Subsoil Constraints for Crop Production

Teogenes Senna de Oliveira
Richard Willian Bell
Editors

Subsoil Constraints for Crop Production

Springer

Editors
Teogenes Senna de Oliveira
Soils Department
Federal University of Viçosa
Viçosa, Minas Gerais, Brazil

Richard Willian Bell
Centre for Sustainable Farming Systems,
Food Futures Institute
Murdoch University
Murdoch, WA, Australia

ISBN 978-3-031-00343-1 ISBN 978-3-031-00317-2 (eBook)
https://doi.org/10.1007/978-3-031-00317-2

This Springer imprint is published by the registered company Springer Nature Switzerland AG
The registered company address is: Gewerbestrasse 11, 6330 Cham, Switzerland

Preface

The productivity of crops is strongly influenced by properties of the subsoil, but surprisingly little attention is paid in agronomy, crop science and soil science to the identification, alleviation, and management of subsoil constraints. The last book on the subject was published 27 years ago by Jayawardene and Stewart (1994). However, a recent survey of high-impact soils journals reveals that in the last three decades 73 % of 1,146 published papers did not even mention soil depth, and of those that did, the average depth of soil investigation was only 27 cm (Yost and Hartemink 2020). Hence, the present literature continues to be dominated by studies of topsoil and a neglect of the subsoil. The present volume is intended to increase the focus of soil and crop research on subsoil constraints. It is intended to inform agronomists and crop scientists of the advances in knowledge about subsoil properties and their management and inspire a new generation of researchers to focus on the subsoil and the alleviation of the subsoil constraints to crop production.

The study of subsoil constraints for crop production depends on a deeper understanding of root systems, and their exploration of the soil profile including the subsoil. The other component of subsoil constraints derives from a knowledge of the subsoil properties themselves. In the present volume, chapters examined acidity, salinity and other salt-related constraints, nutrients, soil physical constraints, sandy and gravel subsoils, and soil biological constraints.

Subsoil constraints are particularly relevant in rainfed agriculture where deep roots are critical for accessing plant-available stored water and nutrients. Indeed, up to 75% of N, 85% of P and 70% of K may be acquired from the subsoil if root growth is not constrained. Subsoil constraints result in water and nutrients resources not being utilised, or being under-utilised, by plants during the growing season. Alleviation of subsoil constraints represents a promising untapped opportunity to increase crop productivity, and in doing so, to enhance food security.

One of the limitations to improved alleviation and management of subsoil constraints is the limited delineation of where such constraints occur. In specific areas, subsoil constraints have been mapped, and there are technologies being developed to alleviate these constraints, for example, in southern Australia. In most parts of the world, such information is lacking. Hence, researchers and policymakers are not in a position to articulate the potential impact of research on subsoil constraints or to identify the best places to undertake research and technology development.

The present volume arose from a short course that was run at the Federal University of Vicosa, Minas Gerais, Brazil, in 2016. The short course was the inspiration of Professor Bob Gilkes, who conducted pioneering research on a range of subsoil constraints in agricultural and post-mining land in south-west Australia, Brazil, and Thailand over many decades. Recognising similarities between the deeply weathered regoliths in southern Australia and in south-east Brazil, he proposed the short course to increase awareness of, and increase research on, subsoil constraints. Following the short course, it was agreed that a book on the subject was needed, and the present editors agreed to undertake that task after Springer agreed to publish this volume.

In both Australia and Brazil, the development of vast areas of land for agriculture has depended on soil science to identify subsoil constraints and develop technologies to alleviate these constraints. Hence, we decided that the book chapters would be written as collaborations between Brazilian and Australian authors, and we have been able to accomplish that with most of the chapters. However, the subject of subsoil constraints has a much greater global relevance. Large areas of deeply weathered regolith occur across tropical regions in Latin America, Asia and Africa and also in many other parts of the world due to past climatic influences. These areas are prospective regions for the identification of subsoil constraints and the application of the technologies reviewed in the current volume. Our aim is to highlight the opportunities for increased crop production by alleviating subsoil constraints and to inspire agronomists, crop scientists and soil scientists to develop programmes of research on subsoil constraints across the globe.

Viçosa, MG, Brazil — Teogenes Senna de Oliveira
Murdoch, WA, Australia — Richard Willian Bell

Contents

Contributors

Walter Antônio Pereira Abrahão Soils Department, Federal University of Viçosa, Viçosa, MG, Brazil

Telmo Jorge Carneiro Amado Agronomy Faculty, Soil Science Department, Santa Maria Federal University, Santa Maria, RS, Brazil

David Backhouse School of Environmental and Rural Science, University of New England, Armidale, NSW, Australia

Richard Willian Bell Centre for Sustainable Farming Systems, Food Futures Institute, Murdoch University, Murdoch, WA, Australia

Clayton Robert Butterly Faculty of Veterinary and Agricultural Sciences, School of Agriculture and Food, University of Melbourne, Parkville, VIC, Australia

Department of Animal, Plant and Soil Science, School of Life Sciences, La Trobe University, Bundoora, VIC, Australia

Guilherme Resende Corrêa Geography Institute, Federal University of Uberlândia, Uberlândia, MG, Brazil

Mirian Cristina Gomes Costa Soil Science Department, Federal University of Ceará, Fortaleza, CE, Brazil

Ítalo Antônio Cotta Coutinho Biology Department, Federal University of Ceará, Fortaleza, CE, Brazil

Ivo Ribeiro da Silva Soils Department, Federal University of Viçosa, Viçosa, MG, Brazil

Claudivan Feitosa de Lacerda Agricultural Engineering Department, Federal University of Ceará, Campus do Pici, Fortaleza, CE, Brazil

Bernard Dell Agriculture and Forest Sciences, Murdoch University, Murdoch, WA, Australia

Fábio Soares de Oliveira Geography Department, Federal University of Minas Gerais, Belo Horizonte, MG, Brazil

Teogenes Senna de Oliveira Soils Department, Federal University of Viçosa, Viçosa, MG, Brazil

Ismail Teodoro de Souza Júnior Plant Protection Department, Federal University of Pelotas, Pelotas, RS, Brazil

José João Lelis Leal de Souza Soils Department, Federal University of Viçosa, Viçosa, MG, Brazil

Raphael Bragança Alves Fernandes Soils Department, Federal University of Viçosa, Viçosa, MG, Brazil

Elpídio Inácio Fernandes Filho Soils Department, Federal University of Viçosa, Viçosa, MG, Brazil

Hans Raj Gheyi Recôncavo of Bahia Federal University, Cruz das Almas, BA, Brazil

César Bauer Gomes Embrapa Temperate Agriculture, Brazilian Agricultural Research Corporation, Pelotas, RS, Brazil

Lucas Carvalho Gomes Soils Department, Federal University of Viçosa, Viçosa, MG, Brazil

Richard Harper Agriculture and Forest Sciences, Murdoch University, Murdoch, WA, Australia

David Henry Murdoch University, Murdoch, WA, Australia

Karen W. Holmes Department of Primary Industries and Regional Development, South Perth, WA, Australia

Paulo Prates Júnior Microbiology Department, Federal University of Viçosa, Viçosa, MG, Brazil

Maria Catarina Megumi Kasuya Agriculture and Forest Sciences, Murdoch University, Murdoch, WA, Australia

Qifu Ma Centre for Sustainable Farming Systems, Food Futures Institute, Murdoch University, Murdoch, WA, Australia

Edson Marcio Mattiello Soils Department, Federal University of Viçosa, Viçosa, MG, Brazil

Stephen P. Milroy Centre for Crop and Food Innovation, Food Futures Institute, Murdoch University, Murdoch, WA, Australia

Bruno Coutinho Moreira Agricultural Engineering College, Federal University of San Francisco Valley, Petrolina, PE, Brazil

Andréa Bittencourt Moura Plant Protection Department, Federal University of Pelotas, Pelotas, RS, Brazil

Pichu Rengasamy School of Agriculture, Food and Wine, Waite Campus, The University of Adelaide, Adelaide, SA, Australia

Podjanee Sangmanee Faculty of Agriculture, Uttaradit Rajabhat University, Uttaradit, Thailand

Craig A. Scanlan Department of Primary Industries and Regional Development, Northam, WA, Australia

Carlos Ernesto Gonçalves Reynaud Schaefer Soils Department, Federal University of Viçosa, Viçosa, MG, Brazil

Hwat Bing So Australian Rivers Institute, Griffith University, Nathan, QLD, Australia

Caixian Tang Department of Animal, Plant and Soil Science, School of Life Sciences, La Trobe University, Bundoora, VIC, Australia

Wendy H. Vance Centre for Sustainable Farming Systems, Food Futures Institute, Murdoch University, Murdoch, WA, Australia

Chapter 1
Introduction to Subsoil Constraints for Crop Production

Teogenes Senna de Oliveira and Richard Willian Bell

Abstract Most investigations of edaphic processes are confined to the topsoil even though plant roots explore a much greater volume of soil than the topsoil. However, plant growth is commonly restricted by adverse physical, chemical, or biological properties of subsoils or combinations of such constraints. Testing surface soils does not reveal these limitations, and hence their significance in crop and soil management is often ignored. There is growing body of research on the impact of subsoil constraints on crop production. To synthesize the insights from this body of knowledge for soil scientists, agronomists, and land managers so that due attention is given to the subsoil in soil, crop, and land management, we have commissioned the 15 following chapters dealing with subsoil constraints for crop production.

Keywords Soil management · Tropical soils · Roots · Alkalinity · Salinity · Nutrient acquisition · Mined areas · Compaction · Sandy subsoils · Gravelly subsoils · Soil water storage · Soilborne pathogens · Water acquisition · Soil biota

1.1 Introduction

There is growing body of research on subsoil constraints for crop production. Many new insights have emerged since the book of Jayawardene and Stewart (1994). We think it is timely to synthesize this body of knowledge for soil scientists, agronomists, and land managers so that due attention is given to the subsoil in soil, crop, and land management.

T. S. de Oliveira (✉)
Soils Department, Federal University of Viçosa, Viçosa, MG, Brazil
e-mail: teo@ufv.br

R. W. Bell
Centre for Sustainable Farming Systems, Food Futures Institute, Murdoch University, Murdoch, WA, Australia
e-mail: r.bell@murdoch.edu.au

T. S. d. Oliveira, R. W. Bell (eds.), *Subsoil Constraints for Crop Production*,
https://doi.org/10.1007/978-3-031-00317-2_1

Agriculture in large tracts of land in both Brazil (e.g., the Cerrado region) and Australia (e.g., the Southwest Wheatbelt) was substantially hampered until research led to the development of technologies to alleviate or manage subsoil constraints. Hence, this book is primarily a joint production of Brazilian and Australian authors and draws much of its material and insight from studies and agriculture in these two countries. However, subsoil constraints are much more widely distributed than just Brazil and Australia. By drawing attention to the consequences of subsoil constraints for crop production, we aim to stimulate researchers worldwide to give this subject greater attention.

The subsoil in this book is considered to be the layers of the root zone below where soil sampling and soil analysis are generally focused. It approximates to below the A1 horizon in soil profile descriptions. In practice, subsoil corresponds to soil below 10, 15, or 20 cm depth, depending on the soil sampling conventions of the region. In much of southern Australia, for example, soil sampling is confined to 0–10 cm. The layer that is normally investigated as the topsoil often coincides with the plough layer also. However, in much of the cropping land in Brazil and southern Australia where zero tillage or minimum soil disturbance is the predominant means for crop establishment (Kassam et al. 2019), the plough layer is no longer a suitable surrogate for defining topsoil. However, even in zero tillage, the placement of fertilizer is rarely more than 10 cm deep, and with pasture lands and perennial crops (coffee, fruits, wood species, etc., irrigated or not), deeper layers than 10 to 20 cm are generally not disturbed or sampled.

1.2 Overview of Subsoil Constraints Chapters

The book was divided in six parts, each one with related chapters. Most of them have Australian and Brazilian authors writing about the same theme and drawing together the research, experience, and realities of the subsoil constraints in the agricultural environments in those countries. The main parts of the book involve the following: (i) Natural Constraints in Subsoils (Part 1); (ii) Chemical Subsoil Constraints (Part 2); (iii) Physical Subsoil Constraints (Part 3); (iv) Soil Biology and Biological Processes in Subsoils (Part 4); Biotic and Root Interactions with Subsoils Constraints (Part 5); and Conclusion and Remarks (Part 6).

In Chap. 2, Schaefer et al. (2022) review the geological, geomorphological, climatic, and hydrological background on the global nature of tropical regoliths and hostile subsoils showing that most regoliths and subsoils in the tropics are polycyclic and their properties may not be directly attributable to the present-day climates because of inherited features from past climatic conditions. Deep regoliths vary in constitution, depending on the parent rocks, but evolve to two basic forms that are dominated by kaolinitic/bauxitic or Fe-rich ferricrete materials. A novel map of the distribution of Brazilian regoliths according to depth to the lithic contact is presented. The authors also discuss key roles that climate and geological structure play in subsoil deepening. To understand the formation of regoliths in the Brazilian landmass, all types of tropical alterations are classified according to the geochemical processes. The authors highlighted Oxisols as the main representative of deeply weathered and saprolite-derived soils, whose chemical, physical, and mineralogical properties reflect prolonged weathering and chemical denudation. Moreover, the nature of saprolites has a definite impact on soil constituents, irrespective of bioclimatic conditions. A framework of the major groups of rocks and their association with soil-landform and regolith attributes was used for a broad discussion on the chemical and physical limitations of hostile subsoils. The importance of subsoils and regoliths, their genesis, and environmental implications were discussed in a broader perspective, incorporating hydrological, pedological, and environmental interactions, focused on practical applications of this knowledge.

Subsoil constraints can be chemical, physical, biological, or combinations of them (Lynch and Wojciechowski 2015), but often combinations of multiple physical and chemical constraints occur in a single soil profile.

The second part of the book deals with chemical constraints which include acidity, high extractable Al, high extractable Mn (Butterly et al. 2022, Chap. 3), alkalinity, salinity, sodicity, boron toxicity (Rengasamy et al. 2022, Chap. 4), low nutrient availability (Ma et al. 2022, Chap. 12), and pyritic subsoils (Souza and Abrahão 2022, Chap. 5).

Butterly et al. (2022) (Chap. 3) note that while many soils are naturally acidic in the subsoil, others have acidified due to land use and fertilization practices that altered the carbon and nitrogen cycles in soils. Acid subsoils (pH_{H_2O} <5.5) which are common in many agricultural systems reduce productivity because crops and pastures fail to achieve water- and nutrient-limited potentials, particularly when topsoils dry out, such as during spring in Mediterranean environments or during summer in tropical environments. The process of acidification is generally linked to nutrient depletion, soil fertilizer reactions, and by root physiological processes involved in nutrient absorption. A key challenge with subsoil acidity is to achieve cost-effective amelioration. Direct incorporation of lime and other amendments into subsoil layers is likely to be effective but is costly and time-consuming and may expose soil to erosion and promote organic matter mineralization. Some machinery is available in the market for deep incorporation, but their use is not common in rural properties. A great deal of research and effort has been focused on finding the most effective materials and methods to ameliorate acid subsoils with minimal or no soil disturbance. Lime, gypsum, calcium oxide, organic amendments, and biological solutions have been tested. Combined incorporation of lime and organic materials could be the most effective, since it provides ameliorative and nutritional benefits as well as improved soil physical conditions. As with many subsoil constraints that are not amenable to economic amelioration with soil treatments, crop breeding appears as an interim solution by developing cultivars with increased tolerance to the subsoil acidity, which has shown positive outcomes. However, acidity covers a range of soil constraints, and correct diagnosis of the specific limitations or combinations of limitations in a subsoil (Al toxicity, Mn toxicity, low P, Ca, Mo, or Mg) is critical to the development of suitable tolerant cultivars, management, or amelioration approaches.

In Chap. 4, Rengasamy et al. (2022) examine processes leading to alkalinity, sodicity, B toxicity, and salinity. These chemical constraints may occur individually in subsoils, but commonly they occur together in combinations in subsoils (McDonald et al. 2012) and can be interrelated: accumulation of salts, salinity, directly affects crop physiology, while the composition of the ions which determines sodicity and alkalinity also affects soil physics and biology, causing direct and indirect interference with crop performance. The potential to accelerate the intensity of these processes through inappropriate management is high and, in some cases, creates outcomes that are not viable to correct. Formation of different types of saline soils, the role of cations in relation to soil structural stability, the iron deficiency and toxicity effects due to sodicity and alkaline pH, and the possible management of these constraints are also discussed in this chapter. The co-occurrence of B toxicity, sodicity, and alkalinity in subsoils has seriously hampered attempts to develop tolerant germplasm with increased yield potential on such soils in southern Australia (McDonald et al. 2010) and northeastern Brazil. Multispecies and perennial crop systems could help to reduce the intensity of the salinity-sodicity-alkalinity processes especially those associated with high water table.

Nutrient acquisition in the subsoils is covered by Ma et al. (2022) in Chap. 12. The widespread emphasis on soil testing in the topsoil (0–10, 0–15, 0–20 cm) means that there is limited understanding of when and where low nutrient levels in the subsoil are a primary cause of deficiencies limiting crop production. While the subsoil is generally lower in available nutrients and organic matter than the topsoil, there is strong evidence that nutrient acquisition from the subsoil can contribute significant amounts of N, P, and K taken up by crops. The subsoil nutrient levels are particularly important in rainfed (cereal or grains) and tree crops (e.g., eucalyptus and coffee). Crops rely on stored subsoil water for growth after topsoils have dried, and, in such cases, the crop may be entirely dependent on the subsoil for uptake of nutrients such as P after the topsoil has dried (e.g., Kabir et al. 2015). There are challenges with amelioration and management of subsoil nutrient limitations. Some farmers in Brazil are using deep subsoiling with nutrient and liming application to improve productivity, which promotes better water and nutrient uptake in droughts or during short periods without rain during the rainy season. The better mechanical support promoted also increases crop resistance to strong wind, which may become more significant as climate changes. However, the crop responses to increased subsoil nutrient levels are not always economic as shown by modeling of K response by Scanlan et al. (2015). Field diagnostics and methods for identification of subsoil nutrient constraints were reviewed. For nutrients such as Zn, Mn, and B, which are immobile in the phloem, a low level of plant available nutrients in subsoil directly limits root growth. Hence, not only is the uptake of that nutrient impaired, but also water uptake from the subsoil may be depressed (Nable and Webb 1993; Grewal et al. 1997). Nutrient acquisition from subsoil will be limited by other constraints such as subsoil acidity, alkalinity, sodicity, toxic B, salinity, anoxia, and the physical impediments to root growth such as compaction, high bulk density, pans, gravels, and deep sand. Ma et al. (2022) focus on plant uptake of N, P, K, S, and micronutrients from subsoils and review the evidence for enhanced subsoil nutrient uptake by fertilizer placement, nutrient localization, rhizosphere modification, and genotypic variation in subsoil nutrient uptake.

The last chapter of Part 3 is about pyritic subsoils in acid sulfate soils and mined areas with sulfidic rocks which represent a special case of subsoil acidity (Souza and Abrahão 2022, Chap. 5). Acid drainage is one of the most dangerous and costly environmental problems that reduce soil and water quality. Souza and Abrahão (2022) review the processes by which drainage of naturally pyritic subsoils or oxidation of sulfidic waste materials produces subsoils with chemical constraints associated with extreme acidity and/or toxicity. The range of constraints and amelioration practices applied to these cases are commonly different to those used in acid agricultural soils. On mine sites, there is capacity to apply much more expensive solutions than would be affordable in agricultural soils. Field diagnostics and methods for identification of subsoil constraints are particularly important for pyritic subsoils since the best management practice for such landforms is often to avoid disturbance or change in soil hydrology and their negative consequences over large areas. Avoidance depends on accurate surveys to delineate the locations of potential acid sulfate soils. Sampling and laboratory analyses can be used to predict the acidification potential so that necessary land management practices can be applied.

Part 4 covers the physical limitations in subsoils which include poor or negligible subsoil structure, compacted layers, or dense pans that restrict root penetration in the subsoil in agricultural, forest, or mined areas (Oliveira and Fernandes 2022, Chap. 6; So et al. 2022, Chap. 7) or very sandy or gravelly subsoils that do not provide adequate water or nutrient storage (Scanlan et al. 2022, Chap. 8). The cited subsoil constraints can be both inherent properties of the soil profile or induced by crop and soil management practices. Examples from agriculture, from forestry (Oliveira and Fernandes 2022), and from mined land (So et al. 2022) are used to illustrate the diversity of natural and induced subsoil constraints for crop production, their origin, and amelioration practices to reduce subsoil physical constraints as well chemical constraints that occur in the same profiles in some cases.

By restricting root growth in subsoils, physical constraints limit plant available water storage and access by roots to both water and nutrients. Oliveira and Fernandes (2022) address the subject in the Chap. 6, considering the origin and pedological conditions that cause natural pans or compaction, as well as the mineralogical and microscopic alterations due to land use that contribute to compaction. Natural and induced processes are compared in terms of how they contribute to the subsoil constraints. Oliveira and Fernandes (2022) also review amelioration methods and plant responses including economic benefits of treatments. Field diagnostics and methods for identification of physical subsoil constraints were reviewed considering agricultural and forest lands.

A specific cause of induced physical subsoil constraints is on mine sites and mineral-processing residues (tailings) that can be attributed to trafficking by heavy machinery especially on the floor of mine pits and associated transport routes or by the deposition of well-sorted materials that pack to high density. In Chap. 7, So et al. (2022) discuss pans and compaction layers and high bulk density materials that limit soil water storage and access by roots on mined land and mineral-processing residues. The range of physical constraints is illustrated by examining contrasting mine substrates from the mining of iron ore and bauxite mining. The chapter examines amelioration practices and how the physical status can change with time as illustrated by a long-term experiment conducted in a bauxite mined area. This experiment tested a range of practices and strategies to rehabilitate the mined area, including surface and subsoil preparation, the return of the original topsoil, and fertilization to determine the most economical form of restoration for family farmers in the important Atlantic rainforest ecological area of Brazil. Unlike agricultural and forestry land use, the rehabilitation of mined areas to remove subsoils constraints, so that the land is fit for the designated post-mining land use, is subject to legal and contractual agreements.

Deep sands and subsoils with a large proportion of gravels (>30%) limit subsoil plant available water storage as well as root density and function, water uptake, and nutrient acquisition from subsoils (Scanlan et al. 2022). Sand and gravel subsoils can both constrain root growth although the mechanisms differ. Root growth is constrained in sand subsoils by factors such compaction and aluminum toxicity that are often induced by land use practices. The impact of gravel subsoils on crop growth depends upon on the penetrability of the gravel layer by crop roots. For impenetrable gravel layers, the properties of the topsoil will have the greatest influence on crop growth. For penetrable layers, root depth or length decreases as gravel content increases. On the other hand, deep sands can have some advantages compared with some other soil classes. They do not strongly adsorb applied nutrients, and normally they also have good conditions for crop mechanization especially when they occur on flat land. Sandy soils are receiving increasing attention in Australia and Brazil, as well around the world. Scanlan et al. (2022) review the use, amelioration, management, and plant responses on deep sands and subsoils with a large proportion of gravels as well as the economic benefits of treatments applied.

Part 4 and 5 cover the biological constraints by harmful soil organisms (Moura et al. 2022, Chap. 9) and the function and interactions of roots and biotic components in the subsoils (Costa and Coutinho 2022, Chap. 10; Moreira et al. 2022, Chap. 11; Vance and Milroy 2022, Chap. 13) as well the interactions between soil fauna and regoliths and their role in subsoil formation (Schaefer and Oliveira 2022, Chap. 15), sometimes with repercussions on the crop productivity. Most studies of soil biology and biological processes are conducted on topsoils, and there has been limited consideration of their significance in subsoils in the literature.

Chapter 9 from Moura et al. (2022) deals with soil pathogens of different groups: bacteria, fungi, oomycetes, and nematodes. It describes the characteristics of the soilborne pathogens, their geographical distribution and host range, favored growth conditions, typical symptoms, and plant damage. They cause diseases on most important species of plants, and almost all crop plants are susceptible to one or more species of soilborne pathogens. These pathogens cause a wide range of symptoms, and the damage caused to the plant can include death of germinating seeds and seedlings, rotting of roots, blocking of xylem, soft rot, deformation, and necrotic lesions on stem bases. The losses caused by these biological constraints are huge, so their mitigation is essential. However, it is necessary to diagnose the specific cause in a given area to take the most appropriate control measure. Control measures, like the use of chemical compounds or products based on biological agents, are reviewed. On the other hand, beneficial soilborne bacteria and fungi are central to the performance of most plants. The review of Moreira et al. (2022) (Chap. 11) describes the main groups of symbiotic and free-living organisms and explores how they contribute to plant and soil health in managed and natural ecosystems. Opportunities for enhancing beneficial plant microbe interactions in the subsoil deserve scrutiny particularly as crop productivity is becoming more dependent on subsoil moisture with declines in rainfall in many parts of the world. Effective rhizobium in subsoil are responsible for nodulation which in some dryland crops like pigeon pea is an important source of symbiotically fixed nitrogen with active nodules formed as deep as 50 cm (Kumar Rao and Dart 1987). There is scant information, however, on whether beneficial bacteria and fungi can persist and enhance root function in subsoils.

Crop root architecture and their response to adverse subsoils (e.g., rhizosphere modification) affect plant growth and tolerance to nutrient and water deficits. Costa and Coutinho (2022) review the depth, distribution, activity, and architecture of roots across a diversity of root systems in agricultural, forestry, and native species. The authors examine how roots are affected by subsoil constraints and how they can contribute to alleviating subsoil constraints. They review the root systems of widely cultivated species and their response to physical and chemical constraints, particularly in subsoils. For most agricultural and horticultural species studied, roots are concentrated in the top 20 cm depth, but 20–30% of the root system commonly grows below this depth. Physical restrictions caused by soil compaction reduce root biomass, density, and length, besides limiting root exploitation depth. Some species such as rapeseed and black mustard may alleviate soil compaction, making the subsoil more suitable for roots of other crops. Plant root systems are negatively affected by constraints that are not always present in the topsoil such as excess of soluble aluminum, heavy metals, and salts. However, deficiencies of N, P, K, as well as micronutrients also impair subsoil root growth. Mechanized tillage systems, more tolerant genotypes, association with soil microorganisms, exudation of organic acids by roots, and adequate nutrient supply are among the strategies that allow for better root development and function, enabling improved crop growth in soil with subsoil constraints.

The acquisition of water from subsoil is critical to crop productivity in rainfed and drought-prone environments. In Chap. 13, Vance and Milroy (2022) review strategies used by plants for water acquisition from subsoils and how physical and chemical constraints and modification of roots architecture and rhizosphere alter the efficiency of water uptake. The physical subsoil constraints impair crop water use either through effects on water availability or the ability of the roots to access the water. The resultant impact on crop water use is modulated by the amount and distribution of rainfall, the soil's water holding capacity, and the depth and severity of the constraint. That subsoil constraints influence crop water uptake is generally well understood, but one aspect that still needs clarification or quantification concerns processes of water transfer in subsoils from the bulk soil to the roots' vascular elements. Also, there are knowledge gaps regarding the link between water uptake at the scale of the crop or the soil profile with the influence of agronomic management on pore size distribution, continuity, and stability in terms of their influence on root system development: these gaps limit understanding of the water acquisition by crops. Simulation models that integrate the effects of variable access to water in differing soil types and climatic zones can help to quantify the agronomic importance of subsoil constraints in the context of interannual variation in rainfall distribution at specific locations.

Root activity is also closely related to the deep storage of soil organic carbon in subsoils (Sangmanee et al. 2022, Chap. 14). Soils with deep roots (meters deep) are located in most continents of the world particularly where there is deep regolith. In some cases, roots up to 40 m have been recovered (Dell et al. 1983). Plant roots are the primary source of carbon in the entire soil profile, and their presence explains in large part the deep stores of soil carbon (Harper and Tibbett 2013). Long after land is deforested for agricultural and other pursuits, significant portions of this root biomass persist. Hence, deforestation, commercial logging, pests, and fires have left behind considerable root biomass at the global scale, but estimation of the deep root biomass and carbon dynamics is challenging in deep soil and regoliths that may reach tens of meters deep (Schaefer et al. 2022). This chapter explores deep soil carbon from the perspectives of definition, source, and persistence of deep soil carbon; methodologies available to study deep soil carbon; and the effect of land use change on deep soil carbon.

The last chapter from Part 5 (Schaefer and Oliveira 2022) reviews the effects of biota on the regolith which operate at scales from the smallest organisms (bacteria) to the largest living ones (big trees). The authors examined how biological agents, especially roots, ants, and termites, play a key role in regolith and subsoil formation, highlighting the processes of pedoturbation, microaggregate genesis, chemical reactions in the rhizosphere, and soil morphological organization. The biological turnover of soils through the action of termites and ants was discussed, emphasizing its importance in the long-term evolution of tropical soils and subsoils.

Finally, all chapters are summarized and synthesized in Chap. 16 by Bell and Oliveira (2022) to highlight the significance of subsoil constraints and their importance in land use. The chapter identifies the key gaps and research questions that remain about subsoil constraints and techniques that can ameliorate subsoil constraints.

References

Bell RW, Oliveira TS. Subsoil constraints for crop production: recent advances, new technologies and priorities for further research. In: Oliveira TS, Bell RW, editors. Subsoil constraints for crop production. Switzerland: Springer; 2022. p. 397–422.

Butterly CR, Amado TJC, Tang C. Soil acidity and acidification. In: Oliveira TS, Bell RW, editors. Subsoil constraints for crop production. Switzerland: Springer; 2022. p. 53–91.

Costa MCG, Coutinho IAC. Root systems of agricultural crops and their response to physical and chemical soil constraints. In: Oliveira TS, Bell RW, editors. Subsoil constraints for crop production. Switzerland: Springer; 2022. p. 225–61.

Dell B, Bartle J, Tacey W. Root occupation and root channels of jarrah forest subsoils. Aust J Bot. 1983;31(6):615–27.

Grewal HS, Lu Z, Graham RD. Influence of subsoil zinc on dry matter production, seed yield and distribution of zinc in oilseed rape genotypes differing in zinc efficiency. Plant Soil. 1997;92:181–9. https://doi.org/10.1023/A:1004228610138.

Harper R, Tibbett M. The hidden organic carbon in deep mineral soils. Plant Soil. 2013;368(1–2):641–8. https://doi.org/10.1007/s11104-013-1600-9.

Jayawardene NS, Stewart BA. Subsoil management techniques. In: Advances in soil science. Boca Raton, FL: Lewis Publishers; 1994.

Kabir E, Johansen CJ, Bell RW. Subsoil rhizosphere modification by chickpea under dry topsoil conditions. J Plant Nutr Soil Sci. 2015;178:904–13. https://doi.org/10.1002/jpln.201500021.

Kassam A, Friedrich T, Derpsch R. Global spread of conservation agriculture. Int J Environ Stud. 2019;76:29–51. https://doi.org/10.1080/00207233.2018.1494927.

Kumar Rao JVDK, Dart PJ. Nodulation, nitrogen fixation and nitrogen uptake in pigeonpea (*Cajanus cajan* (L.) Millsp) of different maturity groups. Plant Soil. 1987;99:255–66. https://doi.org/10.1007/BF02370872.

Lynch JP, Wojciechowski T. Opportunities and challenges in the subsoil: pathways to deeper rooted crops. J Exp Bot. 2015;66:2199–210. https://doi.org/10.1093/jxb/eru508.

Ma Q, Bell RW, Mattiello E. Nutrient acquisition with particular reference to subsoil constraints. In: Oliveira TS, Bell RW, editors. Subsoil constraints for crop production. Switzerland: Springer; 2022. p. 289–321.

McDonald GK, Eglinton JK, Barr AR. Assessment of the agronomic value of QTL on chromosomes 2H and 4H linked to tolerance to boron toxicity in barley (*Hordeum vulgare* L.). Plant Soil. 2010;326:275–90. https://doi.org/10.1007/s11104-009-0006-1.

McDonald GK, Taylor JD, Verbyla A, Kuchel H. Assessing the importance of subsoil constraints to yield of wheat and its implications for yield improvement. Crop Pasture Sci. 2012;263:1043–65. https://doi.org/10.1071/CP12244.

Moreira BC, Jr Prates P, Dell B, Kasuya MCM. Roots and beneficial interactions with soil microbes. In: Oliveira TS, Bell RW, editors. Subsoil constraints for crop production. Switzerland: Springer; 2022. p. 263–87.

Moura AB, Backhouse D, Jr Souza IT, Gomes CB. Soilborne pathogens. In: Oliveira TS, Bell RW, editors. Subsoil constraints for crop production. Switzerland: Springer; 2022. p. 199–224.

Nable RO, Webb MJ. Further evidence that Zn is required throughout the root zone for optimal plant growth and development. Plant Soil. 1993;150:247–53. https://doi.org/10.1007/BF00013021.

Oliveira TS, Fernandes RBA. Physical subsoil constraints of agricultural and forestry land. In: Oliveira TS, Bell RW, editors. Subsoil constraints for crop production. Switzerland: Springer; 2022. p. 125–60.

Rengasamy P, Lacerda CF, Gheyi HR. Salinity, sodicity and alkalinity. In: Oliveira TS, Bell RW, editors. In: Oliveira TS, Bell RW, editors. Subsoil constraints for crop production. Switzerland: Springer; 2022. p. 83–107.

Sangmanee P, Dell B, Harper RJ, Henry DJ. Deep soil carbon – characteristics and measurement with particular bearing on kaolinitic profiles. In: Oliveira TS, p. 347–72. Bell RW, editors. Subsoil constraints for crop production. Switzerland: Springer; 2022.

Scanlan CA, Holmes KW, Bell RW. Sand and gravel subsoils. In: Oliveira TS, Bell RW, editors. Subsoil constraints for crop production. Switzerland: Springer; 2022. p. 179 –98.

Scanlan CA, Huth N, Bell RW. Simulating wheat growth response to potassium availability under field conditions in sandy soils II. Effect of subsurface potassium on grain yield response to potassium fertiliser. Field Crops Res. 2015;178:125–34. https://doi.org/10.1016/j.fcr.2015.03.019.

Schaefer CEGR, Oliveira FS. Live subsoils: tropical regolith and biota interactions. In: Oliveira TS, Bell RW, editors. Subsoil constraints for crop production. Switzerland: Springer; 2022. p. 373–96.

Schaefer CEGR, Souza JJLL, Oliveira FS, Corrêa GR, Fernandes Filho EI. The geological, geomorphological, climatic and hydrological background of tropical regoliths and hostile subsoils: the Brazilian landmass. In: Oliveira TS, Bell RW, editors. Subsoil constraints for crop production. Switzerland: Springer; 2022. p. 11–52.

So HB, Silva IR, Gomes LC, Oliveira TS. Subsoil constraints of mined land and tailings. In: Oliveira TS, Bell RW, editors. Subsoil constraints for crop production. Switzerland: Springer; 2022. p. 161–77.

Souza JJLL, Abrahão WAP. Pyritic subsoils in acid sulfate soils and similar problems in mined areas with sulfidic rocks. In: Oliveira TS, Bell RW, editors. Subsoil constraints for crop production. Switzerland: Springer; 2022. p. 109–23.

Vance WH, Milroy SP. Water acquisition by roots from the subsoil: impact of physical constraints on the dynamics of water capture. In: Oliveira TS, Bell RW, editors. Subsoil constraints for crop production. Switzerland: Springer; 2022. p. 323–45.

Chapter 2
The Geological, Geomorphological, Climatic, and Hydrological Background of Tropical Regoliths and Hostile Subsoils: The Brazilian Landmass

Carlos Ernesto Gonçalves Reynaud Schaefer, José João Lelis Leal de Souza, Fábio Soares de Oliveira, Guilherme Resende Corrêa, and Elpídio Inácio Fernandes Filho

Abstract In this chapter, we first describe the global nature of tropical regoliths, showing the remarkable differences with temperate latitudes counterparts. It is shown that most regoliths and subsoils in the tropics are polycyclic and not directly attributable to the present-day climates because of inherited features from past climatic conditions. The Brazilian land surface has one of the deepest regolith mantles at a planetary scale, resulting from a long-term evolution under terrestrial conditions and warm, tropical climates within a tropical latitudinal belt since the Jurassic. Depending on the parent rocks, this deep regolith is varied in the constitution, but mature pre-weathered sediments are commonly observed throughout the country. The two basic subsoils (kaolinitic/bauxitic and Fe-rich ferricretes) are described, and we offer a fresh perspective on the Brazilian view of these common tropical regoliths. We present a pioneering map of the distribution of Brazilian regoliths according to depth to the lithic contact, estimated from soil surveys. It indicates that the shallowest depths are found on resistant rock belts, especially along the quartzitic Espinhaço Range and scattered areas on granites and quartzites in the southern Amazon. The main area of shallow saprolites is those on granites and migmatites/gneiss from the semiarid northeast. Thus, climate and geological structure are essential for subsoil deepening, and most of the Brazilian landmass is dominated by much deeper subsoils, reaching depths exceeding 20 m of rotten rocks. To understand the formation of regoliths in the Brazilian landmass, we present all types of tropical alterations classified accord-

C. E. G. R. Schaefer · J. J. L. L. de Souza · E. I. Fernandes Filho (✉)
Soils Department, Federal University of Viçosa, Viçosa, MG, Brazil
e-mail: carlos.schaefer@ufv.br; jjlelis@ufv.br; elpidio@ufv.br

F. S. de Oliveira
Geography Department, Federal University of Minas Gerais, Belo Horizonte, MG, Brazil

G. R. Corrêa
Geography Institute, Federal University of Uberlândia, Uberlândia, MG, Brazil

T. S. d. Oliveira, R. W. Bell (eds.), *Subsoil Constraints for Crop Production*,
https://doi.org/10.1007/978-3-031-00317-2_2

ing to the geochemical processes. The allitization, monosiallitization, bisiallitization, and ferrallitization are defined and exemplified in Brazil, showing their relationship with current and past climatic differences. We highlight Oxisols as the main representatives of weathered and deep saprolite-derived soils. Their chemical, physical, and mineralogical properties reflect prolonged weathering and chemical denudation, and the nature of saprolites has a definite impact on soil constituents, irrespective of bioclimatic conditions. The vast distribution of deep regoliths in the tropical domain is associated with extensive aquifers, mainly under sandy formations of the Paleo-Mesozoic basins. For a broad discussion on the chemical and physical limitations of hostile subsoils in Brazil, we used the framework of the major rocks found in the country and their association with soil-landform and regolith attributes. Finally, we believe that the importance of subsoils and regoliths, their genesis, and environmental implications must be discussed in a broader perspective, incorporating hydrological, pedological, and environmental interplays focused on practical applications of this knowledge. Therefore, we argue that regoliths and subsoils are now recognized as an essential part of the critical zone.

Keywords Critical zone · Subsoil definition · Tropical soils · Tropical weathering

2.1 Introduction

Across the entire tropical belt, anybody unfamiliar with their landscapes, upon arrival, will be taken by surprise by the widespread and astonishing exposure of deep, colorful regoliths and associated subsoils, in striking contrast with the common greyish and pale landscape images of the temperate zones. This lively and vivid impression for the newcomer to the tropics betrays one important aspect of this zone: the long-term pre-weathered nature of its tridimensional regolith body and the very ancient, pre-Quaternary memory of its formation, under varying climatic conditions (Fig. 2.1). It is a tropical land print and characteristic signature, and its recognition is key for understanding the far-reaching consequences of these extraordinary, deep subsoils for many practical aspects of everyday life.

The intertropical belt has experienced a long period under warm, seasonally wet-dry climates that left a stable product in the form of a deep lateritic weathering profile, sometimes reaching the most insoluble end product – an $Al(OH)_3$ blanket, which is rich in secondary minerals gibbsite and boehmite and generically named bauxite (Fig. 2.1). Bauxites are, thus, widely distributed across the intertropical belt and may even occur at subtropical latitudes, like in southwestern Australia, where they are inherited from past tropical paleoclimates during the Cenozoic, when Australia was located at lower latitudes, experiencing warmer and wetter climates (Butt et al. 2000). Indeed, paleoweathering conditions and tropical-like paleosols are found in many non-tropical latitudes. With the development of calibrated dating of weathering profiles, several studies point to past weathering formations and paleoenvironments. For example, Cecil et al. (2006) used (U-Th)/He

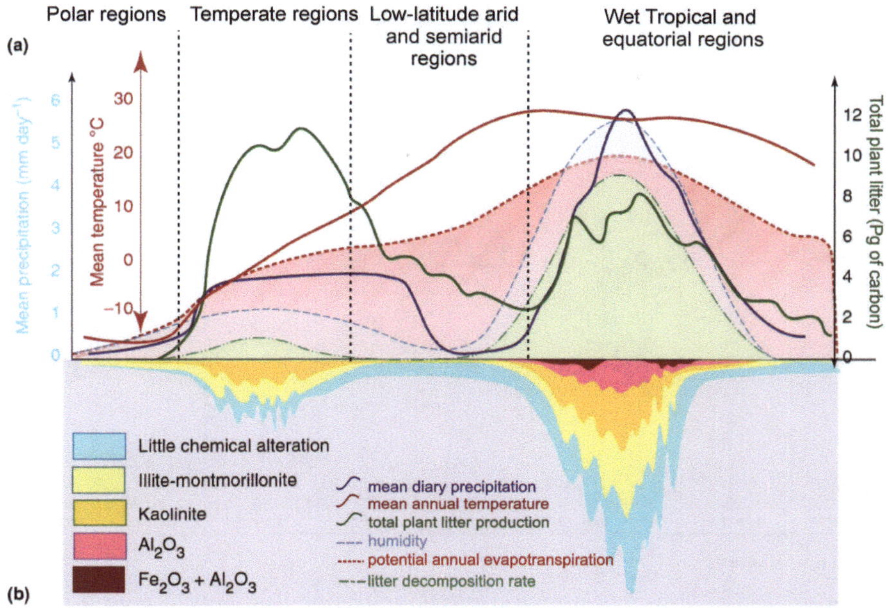

Fig. 2.1 The classical illustration of regolith development along varying latitudes and the mineralogical paragenesis of each layer, with an approximate depth

thermochronology to date the exhumation history of the Sierra Nevada, California, indicating that a lateritic paleosol accompanied a period of tectonic stability during the Eocene when conditions were wetter and warmer (> 5 °C) during this period (Yapp 2008). Along the same line, Retallack (2007) concluded that general wet/warm conditions existed in the northwest and west-central North America based on paleosol features from the Eocene to the Miocene, and later in the Pliocene, just before the Quaternary glaciations.

Climate and biological factors drive the regolith producer reactor, and tropical latitudes have the deepest subsoil development (Fig. 2.1). In contrast, coniferous vegetation releasing organic acids of high acidification potential is key for the temperate deepening of sandy regoliths underneath boreal forests. Deep subsoils in the form of saprolite, either bauxitic or kaolinitic, are typical of all tropical countries located within the belt, from south America to central Africa, from peninsular India to Indochina, finally, to the Indo Malayan region and Australia (Figs. 2.1 and 2.2).

There, classical lateritic profiles are developed, forming very mature secondary minerals (Fe and Al oxides, kaolinite) and abundant quartz. Whenever indurated by long-term seasonal oscillations of oxidizing and reducing conditions, these weathering profiles may evolve into hardened laterites, ironstones, or ferricretes, as indicated by the prominent role of Fe oxides in promoting stable weathered mantles protected by Fe cementation, in some cases. These ironstone crusts occur throughout the tropics, especially in the wet and dry zones, with little formation/

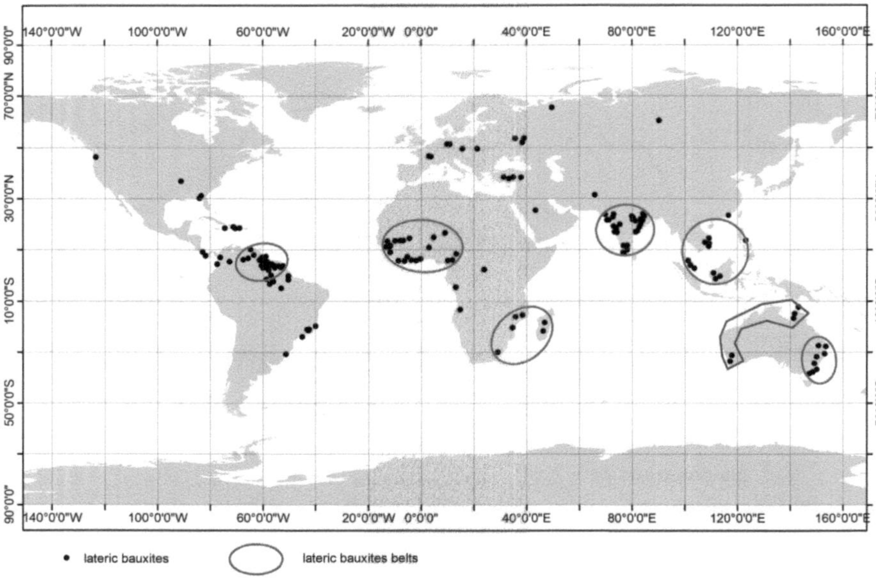

Fig. 2.2 The distribution of highly weathered subsoils containing exploitable Al hydroxides, in the form of bauxite, a key resource for many tropical countries. When found outside the tropical belt, like spots in Russia, Europe, and Southern Australia, they represent paleoweathering profiles, with stable subsoils dating back to Early Cenozoic, or even older (Fig. 2.3), either by climate changes (warm Eocene is a classic example) or tectonic movement towards high latitudes (like Australia moving southwards during the Cenozoic). (After Beckmann 1983)

preservation under more arid climates and progressive destruction under superhumid climates under dissolution conditions. In West Africa, for example, ferricretes, in the form of petroplinthite or "ferruginous courasse," are found along a latitudinal belt, with progressively decreasing depth and degree of Fe cementation towards the Sahara Desert (Fig. 2.3). Therefore, subsoils cemented by Fe oxides are climate-driven features and depend on seasonally wet and dry climates, where they form to the maximum extent.

The Australian landforms and regoliths are controlled by deep weathering on a tectonically stable block and widespread etchplanation under a long-term climate evolution from wet to dry during the Cenozoic (Fig. 2.4). In this illustration, shallow weathering zones (grey areas) only occur where the present-day climate is too dry and arid to allow water infiltration and reaction, and previously weathered substrates from past wet climates have been stripped by erosion. However, most desert landscapes are now on pre-weathered sediments exposed to long-term erosional processes. On the other hand, deep weathering and ferricrete (ironstone) are particularly found on the Yilgarn Craton in the Southwest and on the moderately wet zones of the Northeast (Queensland, mainly). In Fig. 2.5, different types of Australian subsoils were selected, ranging from pre-weathered subsoils to young, poorly drained areas with 2:1 cracking clays of transitional zones of northeastern and

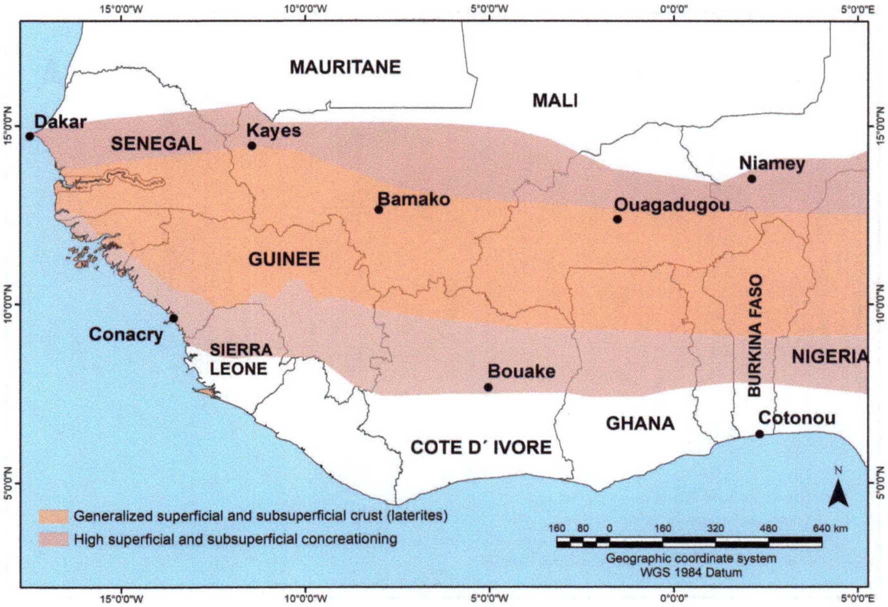

Fig. 2.3 The distribution of a deep lateritic weathering profile with kaolinite and bauxite and the latitudinal range of a continuous blanket of indurated laterites (ferricretes) towards dry savanna areas (southern Sahel), whereas to the south of Conakry-Bouake, ferricretes disappear under wetter climates, being replaced by Ferralsols (Latosols, Oxisols). As aridity increases towards the northern Sahara zone, ferricretes are not formed at all. (After Beckmann 1983)

eastern Australia. These illustrations clearly show many areas of high weathering where the present-day climate does not favor it, so pre-weathering is the rule of most Australian landscapes. In the Quaternary, most Australian landmass has been subjected to arid conditions, and wind erosion became a dominant process across the Australian landscape (Figs. 2.4 and 2.5).

Large areas of the world, especially the broad tropical to the subtropical zone between latitudes 40° north and south, are characterized by a thick regolith cover, with varying mineralogical assemblages. Much of this regolith is residual and consists of intensely weathered bedrock, but there may also be an overlying component of colluvial transported material (Butt and Zeegers 1992), also weathered to varying degrees. The regolith is most extensive in continental regions of low to moderate relief, such as the Precambrian shields and adjacent and overlying Phanerozoic sedimentary basins, of South America, Africa, India, South East Asia, and Australia. It may be present also in old residual mountains and steep slopes from any tropical zone, subjected to gentle/moderate neotectonics.

Consistently, deep regolith is absent from tectonically active and mountainous areas, and thick residual regolith is also generally absent from very arid terrains in the tropics and subtropics, such as the Sahara and Arabian deserts. However,

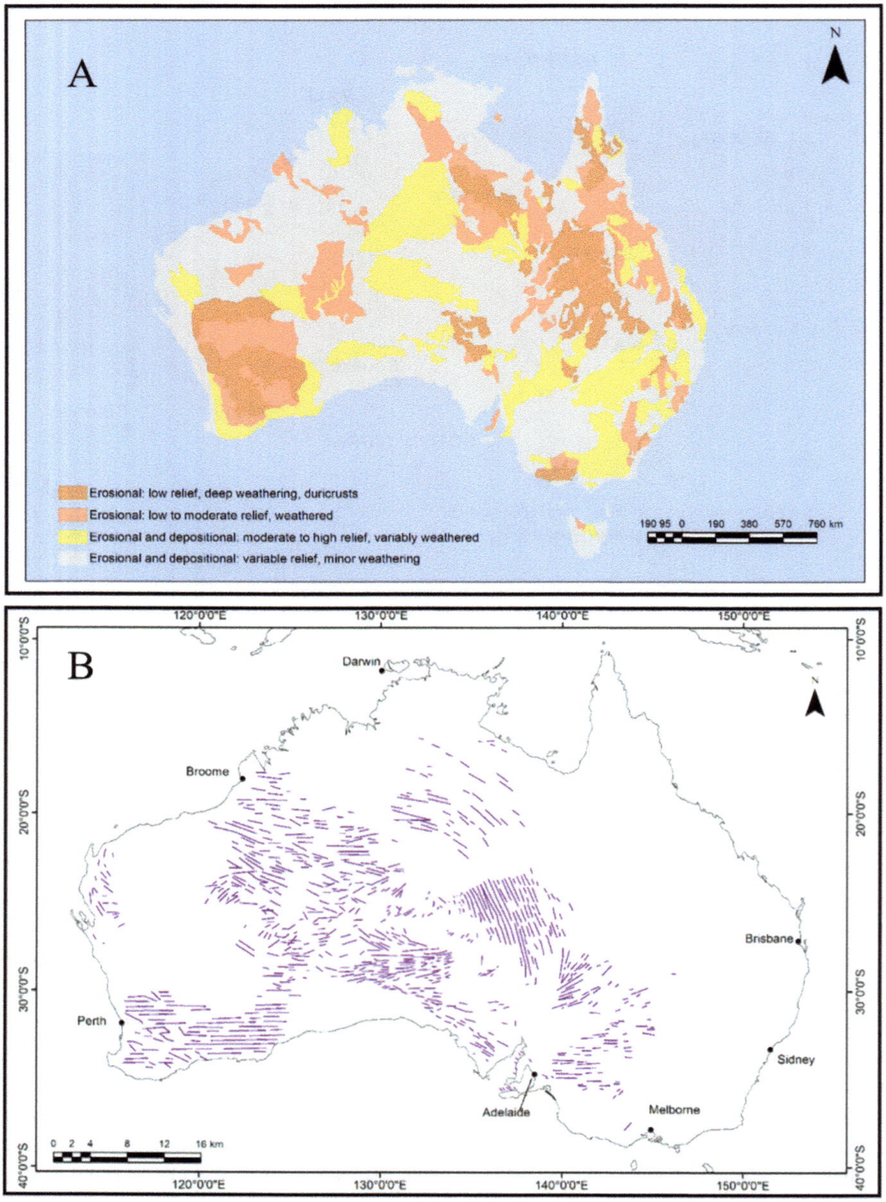

Fig. 2.4 (**a**) A basic general classification of Australian landforms and subsoils (after Hubble et al. 1983). (**b**) The sand dunes and their main directions across Australia indicate strong wind ablation and erosion on the tectonically stable landmass. (After Beckmann 1983)

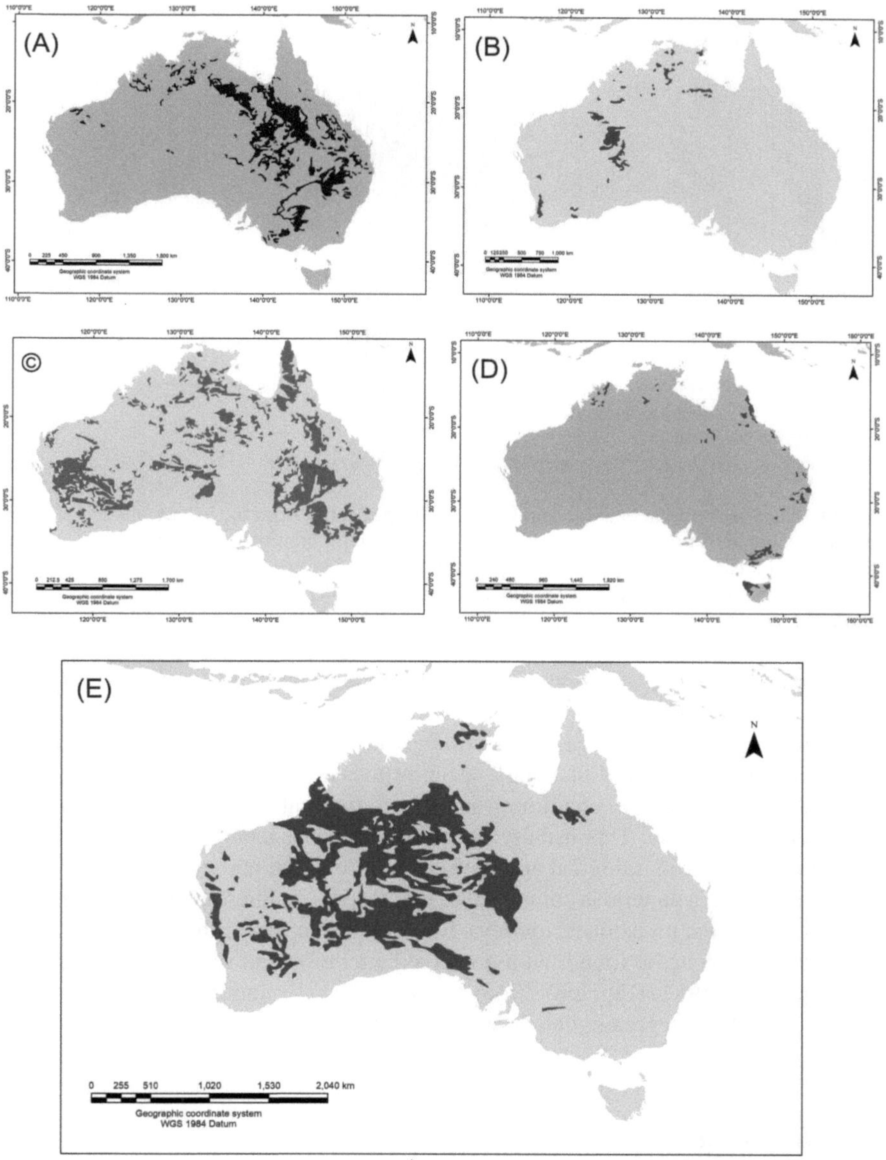

Fig. 2.5 Different types of subsoils in Australia, based on Hubble et al. (1983), selecting and highlighting (**a**) the spatial distribution of 2:1 cracking clay regoliths; (**b**) Fe-rich concretionary soils/ferricrete on pre-weathered regoliths; (**c**) massive, deep kaolinitic regoliths; and (**d**) well-structured, microgranular kaolinitic/oxidic regoliths. In the bottom map (**e**), the deep sandy regoliths of the desert and semi-desert areas indicate the Quaternary (Pleistocene) trend of arenization and desertification of Central and Western Australia. Carbonate-rich subsoils of the southern territory are not shown. (After Beckmann 1983)

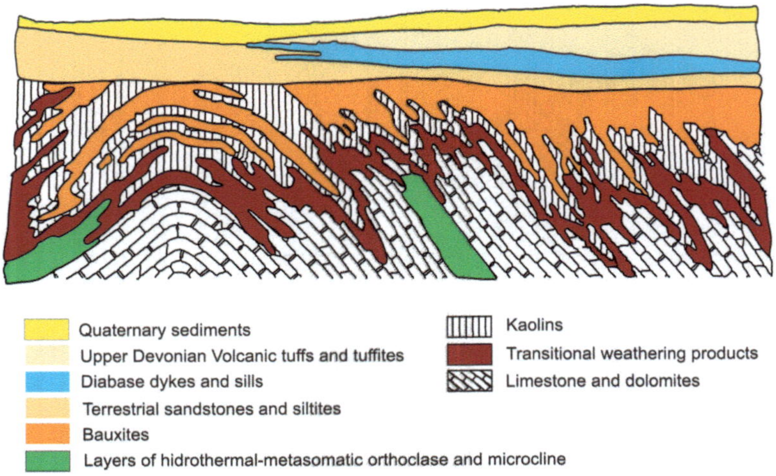

Fig. 2.6 The Precambrian bauxitic and kaolinitic deep regolith from the Timan Mts in Russia is one rare example of pre-Paleozoic deep weathering reaching high bauxite grade (modified after Diomina and Iatskievitsch 1982 quoted in Bardossy 1982). The erosion surface that truncates the bauxite paleosol is certainly pre-Devonian, and one of the oldest worldwide is overlying a deeply weathered terrain

transported materials, including fluvial deposits and sand dunes, are widespread (Butt and Zeegers 1992).

Concerning South America, the Brazilian land surface possesses one of the deepest regolith mantles at the planetary scale, representing the effect of a long-term evolution under terrestrial conditions and warm, tropical climates within a tropical belt since the Jurassic. This usually deep regolith has a much varied constitution, depending on the mineralogical and physical nature of the parent rocks from which they formed, but in general is composed of mature pre-weathered sediments (quartzous sandstones and mudstones, mainly). Elsewhere, deep regoliths dating back from the Precambrian can be found, with preserved alteration and soil features (Nikitina 2016; Bardossy 1982) (Fig. 2.6). From now on, we shall examine the nature of subsoils and regoliths in greater detail, emphasizing Brazil.

2.2 Regoliths and Subsoils

Regolith is broadly defined as the set of unconsolidated or secondarily cemented materials overlying rocks. Its formation is related to surface processes, such as weathering, pedogenesis, erosion, transport, and sedimentation, and therefore, soils, saprolites, and various deposits, such as fluvial, colluvial, glacial, and surface organic accumulations or ashes, can be considered as regoliths. The regolith is in situ or eluvial when the material derives directly from underlying rocks' physical and chemical weathering. If erosion operates upon the unconsolidated materials, they can be transported and deposited in new environments, being classified as

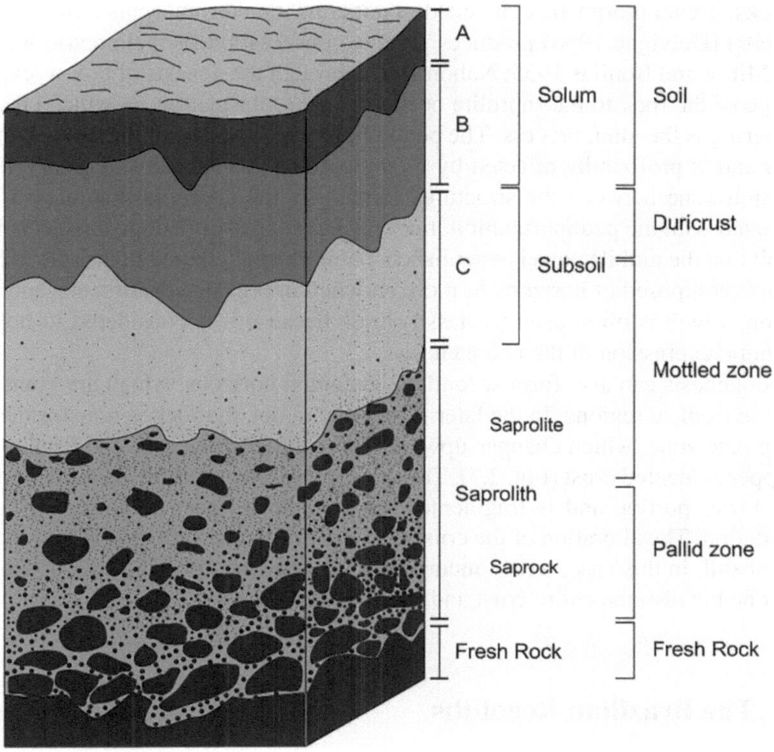

Fig. 2.7 Representative regolith profile, with saprolith (saprock; saprolite) and pedolith (subsoil; *solum*) zones. A, B, and C indicate soil horizons. (Drawing by Henrique Machado)

transported regoliths (Scott and Pain 2009). Regardless of whether it is in situ or transported, regolith is an expression of the interaction between lithosphere, hydrosphere, atmosphere, and biosphere (Taylor and Eggleton 2001).

The soil can be understood as the upper part of regolith, but the two terms are not synonymous. Soil genesis is related to the interaction between several environmental factors (Jenny 1941), resulting in physical, chemical, and biological transformations (weathering) of original materials (rocks, sediments, organic materials) and the formation of new arrangements and structures (pedogenesis). Since pedogenesis always occurs in situ, soils represent a testimony of current and/or past environmental conditions in which they occur.

The concept of subsoil[1] here adopted is understood as part of the pedolith not directly influenced by exposure to the atmosphere and biosphere. The regolith is divided into two main units: saprolite and pedolith (Fig. 2.7). Saprolite incorporates portions in which rock structures are conserved (saprock) and portions in which such structures begin to be disintegrated by the progress of mineral changes. In

[1] This definition varies from the agronomic version defined by Bell and Oliveira (2022) and used in other chapters of this book.

saprocks, pseudomorphism (for euhedral minerals) or alteromorphism (for anhedral minerals) (Delvigne 1998) produces volume conservation (isovolumetric weathering) (Millot and Bonifas 1955; Nahon 1991) through the genesis of box-works. The passage of the rock to the saprolite occurs by alteroplasmation, in which chemical weathering is the main process. The pedolith, in turn, consists of the subsoil and the *solum* and is profoundly affected by biological factors. The subsoil comprises the transition zone between the structures formed by the alteroplasmation and those associated with the pedoplasmation, marking the fringes of pedogenesis. It is in the subsoil that the mobile zone is established, being strongly linked to the role of biota. *Solum* is composed of horizons A, more enriched in organic constituents, and the B horizon, which is often used to classify soils because it is considered to hold the maximum expression of the pedogenesis.

Pedogenesis can also form secondary cemented horizons, which are commonly found in tropical regions. In the lateritic profile model, the rock is transformed into a deep pale zone, which changes upwards to a mottled zone until the formation of the upper cemented crust (Fig. 2.7). The crust (ferricrete, silcrete) is usually massive in its lower portion and is fragmented towards the surface, reflecting its current degradation. The alteration of the crust (ferricrete, silcrete) will give rise to the soil. The subsoil, in this case, would incorporate not only the upper portion of the mottled zone but also the entire crust and its different facies.

2.3 The Brazilian Regoliths

The Brazilian territory is geologically complex, and the regoliths resting upon the parent rocks have varying mineralogical and physical compositions, usually dominated by very stable secondary minerals. The country is in an old and stable continental landmass, representing the main segment of the South American Tectonic Plate (Ab'Saber 1956; Barbosa 1966; Almeida et al. 2000; Schaefer 2013). A rapid examination of the geological structure of the South American Plate (Fig. 2.8) distinguishes two fundamental segments of the continental crust: (i) to the west, the young, Cenozoic Andean orogenetic zone, comprising an elongated north-south, "S" shaped, belt, coinciding with the current collision belt between the Nazca, Pacific, and South American Plates, and (ii) to the east, the much older, vast Brazilian landmass (cratons, mobile belts, and basins), where much older continental crust rocks, and derived sediments, are relatively undisturbed by late tectonics, although the eastern coastal sector represents an active divergence zone from the neighboring African plate (Fig. 2.8).

Thus, the Brazilian territory is the oldest and most stable part of the South American continent and encompasses several distinct structural sub-provinces of great interest for understanding the evolution of regoliths and the Brazilian landscape, in general. The so-called structural provinces, a modified structural model of Brazil, proposed by Almeida et al. (1976, 2000), can be adopted here, assuming

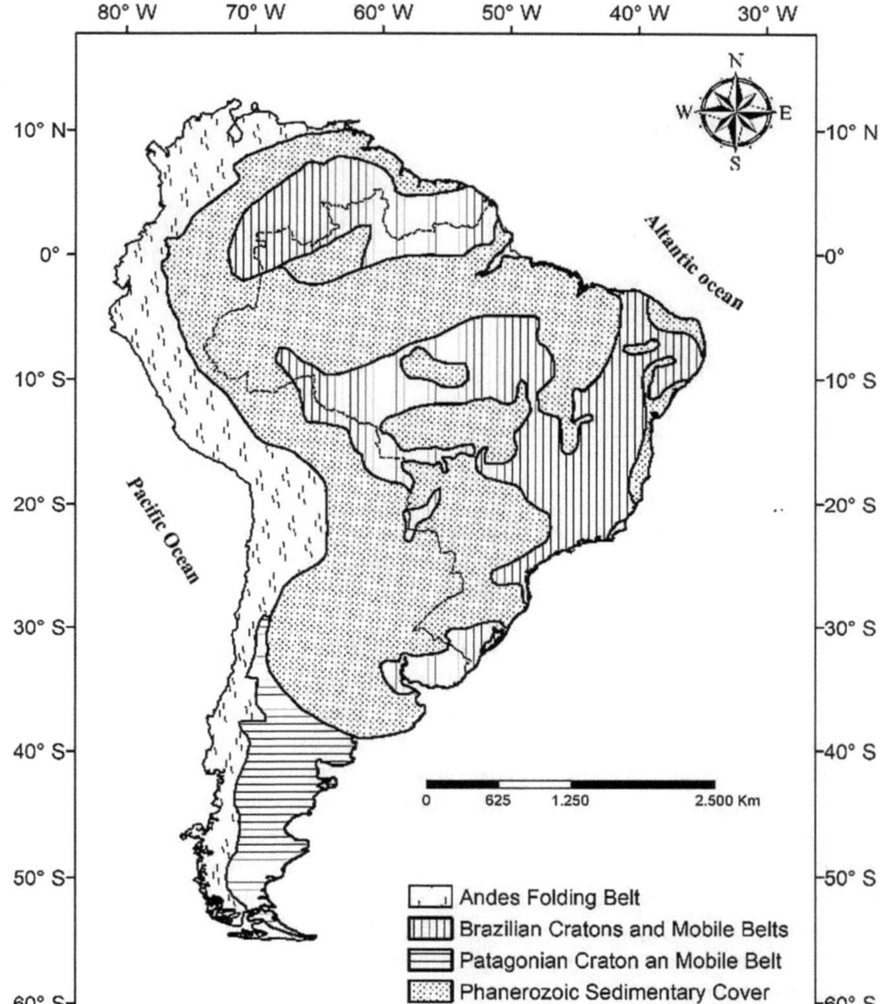

Fig. 2.8 The basic geotectonic framework of South America indicates that deep weathering is concentrated on the Brazilian craton and mobile belts and derived Phanerozoic basins, all connected by sea through the Paleozoic

some generalizations on a small scale adopted, allowing a broad understanding of the nature of subsoils.

Brazil can thus be divided into four distinct structural zones, so-called (i) cratons; (ii) Precambrian mobile belts; (iii) Paleozoic sedimentary basin; and (iv) Meso-Cenozoic sedimentary basins (Fig. 2.9b). As clearly indicated in Fig. 2.9b, most Brazilian highlands are concentrated in the southeastern region, where neotectonic uplift in the Neogene (Miocene to recent) has taken place.

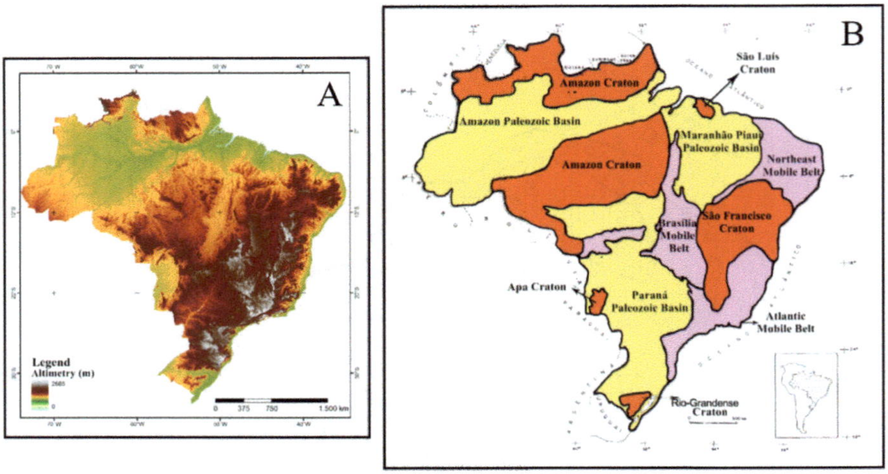

Fig. 2.9 (**a**) A digital elevation model of Brazil, highlighting the high altitudes (> 1000 m) of the corresponding Atlantic mobile belt (Schaefer 2013). (**b**) The basic structural geotectonic of Brazil, showing the framework of structural provinces (cratons, orange; mobile belts, pink; Paleozoic sedimentary basins, yellow); (Data from Schaefer 2013)

According to depth to the lithic contact, the distribution of regoliths in Brazil was estimated by data obtained from soil surveys throughout the country for the first time to produce a general picture of Brazilian subsoil conditions (Fig. 2.10). The shallowest depths are found on resistant rock belts, especially along the quartzitic Espinhaço Range and in some scattered areas on granites and quartzites in southern Amazon. However, the main area of shallow saprolites is those on granites and migmatites/gneiss from the semiarid northeast. Thus, climate and geological structure are the keys for subsoil deepening, but most of the Brazilian landmass is dominated by much deeper subsoils, reaching depths exceeding 20 m of rotten rocks (saprolites). These saprolites reach amazing depths in some places, like in the Carajás (Amazon) and Iron Quadrangle, where more than 300 meters of subsoil are commonly observed.

2.4 Long-Term Weathering and Brazilian Subsoils

The map illustrating the regolith depth to fresh rock, shown in the previous Fig. 2.10, does not distinguish the two main types of weathered mantles that cover the country. The two basic types (kaolinitic and ironstone) are described in the following section. They are part of a broader cover of unconsolidated Late Cenozoic sediments, which geologists have difficulty distinguishing and mapping from a pure pedological cover (Fig. 2.11) and are normally associated with deeper regoliths (Fig. 2.10).

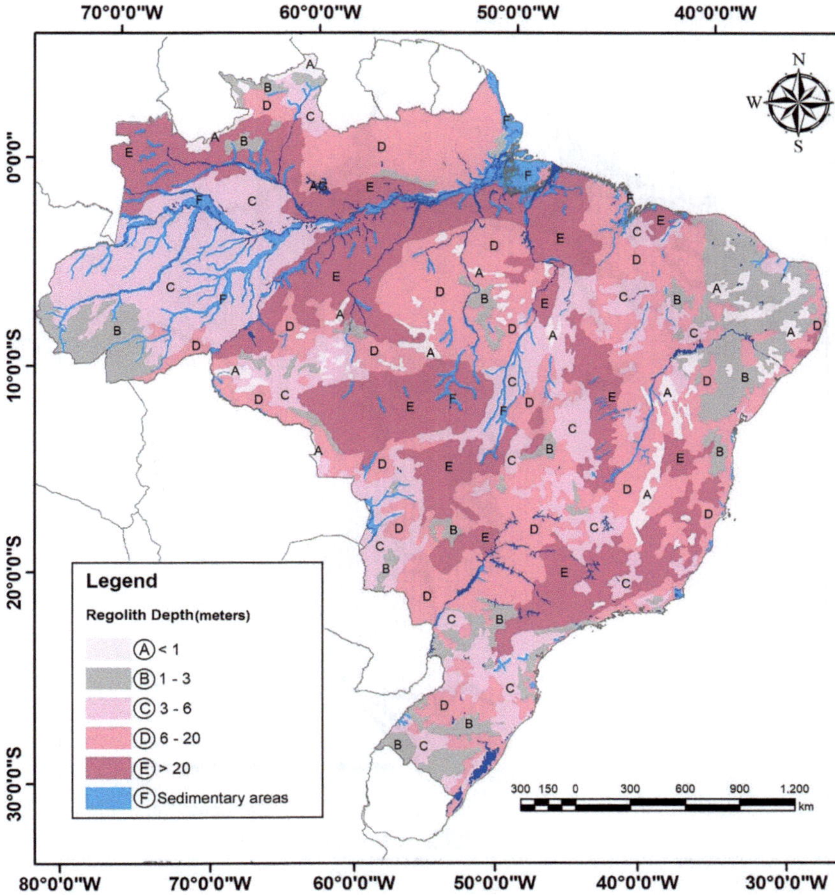

Fig. 2.10 The approximate regolith depth in Brazil is based on data obtained from soil surveys throughout the country and compiled by C. Schaefer

2.4.1 The Kaolinitic Mantle

In the great majority of the Brazilian landmass, a "kaolinitic mantle" (*sensu* Millot 1983) prevails, either on crystalline basement rocks (gneiss, granites, migmatites) or on Cenozoic clayey sedimentary rocks. This so-called lateritic mantle (*Lato sensu*) is composed of kaolinite, quartz, and secondary Fe and Al oxy-hydroxides in lesser amounts. According to Boulet et al. (1998), this kaolinitic mantle in the tropical zone under contrasting seasons (up to 7–8 months of dry season) lacks a pedoclimatic equilibrium under present-day dry (or super humid) climatic extremes, where it is interpreted as inherited from wet and dry tropical climates (Schaefer 2013). Most kaolinitic subsoils under the Atlantic or Amazon rainforest are covered

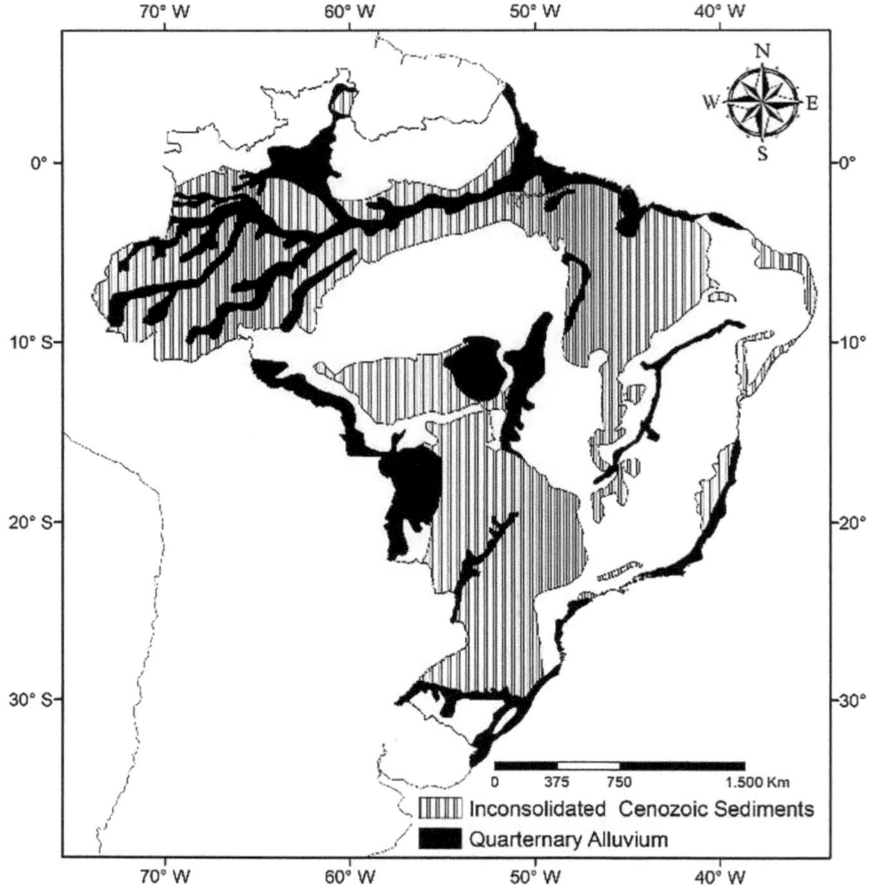

Fig. 2.11 The Cenozoic sedimentary cover (mostly kaolinitic) matches the occurrence of three Paleozoic basins so that a structural control on its distribution is inferred. The recent Quaternary alluvium, however, is either associated with zones under intense post-Miocene tectonic movements (Coastal Tablelands, Pantanal, and Araguaia basins) or to widespread alluvial sedimentation by extraordinary river systems (Amazon, Juruá, Purus, and Madeira rivers) (Schaefer 2013)

by deep Latosols, especially on the formerly forested zone of juxtaposed demi-orange hills and hillocks. Whenever the parent rock is more resistant to weathering, like in coarse-grain plutonic granites, inselbergs emerge from the dissected plateau, a common feature across the entire tropical belt, like in the Manhuaçu region of Minas Gerais (Fig. 2.12).

Most Brazilian soils and subsoils are kaolinitic, but in some cases, they can be degraded. The pedological transformations of the kaolinitic mantle, under superhumid climates, involve strong acidification, clay destruction, and particle migrations, all accompanied by microstructural changes, with a decrease of soil volume (up to 30%), collapse, and formation of the deep residual sandy mantle (Schaefer 2013).

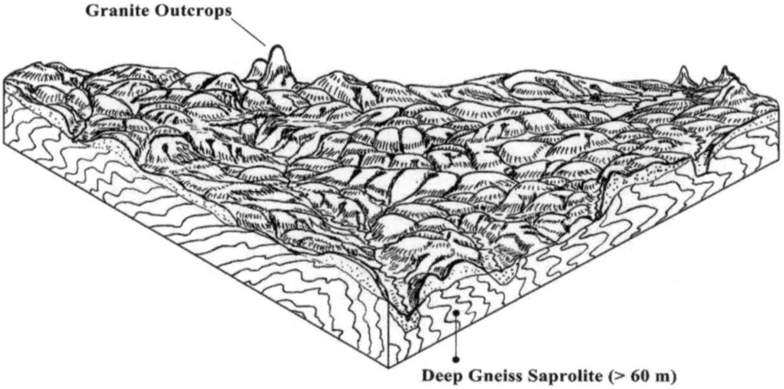

Fig. 2.12 Deep kaolinitic soils and subsoils on deep saprolites and resistant plutonic granites forming rocky crests and outcrops, in a highly dissected terrain at the "Mar de Morros" (Sea of Hills) in the Manhuaçu region, Minas Gerais State (Schaefer 2013). (Drawing by C. Schaefer)

These lateral transformations have been observed in Africa (Senegal and Cameroon) by Chauvel (1977) and Amazon by Andrade et al. (1997) (Figs. 2.13 and 2.14), with dissection of plateaus and development of lateral glacis (Chauvel et al. 1987). The progressive accumulation of skeleton grains and matrix disruption is accompanied by clay hydrolysis. If the climate turns drier, reworking these sandy mantles by eolian erosion gives rise to extensive dunes system and widespread peneplanation (Schaefer et al. 2020; Fig. 2.13).

2.4.2 Ironstone or Ferricrete Mantles

In Cerrado (Savanna) areas, kaolinitic or bauxitic mantles may possess an overlying ironstone cap in the form of hardened laterite (petroplinthite, canga, cuirasse, duricrust, ferricrete, ironstone), particularly if the bedrock is Fe-rich. They are very common on the top of old planation surfaces where soils have been stripped by erosion, exposing the plinthite layer to irreversible induration. They also frequently occur along the steep slopes and scarps of high plateau due to lateral migration of Fe and further oxidation at the edges. In this case, they help preserve the upslope planated surface from erosion and dissection. Under forest (Capões, Fig. 2.15), these ironstones or ferricretes of rupestrian grasslands (Campo Rupestre de Vellozia) are progressively destroyed and form Latosols with loose petroplinthite fragments, called Fe concretions. These are the classical concretionary soils of Brazil. When formed under past or present-day conditions of poor drainage, they have nodular features and pisolites, normally rich in Mn and Fe, and occur on waterlogged bogs under a grassy cover (Campo Brejoso Nodular; Fig. 2.15).

Similar degradation processes of petroplinthite and ferricrete under humid tropical climates have been shown in studies in Upper Volta and Senegal, West Africa

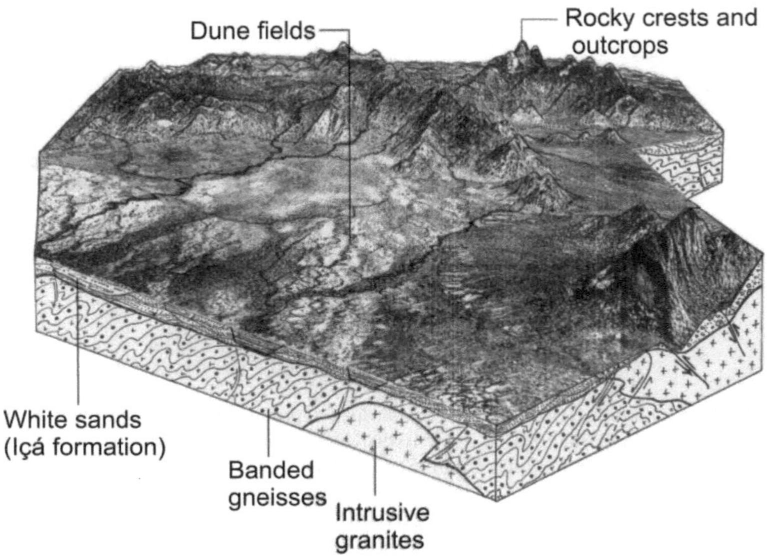

Fig. 2.13 The biggest resistant granitic/gneiss massif in northern Amazon, the Serra da Mocidade, is surrounded by the largest tropical peneplain on Earth: the Rio Negro-Rio Branco white sands peneplain, where deep kaolinitic saprolite mantles are progressively degraded into a quartz-rich complex (Içá Formation), forming sandy hydromorphic plains, sandy alluvial plains, sandy dunes, and large alluvial fans along the mountain footslopes. These mountains possess the largest highland Amazon rainforest preserved, with deep weathering on gneisses and shallow soils on intrusive granites. Down the peneplanated landscape, Spodosols with Campinarana vegetation dominate. (Drawing by C. Schaefer)

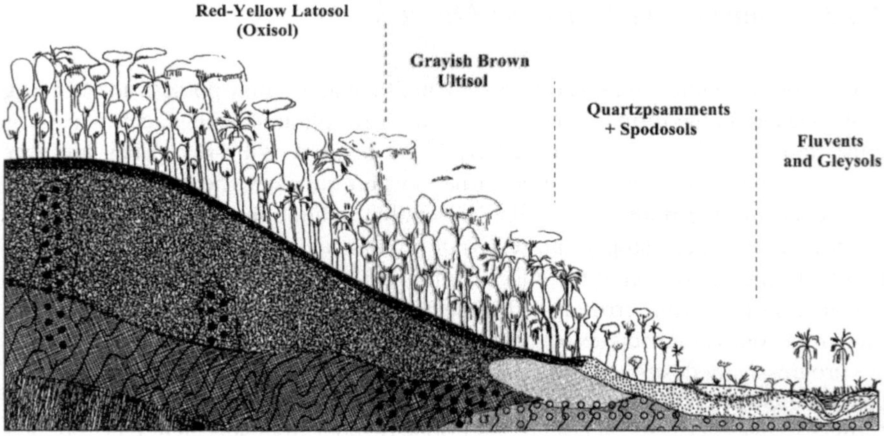

Fig. 2.14 The lateral transformation of the kaolinitic mantle overlying granite/gneiss saprolites into a sandy, hydromorphic domain of Spodosols and Entisols (after Andrade et al. 1997). The black dots indicate the presence of petroplinthite in the subsoils or plinthite with mottling downslope. This pedological transition also forms a vegetation gradient, from Forest to Campinaranas. (Modified from Schaefer 2013. Drawing by C. Schaefer)

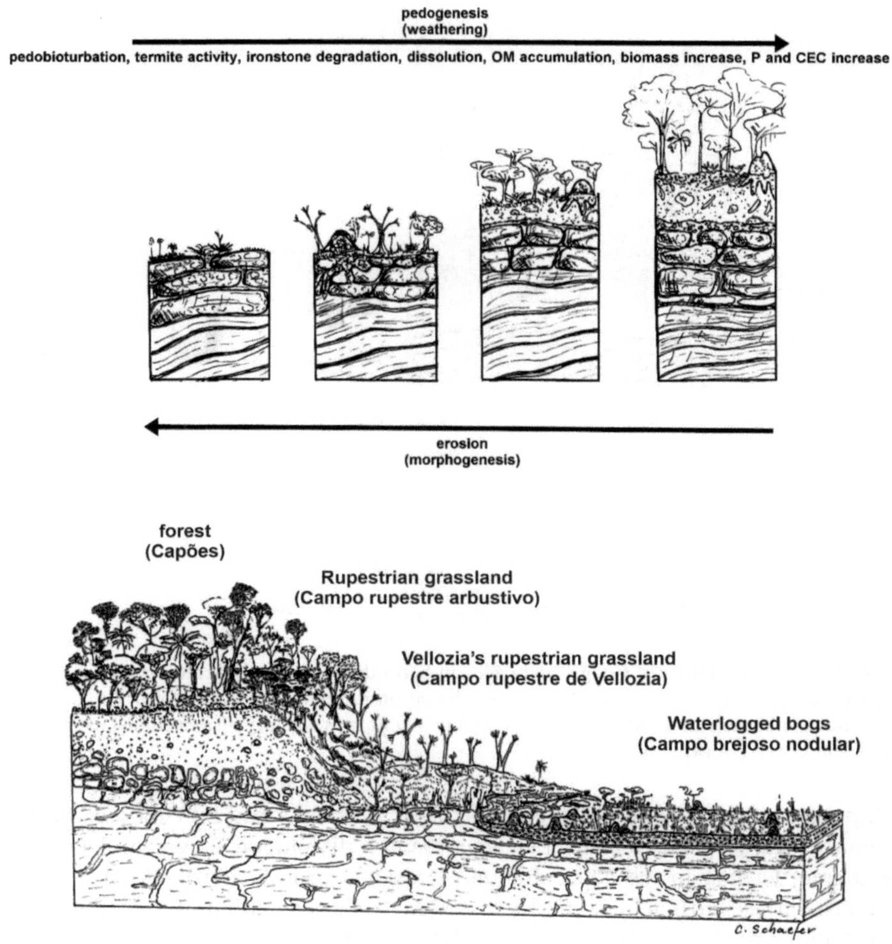

Fig. 2.15 Typical sequence of Canga (ferricrete, ironstone) degradation with accompanying changes in vegetation and soils. These are the richest subsoils in terms of Fe contents, reaching values of >70% Fe_2O_3. (Drawing by C. Schaefer)

(Leprun 1979; Nahon et al. 1977), Brazil (Costa 1991; Schaefer 2013), and Australia (Hubble et al. 1983), illustrating their continuous evolution under contrasting climate conditions, leading to loss by degradation (Figs. 2.16 and 2.17).

Under seasonal climates, the initial iron accumulation does not attack the original saprock structures and develops by filling voids and primary minerals' epigenesis. Further evolution and pedological transformations lead to microstructuration, nodulation, and internal reworking and cementation. In some cases, with high Fe contents, these ferricretes can form meters of "ferruginous canga," similar to a short-range transported agglomerate. Suppose climate (or microclimate) allows water accumulation and pedobioturbation by termites (mostly): in that case, the

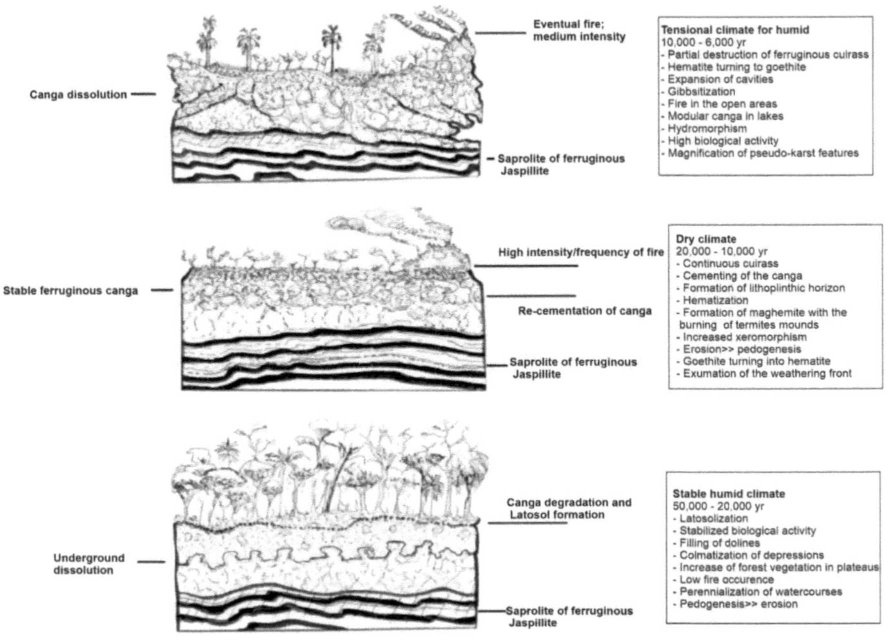

Fig. 2.16 The evolution of Canga landscapes in Serra dos Carajás (ferricrete, ironstone) under contrasting climates in the last 50,000 years, based on different proxies. Latosol formation by intense pedobioturbation followed the onset of humid climates, and dry phases eroded most of the Latosols cover. (Drawing by C. Schaefer)

degradation occurs, and the lithoplinthic layer is transformed into a loose concretionary layer (Figs. 2.16 and 2.17) under the iron dissolution, clay hydrolysis, and leaching, leading to the concentration of residual resistant ferricrete and quartz grains. Surface erosion by rain wash disperses the Fe concretions of colluvial nature in glacis or pediments. An ironstone landscape is leveled by the combined action of weathering, iron accumulation with constant volume, pedological restructuration, pedobioturbation, and development of surface leached horizons with volume losses.

2.4.3 Types of Tropical Alteration and Subsoil Formation

In trying to assemble all types of tropical alterations into classified geochemical types, Pedro and Melfi (1983) proposed the separation of two basic groups: (*i*) those suffering total hydrolysis and desilification and (*ii*) those of partial hydrolysis and desilification (Tables 2.1 and 2.2).

Hence, according to Pedro and Melfi (1983), the evolution of the tropical regolith under humid conditions leads to two main forms of paragenesis (Table 2.2) that represent the two basic types of subsoils in Brazil:

Fig. 2.17 Gondwana ferricrete landscapes: two termite mounds are found on similar pre-Miocene ferricrete – (**a**) one in Gandarela (MG), Brazil, where shallow concretionary soils are transformed into Latosols under wet subtropical climate, and (**b**) at Tom Price, Pilbara (Western Australia), where shallow concretionary soils are transformed into deep Petric Plinthosols by pedobioturbation process. In both cases, the old, pre-weathered Fe-rich saprolites (> 50 m deep) occur on similar Precambrian banded iron formation but very contrasting climates. (Brazil, humid tropical; Western Australia, aridic, desert)

(*i*) The Al hydroxides-ferric hydrates paragenesis, in what is called ferralitization, with low to nill negative liquid charge (CEC) under sensu stricto ultraferralitization and high Fe contents (e.g., Canga).

(*ii*) The ferric hydrates/kaolinite paragenesis, in what is called kaolinite ferralitization.

In many areas of the high altitude Atlantic mobile belts, kaolinitic (monosialitization) and bauxitic (allitization) regoliths are controlled by the fracturing of resistant rocks and degree of desilification leaching, leading to kaolinite and gibbsite occurring side by side in the same profile. One emblematic example, long recognized in the Brazilian highlands, is the Itatiaia massif, where shallow soils occur under subtropical to temperate climates. These soils are developed from alkaline rocks (syenites) of Late Cretaceous age, one of the largest plutonic intrusions in Brazil (Fig. 2.18).

Only under dry semiarid tropical climates do limited areas of sialferritization occur, especially when drainage conditions are not good. In semiarid Brazil, however, kaolinite ferralitization is more common.

Table 2.1 The main geochemical types of tropical alteration and resulting subsoils

Evolution of aluminosilicate minerals	Degree of hydrolysis	Total hydrolysis	Partial hydrolysis	
	Silica geochemistry	Total desilification	Partial desilification	
	Neoformed minerals (general type)	Aluminium hydroxides (allites)	Clay silicates (siallites)	
	Mineralogy	Gibbsite $Al(OH)_3$ Boehmite γ-AlO OH	Kaolinite 1:1	Smectite 2:1
	Geochemical process	Allitization	Monosiallitization	Bisiallitization
	Geochemistry of alkaline and earth alkaline cations	Total desalkalinization		Partial desalkalinization
Evolution of ferromagnesian minerals	Mineralogical types	Amorphous ferric hydrates (ferrihydrite) and crystalline compounds (hematite αFe_2O_3 and goethite αFeO OH); some maghemite		Ferric smectite (Nontronite)
General evolution of parent rocks	Geochemical process	Ferrallitization		Sialferritization
		Ferrallitization Sensu stricto (Ultraferrallitization)	Ferrallitization Sensu *lato* (Kaolinite ferralitization)	
	Characteristic paragenesis	Gibbsite-ferric hydrates (Oxidic type)	Kaolinite-ferric hydrates (Kaolinitic type)	Smectite (Al-Fe)

Modified after Pedro and Melfi (1983)

Table 2.2 The pathways of regolith evolution in ferrallitization in tropical conditions

Evolution pathways	Characteristic mechanisms	Corresponding phenomenon	Genesis conditions		Relations with the pedoclimate
Homogeneous	Close association clay/iron/skeleton – pedoplasmation bonding	Ferrallization	Well drained – Humid		Zonal evolution
Heterogeneous with segregation	Demixture clay-iron skeleton-plasma separation	"Laterization" s.l.	Alternating	Alternating wet and dry seasons	Seasonal evolution
				Alternation of humid drained periods and waterlogged periods (free ground water)	

Modified after Pedro and Melfi (1983)

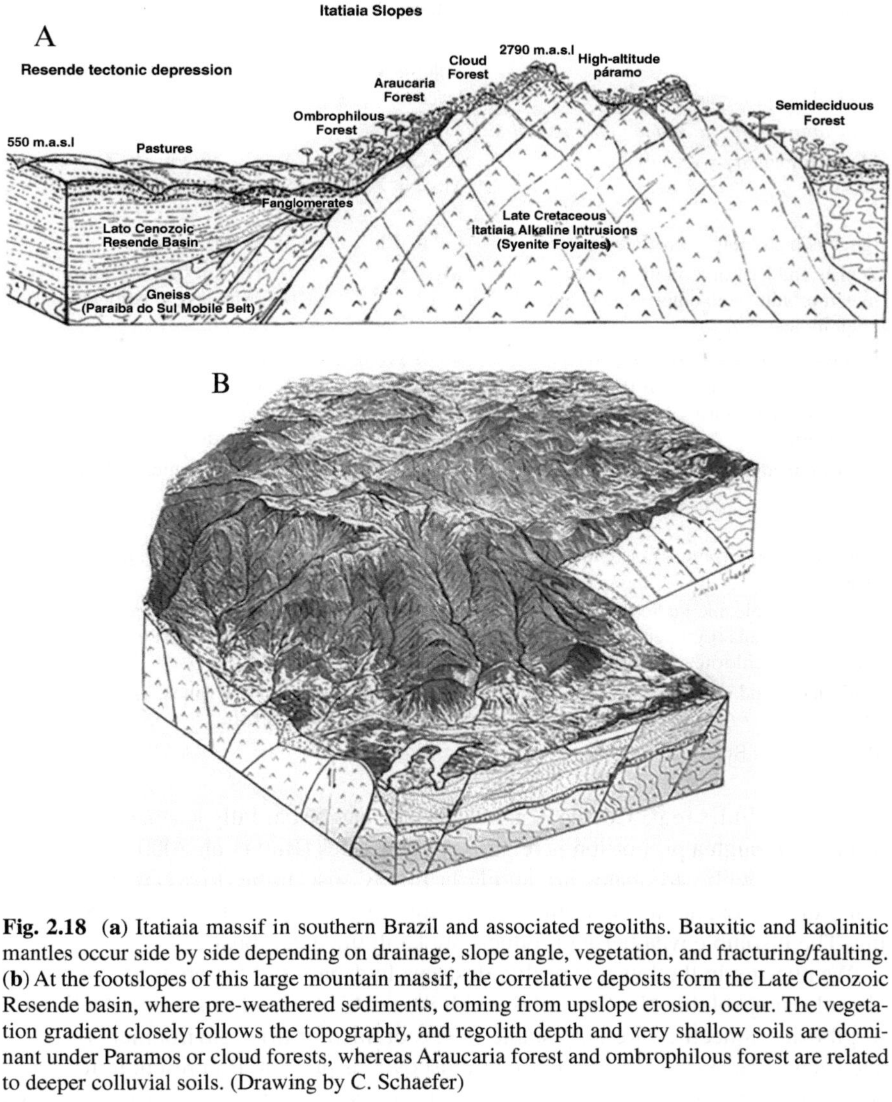

Fig. 2.18 (**a**) Itatiaia massif in southern Brazil and associated regoliths. Bauxitic and kaolinitic mantles occur side by side depending on drainage, slope angle, vegetation, and fracturing/faulting. (**b**) At the footslopes of this large mountain massif, the correlative deposits form the Late Cenozoic Resende basin, where pre-weathered sediments, coming from upslope erosion, occur. The vegetation gradient closely follows the topography, and regolith depth and very shallow soils are dominant under Paramos or cloud forests, whereas Araucaria forest and ombrophilous forest are related to deeper colluvial soils. (Drawing by C. Schaefer)

2.4.4 *The Deep Subsoil Composition*

The behavior of elements in tropical subsoils depends on the host rock's mineral and chemical composition and relative mobility and retention under tropical conditions (Table 2.3.). A good synthesis has been presented by Butt and Zeegers (1992), serving as a framework on which we have based our discussions, as follows.

Rock containing sulfides is one of the most unstable in humid, oxidizing environments, and S is strongly leached from the deep saprolite. Many insoluble elements

Table 2.3 Relative mobility and retention of elements in the regolith, with indication of hazards (bold, underlined)

Rocks/host minerals	Soluble and leached	Partly retained in secondary minerals
Soil/saprolite interface and upper saprolite		
Metasedimentary with sulfides:	As, Cd, Co, Cu, Mo, Ni, Zn, S, Al	As, Cu, Ni, Pb, Sb, Zn (Fe oxides), Al (gibbsite),
Igneous/metamorphic with muscovite	Cs, K, Rb	Si, Al (kaolinite)
Schists and hydrothermally altered metamorphics with chlorite, talc, and amphiboles	Mg, Li	Fe, Ni, Co, Cr, Ga, Mn Ni, Ti, V (Fe, Mn, and Ti oxides)
Pellitic sedimentary with smectites	Ca, Mg, Na	Si, Al (kaolinite)
Ironstone and ferricrete of varying composition (itabirite, jaspilite, and carbonatite)	Fe and trace elements	Al, Cr, Fe, Hf, Ga, Nb, REE, Th, Ti, V, W, Zr (depending on composition)
Limestone and marble with carbonates	Ca, Mg, Mn, Sr	Si, Al (kaolinite, gibbsite), Fe oxides
Lower saprolite		
Igneous/metamorphic – Feldspars and plagioclases	Ca, Mg, Cs, K, Na, Rb	Si, Al, (kaolinite, gibbsite); Ba (barite)
Plutonic or volcanic mafic rocks – ferromagnesians (pyroxene, olivine amphiboles, chlorite, and biotite)	Ca, Mg	Fe, Ni, Co, Cr, Ga, Mn, Ti, V (Fe, Mn, and Ti oxides)
Carbonates and alkaline intrusions with apatite	Ca, Mg, Na, K	U, Th, P (phosphates), Ti, Al (gibbsite)

Modified from Butt et al. (2000)

hosted by sulfides (e.g., Cd, Co, Cu, Mo, Ni, Zn) can be partially leached deep in the profile, although a proportion is retained in Fe oxides (Butt et al. 2000). Limestone and marble with carbonates are similarly highly susceptible to weathering, especially under wet, acidic environments, leading to strong Ca, Mg, Mn, and Sr leaching. The insoluble weathering products are kaolinite and gibbsite.

Weathering in the lower saprolite destroys feldspars and ferromagnesian minerals, releasing soluble Na, Ca, and Sr, with Si and Al, retained as kaolinite/halloysite or gibbsite. Smectites are only stable in less drained or drier environments. K, Rb, and Cs will be leached if orthoclase or muscovite is present. Ba present in feldspars is released and precipitated as stable barite. Less stable ferromagnesian minerals (pyroxene, olivine, amphibole, biotite, etc.) are strongly altered and form Fe oxides, with retention of minor and trace elements (Ni, Co, Cu, Mn, and Ni) and losses of Mg and Si, partially retained in smectite (Mg, Si), kaolinite (Si), or quartz (Si).

Only the most resistant primary minerals persist at the upper saprolite, and less stable secondary minerals (smectites) are destroyed. If present in the host rock, serpentine and chlorite are weathered, but talc and muscovite may persist up to the surface. Ferromagnesian minerals host transition metals (Ni, Co, Cu, and Zn) in mafic and ultramafic rocks which are retained in the saprolite incorporated in secondary Fe-Mn oxides. Limited leaching and redistribution of Ni, Co, and Mn are restricted to periods of poor drainage and reduction.

Most igneous and metamorphic rocks develop a bleached and mottled saprolite, where only quartz persists, whereas all primary minerals are destroyed. The insoluble products are rich in Si, Al, and Fe, as kaolinite, quartz, Fe oxides (hematite and goethite), and gibbsite. The kaolinization of saprolites is most potent over felsic rocks, while mafic and ultramafic rocks usually concentrate Fe oxides and gibbsite.

In physical and morphological terms, the accumulation, partial dissolution, recrystallization, and pedoturbation of Fe oxides result in various features, like pisoliths, nodules, tubules, and concretions. Micro or trace elements (Cr, Co, As, Ga, V, Cr, Ni) closely follow Fe oxides, due to substitution or coprecipitation. The same applies to the distribution of Zr, Hf, Th, Nb, Ta, W, rare earth elements, and Ti due to their relative chemical immobility (e.g., Zr and Hf in zircon; Ti in rutile and anatase).

2.4.5 Latosols and Similar Soils Overly Deep Weathered Saprolites in Brazil

Mostly Latosols (Oxisols), with oxic B horizons, dominate the Brazilian landmass. According to Schaefer (2013), most Latosols and associated Cambisols on deeply weathered saprolites in Brazil have very low CEC, and all charge is pH-dependent. The conventional procedure for measuring cation exchange capacity with neutral, 1 N ammonium acetate, where pH is adjusted to 7, overestimates the real CEC at actual soil pH. In this case, Al^{3+} occupies most negative charged exchangeable sites, and allic soils are the rule (> 50% of Al saturation). As weathering advances towards almost complete leaching, kaolinite is replaced by gibbsite, and pH becomes higher by approaching the zero point charge of Al and Fe oxides. This is the case of the soils on itabirite, one of the oldest and deepest Latosols (Oxisols) worldwide.

Many Latosols have Bw horizons where pH in KCl is higher than pH in water, indicating a positive net charge. This is called "isoelectric weathering," when leaching is so intense that soil pH tends to drift to pH_0. Prolonged leaching under warm and humid conditions leads to "weathering out" of constant surface charge minerals, and the pH of the soil solution drifts to pH_0 of the insoluble residue.

The Latosols are comparable to Oxisols (USA) and "ferrallitic soils" (French system). They are broadly defined as having the following attributes: (*i*) complete weathering of primary minerals with the possible exception of some inherited minerals, mostly quartz and Ti/Zr-resistant minerals (the washing out of alkaline and earth alkaline bases and of the more significant part of silica); (*ii*) presence of the following secondary compounds formed in the soil (1/1 lattice clay minerals and oxy-hydroxides of aluminum and iron, both in crystallized and amorphous forms); and (*iii*) materials which are organized in A-B-C or A-(B)-C profile where the B horizon is usually thick (> 50 cm in depth) and contains the essential secondary

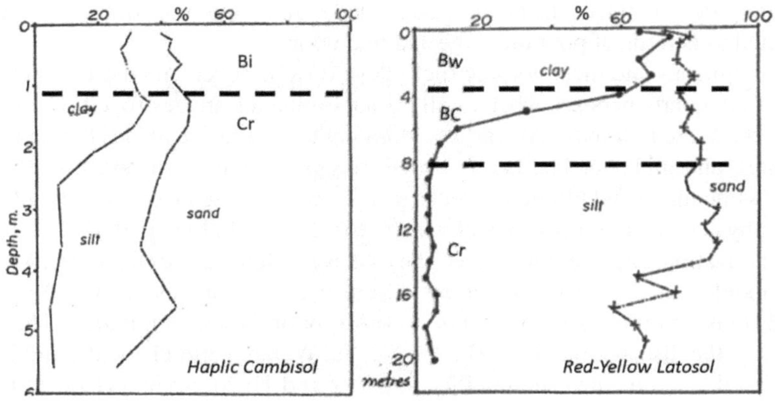

Fig. 2.19 The subsoil/regolith anisotropy in Brazilian highlands: strong textural difference with depth in two distinct soils/saprolites on gneiss (haplic Cambisol and red-yellow Latosol) (after Rezende 1980). The pedoplasmation front (Stoops and Schaefer 2018), where clay is formed by combined chemical, physical, and pedobiological processes, is clearly illustrated at the BC horizon depth (4 m) at the red-yellow Latosol

minerals quoted above, with less than 5% weatherable primary minerals. The A horizon is usually relatively thin with a low organic matter content, except if the climate (past and present) is (was) colder. The C horizon is normally very thick and contains a similar assemblage of secondary, weathered minerals. If igneous or metamorphic, it physically resembles the parent rock but can be easily crushed with the fingers. The isovolumetric weathering is the rule, and (iv) the following characteristics are associated with this morphology: low cations exchange capacity, low sum of exchangeable base cations, medium to high Al saturation, and low pH.

After prolonged weathering and chemical denudation, the fresh parent rocks are generally absent in the tropical belt of Brazil, and the nature of saprolites has a definite impact on soil constituents, irrespective of bioclimatic conditions. Deep mantles of altered rocks are essentially formed by secondary minerals, irrespective of the richness of parent materials (limestone, shales, sandstones, siltstones, gneiss, granites) due to profound long-term losses of all soluble elements and silica. In the case of limestones, where a minor proportion of impurities occur, residual soils are formed by clays inherited from those impurities. On the other hand, quartzites and banded iron formations show a very low rate of weathering and erosion and form prominent landforms.

The clay content rises abruptly to the top, in the boundary between soils and saprolites, across varying lithologies, as illustrated by a classical study in Minas Gerais (Rezende 1980) (Fig. 2.19). Thus, soils and saprolites have contrasting textures, and saprolites are susceptible to erosion if exposed, since they have high silt content and no structural development. This is clearly shown by comparing the microstructure of saprolites and overlying Latosols under thin section, as illustrated in Fig. 2.20.

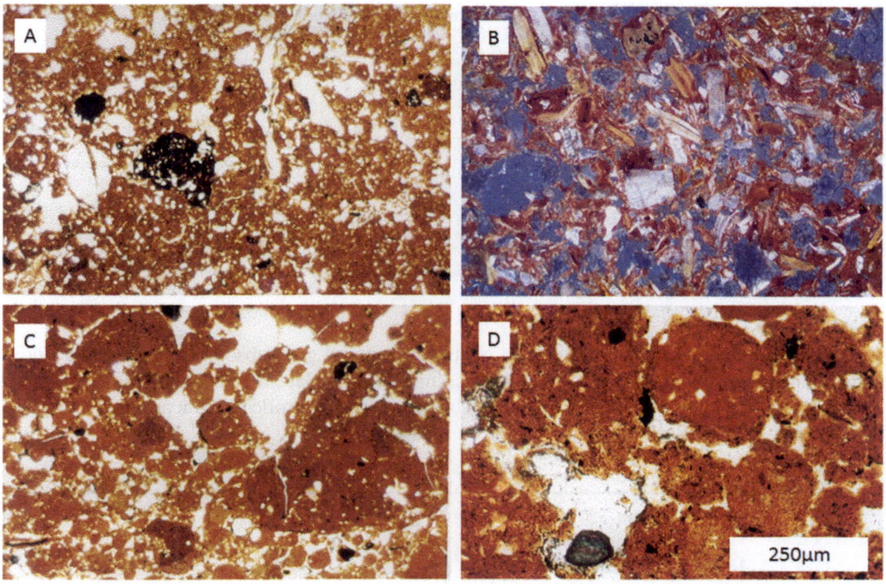

Fig. 2.20 Micromorphology of BA (A, plain light), Bw (C and D, plain light), and Cr (saprolite) horizons (B, polarized light) of a deep weathered red-yellow Latosol from Rio de Janeiro at the Tijuca National Park. The microped structure is clearly seen on the oxic, latosolic horizon (Bw), whereas a welded structure is observed at the BA. In the saprolite, the porous isovolumetric weathering of the gneiss left large silt-sized kaolinite pseudomorphs after biotite, Fe-minute particles dispersed, and quartz grains. The bluish color indicates the porous space formed by intense leaching and deep weathering. (Schaefer 2018, Stoops and Schaefer 2018)

Dissected landforms are typical of the Atlantic Forest morphoclimatic domain where the widespread dissected and polyconvex hill pattern called "Sea of Hills" occurs (Ab'Saber 1996). According Ab'Saber (1996), landforms of the forested environment have (*i*) the general mamelonization (convex slopes) of the low and medium slopes in the mountainous regions forming rounded hills bordered by fluvial terraces and pediments on the bottom valleys; (*ii*) deep saprolites formed by the generalized decomposition of rocks; (*iii*) widespread presence of deep, overlying reddish-to-yellowish Latosols or related soils; (*iv*) buried stone lines formed during Quaternary climatic fluctuation; (*v*) the presence of inselbergs ("sugar loaves") of granitic resistant domes with boulder fields (Fig. 2.21); and (*vi*) presence of alveolar upland plains in mountainous regions, perched valleys, and ill-drained headwaters. The drainage network is typically perennial and with high density of streams. To the west of Brazil, landforms become flatter and planated, as we move towards the transition to the savanna biome. The exceptions are the resistant sandstone or quartzite mountains, where regoliths are poorly developed and shallow soils dominate. The Serra de São José (Fig. 2.22) is a typical example of very shallow soils and regoliths in a humid tropical landscape of resistant quartzites and metasandstones.

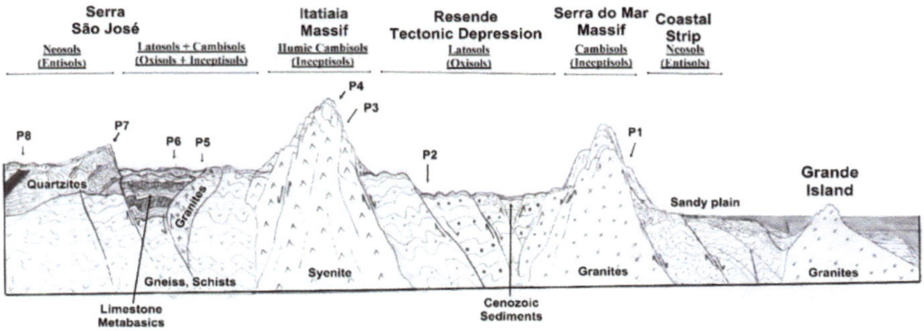

Fig. 2.21 The topographic and structural control of regolith and subsoils in the Atlantic Forest Zone clearly show the important role of tectonics and block displacement in exhuming the resistant cores of granitic plutons, where regolith is much shallow and controlled by fracturing and faulting. (Drawing by C. Schaefer)

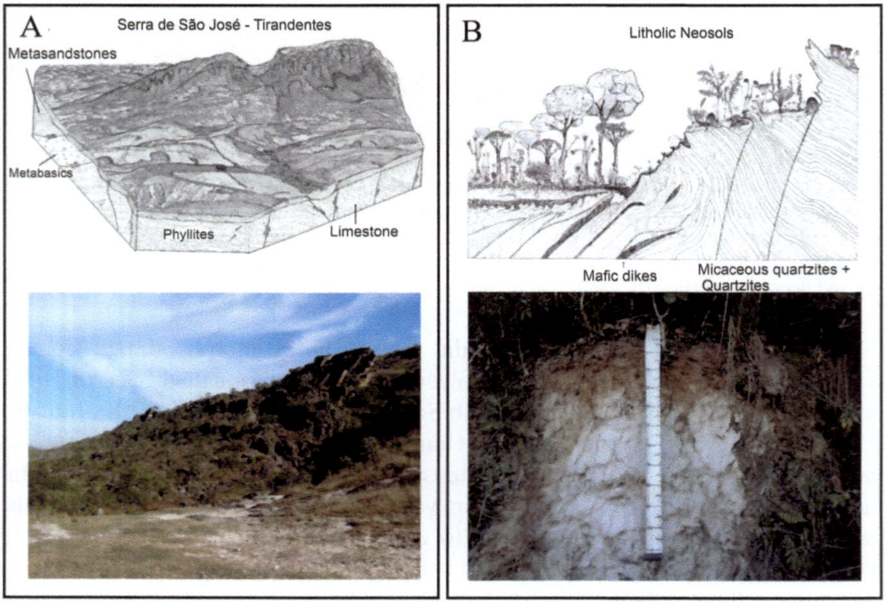

Fig. 2.22 (**a**) The quartzitic landscapes of the Proterozoic at São José Mountain Range (Tiradentes, MG State). The very shallow soils (Litholic Neosols – **b**) are associated with quartzite, and metasandstone outcrops, representing the most resistant rocks under tropical weathering conditions. The regolith is more developed on the metabasic rocks that occur as mafic dikes. Most water infiltration follows the fractures and faults since the overlying soils do not offer good permeability. The vegetation is a typical "Campo Rupestre" (rupestrian grasslands) with many endemic species of high conservation value

2.5 Hydrogeological Characteristics of Deep Subsoils

The relationships between surface and subsurface hydrology of the Brazilian landmass and the nature and depth of associated regolith are diverse and interwoven but remain little explored and interpreted. The hydrogeology of this deeply weathered regolith in tropical landscapes is little understood but has direct practical implications for water management. All long-term processes that led to deep regoliths and chemically hostile subsoil formation have produced extensive groundwater reservoirs worldwide.

The Brazilian subsoils, like others from elsewhere in the tropics, have marked differences in groundwater recharge. In most sedimentary areas, the dominance of sandy substrates and deep strata promoted the formation of the largest aquifers in South America, at three main Paleozoic basins, as follows: (*i*) the Paleozoic Paraná Basin, with a Cretaceous sandstone cover on top; (*ii*) the Maranhão Piaui Basin, also with a similar overlying Cretaceous sandstone; and (*iii*) the Amazon Basin, partially covered with remnants of Cretaceous sandstones but largely capped by a Cenozoic clayey cover, called the Solimões Formation. The Cretaceous cover (Alter do Chão Formation) runs along both margins of the Amazon Valley, in the middle and lower sectors, with a vast groundwater reservoir (Fig. 2.23).

These large areas of deep sandy subsoils allowed the development of very large and extensive subterraneous water reservoirs, named the abovementioned basins. However, the Paraná Basin is best known as the Guarani Aquifer, one of the largest in South America and currently intensively exploited by urban centers and modern farms for irrigation. Other great aquifers in Brazil are the high tablelands of Espigão Mestre (between Tocantins and Bahia) and the Parecis (Mato Grosso), where highly intensive agriculture is taking advantage of groundwater for high productivity under irrigation. Both are associated with sandy subsoils and represent case studies where water uptake needs to be regulated to not exceed long-term water recharge before this groundwater resource is gone.

In fractured aquifers, the highest groundwater recharge is in Minas Gerais (saprolites of Precambrian crystalline rocks) and Goiás and São Paulo State (mostly basalts metapellitic sediments). The highest productivity of the granular aquifers is found on sandstones of the Cretaceous age widely distributed across the continental sedimentary basins (Fig. 2.23a, b). Because of the limited expanse of limestone outcrops and subterranean groundwater karstic aquifers (except the Irecê and Apodi Plateau), we have mapped these with the largest fractured aquifers, since they are closely associated and most limestone are also intensely fractured.

As shown in the earlier section, the tectonically stable Brazilian cratons are usually poor aquifers, except when coated by Late Cenozoic sediments. However, deep chemical weathering has produced thick regoliths (overburden) in which extractable groundwater resources abound (Fig. 2.24). In this case, regional differences in parent materials, climate, and/or spatial variations in weathering depth result in varying characteristics of regolith aquifers. The effects of these factors on the mode of occurrence of the zone of saturation in the weathered mantle and on the relationship

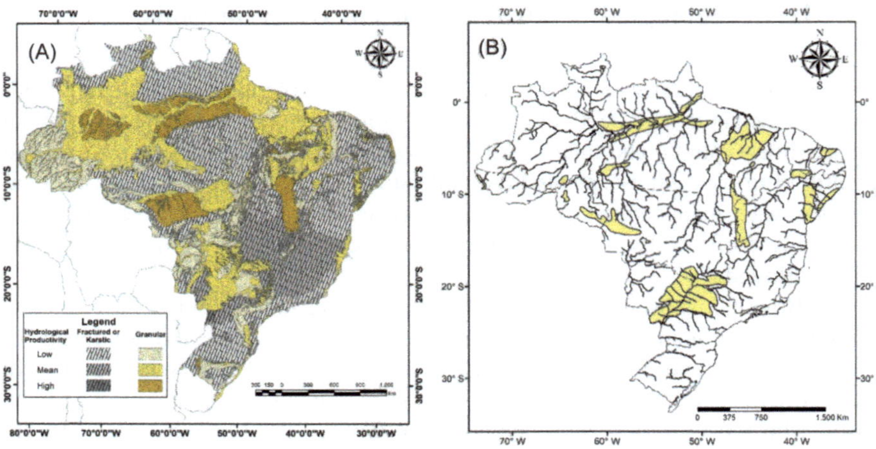

Fig. 2.23 (**a**) The hydrogeological map of Brazil, adapted from the original CPRM (2014) mapping, identifying granular and fractured (together with karstic) aquifers. (**b**) Illustrates the deep quartzose sandstones of Cretaceous age where groundwater productivity of granular aquifers is maximum

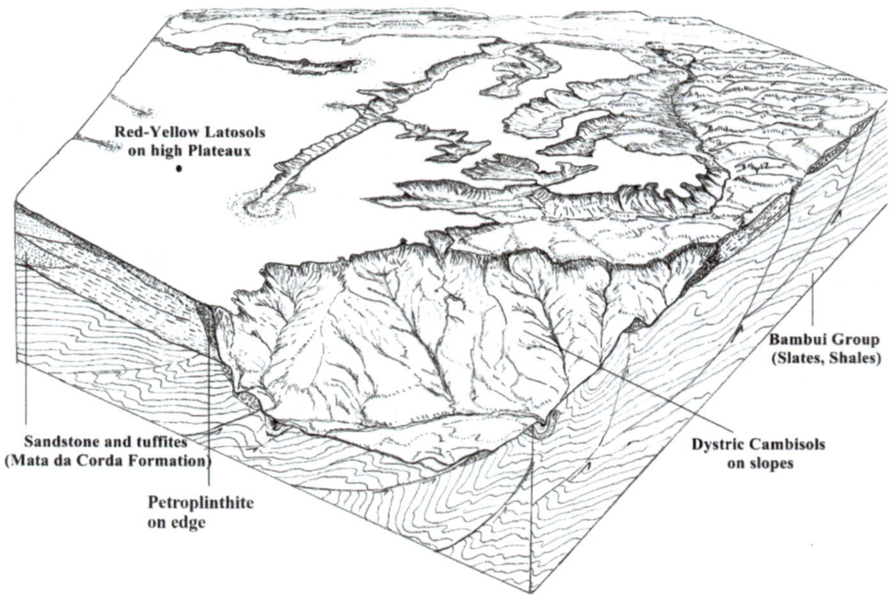

Fig. 2.24 The Upper Paranaíba Basin landscape is a model of a highland plateau of central Brazil with high groundwater recharge due to sedimentary covers. The erosion surface between the Precambrian Bambui Group (slates, shales) and the overlying cover of Cretaceous sandstones and tuffs also marks the level where deeply infiltrated water in a continuous aquifer discharges, forming a myriad of draining watercourses along the steep slopes. Further upslope, a blanket cover of Canga (ironstone, ferricrete, petroplinthite) protects the highland plateau from widespread erosion. Latosols occur on top, while shallow Cambisols are dominant down slope. (Drawing by C. Schaefer)

between saturated zone thickness and weathering depth have been reported for deeply weathered landscapes in Africa (Enslin 1943; Faniran and Omorinbola 1980). While isolated groundwater compartments occurring in discrete basins of decomposition tend to characterize regolith aquifers in semiarid areas (Enslin 1943), the zone of saturation in the regolith overburden is generally widespread or spatially continuous in the more humid low relief areas (Omorinbola 1984; Schaefer 2013). Even in humid areas, local rainfall variations can be used to explain differences in the values of weathering depth threshold for the formation of a groundwater zone in tropical regoliths (Omorinbola 1984).

Probably the greatest challenge posed by the hydrogeology of the tropical belt is to obtain reliable estimates of the groundwater resources in the regoliths. A quantitative evaluation would offer great opportunities for successful groundwater exploitation schemes, especially in the rural areas where piped water is usually rare.

With increasing pressure on groundwater resources by irrigated agriculture in Brazil, there is a big challenge to quantify the groundwater resources in tropical Brazilian regoliths accurately. Limitations of indirect methods for evaluating groundwater storage within regoliths are due to many geohydrological peculiarities of deeply weathered tropical terrains. In this sense, Omorimbola (1984) suggests that direct methods of quantitative groundwater evaluation which utilize data on hydrogeological parameters of regoliths should be applied. The relevant parameters include the spatial pattern of regolith aquifers, the areal extent and thickness of the groundwater zone, and the porosity and specific yield of the saturated zone. Appropriate computational formulae based on these parameters must be used for evaluating the total and the drainable groundwater storage in the regoliths at a given point in time.

Despite the absence of substantial groundwater resources in the semiarid zone of Brazil, the presence of scattered remnants of Cretaceous sandstones, such as the case of Araripe plateau in the Cariri of Ceará state, makes the crucial difference that allows social development and promotes sustainable family agriculture. These are the so-called Brejos do Sertão, where Cretaceous sandstone groundwater is key for the human settlements downslope within the semiarid depression (Fig. 2.25).

2.6 Dating of Brazilian Subsoils, Age of Regoliths, and Rates of Denudation

As previously demonstrated, deep regoliths, saprolites, and soils are seen throughout Brazil. Their ancient age is quite variable when considering the whole *solum* volume. In the upper part, most Latosols are younger than 10,000 years based on 14C dating of soil organic matter, whereas the underlying saprolites have ages in the range of 5–10 MA up to 70–90 MA. One of the oldest regoliths on Earth, the Serra da Moeda and Gandarela (Iron Quadrangle), with a cap of ironstone on banded iron formation, reaches more than 250 meters of depth, and more than 70 MA of age,

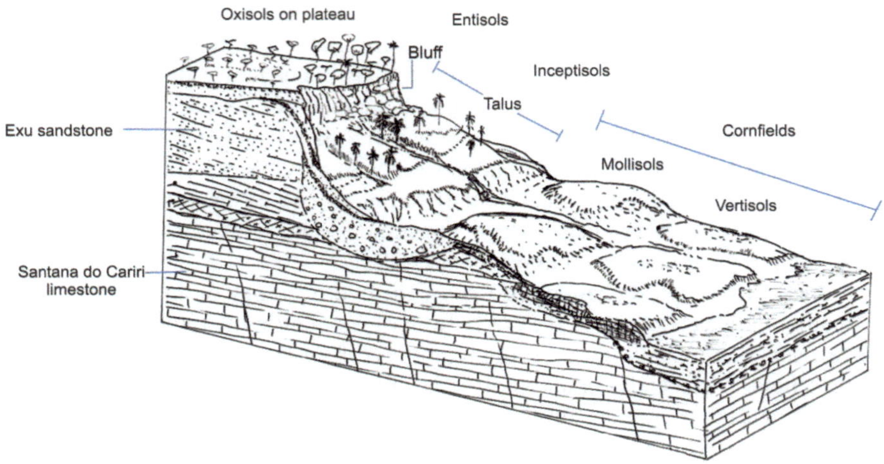

Fig. 2.25 A block diagram of the Araripe plateau and the Cariri depression near Crato (Ceará State). Intensive sugarcane, maize, and beans cultivation on rich Vertisols or Chernosols on limestone are possible only because of the groundwater resources that drain from the extremely poor Exu Sandstone Formation (with Latosols), upslope. (Drawing by C. Schaefer)

dating back to the Late Cretaceous (Fig. 2.26). The estimated erosion rate based on ^{10}Be dating is one the lowest worldwide, so that differential erosion and etchplanation are the two basic processes of landform and landscape sculpturing. Most soils in Brazil are formed on weathered saprolites and do not retain any significant chemical reserves, although they reveal subtle evidence of the lithological inheritance.

On the other hand, the role of differential chemical denudation in different rocks is marked, and the same area may produce strikingly different regoliths (Fig. 2.27). Based on careful measurements of dissolved products of weathering in monitoring sites across river catchments in highlands of Minas Gerais, Salgado et al. (2004) showed a large variation in chemical rates of weathering, varying from high denudation rates in limestone and carbonatic rocks (17–25 m My^{-1}); medium rates for granite and gneisses, schists, and phyllites (4–10 m My^{-1}); and lower rates for banded iron formation and quartzites (<1.5–5 m My^{-1}). This highlights the remarkable role of structural geomorphology and rock resistance in landscape evolution in highly weathered terrains. Mountains, in such cases, are the products of differential erosion, in which prominent reliefs are those with greater resistance to weathering and erosion.

The rate of subsoil formation remains poorly investigated. Geochemical tracers of soil erosion and regolith residence time are still little explored in Brazil. The denudation rates determined by in situ-produced ^{10}Be concentration in sediments and soils vary between 9.0 m My^{-1}, in plateaus of Central Brazil, and 30.2 m My^{-1}, in highlands influenced by recent epeirogenesis. The higher denudation rates in highlands are attributed to by (*i*) humidity (because of orographic rains, the seaward-facing scarps of mountains exhibit mean denudation rates that are approximately

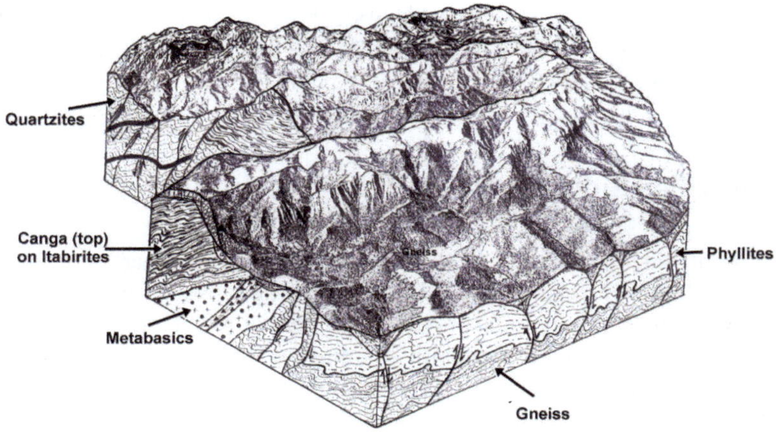

Fig. 2.26 The Gandarela Syncline and the associated deep regolith on banded iron formation, dating ages to Paleogene and Upper Cretaceous (Vasconcelos 1999; Spier et al. 2006). (Drawing by C. Schaefer)

Fig. 2.27 The double planation interfaces (surface and subsoil) in two different regoliths: (**a**) a shallow dystric yellow Latosol developed from quartzose schists from Coronel Xavier Chaves (MG) and (**b**) a deep red Latosol on metabasics from Lagoa Dourada (MG). The presence of abundant quartz veins and layers in the first case promotes the formation of thick lag deposit of stone line made of quartz pebbles and cobbles at the border zone between soil and saprolite, whereas the pedoplasmation front in the metabasics is gradual, without stone lines, due to the complete absence of quartz. The high erodibility of both saprolites shows that road cuts should avoid exposures such as those and revegetation measures must be implemented immediately after the intervention

1.5 times those of the inland-facing scarps (Salgado et al. 2016)); (*ii*) the resistance of rocks (the drainages over more resistant rocks, such as granites and quartzites, had lower average denudation rates than those over schists and gneisses (Salgado et al. 2014)); and (*iii*) neotectonics (the decrease of the denudation rates in upper sediments suggests a reduction in intraplate tectonic activity beginning in the Middle Quaternary or earlier (Varajão et al. 2009)). On the other hand, the landscape of the deeply weathered soils derived from sandstones and mafic rocks in

Paraná Sedimentary Basin is very stable. Very low erosion rates indicate that the plateaus sustained by ferruginized deposits are being dismantled very slowly since the Late Miocene (Braucher et al. 2000).

2.7 Brazilian Subsoils: Hostile for What or Whom?

The nature of Brazilian subsoils and weathered rocks, forming part of the regolith, possess both lithologic and pedologic features. Hostile subsoils for agronomic purposes may not be similarly unfavorable for geological and geotechnical uses and applications. Such is the case of bauxitic or ironstone saprolites, which are valuable assets for producing aluminum or iron ores while completely inhospitable for any agricultural use. Usually, pedogenic and lithological features found in weathered rock substrates can be identified and interpreted according to the specific interest, allowing a broad interpretation of these regoliths' processes and environmental roles.

Graham et al. (1997) pointed out that lithogenic features (rock structure, texture, and composition) strongly influence weathering rates and resulting regolith characteristics. Joint fractures provide rapid access for infiltrating water and roots, assisting weathering. As rock weathering progresses, it develops microporosity, thereby increasing the water-holding capacity of the regolith, which further enhances weathering and water availability for plants. Plant roots can penetrate the saprolite and locally change the nature of the weathered material, as clearly illustrated by bauxite hot spots concentrated around very old Jarrah trees in Western Australia. In less weathered rocks, roots follow fractures and bedding planes, creating high localized organic C concentrations that retain nutrients before they can be leached out, reaching higher amounts than in overlying A horizons. Different organic acids and dissolved CO_2 from dead, decomposing roots further promote weathering, whereas K uptake by living roots causes the degradation of biotite into vermiculite, further increasing weathering along with fractures and bedding planes. Root and faunal exploitation of the young saprolite progressively and eventually reduces the relative importance of lithogenic features, like galleries, channels, and macropores which are formed, helping conduct water more efficiently. Soluble products can be entirely removed by leaching, but suspended colloids only translocate down the regolith, forming argillans in fractures, abandoned root channels, and intergranular pores within the matrix.

These clay infillings and cutans (argillans) can be quite protected from physical disturbances down the regolith. With time, precipitate neoformed mineral phases, such as $CaCO_3$ and opaline silica in semiarid/arid regoliths, can be helpful to proxies to interpret past environmental conditions.

2.7.1 The Nature of Brazilian Subsoils and Their Environmental Limitations

The Brazilian subsoils are directly related to the nature of the parent rock. For a broad discussion on the chemical and physical limitations of hostile subsoils in Brazil, we used the framework of the major rocks found in the country and their association with soil-landform and regolith attributes (Table 2.4). This useful grouping broadly follows the systems proposed by Resende (1988) and Resende et al. (2019), adapted by Schaefer et al. (2000), and finally published by Schaefer (2013).

According to Schaefer (2013), the mineralogical/chemical composition of the rocks is the basic attribute controlling weathering processes in the tropics. However, in many cases, we observe rocks with high vulnerability to chemical change remaining little altered in the soil mass, forming boulders and blocks of granite, limestone, and diabase. In this case, structural compactness is the key factor in explaining the phenomenon since massive rocks resist water penetration, retarding the advance of weathering. The presence of fractures, faults, or banding (alternating minerals having different resistances) leads to preferential deep water flow and the removal of soluble products. On top of that, plant rooting and the role of rhizosphere and soil fauna (termites and ants – see Chap. 15 this volume Schaefer and Oliveira 2022) make new pathways for water penetration through channels and galleries into the regolith, creating a complex network of large, connected pores and channels.

In Group 1 (Table 2.4), the soils developed from acid granitic rocks can be either shallow or deep, depending on the climate in which they form. Granites with smaller grain size when quartz rich are more resistant to weathering and tend to form rocky, bouldery outcrops under steep slopes in the form of rounded peaks. Faults or major fractures control the valleys, and surface erosion and landslides are frequent and severe. Yellow Latosols on top, yellow and red-yellow Argisols on midslopes, and dystric Cambisols occur in close association, the first two on convex slopes, while concave and steep slopes are usually associated with Cambisols under wet climates. In dry areas, Litholic Neosols, Planosols, and red-yellow Argisols are most represented, with lesser Luvisols. K is the only macronutrient in high amounts, whereas all others are very low and limiting. High silt contents make these subsoils prone to erosion. Eutrophic soils and subsoils are very uncommon, except in semiarid areas.

In Group 2, the soils developed from mafic rocks (rich in Fe-Mg minerals) form gentle, smooth landforms. They weather easily, depending on the compactness and fracturing (e.g., very fractured basalts versus compact diabase). Red Nitosols and red Latosols (Oxisols) are very common in the wet and dry zones, but yellowish soils also occur in wetter areas. Chernosols and Vertisols occur sporadically on dry areas. Volcanic rocks are generally far more vulnerable to weathering than plutonic, especially those formed at greater depth. The subsoils are not as deep as those on banded rocks but comparable to Group 1 (granites). Some subsoils attained a high degree of weathering, with bauxite-rich subsoils, but most are kaolinitic (monosialitic) and some even illitic/smectitic (bisialitic). Fresh basalt is uncommon in wet climates, and most regoliths are also very leached.

Table 2.4 Grouping of rocks for pedological purposes with associated nutritional and environmental problems

Rock groups	Characteristics	Nutrient/soil element		Erosion features
		High	Low	
1. Granitic	Granites, granitoids, migmatites. Diorites; SiO_2 content>65%	K, Al, B, Si	Ca, Mg, Fe, Mn, Cu, Zn, P, Co, Se	Rilling erosion common; gullies where deep and exposed Fe-poor saprolites occur (Cambisols)
2. Mafic	Basalts, diabases, gabbros, diorites, tuffites. SiO_2 content between 54 and 65%	Ca, Mg, Fe, Cu, Si, Mn, Co, P (mostly) but depending on weathering degree	K, B, Zn	Rilling erosion is common in basalts and tuffites (red latosols)
3. Pelitic and metapelites	Slates, phyllites, siltstones, muscovite shales, diamictites, mudstone, metassiltites	K, Si, and Al	All other elements in general	Erosion depends on the slope of the sediments strata. Rilling and gullies are common in geological contacts and deeper saprolites, rich in silt. Strong sheet erosion in siltstone and mudstones
4. Arenitic	Sandstones and quartzites	Depends on cement (ferruginous, carbonate, or siliceous); with full dominance of Si	In general, very poor, with very low P and Ca besides other elements	Rill erosion in the most friable Si-cemented sandstones; quartzites are extremely resistant
5. Ferruginous	Banded iron formations, ferruginous laterites	Fe	Other elements in general, especially P	Development of subterranean cavities between the crust and the saprolite; moderate rilling erosion in the deepest and most friable soils; resistant to erosion when cemented
6. Calcareous	Limestone, marble, marl	Ca, Mg	Fe, Cu, Zn, Mn	Common underground erosion by dissolution, soil collapse (dolines)
7. Recent alluvial	Sand, silt, clay, and gravel sediments	Very variable, no trend	Very variable, no trend	Severe sheet erosion; alluvial deposition and burial of topsoil; silting up in the drainage channels

(continued)

Table 2.4 (continued)

Rock groups	Characteristics	Nutrient/soil element High	Nutrient/soil element Low	Erosion features
8. Gneiss	Mesocratic (rich in biotites) or leucocratic gneisses (with muscovite and rich in quartz) form two very distinct groups	K, Mg, Si, Al	P, Ca, B	Deep saprolites retard erosion, but muscovite-rich leucocratic gneisses give rise to very erodible soils and saprolites (rilling and gullying)
9. Conglomerate	Conglomerate, breccia, tillites	They are usually very quartz rich, poor in nutrients	Other elements in general	Gravelly and stony soils are less erodible, but it depends on topography and cementation
10. Organic sedimentary, peat bogs	Peat bogs, lignite	All in general	Other elements in general	Soils susceptible to subsidence by burning and degradation by cultivation and drainage. Extremely fragile systems

Source: Adapted from Schaefer et al. (2000), with data from Resende (1988) and Resende et al. (2019)

Group 3 represents the very common acid and poor pelitic or metapelitic rocks, with shallow or deep soils and subsoils, depending on time, climate, and bedding. Red-yellow and yellow Latosols (Oxisols) are the most developed and leached, but the dystric Cambisols (Inceptisols) are far more frequent. When possessing horizontal stratification, they offer more resistance. With advancing age, deeper soils are less rich in exchangeable Al, making the deep weathered Oxisols in this group less limiting chemically than the young Inceptisols, where exchangeable Al is always high. The subsoils are generally very hostile in terms of chemical properties, especially from toxic levels of soluble Al, and physically limiting. The exposure of saprolites on pelitic rocks makes a kind of chemical desert difficult to rehabilitate. The vegetation growth is reduced, and erosion is usually more severe due to high silt content and little structural development. The only major nutrient in adequate concentrations is potassium, due to the presence of mica, amid the almost universal nutrient depletion in these rocks.

In Group 4, soils developed of quartz-sandstone rocks, landforms tend to be tabular in more friable and horizontally bedded sandstones, with relatively deep, sandy soils that are poorly differentiated in their morphology (weak B horizons development). Litholic Neosols, Quartzarenic Neosols, and Spodosols are particularly common, all very dystric and nutrient-depleted. Landforms are steep and mountainous in more compact and resistant quartzite (metamorphic rock) with very shallow soils. Forest rarely occurs in this group, due to the combined strong deficiencies of water and nutrients.

In Group 5, ferruginous rocks are associated with quartzite in the general high mountain scenery. However, soils are often iron-cemented, petroplinthic or concretionary, and these concretions are stable, helping protect the soil from long-term erosion, under dry or seasonal climates. In humid seasonal, stable climates, these ironstone caps normally degrade, leading to loose concretionary Latosols (Oxisols) or Petric Plinthosols, showing various stages of degradation, ranging from a continuous petroplinthite layer (indurated laterite) (Costa 1991; Schaefer 2013) to scattered residual concretions, common in Fe-rich Oxisols (Latosols), as they form from the ironstone degradation.

In Group 6, the limestone rocks form deep reddish soils, well-drained but with surprising low Fe content (compared to mafic or ferruginous rocks). Drainage tends to be good, with a deep underground network of cavities, in more humid climates. Under dry climates, limestone outcrops may occur with more compact and resistant cores. The dominant soils are red Latosols where weathering is pronounced, but red Nitosols and Argisols also occur. The subsoils are not deep, since the insoluble products that form the shallow saprolites are derived from impurities present in the limestone, usually less than 10% of the total rock material. Under dry climates, Eutrophic Cambisols are common.

In Group 7, with Quaternary alluvial sediments, soils are always young, poorly developed, and little differentiated and with many morphological features (texture, organic matter content) inherited from the original sedimentary deposit. Chemically poor, mature alluvial sediments are the dominant types in Brazil, and Fluvic Neosols and Gleysols are remarkably dystrophic, except in the Amazon River floodplain, with an Andean source.

In Group 8, gneisses represent high-grade metamorphic rocks that cover a large extent in Brazil, with two basic types: the biotite-rich gneisses and schists, deeply altered in the humid regions, and relatively rich in iron, with a pinkish-red, well-drained, and very deep saprolite, covered by red-yellow Latosols (Oxisols) or dystric Cambisols, usually under forest. Dystric red-yellow Argisols also occur. In the second case, leucocratic gneisses (rich in quartz and muscovite), soils are much poorer, yellowish, silt-rich, and prone to erosion. In both cases, however, there is a very deep saprolite development (alteration mantles or regoliths), which makes it difficult to link the current soil formation with rock, so deeply altered. The vegetation is less developed than in biotite gneisses, and dystric Cambisols and Litholic Neosols are more common. Under dry climates, Luvisols and Planosols are predominant, with some Eutric Litholic Neosols.

In Group 9, the conglomerates, depending on the nature (pebbly, gravelly, angular material, breccia, etc.) and the type of cement (Fe, Si, $CaCO_3$), can have both nutrient-rich or nutrient-poor soils and subsoils. Litholic and Regolitic Neosols are the main soils.

In Group 10, the organic sediments in bogs that accumulated in inundated areas (peat bogs) form organic soils, with varying degrees of humification and preservation of the original constitution of the plant material. Deep organic soils are rarely met in Brazil, and their acid nature and high vulnerability to carbon emissions and losses preclude any intensive land use. Organosols and Humic Gleysols are the main soil types.

2.7.2 *Mineral Resistance to Weathering in Tropical Brazil*

In terms of comparative mineral resistance to weathering, quartz is extremely resistant. However, a long time exposure in a free leaching environment can result in strong removal of silica (Loughnan and Bayliss 1961), as evidenced by the deep saprolite of banded iron formation, or Quartzite caves in Ibitipoca State Park, in Minas Gerais (Dias et al. 2003). Feldspars have well-developed cleavage, easing hydrolysis. Plagioclases (Ca > Na) alter more easily than potassium feldspars (orthoclase>microcline). Kaolinite and gibbsite are the most common products of alteration of feldspars and depend basically on the drainage.

Pyroxene has good cleavage among the mafic minerals and is readily weathered, producing clay minerals and oxides. Amphiboles are more resistant than pyroxenes, especially hornblende, which can be found in moderately weathered soils in Brazil's seasonal dry tropical areas (Albuquerque Filho et al. 2008). Micas, which have perfect cleavage in layers, are soft and easily broken, resulting in rapid hydrolysis, generating clay minerals; muscovite is much more resistant than Fe-rich biotite. Although not having cleavage, olivines are easily altered by their network of fractures, exposing the inner structure to water penetration.

Among the carbonates, the most soluble of common minerals, dolomite is more resistant than calcite. More soluble than carbonates, sulfates are not common in soils or rocks in Brazil, requiring dry climate to remain stable. The weathered rock zone is a key, neglected part of the critical zone. Much more effort and research are needed to understand better how it evolves and functions in the environment (Graham et al. 1997).

2.8 The Way Forward: Regolith and Subsoils Within the Critical Zone (CZ) Concept

There is a clear need for a broader perspective on subsoils and regoliths and a methodological shift to address the hydrological, pedological, and environmental interplays focused on practical applications of this knowledge. In this respect, regoliths and subsoils are now recognized as an essential part of the CZ. The CZ is conceptually described as a *feed-through reactor* (Anderson et al. 2007), in which physical and chemical processes, coupled with biological factors, work together to transform and move (upwards, downwards, laterally) materials. The thickness of this reactor, the nature of the weathered product, and the solute leached or washed out are key attributes of the CZ. Also, the rate at which it is formed depends on the rate of multiple chemical, physical, and biological processes in the reactor (Anderson et al. 2007).

The geological memory is fully considered in the CZ concept. Long-term processes operating from tectonic (faulting, jointing, fracturing) below the CZ have a role. Geomorphological processes at the slope scale (soil creep, rilling, gullying)

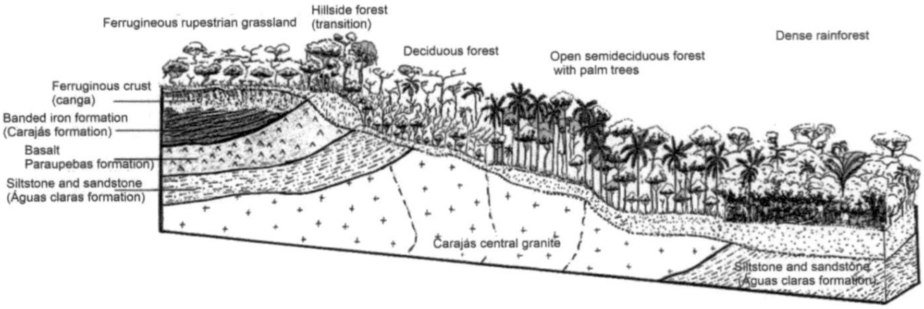

Fig. 2.28 A proposed model of critical zone system at Carajás Region, Amazonia (Schaefer et al. 2015). Vegetation changes according to soils, topographical variations, and lithology, illustrating the close interplay between regolith and vegetation in this ecotonal region of Brazilian Amazonia. The implications for groundwater recharge can be discussed and help management plans for mining operations and environmental conservation measures. (Drawing by C. Schaefer)

down to mineral scales (phase transformations, stresses, microcracks, neoformation) all drive the chemical weathering rate. All these processes combined, besides the key role of living organisms (the biological factor), control the residence time of the resulting material in the CZ. Also, they control the rate of water flow, groundwater recharge, size, and nature of primary or secondary mineral particles in the regolith, all of which influence the chemical weathering flux and, conversely, are influenced by this flux (Fig. 2.28).

It is a challenge to integrate all aspects of the CZ and unravel the interplays operating over a range of scales. This requires a shift in the traditional research approach from one of correlating observations to one of understanding the mechanical and chemical processes, as suggested by Anderson et al. (2007). Departing from these measures and quantifiable parameters, one can model the CZ for different purposes.

There are three dimensions in the CZ volume. With reference to the third dimension, lateral transfers of soluble or suspended material through the CZ affect the nature of regoliths at a given depth. The lowest, deepest parts of the CZ may be quite stable, and subsoils may have reached a kind of steady state, with little changes over time. On the other hand, at the surface, geomorphic (slope) processes are fully recognized in the study of soil catenas, so that this dynamic part of CZ is key for interpreting recent environmental changes, especially those related to human-induced impacts (Fig. 2.29).

Many areas of the tropics have colluvial soils, a kind of pedological cover on the move, by transport processes collectively called soil creep that permanently move soil downslope, intensively affected by biotic or mechanical disturbances. The result of soil creep is the discontinuity between soils and saprolites, and the typical marker is a colluvial lag or stone line. Several models now track regolith production, sediment transport, and chemical weathering on a hillslope (e.g., Mudd and Furbish 2006; Yoo et al. 2007). In a steady state, such models predict that the mean age of the soil column is uniform along the hillslope, so chemical evolution and soil age should not vary much with slope position. These models are not yet sophisticated enough to capture the differences in hydrologic conditions downslope.

Fig. 2.29 Displaced subsoils and regoliths by human action. (**a**) The alluvial terrace of Carmo River just after the deposition of millions of tons of tailings from the Fundão Dam. (**b**) A Technosol profile formed only 5 days after the accident

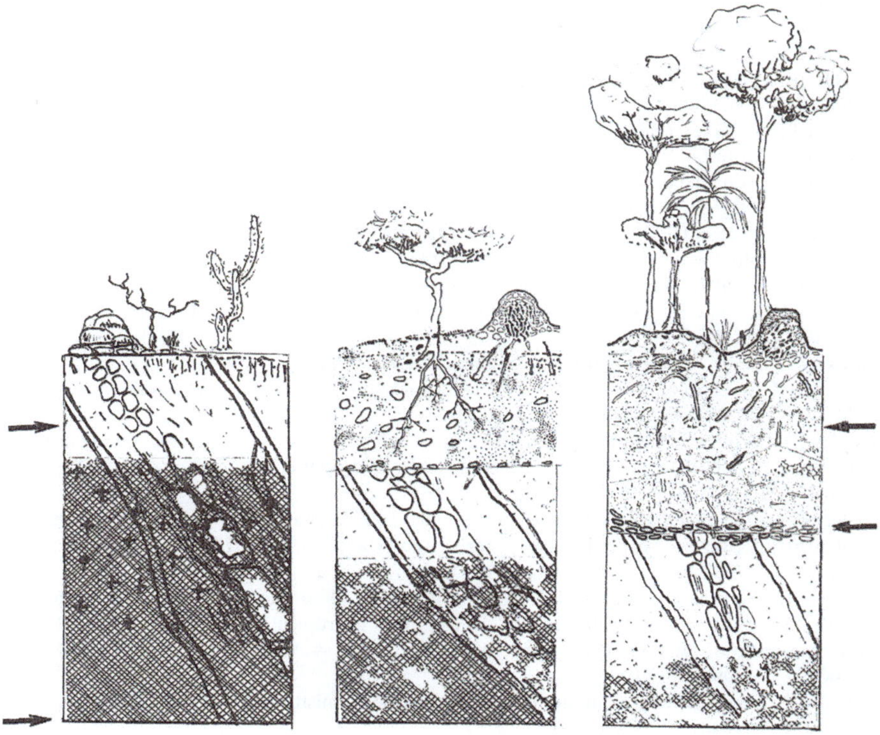

Fig. 2.30 The regolith formation is based on a double planation process, in which the surface evolves under strong pedobiological working, with varying vegetation and climates. In contrast, the subsoil is formed by geochemical processes related to chemical and physical weathering, superimposed on a rock with geotectonic memory in the form of fractures, folds, and faults. (Drawing by C. Schaefer)

Finally, regolith and subsoils are not steady, static entities and are permanently under dynamic stresses from changing climates, so a much more complex picture emerges. In general, it can be synthesized as follows, with shifts of tropical conditions ranging from dry (left) to humid (right), emphasizing the key role of the biological factors, without which the tropical belt would not form its so distinctly singular CZ, regolith, and subsoils (Fig. 2.30).

References

Ab'Saber AN. Relevo, estrutura e rede hidrográfica do Brasil. Bol Geogr. 1956;14(3):225–8.

Ab'Saber AN. Domínios morfoclimáticos e solos do Brasil. In: Alvarez VVH, Fontes LEF, Fontes MPF, editors. Os solos nos grandes domínios morfoclimáticos do Brasil e o desenvolvimento sustentável. Viçosa: Sociedade Brasileira de Ciência do Solo/Universidade Federal de Viçosa; 1996. p. 1–18.

Albuquerque Filho MR, Mugller CC, Schaefer CEGR, et al. Solos com morfologia latossólica e caráter câmbico na região de Governador Valadares, Médio Rio Doce, Minas Gerais: gênese e micromorfologia. Rev Bras Cienc Solo. 2008;32(1):259–70. https://doi.org/10.1590/S0100-06832008000100025.

Almeida FFM, Hasui Y, Brito Neves BB. The Upper Precambrian of South America. Boletim IG. 1976;7(1):45–80.

Almeida FFM, Neves BBB, Carneiro CDR. The origin and revolution of the South American platform. Earth-Sci Rev. 2000;50(1):77–111. https://doi.org/10.1016/S0012-8252(99)00072-0.

Anderson SP, von Blanckenburg F, White A. Physical and chemical controls on the critical zone. Elements. 2007;3(5):315–9. https://doi.org/10.2113/gselements.3.5.315.

Andrade H, Schaefer CEGR, Andrade FV. Pedogeomorfologia e micropedologia de uma sequência Latossolo-Areia Quartzosa hidromórfica sobre rochas cristalinas do Estado do Amazonas. Geonomos. 1997;5(1):55–66. https://doi.org/10.18285/geonomos.v5i1.189.

Barbosa ALM. Síntese da evolução geotectônica da América do Sul. Bol Inst Geol/Rev Esc de Minas. 1966;1(1):91–111.

Bardossy G, editor. Karst bauxites. Amsterdam: Elsevier; 1982.

Beckmann GG. Development of old landscapes and soils. In: Soils: an Australian viewpoint. Melborne: CSIRO Publications; 1983. p. 51–71.

Bell RW, Oliveira TS. Chapter 1: Introduction to subsoil constraints for crop production. In: Oliveira TS, Bell RW, editors. Subsoil constraints for crop production. Switzerland: Springer; 2022. p. 1–10.

Boulet R, Fritsch E, Filizola HF, et al. Iron bands, fragipans and duripans in the northeastern plateaus of Brazil – properties and genesis. Can J Soil Sci. 1998;78(3):519–30. https://doi.org/10.4141/S98-030.

Braucher R, Bourlès DL, Brown ET, et al. Application of in situ-produced cosmogenic ^{10}Be and ^{26}Al to the study of lateritic soil development in tropical forest: theory and examples from Cameroon and Gabon. Chem Geol. 2000;170(1–4):95–111. https://doi.org/10.1016/S0009-2541(99)00243-0.

Butt CRM, Zeegers H, editors. Regolith exploration geochemistry in tropical and subtropical terrains. Amsterdam: Elsevier; 1992.

Butt CRM, Lintern MJ, Anandet RR. Evolution of regolith and landscapes in deeply weathered terrain-implications for geochemical exploration. Ore Geol Rev. 2000;16(3–4):167–83. https://doi.org/10.1016/S0169-1368(99)00029-3.

Butt CRM, Lintern MJ, Anand RR. Evolution of regolith and landscapes in deeply weathered terrain—implications for geochemical exploration. Ore Geology Reviews. 2000;(16):167–183.

Cecil MR, Ducea MN, Reiners PW. Cenozoic exhumation of the northern Sierra Nevada, California, from (U-Th)/He thermochronology. Geol Soc Am Bull. 2006;118(11–12):1481–8. https://doi.org/10.1130/B25876.1.
Chauvel A. Recherches sur la transformation des sols ferrallitiques dans la zone tropicale a saisons contrastées: évolution et réorganisation des sols rouges de moyenne Casamance, Sénégal; 1977.
Costa ML. Aspectos Geológicos dos Lateritos da Amazônia. Rev Bras Geociênc. 1991;21(2):146–60.
CPRM Serviço Geológico do Brasil. Mapa Hidrogeológico no Brasil ao Milionésimo; 2014.
Chauvel A, Lucas Y, Boulet R. On the genesis of the soil mantle of the region of Manaus, Central Amazonia. Experientia. 1987;(43):234–241.
Delvigne J. Atlas of micromorphology of mineral alteration and weathering, Canada; 1998.
Dias HCT, Schaefer CEGR, Fernandes Filho E, et al. Caracterização de solos altimontanos em dois transectos no Parque Estadual do Ibitipoca (MG). Rev Bras Cienc Solo. 2003;27(3):469–81. https://doi.org/10.1590/S0100-06832003000300009.
Enslin JF. Basins of decomposition in igneous rocks: their importance as underground water reserves and their location by the electrical resistivity method. Trans Geol Soc S Afr. 1943;46(1):1–12.
Faniran A, Omorinbola EO. Trend surface analysis and practical implications of weathering depths in Basement Complex rocks of Nigeria. Niger Agric J. 1980;23(1):113–26.
Graham RC, Schoeneberger PJ, Anderson MA, et al. Morphology, porosity, and hydraulic conductivity of weathered granitic bedrock and overlying soils. Soil Sci Soc Am J. 1997;61(2):516–22.
Hubble GD, Isbell RF, Northcote KH. Development of old landscapes and soils. In: Soils: an Australian viewpoint. Melborne: CSIRO Publications; 1983. p. 17–47.
Jenny H. Factors of soil formation: a system of quantitative pedology. McGraw-Hill, New York; 1941.
Leprun JC. Les cuirasses ferrugineuses des pays cristallins de l'Afrique occidentale sèche. Genèse. Transformations. Dégradation; 1979.
Loughnan FC, Bayliss P. The mineralogy of bauxite deposits near Weipa, Queensland. Am Min. 1961;46(1–2):209–17.
Millot G. Planation of continents by intertropical weathering and pedogenetic processes. In: Melfi, AJ, Carvalho A (1982) Paper presented internation seminar on lateritisation processes, São Paulo, 4-12 July 1982; 1983.
Millot G, Bonifas M. Transformations isovolumétriques dans les phénomènes de latéritisation et de bauxitisation. Sci Géologiques Bul Mém. 1955;8(1):3–20.
Mudd SM, Furbish DJ. Using chemical tracers in hillslope soils to estimate the importance of chemical denudation under conditions of downslope sediment transport. J Geophys Res Earth Surf. 2006;111(2):1–18. https://doi.org/10.1029/2005JF000343.
Nahon DB, editor. Introduction to the petrology of soils and chemical weathering. New Jersey: Wiley; 1991.
Nikitina NK. Geoethics: theory, principles, problems, 2nd edn, revised and supplemented. Moscow: Geoinformmark, Ltd.; 2016.
Nahon D, Janot C, Karpoff A M, Paquet H, Tardy Y. Mineralogy,petrography and structures of iron crusts (ferricretes) developed on sandstones in the western part of Senegal. Geoderma. 1977;(19):263–277.
Omorinbola EO. Groundwater resources in tropical African regoliths. IAHS-AISH. 1984;144(1):15–23.
Pedro P, Melfi AJ. The superficial alteration in tropical region and the laterization phenomena. In: Lateritisation process. 2nd ed. São Paulo: Instituto Astronômico e Geofísico; 1983. p. 3–13.
Retallack GJ. Cenozoic paleoclimate on land in North America. J Geol. 2007;115(3):271–94. https://doi.org/10.1086/512753.
Rezende SB. Geomorphology, mineralogy and genesis of four soils on gneiss in southeastern Brazil. Thesis Purdue University; 1980.
Resende M. Caracterização de solos tropicais brasileiros. Brasília, ABEAS, 1988.

Resende M, Curi N, Rezende SB, Silva HG. Da rocha ao solo: enfoque ambiental. Lavras: EDUFLA, 2019.

Salgado AR, Colin F, Nalini Júnior HA, et al. O papel da denudação geoquímica no processo de erosão diferencial no Quadrilátero Ferrífero/MG. Rev Bras de Geomorfol. 2004;5(1):55–69. https://doi.org/10.20502/rbg.v5i1.32.

Salgado AR, Marent BR, Cherem LFS, et al. Denudation and retreat of the Serra do Mar escarpment in southern Brazil derived from in situ-produced ^{10}Be concentration in river sediment. Earth Surf Process Landf. 2014;39(3):311–9. https://doi.org/10.1002/esp.3448.

Salgado AAR, Rezende EA, Bourlès D, et al. Relief evolution of the Continental Rift of Southeast Brazil revealed by in situ-produced ^{10}Be concentrations in river-borne sediments. J S Am Earth Sci. 2016;67(2):89–99. https://doi.org/10.1016/j.jsames.2016.02.002.

Schaefer CEGR (2013) Bases físicas da paisagem brasileira: estrutura geológica, relevo e solos. In: Araújo AP, Alves, BJR (ed) Tópicos em ciência do solo Sociedade Brasileira de Ciência do Solo: Viçosa 8(1):1–69.

Schaefer C, Oliveira FS. Chapter 15: Live subsoils: tropical regolith and biota interactions. In: Oliveira TS, Bell RW, editors. Subsoil constraints for crop production. Switzerland: Springer; 2022. p. 373–96.

Schaefer CEGR, Albuquerque MA, Charmelo LL, et al. Elementos da Paisagem e Gestão da Qualidade Ambiental. Informe Agropecuário. 2000;21(202):20–44.

Schaefer CEGR, Cândido HG, Corrêa GR, et al. Solos desenvolvidos sobre canga ferruginosa no Brasil: uma revisão crítica e papel ecológico de termiteiros. In: Do Carmo FF, Kamino LH, editors. Geossistemas ferruginosos do Brasil. Belo Horizonte: Instituto Prístino; 2015. p. 77–102.

Schaefer CEGR. Brazilian highland soils and Oxisols from deep weathered saprolites.In: 21st World Congress of Soil Science; August 2018; Rio de Janeiro. Rio de Janeiro: Sociedade Brasileira deCiência do Solo; 2018.

Schaefer CEGR, Campos PV, Candido HG, et al. Serras e pantanais arenosos: solos e geoambientes em unidade de conservação da Amazônia. Brasil Neotrop Biol Conserv. 2020;15(1):43–69. https://doi.org/10.3897/neotropical.15.e49221.

Scott K, Pain C, editors. Regolith science. CSIRO Publishing; 2009.

Spier CA, Vasconcelos PM, Oliviera SMB. ^{40}Ar/^{39}Ar geochronological constraints on the evolution of lateritic iron deposits in the Quadrilátero Ferrífero, Minas Gerais, Brazil. Chem Geol. 2006;234(1–2):79–104. https://doi.org/10.1016/j.chemgeo.2006.04.006.

Stoops G, Schaefer CEGR. Pedoplasmation: formation of soil material. In: Taylor G, Eggleton RA, editors. Interpretation of micromorphological features of soils and regoliths. 2nd ed. Amsterdam: Elsevier; 2018. p. 59–7. https://doi.org/10.1016/B978-0-444-53156-8.00005-2.

Taylor G, Eggleton RA. Regolith geology and geomorphology. Chichester/New York/Weinheim/Brisbane,/Singapore/Toronto: Wiley; 2001.

Varajão CAC, Salgado AAR, Varajão AFDC, et al. Estudo da evolução da paisagem do Quadrilátero Ferrífero (Minas Gerais, Brasil) por meio da mensuração das taxas de erosão (^{10}Be) e da pedogênese. Rev Bras Cienc Solo. 2009;33(5):1409–25. https://doi.org/10.1590/S0100-06832009000500032.

Vasconcelos P. K-Ar and ^{40}Ar/^{39}Ar geochronology of weathering processes. Ann Rev Earth Sci. 1999;27(1):183–222. https://doi.org/10.1146/annurev.earth.27.1.183.

Yapp CJ. ^{18}O/^{16}O and D/H in goethite from a North American Oxisol of the early Eocene climatic optimum. Geochim Cosmochim Acta. 2008;72(1):5838–51. https://doi.org/10.1016/j.gca.2008.09.002.

Yoo K, Heimsath RA, Dietrich WE, et al. Integration of geochemical mass balance with sediment transport to calculate rates of soil chemical weathering and transport on hillslopes. J Geophys Res Planets. 2007;112(2):1–15. https://doi.org/10.1029/2005JF000402.

Chapter 3
Soil Acidity and Acidification

Clayton Robert Butterly, Telmo Jorge Carneiro Amado, and Caixian Tang

Abstract Acid subsoils ($pHH_2O < 5.5$), which are common in many agricultural systems, reduce productivity because crops and pastures fail to achieve water- and nutrient-limited potentials, particularly when topsoils dry out, such as during spring in Mediterranean environments or during summer in tropical environments. Depending on soil type and management, acid soil horizons can occur in the subsurface (8–20 cm) or subsoils (>20 cm) making these difficult, slow and expensive to ameliorate. The mechanisms which cause soil acidification or intensify existing acidity in managed agricultural production systems and lead to the development of acid subsurface and subsoil layers are presented. Unless there are visible effects on root development, soil tests are generally required to identify and diagnose subsoil acidity. Growing acid-tolerant crops can overcome some of the limitations of soil acidity but does not solve the problem. Hence, the use of lime ($CaCO_3$) is invariably required. Lime applied on the soil surface moves very slowly through the soil profile and is not suitable to ameliorate deep acid subsoil layers (>10 cm) in the short term. Surface applications of liming materials can create abrupt vertical fertility gradients and increase the susceptibility of plants to water stress, particularly in systems that use conservation agriculture. Consequently, direct incorporation of lime and other amendments into subsoil layers is likely to be the most effective but is costly and time-consuming and may expose soil to erosion and promote organic matter

C. R. Butterly (✉)
Faculty of Veterinary and Agricultural Sciences, School of Agriculture and Food, University of Melbourne, Parkville, VIC, Australia

Department of Animal, Plant and Soil Science, School of Life Sciences, La Trobe University, Bundoora, VIC, Australia
e-mail: clayton.butterly@unimelb.edu.au

T. J. C. Amado
Agronomy Faculty, Soil Science Department, Santa Maria Federal University, Santa Maria, RS, Brazil

C. Tang
Department of Animal, Plant and Soil Science, School of Life Sciences, La Trobe University, Bundoora, VIC, Australia
e-mail: c.tang@latrobe.edu.au

T. S. d. Oliveira, R. W. Bell (eds.), *Subsoil Constraints for Crop Production*,
https://doi.org/10.1007/978-3-031-00317-2_3

mineralisation. A great deal of research and effort has been focussed on finding the most effective materials and methods to ameliorate acid subsoils with minimal or no soil disturbance. Lime, gypsum, calcium oxide, organic amendments and biological solutions have been tested. Combined incorporation of lime and organic materials could be the most effective, since it provides ameliorative and nutritional benefits as well as improved soil physical conditions. Establishing cover crops on amended soils is advocated in Brazil to improve the physical condition of the soil profile. The continuous biopores formed by roots or macrofauna increase macroporosity and maximise water infiltration which is critical to achieve amelioration of subsoil layers. However, ameliorating acid subsoils may not always be practical and/or profitable, and it is better to take actions to minimise the rate at which subsoil acidity develops. Future development of cost-effective and integrative approaches is still needed.

Keywords Amelioration · Biopores · Cation/anion balance · Organic amendments · Subsoil acidity · Subsurface acidity

3.1 Introduction

Soil acidity combined with low plant nutrient availability is a major constraint for crop production in many parts of the world, particularly throughout the Asia-Pacific, Africa and South America. Acid soils ($pHH_20 < 5.5$ or $pHCaCl_2 < 4.8$) are estimated to cover 30% of the world's land area and about 50% of that which can be used for food production (von Uexküll and Mutert 1995; Wang et al. 2013). Importantly, many of these soils are also acid in subsurface (8–20 cm) or subsoil layers (>20 cm) (Paul et al. 2003; Scott et al. 2007). In most cases, acid soils have developed naturally from either base cation-poor parent materials or in areas which have undergone extensive weathering and leaching (Sumner and Noble 2003). However, intensive management of agricultural systems has exacerbated soil acidification and ongoing pressures on land-use mean that many are becoming highly acid mainly in the subsoil. Assessing the extent, rate of acidification and severity of soil acidity is often difficult, particularly at depths >10 cm, due to the lack of monitoring. Also, traditional soil sampling depths (0–10, 10–20 cm, etc.) fail to sufficiently detect pH stratification within soil profiles (Condon et al. 2021). Furthermore, the rate of acidification is highly variable depending on the soil type, farming system, fertiliser inputs, management and net primary productivity. In Australia, it is estimated that around 50% of agricultural land (50 M ha) has an acid topsoil and 23 M ha an acid subsoil (NLWRA 2001) occurring predominantly in areas used for crop production (de Caritat et al. 2011). Around 70% of Brazilian agricultural soils are also naturally acid, and while four decades of lime use has improved topsoil pH, extensive areas of subsoil acidity remain. Soil acidity is a critical issue in the Brazilian Cerrado, an important crop production area that expanded from 7.4 M ha in 2000 to 20.5 M ha in 2016 and has a total potential agriculture/pasture area of about 205 M ha (Fageria and Nascente 2014). The cost associated with lost agricultural productivity by soil acidity in Australia was estimated at $1.6 G per annum (1997–2001) (Lockwood et al. 2003) and is greater in recent years (Orton et al. 2018). Reductions in the choice and performance of crop and pasture species ultimately impact the profitability and sustainability of these systems.

Constraints on crop growth and performance are generally apparent at $pHH_20 \leq 5.5$. Reductions in soil pH are associated with a number of important chemical changes (Meng et al. 2019). A fundamental change is the sharp increase in solubility and availability of Al^{3+} (below pHH_20 5.5) and Mn^{2+} (below pHH_20 5.5) which are phytotoxic. Aluminium binds to the surface of root tips, preventing elongation and causing roots to rupture (Kopittke and Blamey 2016), and eventually inhibits cell division (Kochian et al. 2004). This Al^{3+} mostly occurs in the apoplast, with some moving to the symplast, but little is translocated to shoots in annual crop plants (Kopittke et al. 2016; Kochian et al. 2004). Manganese, an essential plant nutrient, accumulates in plant tissues in acid soils and interrupts key biochemical processes such as photosynthesis (Hue et al. 2001). Moreover, as soils acidify, their effective cation exchange capacity is decreased, and basic cations (Ca^{2+}, Mg^{2+} and K^+) are displaced into the soil solution making these susceptible to leaching. Hence, greater availability of Al^{3+} is also associated with it replacing Ca^{2+} as the dominant cation contributing to the effective cation exchange capacity. Crop growth in acid soils can also be limited by low availability of nutrients. Plant-available P is reduced via the formation of stable complexes with Al and Fe oxides. Low availability of Mo often limits legume performance. In tropical soils, low calcium and high aluminium levels constrain root and crop growth, decreasing nutrient uptake and water use efficiency with negative consequences for plant vigour.

Subsoil acidity primarily reduces the amount, distribution and function of plant roots. The uptake of water and nutrients from subsoil layers is restricted by shallow root growth, and plants do not achieve their water- and nutrient-limited potentials. This is particularly important in grain production systems and during dry periods when grain filling occurs in Mediterranean, temperate, tropical and subtropical environments. In addition to drought, symptoms of subsoil acidity at the paddock level are difficult to distinguish from other nutritional disorder such as P deficiencies. Closer examination of roots often reveals poor root development, characterised by stunted and thick roots, and poor nodulation of legumes. Reducing the development of subsoil acidity is of paramount importance as it is orders of magnitude cheaper than trying to reverse the problem once it has developed. Considerable effort has been placed on the selection and development of acid-tolerant crops and pasture species (Ryan 2018). However, this approach is essentially adaptation to the problem rather than a treatment or long-term solution.

The processes contributing to soil acidification have largely been known for some time (Helyar and Porter 1989), and these have been reviewed for a number of regions, including the United Kingdom (Goulding 2016), South America (Fageria and Nascente 2014) and for dryland cropping systems of Australia (Slattery and

Helyar 2004). However, acidification and the development of acid subsurface or subsoil layers have gained less attention (Paul et al. 2003; Tang et al. 2013). In all cases, acidification of soil layers occurs via the spatial separation of acid-producing and acid-consuming processes in the soil profile. It is important to differentiate between the acidification of subsurface soil layers and further acidification of naturally occurring acid subsoils. The later occurs deeper within the soil profile, remains acid for the entire root zone and is inherently more difficult to manage. In contrast, acid subsurface layers that have developed under agricultural management are generally shallower, occupy only a narrow soil layer and importantly often overlay subsoils which are not sufficiently acid to constrain root growth.

The benefits of lime ($CaCO_3$) application for crop productivity are well established (Liu et al. 2004), and this practice has been used for centuries to correct acid soils (Goulding 2016). However, the solubility of lime in water is low, and the dissolution of lime may be slow in climates with low rainfall and long dry periods. Less mechanical mixing of topsoils occurs under conservation agriculture (ca. 180 M ha globally in 2015/16) which means reduced incorporation and slower dissolution of lime (Kassam et al. 2015). Residual surface-applied lime can only move down the soil profile once most acid in the topsoil is neutralised ($pHCaCl_2 > 5.5$), and this requires high lime rates that are often agronomically unrealistic (Conyers and Scott 1989; Scott et al. 2000). In addition, it is likely to take a number of years (2–10) to achieve some amelioration of subsoil (Whitten et al. 2000). Surface applications of lime can effectively maintain adequate pH levels in the topsoil but lead to the development of abrupt transition between this layer and the non-ameliorated subsurface or subsoil layers. The stratification of pH is exacerbated in farming systems with reduced or no-tillage (Dalla Nora et al. 2017b), and consequently the agricultural system is more vulnerable to climate change and variability, i.e. drought and high temperatures.

In general, further acidification of subsurface and subsoils has occurred in recent times due to insufficient lime application, increased crop yields and export of alkalinity, soil erosion due to agricultural intensification and possibly the reliance on acid-tolerant crop varieties. Rates of lime application for many parts of Australia and Brazil are still well below what is needed to arrest the problem and prevent re-acidification (Lockwood et al. 2003). Many soils have become sufficiently acidic to limit productivity and represent an important threat to the sustainability of these systems. Traditionally, the amelioration of subsoil acidity (below 20 cm) was considered too difficult and too expensive (Cregan and Scott 1998). While this may still hold true in some cases, Wong and Asseng (2007) showed that yield and environmental benefits exist from amelioration of subsurface or subsoil acidity in Australian systems in areas with higher annual rainfall and in other areas when adequate rainfall occurs. Similarly, Dalla Nora et al. (2017a) reported in Brazil that high-yielding wheat, corn and soybean performed better when acid subsoil was alleviated, which was attributed to reducing water stress of these modern, short-season crops with high water demand. Recently, attention has again focussed on innovative approaches to manage acid subsoils including the placement of lime, fertiliser and other amendments directly into acid subsurface layers via deep ripping (chisel) in a minimum tillage system. New machinery and precision agriculture technologies such as soil sensors (apparent electrical conductivity; ECa; gamma radiation) may enable the right amount of amendment to be placed where it is needed. Areas at risk of becoming acidic can be identified with these technologies (Wong et al. 2008) and their pH buffering capacity (pHBC) and hence lime requirements determined using pedotransfer functions based on more readily available soil properties such as organic matter and clay contents (Wong et al. 2013; Wang et al. 2015). Developments in soil sensors that help to quantify pH, SOM and ECa and estimate pHBC, using parameters such as clay content, will enable lime rates to be calculated for the amelioration of specific areas of acid soils (Corassa et al. 2016; Schirrmann et al. 2011). In particular, amelioration of subsurface layers in cropping soils is likely to be easier and cheaper than subsoils and could be beneficial if productivity gains exceed the cost of amelioration. Pressures on land-use and food security mean that areas affected by subsoil acidity can no longer be viewed as permanently damaged and permanent areas of low productivity. This chapter provides an overview of the development of acid subsoils in cropping systems, their identification and management options including reducing the rate of acidification and the amelioration of acid subsoil layers with lime and other amendments.

3.2 Soil Acidification and the Development of Acid Subsoils (Soil pH Gradients)

Soil acidification is a natural process that is prevalent in medium- to high-rainfall areas where accelerated leaching of basic cations has occurred and in many highly weathered soils which have had a long history of pedological development. Nitrate is weakly adsorbed by soils and is highly mobile. Its leaching causes the leaching of an equivalent amount of cations to maintain charge neutrality. The cations accompanying nitrate (and sulphate, bicarbonate and chloride) are primarily the base cations K^+, Ca^{2+} and Mg^{2+} which become depleted in acid soils (Wong et al. 1992). Leaching of bicarbonates, although more slowly than cations, can cause significant pH changes (Helyar and Porter 1989). Rates of soil acidification that occur under natural conditions are very slow and take many decades or hundreds of years to develop. In agricultural systems, many farming practices have resulted in accelerated soil acidification, and these can be attributed to imbalances in the C, N and S cycles (Bolan and Hedley 2003; Helyar and Porter 1989). These processes may or may not subsequently lead to the development of pH gradients down the soil profile and importantly to the acidification of subsoils. The effects of the C and N cycles on soil pH change depend on the amounts and forms of C and N entering a specific soil layer and whether it is either removed or transported to subsequent layers down the soil profile.

During normal growth and development, plants obtain nutrients from the soil solution. The ratio in which cations and anions are taken up by plants is critical to the development of soil pH gradients (Tang and Rengel 2003). Excess cation uptake through the plasma membrane and into the root is often associated with the net release of H^+. Conversely, plants maintain overall charge neutrality when excess anion uptake occurs with the export of OH^- / HCO_3^-. Hence, net release of H^+ occurs when plants take up more cations than anions, or net consumption of H^+ occurs when plants take up more anions than cations, and OH^- / HCO_3^- are released into the rhizosphere. Indeed, legume crops generally take up an excess of cations over anions under current farming systems, with the amount and distribution of H^+ production being proportional to root length and their allocation in the soil profile (Tang and Rengel 2003). Of course, the potential impact of plants on soil pH during their growth is dependent on the species (Loss et al. 1993a; McLay et al. 1997; Tang et al. 1997a, b), whether they are legumes or non-legumes and the amount and forms of N present or added to the soil during plant growth. The greatest excess cation uptake occurs by legumes that are actively fixing atmospheric N_2 (Coventry and Slattery 1991; Loss et al. 1993b) and can ultimately lead to the development of acid subsurface layers (5–20 cm) (Chan and Heenan 2005; Dolling 1995; Conyers et al. 1996). Clearly, the proportion of N derived from N_2 fixation will likely be reduced in soils with a high N content (Tang et al. 1999b). For all plants, N form is also critical for the cation-anion balance, with NH_4^+ fertilisers inducing net acidification and NO_3^- fertilisers net alkalisation of the rhizosphere. Nitrogen form is important for soil acidification considering that N is a major plant nutrient (10–50 g N kg^{-1}). Over time the efflux of H^+ into plant rhizospheres can influence the pH of the bulk soil, particularly if the pHBC is low.

In productive agricultural systems, much of the resulting biomass is either removed directly (hay, grain and crop residues) or indirectly (meat, wool and animal waste). The export of these products represents a substantial net loss of alkalinity from the plant-soil system. Large differences in the amounts exported with each product can be attributed to the differences in the alkalinity content of each material and the relative quantities that are exported (Slattery et al. 1991). The alkalinity contents of plant and animal products are expressed relative to their organic anion content. Plants synthesise or catabolise organic acid anions, such as malate and citrate, to balance internal electrical charge due to excess cation uptake (Noble et al. 1996; Tang et al. 1999a). It is the export of organic anions with base cation pairs from agricultural systems which are critical for loss of alkalinity (net acidification). Accounting for the greater organic anion contents that occur in legumes than non-legumes and the relative yields, alkalinity exported in legume hay is ~2 times greater than grass hay, 12–48 times that of legume grain and 49–66 times more than from cereal grains (Slattery et al. 1998). Alkalinity removed in animal products varies widely depending on the age of the animal but can be considerable when animals are large or if successive growing seasons are required to produce each animal (Slattery et al. 1991). Notably, the return of animal waste minimises the net loss of alkalinity from an animal production system, but invariably these materials are returned to specific areas of the paddock and constitute an important loss of alkalinity from other areas.

The processes whereby the remaining above-ground plant biomass enters the soil and is decomposed along with root residues are particularly important for the development of pH gradients (Paul et al. 2001b). Both the rate and depth of incorporation of shoot residues into topsoil are likely to be restricted in farming systems which employ minimum tillage or no-tillage. Increases in pH can occur quite rapidly upon mixing with soil, and this is attributed to abiotic association reactions between H^+, organic anions and other chemical constituents present within the materials (Tang and Yu 1999). However, the liming effect of plant materials is predominantly a biological process and often highly correlated with soil microbial activity (CO_2 release) (Paul et al. 2001a). During decomposition by soil microbes, organic anions within plant materials undergo decarboxylation, a process that consumes H^+ (Yan et al. 1996). Accumulation of organic matter within the topsoil is often considered as a cause of soil acidification. However, in agricultural systems acidification mainly occurs in subsurface layers, that is, below the topsoil layer where organic matter has accumulated. Hence, acidification is not likely to be due to build-up of organic material per se but the redistribution of alkalinity to the topsoil at the expense of net acidification in the root zone below (Ritchie and Dolling 1985). In fact, the acid-neutralising capacity of the plant materials can be measured explicitly to estimate their effect on soil pH (Wong et al. 2000; Noble et al. 1996).

Some studies have demonstrated that alkalinity may be generated below the layer in which shoot residues are incorporated. Leaching and subsequent decomposition of soluble components from plant materials, including organic acid anions, are thought to be the main process (Butterly et al. 2011). The experimental addition of organic acids, including citric and malic acid, to soil solutions has been shown to effectively detoxify Al and increase cotton root growth (Hue et al. 1986). While the formation of organic Al complexes by soluble components of crop residues is likely to be important for acid soils, it is unclear whether they influence subsoil pH. Considering that only a portion of the alkalinity in plant materials is potentially soluble and these organic compounds are likely to be rapidly decomposed by soil microbes, they may not leach very deep within the soil profile (Butterly et al. 2013). However, the leaching and decarboxylation of organic anions are probably an important process in subsurface soil layers. This is particularly important since root residue decomposition is generally not sufficient to neutralise acidity generated by the root system due to their lower organic anion content compared with shoots (Tang et al. 1999a).

Nitrogen cycle processes that occur during production, removal and subsequent decomposition of organic matter are also critical for changes in soil pH. Theoretically, there is no net effect of N transformation on pH (H^+ concentration) when plant material is produced and decomposed in the same location, but this is rarely the case in agricultural production systems. Net changes in pH induced by the N cycle are a result of N transformation between different pools and their movement between soil layers. Inorganic N forms, NO_3^- and NH_4^+, in plant and animal residues are negligible (Tang and Yu 1999; Butterly et al. 2011). Furthermore, NH_4^+ movement in soil is minor and is rapidly converted to NO_3^- in the topsoil unless nitrification is inhibited by low pH. The leaching of NH_4-N from urine patches is an exception (Condon et al. 2004). Hence, N cycle processes are dominated by the production, movement and conversion of organically bound N, and its subsequent ammonification and nitrification mainly occur in the topsoil due to the concentration of organic matter and soil microbial biomass. Here, the charge associated with ammonification ($-H^+$) and nitrification ($+2H^+$) is balanced if NO_3^- is taken up within the topsoil by plants and microorganisms ($-H^+$). However, NO_3^- is highly mobile and often leached to the subsoil resulting in net acidification ($+H^+$) of the topsoil. This process is amplified by high rainfall as occurs in tropical and subtropical regions, mainly during the wet season.

It is well established that NO_3^- leaching is a dominant process for acidification of surface soils (Dolling 1995; Scott et al. 2000; Burle et al. 1997). Legume-based systems, particularly permanent pastures, have the greatest potential for NO_3^- leaching due to the high N concentration of the plant residues and the relatively small root systems which are unable to capture NO_3^- before it is leached (Bolan et al. 1991). In addition, NO_3^- leaching is more prone to occur in cropping systems with abrupt chemical gradients between topsoil and subsurface layers due to the concentration of crop roots in shallow topsoil. Systems that use high-yielding cultivars that demand high mineral N fertilisation rates (e.g. corn and wheat) but only achieve low N use efficiency can also be a source of NO_3^- leaching. Conversely, transformation of NO_3^- that has moved to subsurface or subsoil layers such as denitrification, immobilisation or its uptake by plants would result in alkalinisation rather than acidification (Tang et al. 2000). These N cycle processes in the subsoil will therefore reduce net acid production. However, organic N is only one component of the soil N pool. Application of NH_4^+-based fertilisers, MAP ($NH_4H_2PO_4$) and DAP ($(NH_4)_2HPO_4$) will result in net acidification of the topsoil even if all the N is taken up by plants. Nitrification of NH_4 in NH_4NO_3 or derived from urea hydrolysis also acidifies. The effects of these fertilisers will be worsened if subsequent nitrification of NH_4^+ in the topsoil is followed by NO_3^- leaching (Sumner 2009). In contrast, NO_3^--based fertilisers will result in net alkalinisation of the soil following NO_3^- uptake into plants and soil microbes. Hence, NO_3^- uptake by plants can generate alkalinity in the subsoil (Weligama et al. 2008, 2010) or reduce net acidification (Tang et al. 1999b).

Overall, it is evident that C cycle rather than N cycle processes mainly contribute to subsurface and subsoil acidity. Consequently, even for cropping systems where only grain is removed, the return of alkalinity from residues produced in situ is highly stratified. Acidity produced by the roots in the subsoil will remain, and this effect will be quantitatively greater for deep-rooted plant species (grain legumes) than shallow-rooted species (pasture legumes). Since the greatest net acidification occurs during a legume phase, the severity of pH gradients will depend on their frequency in the cropping rotation and also the duration of the farming system. Thus, differences in pH gradients between surface and deeper soil layers will be stronger when crop residues have greater organic anion content (residue quality), in mildly acid systems (higher initial pH) with greater productivity (residue quantity) and for soils with a low pHBC, low bulk density and greater potential for leaching.

3.3 Identifying and Diagnosing Subsoil Acidity

Identification and diagnosis of subsoil acidity in the field are challenging. Above ground, the symptoms of subsoil acidity are difficult to distinguish from common nutritional deficiencies, due to the negative impacts of acidity on plant roots. Nevertheless, uneven crop and pasture growth, small leaves often with abnormal colouration (intense green and purple colours) and yellowing and necrosis of leaf tips can all occur. Leaf symptoms are generally due to Mn toxicity and P deficiency and in some cases Ca and Mn deficiencies. Below-ground effects of acidity are more easily identifiable. Restricted root growth, often with stunted and thickened roots, especially root tips, is common due to the presence of Al^{3+} as is the lack of fine branching and poor nodulation of legumes (Araújo et al. 2016; Adams and Adams 1983). Consequently, grain yield is often reduced due to water stress and is more severe in years with low rainfall during critical flowering and reproductive stages (Joris et al. 2013; Dalla Nora et al. 2017a). In fact, the combined stresses of Al toxicity and drought on plant growth are likely to be greater than by either stress alone. For example, there was greater callose formation and reduced root elongation of bean (*Phaseolus vulgaris* L.) with combined drought and Al toxicity (Yang et al. 2012) and a detrimental interaction of these factors on all plant traits in soybean (*Glycine max* L.) particularly leaf water status (Goldman et al. 1989).

Emergent precision agriculture techniques that use on-the-go sensors are available to measure apparent soil pH, electrical conductivity and soil organic matter directly in the field (Doolittle and Brevik 2014). Combining different techniques such as electromagnetic induction (EMI) and gamma-ray emission spectrometry with the farmer's paddock data (soil types, yield maps) was effective at identifying subsoil acidity and was superior to using EMI alone (Wong et al. 2008). Alternatively, sensors may be mounted to farm implements. A study by Kweon (2012) utilised six coulter electrodes for EC measurements, a specially configured row unit for optical measurements of organic matter and a soil sampling shoe and ion-selective electrodes for pH measurements. Soil cores (0–15 cm) extracted on the go were brought into direct contact with two antimony ion-selective pH electrodes for 7–20 s. Correlation between laboratory measured values and those estimated by the equipment was high for soil organic matter ($R^2 = 0.95$), CEC ($R^2 = 0.88$) and pH ($R^2 = 0.85$) with the datasets having low root mean square errors of prediction for mean values of 27 g kg^{-1}, 22 $cmol_c$ kg^{-1} and pH 6.4, respectively (Kweon 2012). Although the use of ion-selective electrodes for determining soil pH is promising, the depth of sampling may be limited since soil needs to be brought up from the subsoil to physically contact with the sensor.

Soil analysis is often the only reliable way to diagnose soil acidity. In the field, soil pH is easily measured using colour indicators via commercially available and inexpensive kits. However, results from these kits are highly variable among manufacturers, and data are given as pH classes (often increments of 0.5 pH units). In addition, the matching of colours between standard charts and saturated pastes by the user is subjective (Benke and Robinson 2017). The use of more sensitive portable pH meters with glass electrodes is recommended. Due to convenience, water rather than $CaCl_2$ (and less so KCl) is often used as the extractant in the field. Irrespective of the method, users should be mindful of the extractant used or the relationship of pH data obtained with extractant versus saturated paste methods from commercial kits. It is advisable that data from field tests be confirmed in an appropriate laboratory where possible.

In the laboratory, soil pH is commonly measured following extraction with either water or 0.01 M $CaCl_2$ (Miller and Kissel 2010; Goulding 2016). Analysis of extracts is an accepted compromise with the direct analysis of the soil solution due to the difficulty of extracting soil solution (Rengel 2011). The use of $CaCl_2$ as the extractant in research and commercial laboratories is common in many countries, including Australia (Rayment and Lyons 2011). In particular, pH of $CaCl_2$ extracts is less affected by differences in soil water content, ionic strength of the soil solution and the liquid junction potential of glass pH electrodes than water extracts (Kissel et al. 2009). As a general rule, soil pH measured in $CaCl_2$ is about 0.8 pH units lower than pH in water in highly weathered soils of Australia ($pHCaCl_2$ 4–5) (Slattery et al. 1999). The actual difference between the two extractants can range from 0.5 to 1.2 pH units, and the relationship is non-linear for soils with variable charge components (Minasny et al. 2011; Henderson and Bui 2002). The relative differences in pH between the two extracts are less in acid soils and diminish further as the pH approaches the point of zero charge for the soil (Gillman and Bell 1978; Aitken and Moody 1991). In order to standardise methods in Australia, soil pH is determined using air-dried soil following extraction with 0.01 M $CaCl_2$ 1:5 (soil-to-extract) by shaking end-over-end for 1 h, allowing to settle for 30 min and measuring the pH of the supernatant using a glass electrode without stirring (Rayment and Lyons 2011). The use of fresh buffer solutions and reference soils is essential.

Determining the amounts and forms of Al in the soil is critical to diagnose acid subsoils. However, the total soil Al concentration is not suitable to indicate the potential for constraints on plant growth. Solution culture experiments have identified that monomeric Al species (especially Al^{3+}) are predominantly responsible for phytotoxicity. The availability of Al^{3+} in the soil solution and its activity on exchange sites increases sharply when the soil pH drops below a critical threshold, approximately pHH_2O 5.5 or $pHCaCl_2$ 4.8. However, Al exists in many soluble forms with OH^- which vary according to pH, and quantifying phytotoxic Al concentrations in the soil solution is inherently difficult. Firstly, soluble Al can complex with inorganic anions, e.g. SO_4^{2-}, NO_3^- and PO_4^-, with organic ligands, e.g. organic acid anions, and form polynuclear hydroxyl-Al compounds that are not toxic to plants (Soon et al. 2007). Secondly, the methods used to obtain the soil solution may unintentionally alter the speciation of toxic Al species or contaminate or precipitate Al in the sample. Though ideal, the direct measurement of Al^{3+} in the soil solution is not considered appropriate for routine measurements and diagnostic purposes.

Soil Al is determined using various extraction procedures. Since the Al in the soil solution is a function of the exchangeable Al pool, soil Al is commonly assessed by the amount of Al that is exchanged by an unbuffered salt, usually 1 M KCl (Bertsch and Bloom 1996). This method has also commonly been used to calculate lime requirements. Briefly, soil is extracted with 1 M KCl 1:10 (soil-to-extract) by shaking end-over-end for 1 h, and the filtered (Whatman #2) extract (diluted 1:1 with water) is titrated to pH 8 to obtain exchangeable acidity and then reacted with 5 ml NaF (4% w v^{-1}) and titrated back to pH 8 to obtain exchangeable Al^{3+} ($cmol^{+}$ kg^{-1}) c.f. (Rayment and Lyons 2011). Alternatively, Al in acidified KCl extracts can be determined colourimetrically (Bertsch and Bloom 1996). However, the use of high-ionic-strength extractant raises doubts about whether this pool of Al is only the exchangeable form (Menzies 2003).

Another common approach is to express soil Al as a function of the cation exchange capacity (CEC). This is particularly important because in addition to the absolute concentration of Al in the soil solution, when Al is the dominant exchangeable cation, it reduces or excludes the uptake of other important cations by plants (Ritchie 1989). As for many Al measurements, the extractant, its ionic strength and the duration of extraction vary extensively among laboratories. However, the compulsive exchange method of Gillman and Sumpter (1986) that estimates CEC using unbuffered solutions while maintaining ionic strength and soil pH close to field conditions has been widely adopted in the United States and Australia. Briefly, soil is extracted (1:10) with 0.1 M $BaCl_2/NH_4Cl$ for 2 h, NH_4^+ removed and the ionic strength reduced to 0.006 M (approximating the soil solution) before the compulsive exchange of Ba^{2+} by Mg^{2+} for 1 h. The combined concentrations of base cations Ca^{2+}, K^{+}, Mg^{2+} and Na^{+} (CEC_B) plus Al^{3+} ($cmol_c$ kg^{-1}) in the $MgSO_4$ supernatant are determined by atomic absorption spectroscopy (AAS) or inductively coupled plasma optical emission spectroscopy (ICP-OES) (Gillman and Sumpter 1986; Rayment and Lyons 2011). Alternatively, CEC_B value measured by this method can be added to the exchange acidity (Al^{3+}+H^{+}; 1 M KCl exchangeable). In both cases, CEC_B plus acidic cations is termed the effective CEC (eCEC). These approaches are useful as they allow Al and base cations (e.g. Ca^{2+}+ Mg^{2+}) to be calculated as a % of eCEC. However, neither Al saturation nor exchangeable Al have proved to be reliable indicators of Al toxicity over a wide range of soil types, and as such critical levels for phytotoxicity are not easily transferable among soils.

Various extractants with lower ionic strength have also been used to quantify toxic levels of soil Al. Among these, 0.01 M $CaCl_2$ has been popular given the efficiencies of using the same extract for both soil pH and Al measurement. Following pH determination, whereby samples have been standing for 1–2 h, an aliquot of the extract is taken, filtered and the Al concentration determined colorimetrically or by ICP-OES. Pyrocatechol violet (PCV) is a common colorimetric reagent and has been recommended due to its simplicity, and Al concentrations in filtered extracts (0.8 μm, Whatman grade 40) are determined spectrophotometrically at 580 nm following a 1-h colour development (Conyers et al. 1991). Although the extracts may contain both soluble toxic inorganic Al forms (Al^{3+}) and non-toxic organically complexed and particulate forms of Al (Menzies et al. 1992), the use of PCV is thought to discriminate against micro-particulates and, to some extent, organically complexed Al (Kerven et al. 1989; Menzies et al. 1992). The use of 0.22-μm filters is recommended, and generally a 1:1 relationship between the Al concentrations determined by PCV and ICP-OES is observed for agricultural soils with relatively low organic matter (<2% total C), suggesting that the ICP-OES may be suitable for rapid Al determination in some cases. Some studies have used shorter periods of colour development to determine labile monomeric Al^{3+} (1 min) (Kerven et al. 1989) and total monomeric Al^{3+} (4 min) (Wang et al. 1999a) to discriminate against organically bound Al which reacts more slowly with PCV. This approach is made easier with modern plate readers. Although PCV-based methods for the determination of Al in the field have been developed (Barton and Carr 1996), these are not widely adopted.

Given the effort and cost of soil analysis, careful consideration needs to be given to field sampling methodology. Soil pH and Al vary considerably both spatially and temporally. Where possible, geospatial information, yield maps, remotely sensed imagery, etc. should be used for targeted sampling, and in semi-arid environments, the collection of samples from dry soils, most likely during summer periods, is recommended to minimise seasonal variability (Oliver et al. 2015). Furthermore, traditional sampling depths (e.g. 0–10, 10–20 cm) may be too coarse or too shallow to accurately determine the severity of subsoil acidity and the chemical gradients faced by plant roots. Particularly for subsurface acidity, acid layers that occur at 5–15 cm may be overlooked by traditional sampling (Scott et al. 2017).

3.4 Minimisation of Subsoil Acidification

Minimising the rate at which farming systems acidify is of primary importance, since amelioration of subsoil acidity is difficult and may not always be practical or economically feasible. There have been a number of excellent reviews on reducing soil acidification in cropping systems (Coventry et al. 2003; Helyar 1991; Fageria and Nascente 2014; Conyers et al. 1996). Approaches include reducing rates of ammonium-based fertilisers, slowing the rate of nitrification, reducing legume pasture or fallow periods that increase the risk of NO_3^- leaching, fertiliser management to maximise N use efficiency, selecting crops and their sequence to maximise NO_3^- uptake and minimising the export of alkalinity as farm products. However, many of these target N-cycle processes which acidify topsoils but have the opposite or no effect in the subsoil. Approaches to minimise subsurface or subsoil acidification are inherently different from those that treat surface soils and are limited to designing cropping systems which have low excess cation uptake or the manipulation of the biological cation/anion balance in the subsoil by expediting anion uptake.

Designing cropping systems with reduced excess cation uptake is technically possible but may not be achievable. Essentially it requires eliminating or reducing the frequency of crops with high excess cation uptake (that acidify the greatest) and legumes that have deep roots systems (that acidify the deepest). While the acidifying potential of crops is highly variable, particularly for legumes (McLay et al. 1997), crop choice is largely determined by profitability, soil and climate constraints, seed availability and cost and other economic factors. Given their importance in crop rotations, replacement of legumes is unlikely, but options that acidify less may exist. In addition, reducing the net export of alkalinity also potentially reduces the net acidification of the entire soil profile. Modern crop varieties generally have higher harvest indices, and the retention of crop residues is common practice, so there may be little scope for improvements in this regard. However, minimising the removal of hay is important since the alkalinity in shoots is much higher than in grains. The critical challenge is getting the above-ground residues back into the soil where the acidity was generated during their production. Strategic tillage to incorporate crop residues in the subsurface soil may be an option to accelerate this process (Azam and Gazey 2021).

Expediting the biological uptake of anions in the deeper soil layers could be an important approach to reduce subsoil acidification. Supplying plants with N in the form of NO_3 promotes greater anion uptake, results in the release of OH^- ions into the rhizosphere (or co-transport of H^+ into the root) and increases the pH of the rhizosphere soil (Tang et al. 2013; Jarvis and Robson 1983) with the effect differing among crop species (Weng et al. 2021). Root-induced alkalinisation of plant rhizospheres can ultimately increase pH of the bulk soil (Hinsinger et al. 2003). In soil columns, NO_3^- taken up by lupin and subterranean clover roots reduced the net acidification of these legumes by 14% in a 105-day leaching study (Tang et al. 2000). Similarly, surface-applied NO_3^- taken up by deep wheat roots in the subsoil achieved pH increases of 0.8 and 0.3 units of the rhizospheres and bulk soils, respectively, when soil water content and N rates were high (240 kg N ha^{-1}) (Weligama et al. 2010). Wheat grown at two field sites and supplied with $Ca(NO_3)_2$ (200 kg N ha^{-1}) showed increases in pH of 0.5 units in the rhizosphere and 0.3 units in bulk soil with maximal effects occurring at the depth of N placement (5–10 cm) after 3 years of treatment (Tang et al. 2011). Importantly, in two consecutive growing seasons at the same field sites, it was confirmed that the alkalinisation effect of $Ca(NO_3)_2$ could be achieved while maintaining similar N use efficiency and grain yield to that of urea (Conyers et al. 2011). This approach can induce net alkalinisation of soil under non-legumes and is therefore likely to be important for increasing or at least maintaining subsoil pH (Butterly et al. 2021b).

The effectiveness of the biological manipulation of the cation/anion balance relies on a number of soil and plant factors. First, NO_3^- would need to be placed either directly into the acid layer or above it with potential to leach into the acid layer. Leaching should not be excessive and move the NO_3^- past the acid layer or out of the root zone. Root growth in the acid layer must be sufficient for the complete uptake of NO_3^- and subsequent release of OH^-. Hence, root function must not be impaired, nor plant growth limited by nutrient availability which may occur in acid subsoils. This approach will be most effective on moderately acidic soils with lower pHBC and using acid (Al)-tolerant crop species (Conyers et al. 2011). The impact of initial pH is likely to depend on crop species and soil nutrient status. For example, greater alkalinity was generated by maize at initial pH 4 than pH 5 in solution culture, and greater NO_3^- uptake was attributed to enhanced enzyme activity (Masud et al. 2014). Generally, plants take up more NH_4^+ than NO_3^- when supplied at similar concentrations. However, NO_3^- may be preferentially taken up over NH_4^+ at lower pH due its co-transport with H^+ through the plasma membrane. Also, higher concentrations of other nutrients in the subsoil, such as P, are likely to enhance root growth and the relative impact of NO_3^- uptake (Weligama et al. 2008; Hansel et al. 2017). Conversely, decomposing legume residues or high soil N status will reduce the uptake of NO_3^- applied as fertiliser (Caires et al. 2015). Plants that have a greater depth and density of roots or are active for a longer duration will take up more NO_3^- and other anions (e.g. SO_4^-) and therefore have the largest impact (e.g. *Brachiaria* reaching up to 2 m deep). This approach should be practical if NO_3^- supply in the subsurface soils can be synchronised with plant demand.

3.5 Ameliorating Subsoil Acidity Using Lime and Other Amendments

Lime is the most effective and common approach to ameliorate acid topsoils. Lime applied to soil is predominantly derived from naturally occurring limestone or chalk ($CaCO_3$) and dolomitic limestone [$CaMg(CO_3)_2$] deposits, but burnt lime (CaO) (quicklime) and slaked lime [$Ca(OH)_2$] (hydrated lime) produced as industrial by-products are also used (Goulding 2016). The ameliorative effects of lime on crop growth occur via increases in pH and subsequent reductions in Al and Mn concentrations but also via reduced Al activity (increased Ca^{2+} saturation). However, lime is mainly broadcast on the soil surface, and without mechanical incorporation, movement of carbonate (CO_3^{2-}) and bicarbonate (HCO_3^-) through the soil profile from the initial zone of placement is very slow (Scott et al. 2007). Especially in variable charge soils, the effects of surface-applied lime on pH change may be restricted to a few centimetres depth, and the added Ca^{2+} may not be available to plants since it is adsorbed onto pH-dependent exchange sites (Ernani et al. 2004; Sumner 1994). Nevertheless, surface-applied lime can eventually ameliorate subsoil acidity over many years (Li et al. 2019). The relative impact of lime below its zone of incorporation is a function of the amount and quality of lime added, the initial soil pH and pHBC as well as soil texture and rainfall which influence leaching. Whitten et al. (2000) showed that 62% and 24% of lime trials on sandy soils had soil pH increases at 10 and 20 cm below the limed layer, respectively. Furthermore, these increases in pH of at least 0.2 pH units occurred within 4–7 years depending on the amount of lime applied (2.5–5 t ha^{-1}) (Whitten et al. 2000). Similarly, soil pH below 10 cm had increased over 5 years after surface application (Tang et al. 2003; Wang et al. 2016c; Aye et al. 2016). The importance of lime rate has been highlighted in numerous field studies. At three sites, 5 years after lime application, increases in soil pH were detected up to 6 cm below the layer of incorporation for 8 t ha^{-1} but not 2 t ha^{-1} (Conyers and Scott 1989). Annual rates of subsoil (15–20 cm) amelioration at 0.044 pH units $year^{-1}$ (1 unit over 22 years) were achieved with surface liming to maintain the $pHCaCl_2$ above 5.5 (Li et al. 2010). Generally, a surface soil $pHCaCl_2 > 5.5$ is needed to enable leaching of alkaline species and neutralise acidity generated below the limed layer (Slattery and Helyar 2004). However, the lime required to achieve

$pHCaCl_2$ above 5.5 commonly exceeds that used by farmers (<2.5 t ha^{-1}) and may not be agronomically realistic (Conyers and Scott 1989; Scott et al. 2000). Adding high rates of lime to surface soils to expedite downward movement could have a negative impact on crops and induce micronutrient deficiencies (Zn and B), nutritional imbalances and reduced growth.

Agricultural lime products are insoluble and can only move from the point of application via water movement through soil pores (Ernani et al. 2004). The relative effectiveness of surface-applied lime on ameliorating acid subsoils is slightly greater in areas with high rainfall and more permeable soils, such as Brazil where rainfall could range from 1200 to 1800 mm $year^{-1}$ in many agricultural areas, and lime particles can move down before they are completely reacted at the point of placement (Caires et al. 2015; Castro et al. 2017). Lime movement is particularly slow in no-till systems compared with conventional tillage (2–4 years longer) (Conyers et al. 2003). These issues have been known for some time, and various attempts have been made to either mix lime within a larger volume of soil or place it deeper within the profile. Incorporating lime in the 0–20 cm layer had a greater impact on cereal growth and yield than surface (0–10 cm) incorporation, particularly in dry seasons and for acid-sensitive crops (Scott et al. 1997). Although the technology is commonly available, ploughing to mix such a large volume of the soil profile is arduous, labour, time and fuel demanding. Also, intensification of cropping systems means that farmers are either not willing to consider ploughing or only have short periods where ploughing is suitable such as in tropical regions of Central Brazil with short dry seasons. Hence, approaches to partially ameliorate acid soil profiles with lime have been developed using deep tillage. Slotting uses a wider (e.g. 15 cm) cut than traditional deep ripping and achieves greater mixing. Maximal wheat yields were achieved when 6.5 t of lime ha^{-1} was incorporated by slotting (30 cm deep) and one-third of the soil was treated (Kauffman and Gardner 1978). Similarly, slotting achieved almost half the benefit of full lime incorporation (0–80 cm deep) with less than 17% of the lime and 25% of the soil being ameliorated (Jayawardane et al. 1995). However, this approach is very expensive and not widely adopted. In some cases, slotting has not had any advantage over modified deep ripping and conventional moldboard plough incorporation (Farina and Channon 1988a). Similarly, no difference in subsoil amelioration (10–20 cm) via ploughing and deep ripping (subsoiling) to incorporate dolomitic lime (15 t ha^{-1}) was observed (Auler et al. 2017). In the long term (over 10 years), soil chemical properties were similar between mechanical practices (Farina et al. 2000b), but costs associated with amending the subsoil were rapidly recovered (Farina et al. 2000a). Mechanical approaches to incorporate lime are associated with increased risk of erosion and damage to soil structure.

Ameliorating subsoils without direct incorporation requires the amendments, their soluble components or their decomposition products to be more mobile than lime. Gypsum is often advocated to ameliorate subsoils with minimal disturbance (Sumner 1994), and its use is increasing due to improvements in water use efficiency and yield stability (Dalla Nora et al. 2017b). Gypsum reduces the toxic effects of Al^{3+} on roots, albeit without increasing the soil pH, and has been shown to be superior compared to deep incorporation of lime in some circumstances (Farina et al. 2000a). Large increases in yield observed in field studies have confirmed its economic benefit (Farina and Channon 1988b). A number of possible mechanisms have been proposed for the improvement in plant growth once gypsum leaches into an acid subsoil. These include reduced toxic Al^{3+} by forming $Al_2(SO_4)_3$ ion pairs, facilitating Al^{3+} leaching, increasing Ca^{2+} availability and base saturation and the displacement of OH^- by SO_4^- (self-liming) (Wang et al. 1999a; Dalla Nora et al. 2017a). Although polymerisation of Al into insoluble forms was suggested to occur since no evidence for Al leaching was detected in an oxisol (Pavan et al. 1984), other research has shown that considerable Al is leached in gypsum-amended soils creating a deep soil layer (up to 100 cm) that has very low exchangeable Al (<5% CEC), but this requires high rates of lime (1–1.5 t ha^{-1} $year^{-1}$) and gypsum application (50% of lime). Importantly, the reduction in phytotoxic Al in the subsurface layer must be sufficient to enable plant roots to effectively access subsoil water and nutrients. A number of studies have shown that improvements in plant growth are greater with combined use of gypsum and lime (Dalla Nora 2017; McLay et al. 1994a, b; Dalla Nora et al. 2014). Lime dissolution is thought to occur faster in the presence of gypsum. Also, leaching of Ca into the subsoil occurs more rapidly presumably moving as the $CaSO_4$ ion pair (Dalla Nora et al. 2017b). Increases in grain yield from the combined ameliorants were superior in dry seasons, highlighting the importance of better root growth and access to subsoil water (Dalla Nora et al. 2017a) and consequently enhanced water use efficiency (Dalla Nora 2017). However, better root growth and plant performance in gypsum-amended soils is not always achieved (Wang et al. 1999b; McLay et al. 1994a). The relative differences between studies are likely due to species variation and the rate of gypsum application and the edaphic and climatic effects that control leaching (Dalla Nora and Amado 2013; Tiecher et al. 2018).

Organic materials have been used to ameliorate acid soils for some time. The mechanisms whereby organic materials, including crop residues, alter soil pH and lead to the development of pH gradients have been summarised earlier. In general, plant residues have a liming effect in the layer in which they are added (Wang et al. 2013; Tang et al. 1999a; Xu et al. 2006). Consequently, Al toxicity is reduced, and the processes involved have been reviewed (Haynes and Mokolobate 2001). Under field conditions, surface-applied crop residues can generate alkalinity below the amended layer (Butterly et al. 2013). This is attributed to the leaching and subsequent decomposition of organic compounds, including organic acid anions, that move into the soil profile with cation pairs (Miyazawa et al. 2002; Franchini et al. 2001; Butterly et al. 2011). Soluble components may therefore reduce the constraints of subsurface acidity on plant growth by pH-induced decreases in Al^{3+} concentration, forming non-toxic organo-Al complexes or decreasing Al saturation (increasing base saturation). It is not clear whether this could also facilitate the leaching of Al^{3+} out of acid subsurface layers and into less acid subsoils thereby reducing the Al^{3+} concentration. Nevertheless, because of ongoing acidification, amelioration of acid subsurface and subsoil layers will indefinitely require the direct incorporation of organic materials. It is imperative to make the distinction between those materials produced in situ and those from offsite. Imported organic materials represent a net increase in alkalinity to the soil system. However, alkalinity generated from the incorporation of organic materials (e.g. crop residues) produced in situ represents the redistribution of alkalinity, since its production is associated with acidification in some part of the soil profile. Furthermore, alkalinity generated by organic materials with low C:N may diminish over time due to net nitrification and the production of NO_3^- that is not taken up by plant roots or immobilised by microbes and leaches down the soil profile (Vanzolini et al. 2017; Butterly et al. 2013).

A diverse range of materials are used to ameliorate acid soils, and these are often employed when lime is not available or too expensive. These include bone meal, rapeseed cake, alkaline slag, dolomite, phosphogypsum, chicken manure, poultry litter, sewage sludge and composts (Hue and Licudine 1999; Li et al. 2015a; Liu and Hue 2001; Shi et al. 2017; Wang et al. 2016a; Lauricella et al. 2020). The potential for these materials to ameliorate acid soils is highly variable (Wong and Swift 2001, 2003). With the exception of alkaline slag (Masud et al. 2015), the alternative materials are generally less effective than lime at increasing soil pH. However, long-term application of organic materials, such as farmyard manure, has been successfully shown to maintain soil pH (Cai et al. 2015; Kidd et al. 2017). For all organic amendments, crop responses may be due to combined ameliorative and nutritional benefits, and reductions in Al availability and activity may occur independently of pH (Wong and Swift 2003). Organic and inorganic materials are often applied in combination, in some cases having additive effects (Shi et al. 2017; Wang et al. 2016a; Butterly et al. 2021a). Mixing crop residues with lime facilitated the leaching of Ca and Mg and increased soil pH in a 24-week column study, with black oat residue being the most effective (Miyazawa et al. 2002). However, this was not observed under field conditions, which is likely due to lower input of organic material (Caires et al. 2008). Canola straw was more effective than peanut straw when combined with alkaline slag, and this was attributed to higher S and Cl contents that facilitate the leaching of cations (Li et al. 2015b). It was suggested that crop residues with higher C:N ratio be used with alkaline slag to minimise re-acidification via nitrification and maximise the liming effect (Wang et al. 2012). Similarly, compost with high gypsum (SO_4^{2-}) content was shown to increase the effectiveness of lime below the layer of mixing (Liu and Hue 2001). Amendments such as manures and sewage sludge are able to ameliorate acidity deep within the soil profile due to high alkalinity and nutrient contents that are able to be leached (Hue and Licudine 1999; Wang et al. 2016b). Importantly, there is evidence of decreases in exchangeable Al, Al saturation as well as expedited leaching of Al, all of which reduce its phytotoxicity (Li et al. 2015b). However, some studies have shown that manures can increase H^+ and Al^{3+} concentrations in the subsoil via displacement, due to the salt effect of the added material (Haynes and Judge 2008). Many of the organic materials may be pyrolysed into biochar before they are added to soil (Dai et al. 2017; Steiner et al. 2007). Biochar is generally very effective at increasing soil pH, but the effects are highly variable and depend on feedstock and the temperature and residence time of pyrolysis (Hass et al. 2012; Buss et al. 2018). Lower rates (10 g kg^{-1}) of wheat, rice and peanut biochar achieved the same pH increases as higher (20 g kg^{-1}) rates (Wang et al. 2014). Also, the ameliorative effects of canola and peanut straw biochar were enhanced by the addition of $Ca(NO_3)_2$ (Mehmood et al. 2015). Biochar with high alkalinity and nutrient contents, e.g. poultry litter biochar, successfully ameliorates acid soils (Lauricella et al. 2021b). Although effective, limited availability of many of these products is likely to restrict the locations or areas of land in which they are used.

3.6 Future Farming Practices to Combat Subsoil Acidity

Future farming practices will need to focus on overcoming the economic and practical limitations of ameliorating acid subsoils (Shoghi Kalkhoran et al. 2019). Since minimising subsoil acidification is limited to practices and crop choices in cropping systems that have low excess cation uptake, which biologically manipulate the cation/anion balance and ensure low removal of alkalinity in farm products, these approaches are not likely to eliminate the need to use lime or other amendments. Nevertheless, future farming practices aimed at minimising nitrate leaching and supplying crops with N in the form of $Ca(NO_3)_2$ are likely to be important approaches to minimise subsoil acidification. In the Brazilian Midwest (Cerrado), cover crops such as *Brachiaria* spp. and pearl millet with deep roots (up to 2 m) are used to maximise nitrate uptake from the subsoil. For example, corn intercropped with *Brachiaria* is grown after soybean (it is possible to grow two crops per agricultural year in some regions of the Brazilian Midwest), and after harvesting corn, the *Brachiaria* is well established with deep roots (this practice is called the Santa Fe system). Variable rate technology, split applications of nitrate and using acid-tolerant crops will expedite anion uptake and N use efficiency and have the greatest effect in slowing subsoil acidification. Hence, the development of acid-tolerant crop species and their adoption is important to complement other management strategies (Ryan 2018). The nutrition of these crops will need to be managed appropriately, and in particular, subsurface placement of P may help reduce Al toxicity and maximise NO_3^- uptake.

The greatest advances are likely to come with future development of cost-effective approaches to ameliorate acid subsoils with lime and other amendments. This will invariably require greater detail of soil pH in biopores and its gradient within the profile by combining new precision agriculture technologies including on-the-go smart soil sensors and soil and yield data collected by farmers together with simulation modelling (Wong et al. 2008; Filippi et al. 2018). In addition to pH, these approaches could also help to predict pHBC and enable better calculations of the rates of lime and other materials required to ameliorate subsurface soil acidity (Wong et al. 2013). Hence, efficiencies could be improved via reductions in the quantities of materials used or targeting the areas to which these materials are applied (Oliver et al. 2015), compared with current practices that apply blanket rates of lime. Also, lime could be combined with other materials to facilitate alkalinity movement and acidity amelioration down the soil profile (Butterly et al. 2021a; Lauricella et al. 2021a). Even without new technology, some farmers in Australia have realised the need to better characterise soil pH within their paddocks and have embarked on obtaining these data themselves using handheld pH electrodes (Baxter 2018).

Further development of machinery may also be required to incorporate lime and other materials into subsoils. For land areas still being converted to agriculture, such as the Brazilian 'Cerrado' region, it is advocated that high rates of lime (~5 t ha^{-1}) be incorporated up to 30 cm deep with heavy disk harrows, before planting with fast-growing crops (e.g. pearl millet and *Brachiaria*) to limit soil erosion and restore soil physical attributes affected by tillage. In other areas of the world, however, amelioration of acid subsoils already under agriculture is more difficult because such cultivation is often not desirable. Opportunistic tillage (ploughing) may be an option to reduce the stratification of existing pH within the soil profile or facilitate the incorporation of soil amendments (Dalla Nora et al. 2017b). Depending on the intensity, strategic tillage has ancillary benefits including improved P distribution, burying weed seeds, correcting hydrophobicity and reducing compaction. However, farmers are likely to opt for less aggressive approaches for deep liming and subsoil injection via deep ripping (chisel), which will require new machinery. Depending on the materials, alternatives to lime may require high rates of application (e.g. 15 t ha^{-1}), and pelletisation of these materials will facilitate their handling and delivery into the soil. The choice of these materials will depend on the severity and depth of acid soil layers.

Soils will need to be managed properly to maximise the longevity of the subsoil amendments. It is likely that traditional surface applications of lime will complement new approaches. Deep-rooted crops and pasture species have been shown to facilitate lime movement through compacted layers (12–18 cm) that develop under no-till monocropping systems in Central Brazil, with minimal disturbance. The impact of grazing will also be important since large areas of acid soils are managed as integrated crop-livestock systems. However, in Brazil grazing intensity did not reduce the effectiveness of surface-applied lime (Martins et al. 2014). Greater use of lime to correct soil acidity may also have complementary effects on the sustainability of farming systems. In subtropical systems, studies have reported that lime application (Briedis et al. 2012) and combined application of lime and gypsum (Inagaki et al. 2017) can increase soil carbon accumulation under minimum tillage (conservation agriculture) due to increases in above-ground and root biomass and the presence of Ca^{2+} ions which increase soil aggregation and chemically protect soil C (Ferreira et al. 2018). However, increases in soil C following long-term lime use have not been observed in Australian cropping systems (Wang et al. 2016c). In fact, some studies suggest that liming can promote native soil C mineralisation, i.e. have a priming effect (Aye et al. 2017; Grover et al. 2017, 2021).

3.7 Conclusions

Subsoil acidity is an insidious problem, and innovative ways to minimise acidification and ameliorate acid subsoils will be required to maximise future crop production and ensure food security. This drive to remove the constraints of subsoil acidity on crop production in order to maximise water and nutrient-limited yield potentials will be particularly important in areas with increased climate variability and declining rainfall. Although the mechanisms that lead to the development of acid subsoils and pH gradients are generally well understood, lime application and other ameliorative approaches have not been sufficient to offset rates of acidification. Traditionally, deep placement of lime with or without other organic amendments was considered not practical or profitable, but this is being revisited. For acid subsoil amelioration, the treatments need to be able to improve a large volume of soil; otherwise, there will be no real benefit in terms of root growth and crop performance. In contrast, the key to ameliorating acid subsurface layers that overlay non-acidic subsoils is improving root access to the subsoil where they can proliferate, and this may not require amelioration of large parts of the soil profile but instead zones of access where roots can pass through the acid layer. The effectiveness and longevity of any approach to ameliorate acid subsoils need to be considered in terms of crop response (low pH, Al and Mn toxicities and nutrient deficiencies) and re-acidification. Future development of cost-effective methods is still needed.

Acknowledgements Caixian Tang was supported by the Grains Research and Development Corporation through GRDC project (DAN00206) investigating innovative approaches to managing subsoil acidity in the southern grain region of Australia.

References

Adams TM, Adams SN. The effects of liming and soil-pH on carbon and nitrogen contained the soil biomass. J Agric Sci. 1983;101(3):553–8. https://doi.org/10.1017/S0021859600038570.

Aitken R, Moody P. Interrelations between soil pH measurements in various electrolytes and soil solution pH in acidic soils. Soil Res. 1991;29(4):483–91. https://doi.org/10.1071/SR9910483.

Araújo LG, de Figueiredo CC, de Sousa DMG, de Souza Nunes R, Rein TA. Influence of gypsum application on sugarcane yield and soil chemical properties in the Brazilian Cerrado. Aust J. Crop Sci. 2016;10(11):1557–63. https://doi.org/10.21475/ajcs.2016.10.11.PNE156.

Auler AC, Pires LF, dos Santos JAB, Caires EF, Borges JAR, Giarola NFB. Effects of surface-applied and soil-incorporated lime on some physical attributes of a Dystrudept soil. Soil Use Manag. 2017;33(1):129–40. https://doi.org/10.1111/sum.12330.

Aye NS, Sale PWG, Tang C. The impact of long-term liming on soil organic carbon and aggregate stability in low-input acid soils. Biol Fertil Soils. 2016;52(5):697–709. https://doi.org/10.1007/s00374-016-1111-y.

Aye NS, Butterly CR, Sale PWG, Tang C. Residue addition and liming history interactively enhance mineralization of native organic carbon in acid soils. Biol Fertil Soils. 2017;53(1):61–75. https://doi.org/10.1007/s00374-016-1156-y.

Azam G, Gazey C. Slow movement of alkali from surface-applied lime warrants the introduction of strategic tillage for rapid amelioration of subsurface acidity in South-Western Australia. Soil Res. 2021;59(1):97–106. https://doi.org/10.1071/SR19329.

Barton L, Carr S. Development and verification of a quick test for the diagnosis of aluminium toxicity on the deep yellow earths of Western Australia. Soil Res. 1996;34(5):781–8. https://doi.org/10.1071/SR9960781.

Baxter N. Intensive pH tests set up precision liming. GroundCover, Issue 133; 2018, March–April. https://grdc.com.au/resources-and-publications/groundcover/groundcover-133-march-april-2018/intensive-ph-tests-set-up-precision-liming

Benke KK, Robinson NJ. Quantification of uncertainty in mathematical models: the statistical relationship between field and laboratory pH measurements. Appl Environ Soil Sci. 2017;2017:1–12. https://doi.org/10.1155/2017/5857139.

Bertsch PM, Bloom PR. Aluminum. In: Sparks DL, Page PA, Helmke RH, Loeppert RH, editors. Methods of soil analysis part 3 – chemical methods. Madison: SSSA, ASA; 1996. p. 517–50. https://doi.org/10.2136/sssabookser5.3.

Bolan NS, Hedley MJ. Role of carbon, nitrogen, and sulfur cycles in soil acidification. In: Rengel Z, editor. Handbook of soil acidity. New York: Marcel Dekker; 2003. p. 29–56. https://doi.org/10.1201/9780203912317.

Bolan NS, Hedley MJ, White RE. Processes of soil acidification during nitrogen cycling with emphasis on legume based pastures. Plant Soil. 1991;134(1):53–63. https://doi.org/10.1007/BF00010717.

Briedis C, Sá JCM, Caires EF, Navarro JF, Inagaki TM, Boer A, Neto CQ, Ferreira AO, Canalli LB, Santos JB. Soil organic matter pools and carbon-protection mechanisms in aggregate classes influenced by surface liming in a no-till system. Geoderma. 2012;170:80–8. https://doi.org/10.1016/j.geoderma.2011.10.011.

Burle ML, Mielniczuk J, Focchi S. Effect of cropping systems on soil chemical characteristics, with emphasis on soil acidification. Plant Soil. 1997;190(2):309–16. https://doi.org/10.1023/a:1004266831343.

Buss W, Shepherd JG, Heal KV, Mašek O. Spatial and temporal microscale pH change at the soil-biochar interface. Geoderma. 2018;331:50–2. https://doi.org/10.1016/j.geoderma.2018.06.016.

Butterly CR, Kaudal BB, Baldock JA, Tang C. Contribution of soluble and insoluble fractions of agricultural residues to short-term pH changes. Eur J Soil Sci. 2011;62(5):718–27. https://doi.org/10.1111/j.1365-2389.2011.01387.x.

Butterly CR, Baldock JA, Tang C. The contribution of crop residues to changes in soil pH under field conditions. Plant Soil. 2013;366(1):185–98. https://doi.org/10.1007/s11104-012-1422-1.

Butterly CR, Costello B, Lauricella D, Sale P, Li G, Tang C. Alkalinity movement down acid soil columns was faster when lime and plant residues were combined than when either was applied separately. Eur J Soil Sci. 2021a;72(1):313–25. https://doi.org/10.1111/ejss.12945.

Butterly CR, Wang X, Sale P, Li G, Tang C. Liming effect of non-legume residues promotes the biological amelioration of soil acidity via nitrate uptake. Plant Soil. 2021b;464(1):63–73. https://doi.org/10.1007/s11104-021-04937-6.

Cai Z, Wang B, Xu M, Zhang H, He X, Zhang L, Gao S. Intensified soil acidification from chemical N fertilization and prevention by manure in an 18-year field experiment in the red soil of southern China. J Soils Sediments. 2015;15(2):260–70. https://doi.org/10.1007/s11368-014-0989-y.

Caires EF, Pereira PRS, Zardo R, Feldhaus IC. Soil acidity and aluminium toxicity as affected by surface liming and cover oat residues under a no-till system. Soil Use Manag. 2008;24(3):302–9. https://doi.org/10.1111/j.1475-2743.2008.00166.x.

Caires EF, Haliski A, Bini AR, Scharr DA. Surface liming and nitrogen fertilization for crop grain production under no-till management in Brazil. Eur J Agron. 2015;66:41–53. https://doi.org/10.1016/j.eja.2015.02.008.

Castro GSA, Crusciol CAC, Rosolem CA, Calonego JC, Brye KR. Surface lime and silicate application and crop production system effects on physical characteristics of a Brazilian Oxisol. Soil Res. 2017;55(8):778–87. https://doi.org/10.1071/SR16247.

Chan KY, Heenan DP. The effects of stubble burning and tillage on soil carbon sequestration and crop productivity in southeastern Australia. Soil Use Manag. 2005;21(4):427–31. https://doi.org/10.1079/sum2005357.

Condon JR, Black AS, Conyers MK. The role of N transformations in the formation of acidic subsurface layers in stock urine patches. Soil Res. 2004;42(2):221–30. https://doi.org/10.1071/SR03109.

Condon J, Burns H, Li G. The extent, significance and amelioration of subsurface acidity in southern New South Wales, Australia. Soil Res. 2021;59(1):1–11. https://doi.org/10.1071/SR20079.

Conyers M, Scott B. The influence of surface incorporated lime on subsurface soil acidity. Aust J Exp Agric. 1989;29(2):201–7. https://doi.org/10.1071/EA9890201.

Conyers MK, Poile GJ, Cullis BR. Lime responses by barley as related to available soil aluminium and manganese. Aust J Agric Res. 1991;42(3):379–90. https://doi.org/10.1071/ar9910379.

Conyers MK, Heenan DP, Poile GJ, Cullis BR, Helyar KR. Influence of dryland agricultural management practices on the acidification of a soil profile. Soil Tillage Res. 1996;37(2–3):127–41. https://doi.org/10.1016/0167-1987(95)01005-x.

Conyers MK, Heenan DP, McGhie WJ, Poile GP. Amelioration of acidity with time by limestone under contrasting tillage. Soil Tillage Res. 2003;72(1):85–94. https://doi.org/10.1016/S0167-1987(03)00064-3.

Conyers MK, Tang C, Poile GJ, Liu D, Chen DL, Nuruzzaman Z. A combination of biological activity and the nitrate form of nitrogen can be used to ameliorate subsurface soil acidity under dryland wheat farming. Plant Soil. 2011;348(1–2):155–66. https://doi.org/10.1007/s11104-011-0827-6.

Corassa GM, Amado TJC, Tabaldi FM, Schwalbert RA, Reimche GB, Nora DD, Alba PJ, Horbe TAN. High-resolution spacialization of an Oxisol's acidity attributes by on-the-go sensing. Presqui Agropecu Bras. 2016;51(9):1306–16. https://doi.org/10.1590/s0100-204x2016000900030.

Coventry DR, Slattery WJ. Acidification of soil associated with lupins grown in a crop-rotation in north-eastern Victoria. Aust J Agric Res. 1991;42(3):391–7. https://doi.org/10.1071/AR9910391.

Coventry DR, Farhoodi A, Xu RK. Managing soil acidification through crop rotations in southern Australia. In: Rengel Z, editor. Handbook of soil acidity. New York: Marcel Dekker; 2003. p. 407–29. https://doi.org/10.1201/9780203912317.

Cregan PD, Scott BJ. Soil acidification. In: Pratley JE, Robertson AI, editors. Agriculture and the environmental imperative. Melbourne: CSIRO Publishing; 1998. p. 98–128. https://doi.org/10.1071/9780643104877.

Dai Z, Zhang X, Tang C, Muhammad N, Wu J, Brookes PC, Xu J. Potential role of biochars in decreasing soil acidification – a critical review. Sci Total Environ. 2017;581:601–11. https://doi.org/10.1016/j.scitotenv.2016.12.169.

Dalla Nora D. Improvement of chemical attributes of the root layer and its effect in crop grain yield under no-tillage. PhD, Federal University of Santa Maria, Santa Maria, Rio Grande do Sul, Brazil; 2017.

Dalla Nora D, Amado TJC. Improvement in chemical attributes of Oxisol subsoil and crop yields under no-till. Agron J. 2013;105(5):1393–403. https://doi.org/10.2134/agronj2013.0031.

Dalla Nora D, Amado TJC, Bortolotto RP, Ferreira AO, Reichardt K, Santi AL. Subsoil chemical amelioration and crop yields under continuous long-term no-till in a subtropical Oxisol. Afr J Agric Res. 2014;9(45):3338–49. https://doi.org/10.5897/AJAR2013.8283.

Dalla Nora D, Amado TJC, Nicoloso RS, Gruhn EM. Modern high-yielding maize, wheat and soybean cultivars in response to gypsum and lime application on no-till Oxisol. Rev Bras Cienc Solo. 2017a;41:1–21. https://doi.org/10.1590/18069657rbcs20160504.

Dalla Nora D, Amado TJC, Nicoloso RS, Mazuco ACB, Piccin M. Mitigation of the gradient of chemical properties in the rooting zone of dystrophic Oxisols by gypsum and lime inputs under a no-till system. Rev Bras Cienc Solo. 2017b;41:1–22. https://doi.org/10.1590/18069657rbcs20150541.

de Caritat P, Cooper M, Wilford J. The pH of Australian soils: field results from a national survey. Soil Res. 2011;49(2):173–82. https://doi.org/10.1071/SR10121.

Dolling PJ. Effect of lupins and location on soil acidification rates. Aust J Exp Agric. 1995;35(6):753–63. https://doi.org/10.1071/EA9950753.

Doolittle JA, Brevik EC. The use of electromagnetic induction techniques in soils studies. Geoderma. 2014;223–225:33–45. https://doi.org/10.1016/j.geoderma.2014.01.027.

Ernani PR, Ribeiro MFS, Bayer C. Chemical modifications caused by liming below the limed layer in a predominantly variable charge acid soil. Commun Soil Sci Plan. 2004;35(5–6):889–901. https://doi.org/10.1081/css-120030365.

Fageria NK, Nascente AS. Management of soil acidity of South American soils for sustainable crop production. In: Sparks DL, editor. Advances in agronomy, vol. 128. Academic Press; 2014. p. 221–75. https://doi.org/10.1016/B978-0-12-802139-2.00006-8.

Farina MPW, Channon P. Acid-subsoil amelioration: I. A comparison of several mechanical procedures. Soil Sci Soc Am J. 1988a;52(1):169–75. https://doi.org/10.2136/sssaj1988.03615995005200010030x.

Farina MPW, Channon P. Acid-subsoil amelioration: II. Gypsum effects on growth and subsoil chemical properties. Soil Sci Soc Am J. 1988b;52(1):175–80. https://doi.org/10.2136/sssaj1988.03615995005200010031x.

Farina MPW, Channon P, Thibaud GR. A comparison of strategies for ameliorating subsoil acidity: I. Long-term growth effects. Soil Sci Soc Am J. 2000a;64(2):646–51. https://doi.org/10.2136/sssaj2000.642646x.

Farina MPW, Channon P, Thibaud GR. A comparison of strategies for ameliorating subsoil acidity: II. Long-term soil effects. Soil Sci Soc Am J. 2000b;64(2):652–8. https://doi.org/10.2136/sssaj2000.642652x.

Ferreira AO, Amado TJC, Rice CW, Ruiz Diaz DA, Briedis C, Inagaki TM, Gonçalves DRP. Driving factors of soil carbon accumulation in Oxisols in long-term no-till systems of South Brazil. Sci Total Environ. 2018;622–623:735–42. https://doi.org/10.1016/j.scitotenv.2017.12.019.

Filippi P, Cattle SR, Bishop TFA, Odeh IOA, Pringle MJ. Digital soil monitoring of top- and sub-soil pH with bivariate linear mixed models. Geoderma. 2018;322:149–62. https://doi.org/10.1016/j.geoderma.2018.02.033.

Franchini JC, Gonzalez-Vila FJ, Cabrera F, Miyazawa M, Pavan MA. Rapid transformations of plant water-soluble organic compounds in relation to cation mobilization in an acid Oxisol. Plant Soil. 2001;231(1):55–63. https://doi.org/10.1023/A:1010338917775.

Gillman GP, Bell LC. Soil solution studies on weathered soils from tropical North Queensland. Aust J Soil Res. 1978;16(1):67–77. https://doi.org/10.1071/SR9780067.

Gillman G, Sumpter E. Modification to the compulsive exchange method for measuring exchange characteristics of soils. Soil Res. 1986;24(1):61–6. https://doi.org/10.1071/SR9860061.

Goldman IL, Carter TE, Patterson RP. A detrimental interaction of subsoil aluminum and drought stress on the leaf water status of soybean. Agron J. 1989;81(3):461–3. https://doi.org/10.2134/agronj1989.00021962008100030013x.

Goulding KWT. Soil acidification and the importance of liming agricultural soils with particular reference to the United Kingdom. Soil Use Manag. 2016;32(3):390–9. https://doi.org/10.1111/sum.12270.

Grover SP, Butterly CR, Wang X, Tang C. The short-term effects of liming on organic carbon mineralisation in two acidic soils as affected by different rates and application depths of lime. Biol Fertil Soils. 2017:1–13. https://doi.org/10.1007/s00374-017-1196-y.

Grover SP, Butterly CR, Wang X, Gleeson DB, Macdonald LM, Tang C. Liming and priming: the long-term impact of pH amelioration on mineralisation may negate carbon sequestration gains. Soil Secur. 2021;3:100007. https://doi.org/10.1016/j.soisec.2021.100007.

Hansel FD, Amado TJC, Ruiz Diaz DA, Rosso LHM, Nicoloso FT, Schorr M. Phosphorus fertilizer placement and tillage affect soybean root growth and drought tolerance. Agron J. 2017;109(6):2936–44. https://doi.org/10.2134/agronj2017.04.0202.

Hass A, Gonzalez JM, Lima IM, Godwin HW, Halvorson JJ, Boyer DG. Chicken manure biochar as liming and nutrient source for acid Appalachian soil. J Environ Qual. 2012;41(4):1096–106. https://doi.org/10.2134/jeq2011.0124.

Haynes RJ, Judge A. Influence of surface-applied poultry manure on topsoil and subsoil acidity and salinity: a leaching column study. J Plant Nutr Soil Sci. 2008;171(3):370–7. https://doi.org/10.1002/jpln.200700167.

Haynes RJ, Mokolobate MS. Amelioration of Al toxicity and P deficiency in acid soils by additions of organic residues: a critical review of the phenomenon and the mechanisms involved. Nutr Cycl Agroecosys. 2001;59(1):47–63. https://doi.org/10.1023/a:1009823600950.

Helyar KR. The management of acid soils. In: Wright RJ, Baligar VC, Murrmann RP, editors. Plant-soil interactions at low pH: proceedings of the second international symposium on plant-soil interactions at low pH, 24–29 June 1990, Beckley West Virginia, USA. Springer, Dordrecht; 1991. p. 365–82. https://doi.org/10.1007/978-94-011-3438-5_42.

Helyar KR, Porter WM. Soil acidification, its measurement and the processes involved. In: Robson AD, editor. Soil acidity and plant growth. Sydney: Academic Press; 1989. p. 61–101.

Henderson BL, Bui EN. An improved calibration curve between soil pH measured in water and $CaCl_2$. Soil Res. 2002;40(8):1399–405. https://doi.org/10.1071/SR01098.

Hinsinger P, Plassard C, Tang CX, Jaillard B. Origins of root-mediated pH changes in the rhizosphere and their responses to environmental constraints: a review. Plant Soil. 2003;248(1–2):43–59. https://doi.org/10.1023/a:1022371130939.

Hue NV, Licudine DL. Amelioration of subsoil acidity through surface application of organic manures. J Environ Qual. 1999;28(2):623–32. https://doi.org/10.2134/jeq1999.00472425002800020028x.

Hue NV, Craddock GR, Adams F. Effect of organic-acids on aluminum toxicity in subsoils. Soil Sci Soc Am J. 1986;50(1):28–34. https://doi.org/10.2136/sssaj1986.03615995005000010006x.

Hue NV, Vega S, Silva JA. Manganese toxicity in a Hawaiian oxisol affected by soil pH and organic amendments. Soil Sci Soc Am J. 2001;65(1):153–60. https://doi.org/10.2136/sssaj2001.651153x.

Inagaki TM, de Moraes Sá JC, Caires EF, Gonçalves DRP. Why does carbon increase in highly weathered soil under no-till upon lime and gypsum use? Sci Total Environ. 2017;599–600:523–32. https://doi.org/10.1016/j.scitotenv.2017.04.234.

Jarvis S, Robson A. The effects of nitrogen nutrition of plants on the development of acidity in Western Australian soils. II. Effects of differences in cation/anion balance between plant species grown under non-leaching conditions. Aust J Agric Res. 1983;34(4):355–65. https://doi.org/10.1071/AR9830355.

Jayawardane N, Barrs H, Muirhead W, Blackwell J, Murray E, Kirchhof G. Lime-slotting technique to ameliorate subsoil acidity in a clay soil II. Effects on medic root-growth, water extraction and yield. Soil Res. 1995;33(3):443–59. https://doi.org/10.1071/SR9950443.

Joris HAW, Caires EF, Bini AR, Scharr DA, Haliski A. Effects of soil acidity and water stress on corn and soybean performance under a no-till system. Plant Soil. 2013;365(1):409–24. https://doi.org/10.1007/s11104-012-1413-2.

Kassam A, Friedrich T, Derpsch R, Kienzle J. Overview of the worldwide spread of conservation agriculture. Field Actions Sci Rep. 2015;8:1–11.

Kauffman MD, Gardner EH. Segmental liming of soil and its effect on the growth of wheat. Agron J. 1978;70(2):331–6. https://doi.org/10.2134/agronj1978.00021962007000020027x.

Kerven GL, Edwards DG, Asher CJ, Hallman PS, Kokot S. Aluminum determination in soil solution 2. Short-term colorimetric procedures for the measurement of inorganic monomeric aluminum in the presence of organic-acid ligands. Aust J Soil Res. 1989;27(1):91–102. https://doi.org/10.1071/sr9890091.

Kidd J, Manning P, Simkin J, Peacock S, Stockdale E. Impacts of 120 years of fertilizer addition on a temperate grassland ecosystem. PLoS One. 2017;12(3):e0174632. https://doi.org/10.1371/journal.pone.0174632.

Kissel DE, Sonon L, Vendrell PF, Isaac RA. Salt concentration and measurement of soil pH. Commun Soil Sci Plan. 2009;40(1–6):179–87. https://doi.org/10.1080/00103620802625377.

Kochian LV, Hoekenga OA, Piñeros MA. How do crop plants tolerate acid soils? Mechanisms of aluminum tolerance and phosphorous efficiency. Annu Rev Plant Biol. 2004;55(1):459–93. https://doi.org/10.1146/annurev.arplant.55.031903.141655.

Kopittke PM, Blamey FPC. Theoretical and experimental assessment of nutrient solution composition in short-term studies of aluminium rhizotoxicity. Plant Soil. 2016;406(1–2):311–26. https://doi.org/10.1007/s11104-016-2890-5.

Kopittke PM, Menzies NW, Wang P, Blamey FPC. Kinetics and nature of aluminium rhizotoxic effects: a review. J Exp Bot. 2016;67(15):4451–67. https://doi.org/10.1093/jxb/erw233.

Kweon G. Toward the ultimate soil survey: sensing multiple soil and landscape properties in one pass. Agron J. 2012;104(6):1547–57. https://doi.org/10.2134/agronj2012.0156.

Lauricella D, Butterly CR, Clark GJ, Sale PWG, Li G, Tang C. Effectiveness of innovative organic amendments in acid soils depends on their ability to supply P and alleviate Al and Mn toxicity in plants. J Soils Sediments. 2020;20(11):3951–62. https://doi.org/10.1007/s11368-020-02721-0.

Lauricella D, Butterly CR, Weng Z, Clark GJ, Sale PWG, Li G, Tang C. Impact of novel materials on alkalinity movement down acid soil profiles when combined with lime. J Soils Sediments. 2021a;21(1):52–62. https://doi.org/10.1007/s11368-020-02747-4.

Lauricella D, Weng Z, Clark GJ, Butterly CR, Li G, Gazey C, Sale PWG, Tang C. Biochars and their feedstocks differ in their short-term effects in ameliorating acid soils grown with aluminium-sensitive wheat. J Soils Sediments. 2021b;21(8):2805–16. https://doi.org/10.1007/s11368-021-03001-1.

Li G, Conyers M, Cullis BR. Long-term liming ameliorates subsoil acidity in high rainfall zone in south-eastern Australia. Paper presented at the 19th world congress of soil science: soil solutions for a changing world, Brisbane; 2010.

Li JY, Liu ZD, Zhao WZ, Masud MM, Xu RK. Alkaline slag is more effective than phosphogypsum in the amelioration of subsoil acidity in an Ultisol profile. Soil Tillage Res. 2015a;149:21–32. https://doi.org/10.1016/j.still.2014.12.017.

Li JY, Masud MM, Li ZY, Xu RK. Amelioration of an Ultisol profile acidity using crop straws combined with alkaline slag. Environ Sci Pollut Res. 2015b;22(13):9965–75. https://doi.org/10.1007/s11356-015-4176-5.

Li GD, Conyers MK, Helyar KR, Lisle CJ, Poile GJ, Cullis BR. Long-term surface application of lime ameliorates subsurface soil acidity in the mixed farming zone of south-eastern Australia. Geoderma. 2019;338:236–46. https://doi.org/10.1016/j.geoderma.2018.12.003.

Liu J, Hue NV. Amending subsoil acidity by surface applications of gypsum, lime, and composts. Commun Soil Sci Plan. 2001;32(13–14):2117–32. https://doi.org/10.1081/CSS-120000273.

Liu DL, Helyar KR, Conyers MK, Fisher R, Poile GJ. Response of wheat, triticale and barley to lime application in semi-arid soils. Field Crop Res. 2004;90(2–3):287–301. https://doi.org/10.1016/j.fcr.2004.03.008.

Lockwood PV, Wilson BR, Daniel H, Jones MJ. Soil acidification and natural resource management directions for the future. Armidale: NSW Agriculture; 2003.

Loss S, Ritchie G, Robson A. Effect of lupins and pasture on soil acidification and fertility in Western Australia. Aust J Exp Agric. 1993a;33(4):457–64. https://doi.org/10.1071/EA9930457.

Loss SP, Robson AD, Ritchie GGP. H^+/OH^- excretion and nutrient uptake in upper and lower parts of Lupin (*Lupinus angustifolius* L.) root systems. Ann Bot. 1993b;72(4):315–20. https://doi.org/10.1006/anbo.1993.1113.

Martins AP, Anghinoni I, Costa SEVGA, Carlos FS, Nichel GH, Silva RAP, Carvalho PCF. Amelioration of soil acidity and soybean yield after surface lime reapplication to a long-term no-till integrated crop-livestock system under varying grazing intensities. Soil Tillage Res. 2014;144:141–9. https://doi.org/10.1016/j.still.2014.07.019.

Masud MM, Guo D, Li JY, Xu RK. Hydroxyl release by maize (*Zea mays* L.) roots under acidic conditions due to nitrate absorption and its potential to ameliorate an acidic Ultisol. J Soils Sediments. 2014;14(5):845–53. https://doi.org/10.1007/s11368-013-0837-5.

Masud MM, Li JY, Xu RK. Application of alkaline slag and phosphogypsum for alleviating soil acidity in an Ultisol profile: a short-term leaching experiment. J Soils Sediments. 2015;15(2):365–73. https://doi.org/10.1007/s11368-014-0993-2.

McLay C, Ritchie G, Porter W. Amelioration of subsurface acidity in sandy soils in low rainfall regions. 1. Responses of wheat and lupins to surface-applied gypsum and lime. Soil Res. 1994a;32(4):835–46. https://doi.org/10.1071/SR9940835.

McLay C, Ritchie G, Porter W, Cruse A. Amelioration of subsurface acidity in sandy soils in low rainfall regions. 2. Changes to soil solution composition following the surface application of gypsum and lime. Soil Res. 1994b;32(4):847–65. https://doi.org/10.1071/SR9940847.

McLay CDA, Barton L, Tang C. Acidification potential of ten grain legume species grown in nutrient solution. Aust J Agric Res. 1997;48(7):1025–32. https://doi.org/10.1071/A96174.

Mehmood K, Li JY, Jiang J, Masud MM, Xu RK. Effect of low energy-consuming biochars in combination with nitrate fertilizer on soil acidity amelioration and maize growth. J Soils Sediments. 2015:1–10. https://doi.org/10.1007/s11368-015-1219-y.

Meng C, Tian D, Zeng H, Li Z, Yi C, Niu S. Global soil acidification impacts on belowground processes. Environ Res Lett. 2019;14(7):074003. https://doi.org/10.1088/1748-9326/ab239c.

Menzies NW. Toxic elements in acid soils: chemistry and measurement. In: Rengel Z, editor. Handbook of soil acidity. New York: Marcel Dekker; 2003. p. 267–96. https://doi.org/10.1201/9780203912317.

Menzies NW, Kerven GL, Bell LC, Edwards DG. Determination of total soluble aluminum in soil solution using pyrocatechol violet, lanthanum and iron to discriminate against microparticulates and organic-ligands. Commun Soil Sci Plan. 1992;23(17–20):2525–45. https://doi.org/10.1080/00103629209368754.

Miller RO, Kissel DE. Comparison of soil pH methods on soils of North America. Soil Sci Soc Am J. 2010;74(1):310–6. https://doi.org/10.2136/sssaj2008.0047.

Minasny B, McBratney AB, Brough DM, Jacquier D. Models relating soil pH measurements in water and calcium chloride that incorporate electrolyte concentration. Eur J Soil Sci. 2011;62(5):728–32. https://doi.org/10.1111/j.1365-2389.2011.01386.x.

Miyazawa M, Pavan MA, Franchini JC. Evaluation of plant residues on the mobility of surface applied lime. Braz Arch Biol Technol. 2002;45(3):251–6. https://doi.org/10.1590/s1516-89132002000300001.

NLWRA. Australian agricultural assessment 2001. Canberra: National Land & Water Resources Audit. CSIRO; 2001.

Noble AD, Zenneck I, Randall PJ. Leaf litter ash alkalinity and neutralisation of soil acidity. Plant Soil. 1996;179(2):293–302. https://doi.org/10.1007/BF00009340.

Oliver YM, Wong M, Robertson M. Targeting the subsoil to better manage acidity spatially. In: Proceedings of the 17th ASA conference, Hobart, Australia. Australian Society of Agronomy; 2015. p 20–4.

Orton TG, Mallawaarachchi T, Pringle MJ, Menzies NW, Dalal RC, Kopittke PM, Searle R, Hochman Z, Dang YP. Quantifying the economic impact of soil constraints on Australian agriculture: a case-study of wheat. Land Degrad Dev. 2018;29(11):3866–75. https://doi.org/10.1002/ldr.3130.

Paul KI, Black AS, Conyers MK. Effect of plant residue return on the development of surface soil pH gradients. Biol Fertil Soils. 2001a;33(1):75–82. https://doi.org/10.1007/s003740000293.

Paul KI, Black AS, Conyers MK. Influence of fallow, wheat and subterranean clover on pH within an initially mixed surface soil in the field. Biol Fertil Soils. 2001b;33(1):41–52. https://doi.org/10.1007/s003740000288.

Paul KI, Black AS, Conyers MK. Development of acidic subsurface layers of soil under various management systems. In: Sparks DL, editor. Advances in agronomy, vol. 78. San Diego: Academic Press; 2003. p. 187–214. https://doi.org/10.1016/S0065-2113(02)78005-X.

Pavan MA, Bingham FT, Pratt PF. Redistribution of exchangeable calcium, magnesium, and aluminum following lime or gypsum applications to a Brazilian Oxisol. Soil Sci Soc Am J. 1984;48(1):33–8. https://doi.org/10.2136/sssaj1984.03615995004800010006x.

Rayment GE, Lyons DJ. Soil chemical methods – Australasia. Melbourne: CSIRO Publishing; 2011.
Rengel Z. Soil pH, soil health and climate change. In: Singh BP, Cowie AL, Chan KY, editors. Soil health and climate change. Berlin/Heidelberg/: Springer; 2011. p. 69–85. https://doi.org/10.1007/978-3-642-20256-8_4.
Ritchie GSP. The chemical behavior of aluminium, hydrogen and manganese in acid soils. In: Robson AD, editor. Soil acidity and plant growth. Sydney: Academic Press; 1989. p. 1–60.
Ritchie GSP, Dolling PJ. The role of organic matter in soil acidification. Aust J Soil Res. 1985;23(4):569–76. https://doi.org/10.1071/SR9850569.
Ryan PR. Assessing the role of genetics for improving the yield of Australia's major grain crops on acid soils. Crop Pasture Sci. 2018;69(3):242–64. https://doi.org/10.1071/CP17310.
Schirrmann M, Gebbers R, Kramer E, Seidel J. Soil pH mapping with an on-the-go sensor. Sensors. 2011;11(1):573. https://doi.org/10.3390/s110100573.
Scott BJ, Conyers MK, Poile GJ, Cullis BR. Subsurface acidity and liming affect yield of cereals. Aust J Agric Res. 1997;48(6):843–54. https://doi.org/10.1071/A96140.
Scott BJ, Ridley AM, Conyers MK. Management of soil acidity in long-term pastures of south-eastern Australia: a review. Aust J Exp Agric. 2000;40(8):1173–98. https://doi.org/10.1071/EA00014.
Scott BJ, Fenton IG, Fanning AG, Schumann WG, Castleman LJC. Surface soil acidity and fertility in the eastern Riverina and Western slopes of southern New South Wales. Aust J Exp Agric. 2007;47(8):949–64. https://doi.org/10.1071/EA05155x.
Scott BJ, Conyers MK, Burns HM, Evans CM, Fettell NA. Stratification of acidity in the shallow soil surface – experiences in the cropping areas of southern and central NSW. Paper presented at the Doing more with Less, Ballarat, Australia; 2017.
Shi R-y, Li J-y, Ni N, Mehmood K, Xu R-k, Qian W. Effects of biomass ash, bone meal, and alkaline slag applied alone and combined on soil acidity and wheat growth. J Soils Sediments. 2017;17(8):2116–26. https://doi.org/10.1007/s11368-017-1673-9.
Shoghi Kalkhoran S, Pannell DJ, Thamo T, White B, Polyakov M. Soil acidity, lime application, nitrogen fertility, and greenhouse gas emissions: optimizing their joint economic management. Agric Syst. 2019;176:102684. https://doi.org/10.1016/j.agsy.2019.102684.
Slattery WJ, Helyar KR. Acidification and its evolution under Australian dryland cropping systems. In: Rao SC, Ryan J, editors. Challenges and strategies of dryland agriculture, CSSA special publication, vol. 32. Crop Science Society of America and American Society of Agronomy: Madison, WI; 2004. p. 335–58. https://doi.org/10.2135/cssaspecpub32.c21.
Slattery W, Ridley A, Windsor S. Ash alkalinity of animal and plant products. Aust J Exp Agric. 1991;31(3):321–4. https://doi.org/10.1071/EA9910321.
Slattery WJ, Edwards DG, Bell LC, Coventry DR, Helyar KR. Soil acidification and the carbon cycle in a cropping soil of north-eastern Victoria. Aust J Soil Res. 1998;36(2):273–90. https://doi.org/10.1071/S96095.
Slattery W, Conyers M, Aitken R. Soil pH, aluminium, manganese and lime requirement. In: Peverill KI, Sparrow LA, Reuter DJ, editors. Soil analysis: an interpretation manual. Melbourne: CSIRO Publishing; 1999. p. 103–28.
Soon YK, Belanger N, Hendershot WH. Readily soluble aluminum and manganese in acid soils. In: Carter MR, Gregorich EG, editors. Soil sampling and methods of analysis. 2nd ed. Boca Raton: CRC Press; 2007. p. 121–8. https://doi.org/10.1201/9781420005271.
Steiner C, Teixeira WG, Lehmann J, Nehls T, de Macedo JLV, Blum WEH, Zech W. Long term effects of manure, charcoal and mineral fertilization on crop production and fertility on a highly weathered Central Amazonian upland soil. Plant Soil. 2007;291(1–2):275–90. https://doi.org/10.1007/s11104-007-9193-9.
Sumner ME. Amelioration of subsoil acidity with minimum disturbance. In: Stewart BA, Jayawardane NS, editors. Subsoil management techniques. Boca Raton: CRC Press; 1994. p. 147–85.
Sumner ME. Soil acidification. In: Verheye WH (ed) Land use, land cover and soil sciences – volume VII: soils and soil sciences – 2. vol pt. 2. EOLSS Publishers Company Limited, Oxford; 2009. p. 187–201.

Sumner ME, Noble AD. Soil acidification: the world story. In: Rengel Z, editor. Handbook of soil acidity. New York: Marcel Dekker; 2003. https://doi.org/10.1201/9780203912317.
Tang C, Rengel Z. Role of plant cation/anion uptake ratio in soil acidification. In: Rengel Z, editor. Handbook of soil acidity. New York: Marcel Dekker; 2003. https://doi.org/10.1201/9780203912317.
Tang C, Yu Q. Impact of chemical composition of legume residues and initial soil pH on pH change of a soil after residue incorporation. Plant Soil. 1999;215(1):29–38. https://doi.org/10.1023/A:1004704018912.
Tang C, Barton L, Raphael C. Pasture legume species differ in their capacity to acidify soil. Aust J Agric Res. 1997a;49(1):53–8. https://doi.org/10.1071/A97080.
Tang C, McLay CDA, Barton L. A comparison of proton excretion of twelve pasture legumes grown in nutrient solution. Aust J Exp Agric. 1997b;37(5):563–70. https://doi.org/10.1071/EA96151.
Tang C, Sparling GP, McLay CDA, Raphael C. Effect of short-term legume residue decomposition on soil acidity. Aust J Soil Res. 1999a;37(3):561–73. https://doi.org/10.1071/S98104.
Tang C, Unkovich MJ, Bowden JW. Factors affecting soil acidification under legumes. III. Acid production by N_2-fixing legumes as influenced by nitrate supply. New Phytol. 1999b;143(3):513–21. https://doi.org/10.1046/j.1469-8137.1999.00475.x.
Tang C, Raphael C, Rengel Z, Bowden JW. Understanding subsoil acidification: effect of nitrogen transformation and nitrate leaching. Aust J Soil Res. 2000;38(4):837–49. https://doi.org/10.1071/SR99109.
Tang C, Rengel Z, Diatloff E, Gazey C. Responses of wheat and barley to liming on a sandy soil with subsoil acidity. Field Crop Res. 2003;80(3):235–44. https://doi.org/10.1016/S0378-4290(02)00192-2.
Tang C, Conyers MK, Nuruzzaman M, Poile GJ, Liu DL. Biological amelioration of subsoil acidity through managing nitrate uptake by wheat crops. Plant Soil. 2011;338(1–2):383–97. https://doi.org/10.1007/s11104-010-0552-6.
Tang C, Weligama C, Sale P. Subsurface soil acidification in farming systems: its possible causes and management options. In: Xu J, Sparks DL, editors. Molecular environmental soil science; 2013. p. 389–412. https://doi.org/10.1007/978-94-007-4177-5_13.
Tiecher T, Pias OHC, Bayer C, Martins AP, Denardin LGO, Anghinoni I. Crop response to gypsum application to subtropical soils under no-till in Brazil: a systematic review. Rev Bras Cienc Solo. 2018;42 https://doi.org/10.1590/18069657rbcs20170025.
Vanzolini JI, Galantini JA, Martínez JM, Suñer L. Changes in soil pH and phosphorus availability during decomposition of cover crop residues. Arch Agron Soil Sci. 2017:1–11. https://doi.org/10.1080/03650340.2017.1308493.
von Uexküll HR, Mutert E. Global extent, development and economic impact of acid soils. Plant Soil. 1995;171(1):1–15. https://doi.org/10.1007/bf00009558.
Wang HL, Hedley MJ, Bolan NS, Horne DJ. The influence of surface incorporated lime and gypsiferous by-products on surface and subsurface soil acidity. I. Soil solution chemistry. Aust J Soil Res. 1999a;37(1):165–80. https://doi.org/10.1071/s97057.
Wang HL, Hedley MJ, Bolan NS, Horne DJ. The influence of surface incorporated lime and gypsiferous by-products on surface and subsurface soil acidity. II. Root growth and agronomic implications. Aust J Soil Res. 1999b;37(1):181–90. https://doi.org/10.1071/s97058.
Wang L, Butterly CR, Yang XL, Wang Y, Herath HMSK, Jiang X. Use of crop residues with alkaline slag to ameliorate soil acidity in an Ultisol. Soil Use Manag. 2012;28(2):148–56. https://doi.org/10.1111/j.1475-2743.2012.00396.x.
Wang Y, Liu X, Butterly C, Tang C, Xu J. pH change, carbon and nitrogen mineralization in paddy soils as affected by Chinese milk vetch addition and soil water regime. J Soils Sediments. 2013;13(4):654–63. https://doi.org/10.1007/s11368-012-0645-3.
Wang L, Butterly CR, Wang Y, Herath HMSK, Xi YG, Xiao XJ. Effect of crop residue biochar on soil acidity amelioration in strongly acidic tea garden soils. Soil Use Manag. 2014;30(1):119–28. https://doi.org/10.1111/sum.12096.

Wang X, Tang C, Mahony S, Baldock JA, Butterly CR. Factors affecting the measurement of soil pH buffer capacity: approaches to optimize the methods. Eur J Soil Sci. 2015;66(1):53–64. https://doi.org/10.1111/ejss.12195.

Wang L, Butterly CR, Chen Q, Mu Z, Wang X, Xi Y, Zhang J, Xiao X. Surface amendments can ameliorate subsoil acidity in tea garden soils of high-rainfall environments. Pedosphere. 2016a;26(2):180–91. https://doi.org/10.1016/s1002-0160(15)60031-5.

Wang L, Butterly CR, Tian W, Herath HMSK, Xi Y, Zhang J, Xiao X. Effects of fertilization practices on aluminum fractions and species in a wheat soil. J Soils Sediments. 2016b;16(7):1933–43. https://doi.org/10.1007/s11368-016-1380-y.

Wang X, Tang C, Baldock JA, Butterly CR, Gazey C. Long-term effect of lime application on the chemical composition of soil organic carbon in acid soils varying in texture and liming history. Biol Fertil Soils. 2016c;52(3):295–306. https://doi.org/10.1007/s00374-015-1076-2.

Weligama C, Tang C, Sale PWG, Conyers MK, Liu DL. Localised nitrate and phosphate application enhances root proliferation by wheat and maximises rhizosphere alkalisation in acid subsoil. Plant Soil. 2008;312(1–2):101–15. https://doi.org/10.1007/s11104-008-9581-9.

Weligama C, Sale PWG, Conyers MK, Liu DL, Tang C. Nitrate leaching stimulates subsurface root growth of wheat and increases rhizosphere alkalisation in a highly acidic soil. Plant Soil. 2010;328(1–2):119–32. https://doi.org/10.1007/s11104-009-0087-x.

Weng Z, Butterly CR, Sale P, Li G, Tang C. Combined nitrate and phosphorus application promotes rhizosphere alkalization and nitrogen uptake by wheat but not canola in acid subsoils. J Soils Sediments. 2021;21(9):2995–3006. https://doi.org/10.1007/s11368-021-03000-2.

Whitten MG, Wong MTF, Rate AW. Amelioration of subsurface acidity in the south-west of Western Australia: downward movement and mass balance of surface-incorporated lime after 2–15 years. Aust J Soil Res. 2000;38(3):711–28. https://doi.org/10.1071/sr99054.

Wong MTF, Asseng S. Yield and environmental benefits of ameliorating subsoil constraints under variable rainfall in a Mediterranean environment. Plant Soil. 2007;297(1):29–42. https://doi.org/10.1007/s11104-007-9316-3.

Wong MTF, Swift RS. Application of fresh and humified organic matter to ameliorate soil acidity. In: Swift RS, Spark KM, editors. Understanding and managing organic matter in soils, sediments and water. International Humic Substance Society, Adelaide; 2001. p. 235–42.

Wong MTF, Swift RS. Role of organic matter in alleviating soil acidity. In: Rengel Z, editor. Handbook of soil acidity, CRC Press, vol. 94. USA: New York; 2003. p. 337–58. https://doi.org/10.1201/9780203912317.

Wong MTF, van der Kruijs ACBM, Juo ASR. Leaching loss of calcium, magnesium, potassium and nitrate derived from soil, lime and fertilizers as influenced by urea applied to undisturbed lysimeters in south-east Nigeria. Fertil Res. 1992;31(3):281–9. https://doi.org/10.1007/bf01051280.

Wong MTF, Gibbs P, Nortcliff S, Swift RS. Measurement of the acid neutralizing capacity of agroforestry tree prunings added to tropical soils. J Agric Sci. 2000;134:269–76. https://doi.org/10.1017/S0021859699007388.

Wong MTF, Asseng S, Robertson MJ, Oliver Y. Mapping subsoil acidity and shallow soil across a field with information from yield maps, geophysical sensing and the grower. Precis Agric. 2008;9(1):3–15. https://doi.org/10.1007/s11119-008-9052-6.

Wong MTF, Webb MJ, Wittwer K. Development of buffer methods and evaluation of pedotransfer functions to estimate pH buffer capacity of highly weathered soils. Soil Use Manag. 2013;29(1):30–8. https://doi.org/10.1111/sum.12022.

Xu JM, Tang C, Chen ZL. The role of plant residues in pH change of acid soils differing in initial pH. Soil Biol Biochem. 2006;38(4):709–19. https://doi.org/10.1016/j.soilbio.2005.06.022.

Yan F, Schubert S, Mengel K. Soil pH increase due to biological decarboxylation of organic anions. Soil Biol Biochem. 1996;28(4–5):617–24. https://doi.org/10.1016/0038-0717(95)00180-8.

Yang Z-B, Eticha D, Albacete A, Rao IM, Roitsch T, Horst WJ. Physiological and molecular analysis of the interaction between aluminium toxicity and drought stress in common bean (*Phaseolus vulgaris*). J Exp Bot. 2012;63(8):3109–25. https://doi.org/10.1093/jxb/ers038.

Chapter 4
Salinity, Sodicity and Alkalinity

Pichu Rengasamy, Claudivan Feitosa de Lacerda, and Hans Raj Gheyi

Abstract Many of the subsoil constraints to crop productivity are due to the chemical components of soil water influenced by salinity, sodicity and soil pH. Accumulation of salts above a threshold level, commonly described as 'salinity', affects plants by osmotic and ionic effects. When these salts are dominated by sodium, increased level of soil adsorbed (exchangeable) sodium, known as 'sodicity', affects soil structural stability leading to poor soil physical conditions limiting water and air movement in the root zone layers. Domination of bicarbonate and carbonate anions in soil water increases soil pH above 8 resulting in soil alkalinity and pH-related ion deficiency and toxicity. This chapter focusses on subsoil constraints caused by salinity, sodicity and alkalinity. These three phenomena are interrelated: accumulation of salts, salinity, directly affects crop physiology, while the composition of the ions determines sodicity and alkalinity, affects soil physics and biology and indirectly interferes with crop performance. Formation of different types of saline soils, the role of cations in relation to soil structural stability, the ion deficiency and toxicity effects due to alkaline pH and the possible management of these constraints are also discussed.

Keywords Salt-affected soils · Subsoil salinity · Salt leaching · Soil pH · Salt tolerance · Exchangeable sodium

---

P. Rengasamy (✉)
School of Agriculture, Food and Wine, Waite Campus, The University of Adelaide, Adelaide, SA, Australia
e-mail: pichu.rengasamy@adelaide.edu.au

C. F. de Lacerda
Agricultural Engineering Department, Federal University of Ceará, Campus do Pici, Fortaleza, CE, Brazil
e-mail: cfeitosa@ufc.br

H. R. Gheyi
Recôncavo of Bahia Federal University, Cruz das Almas, BA, Brazil

T. S. d. Oliveira, R. W. Bell (eds.), *Subsoil Constraints for Crop Production*,
https://doi.org/10.1007/978-3-031-00317-2_4

4.1 Introduction

Plant growth is highly influenced by the chemical components of soil solutions at different water contents in root zone soil layers under field conditions. The composition and concentration of dissolved ions determine the multiphase equilibria involving the solid phase containing inorganic and organic compounds, the liquid phase comprising soil water and the gaseous phase dominated by oxygen, nitrogen, water vapour and carbon dioxide. Plant roots need a well-hydrated and well-aerated soil environment of low osmotic pressure, with adequate nutrients and low toxicity. Changes in soil chemistry impact on most of the plant requirements: soil structure, osmotic potential of soil water and presence of toxic ions. Accumulation of salts in subsoil above a threshold level leads to 'salinity' impacts on crops by osmotic and ion toxicity effects. Increased concentrations of sodium in soil water lead to increased levels of exchangeable sodium, known as 'sodicity', which affects the soil structural stability. Similarly, accumulation of bicarbonate and carbonate ions in soil water increases soil pH above 8, leading to soil 'alkalinity' and pH-related ion deficiency and toxicity.

While this book deals with most aspects of subsoil constraints, this chapter will deal with the constraints caused by soil chemistry related to salinity, sodicity and alkalinity. These three phenomena are interrelated: accumulation of salts causes salinity effects, while the composition of the accumulated ions determines sodicity and alkalinity. These chemical factors, in addition to directly affecting crop physiology, indirectly influence physical and biological properties of soils, also interfering with plant performance. Although the focus is on subsoils, it is important to note that the interactive processes between top soil and subsoil layers influence the environment in both layers.

4.2 Salinisation Processes: Subsoil Accumulation of Salts

Commonly, salt accumulation over long periods of time through natural process is termed 'primary salinity', while salinisation of soil layers as a consequence of mismanagement of natural resources is termed 'secondary salinity'. Salinity is a growing problem worldwide, especially in regions with arid and semi-arid climates, where the annual total precipitation is not sufficient to leach the salts present in the root zone. Overall, it is estimated that about 1.0 billion hectares is affected by salinity and sodicity (Sharma and Singh 2015), including primary and secondary salinisation. These areas are mainly in the Middle East, Australia and North Africa. In South Asia, including India, about 52 million hectares are affected by salts (Sharma and Singh 2015), and extensive areas of salt-affected soils are also found in Europe and South America (Beltrán 2016). According to Wicke et al. (2011), most of the salt-affected areas (~85%) are only slightly to moderately affected, while the remaining 15% suffer from severe to extreme constraints to plant development.

According to Ribeiro et al. (2016), based on the soil map of Brazil, saline and/or sodic soils occupy about 160,000 km^2 or 2% of the national territory. Most of these salt-affected areas are located in the Brazilian semi-arid region, and the accumulation of salts and sodium is mainly due to natural causes (primary salinity). However, in this region there are also problems of secondary salinity, where about 25% of the irrigated areas present varying degrees of soil salinisation. This has resulted in serious socioeconomic problems related to loss of soil productivity and abandonment of area by farmers.

In general, these processes involve several sources of salts, such as weathering of rocks and minerals, wind transportation of aeolian dust, rainfall containing varying amounts of salts from oceanic spray, discharge of saline groundwater, sea water intrusion, agronomic input of fertilisers and pesticides and irrigation using brackish water or industrial waste water. The particular processes contributing salt, combined with the influence of climatic and landscape features, and the effects of human activities, determine where and what type of salt is likely to accumulate in soil layers. Rengasamy (2006) identified three major types of salinity based on soil and groundwater processes found globally.

4.2.1 Groundwater-Associated Salinity

Leaching of salts from the permeable soil layers leads to salt storage in deep layers and also in the water table over geological time scales. When the equilibrium levels of the water table have been changed due to clearing of native vegetation and introduction of agricultural activities, the water table levels may rise to close to the root zone soil layers. Capillary rise of salty groundwater leads to salt deposition in top and subsoil layers, the degree of salinity being dependant on the quality of groundwater and rate of capillary rise which depends on soil texture and depth of water table.

4.2.2 Transient Salinity Not Associated with Groundwater Processes

Even in landscapes where the water table is very deep (below 10 m), salt has been accumulating in substrata over thousands of years. Under semi-arid conditions, the rainfall has not been sufficient to leach all the salts accumulated in the root zone to the layers below. The clay layers below root zone have restricted water and salt movement. In salt-affected soils dominated by sodium compounds, soil layers become sodic and highly dispersive. Sodic subsoils with degraded soil structure prevent water flow and restrict salt leaching leading to salinisation of root zone layers which affects plant growth. This 'transient salinity' in dispersive subsoils (Rengasamy 2002a; Fig. 4.1) varies in depth and concentration, and its effect on

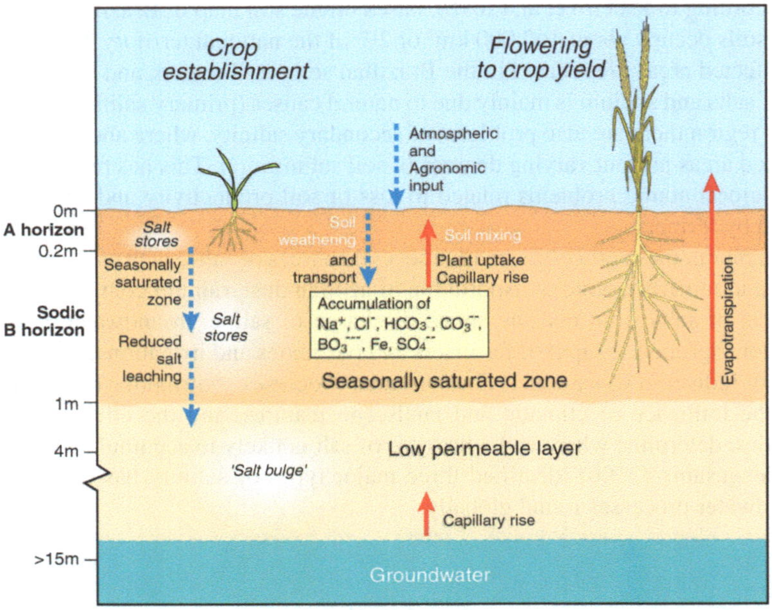

Fig. 4.1 Schematic diagram of accumulation of salt in the root zone layers (transient salinity) in a sodic subsoil. (Reproduced from Rengasamy 2002a with permission from CSIRO Publishing)

plant growth changes with seasonal factors including rainfall, leaching, water use by plants and water evaporation from the soil surface. Generally transient salinity is not influenced by groundwater processes. It is estimated that about 67% of dryland cropping area in Australia are affected by this type of salinity (Rengasamy 2002a; Barrett-Lennard et al. 2016). The problem of transient salinity is not, however, confined to Australia. About 5.8×10^6 km^2 of soils around the world are sodic with dispersive soil layers (Bui et al. 1998) and have the potential for transient salinity.

The salt accumulation, in addition to increasing the osmotic pressure of soil water, can also introduce high concentrations of soluble boron, carbonates and microelements such as Al, Mn and Fe, which can be toxic to crops. As a result of future climate changes, regions experiencing high temperature and low rainfall can expect an increase in the occurrence of transient salinity. By contrast, in high rainfall regions, salinity, induced by groundwater fluctuations, may increase.

4.2.3 Salinity Induced by Irrigation

In soils with low hydraulic conductivity (i.e. low leaching fraction), salts introduced by poor quality irrigation water are stored in the root zone layers. Clayey dispersive (sodic) soils and high evaporative conditions accelerate salinisation of soil layers.

Irrigation waters from groundwater sources usually contain high to moderate amounts of salts and, unless appropriately used with proper drainage, can induce salinity of soil layers. The dominance of sodium and/or potassium ions in irrigation waters, as well as in industrial effluents and recycled waters (e.g. Arienzo et al. 2012), may lead to deterioration of soil structure. In addition to the quality of irrigation water, improper drainage and methods of application of water play important roles in irrigation-induced salinity. FAO (2015) reported that globally, about 20–30 million hectares of irrigated lands were severely affected by salinity and about 80 million hectares were affected moderately.

Salinity and sodicity are interrelated, and the three processes discussed above can occur together in many instances. It is difficult to state which process contributes more in salt-affected regions globally, because surveys conducted by national and international agencies are not systematic and also not appropriately coordinated.

4.3 Physical and Chemical Processes Influencing Salinity and Sodicity

Irrespective of the sources of salts, discussed above, the total concentration and composition of the accumulated salts can vary widely in a given landscape or even between horizons of a soil profile within a given location in a paddock. The interactions between root-zone environments and plant responses to salinity in the field are complicated by many soil processes such as soil water dynamics, soil structural stability, solubility of compounds in relation to pH and pE and movement of nutrient and water in soil.

Because of poor soil physical conditions, restricted water movement in a given soil layer leads to waterlogging in the overlying soil layer. As the water evaporates, the dissolved salts are also stored in that layer. Thus, subsoil salinity (or transient salinity) is a result of the condition of the layers below, usually with high clay content and sodicity. Seasonal changes in rainfall pattern and evaporation and subsoil sodicity (reflecting the reduced drainage) have been found to be highly correlated with subsoil salinity in Australia (Rengasamy 2010).

When the accumulated salts are predominantly sodium compounds, soil particles adsorb Na^+, and the soil becomes saline-sodic. When salinity is in excess of a threshold value (point of zero dispersion in Fig. 4.2), soil structural problems such as clay dispersion do not occur, but osmotic and ionic effects of salinity will prevail. Once the salt levels are leached below the threshold, soil becomes sodic and dispersive, and the limitations on plant growth are imposed primarily by the soil physical condition. Therefore, in a landscape dominated by sodium salts, salinity and sodicity are interrelated.

Soil physical conditions influence soil water retention, while changes in water content largely affect salinity dynamics in a soil profile. As the soil dries due to evapotranspiration, the salt concentration increases in the soil water, as does the

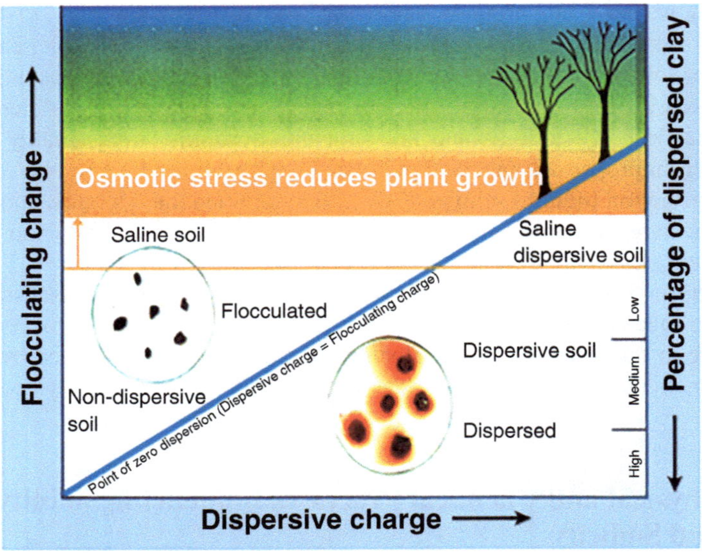

Fig. 4.2 Distinction between saline and sodic (dispersive) soils based on dispersive charge, flocculating charge and percentage of dispersed clay. (After Rengasamy 2016a with permission from MDPI CC BY 4.0)

Table 4.1 Percentage of stored water not taken by plants in soils affected by salinity in southern Australia

Average EC (dS m^{-1}) of soil water[a]	% of stored water not taken by plants
0.7	0
2.8	5.1
10.6	5.9
22.6	50.2
30.8	59.6
41.0	84.8
63.9	95.4

Source: Rengasamy (2010)
[a]The values are average of several observations from seven locations in Southern Australia. By comparison, sea water has an EC of approximately 50 dS m^{-1}

osmotic pressure of soil water. Concomitant changes in matric and osmotic potentials of soil water determine plant water uptake in the field. In dryland cropping, fluctuating soil water content during the growing season is an important factor in considering the effects of salinity on crops. The percentage of water in soil that is apparently available based on matric potential but not used by plants in soils affected by subsoil salinity in southern Australia is given in Table 4.1. The unused soil water stored in the subsoil is also termed 'green water' (Sposito 2014).

Many soil chemical processes affect the ionic composition and chemical reactivity of the water in salt-affected soils (Naidu and Rengasamy 1993). For example, ion exchange reactions between solid components and the dissolved ions in soil water determine the distribution of cations and anions in each of these phases. Dissolution and precipitation of soil compounds induced by soil pH, pE, CO_2 and organic matter alter the ionic composition and control ion activities including that of protons (H_3O^+). Toxicity of elements depends on chemical speciation (e.g. different species of Al, Mn, Fe and hydroxy complexes) which is a function of pH and pE. Biochemical reactions due to soil microbes also control the chemical reactivity of soil solutions. Further, under field conditions, soil water content fluctuates from high to very low levels, and the chemical equilibria become more complex as soils dry (Rengasamy et al. 2016). Thus, chemistry in soil layers is an intricate phenomenon, and appropriate soil management is necessary to maintain a suitable soil chemical environment.

4.4 Mechanisms of Salinity-Reducing Crop Production

Salt tolerance varies widely among plant species, ranging from those growing at salinity levels similar to seawater to those that do not survive even at low salt concentrations. This amplitude of response is due to differences in the efficiency of the mechanisms that can minimise the effects of salinity, which are divided into three types: osmotic, toxic and nutritional effects. The osmotic effect, associated with the accumulation of soluble salts in the root environment, reduces the gradient of water potential between the roots and the external soil solution, resulting in a decrease in the rate of water absorption by the cells. The initial reductions in plant growth and stomatal conductance, for example, are mainly due to this type of effect of salts (Munns and Tester 2008).

If the plant continues growing in a salt-affected soil or saline solution, then an unavoidable consequence is the accumulation of the predominant dissolved ions in the rhizosphere. This may cause toxicity to the plant cells, which is expressed by direct effects on metabolism and can lead to death of cells and necrosis in the leaves. In most waters and saline soils of the world, Na^+ and Cl^- ions predominate, although in some cases high concentrations of Mg^{2+}, Ca^{2+} and SO_4^{2-} may also be found. The low capacity of the plants to compartmentalise these ions in the vacuole, especially Na^+ and Cl^-, accelerates toxicity problems, being a strong indicator of the low capacity of the species to grow in salt-affected areas.

The nutritional effect or imbalance of nutrients occurs when the excess of one ion in solution inhibits the absorption of another. For example, excess of Na^+ can inhibit K^+ and/or Ca^{2+} uptake, while excess of Cl^- can reduce the uptake of nitrate. This effect is certainly the most complex and also the least relevant. The nutritional effects of salinity are complex because of the large number of plant species (more than 300,000) that can be subjected to many combinations of salinity conditions, and these species vary in nutrient demand and degrees of salt tolerance. However, in general the reduction in growth is primarily caused by the osmotic and toxic effects of salinity, so that the nutritional demand of the whole plant is reduced (Lacerda et al. 2016) and the reduction in growth itself may lead to changes in the foliar concentration of some nutrients (Lacerda et al. 2006). The nutritional imbalance could be more severe under sodic soil conditions, due to the high percentage of exchangeable sodium and high pH. However, in these soils the negative effects on the physical properties are mainly responsible for limiting plant growth.

In plants growing under field conditions, it is possible that osmotic, toxic and nutritional effects are not uniform throughout the root zone, since neither salinity nor sodicity is uniform in the root zone (Nuttall and Armstrong 2010; He et al. 2014). For annual irrigated crops, most roots concentrate in the upper 20 or 30 cm of the soil. In this case, a common practice is to promote the leaching of salts into the subsoil, thereby reducing impacts on plant growth and yield. However, the lack of adequate drainage can cause an increase in subsoil salinity over time (Roux et al. 2007), and subsequently salinity enters the surface layer, rendering the area unproductive in the future.

Under rainfed farming with annual and/or perennial crops and for native vegetation, the response to salinity tends to be different to well-watered conditions, as the roots tend to explore deeper layers, due to periods of water deficit and reduction of water content in the surface soil layer. Thus, it is possible for roots to find a higher water content in the subsoil but also to encounter higher concentration of salts and sodium (Rengasamy et al. 2003; He et al. 2014). High soil salinity and sodicity in the subsoil, therefore, are limiting factors for plant yield (Dixit and Chen 2010), because they restrict the ability of the plant roots to absorb water and nutrients stored in subsoils, in areas of rainfed agriculture.

Salinity in the subsoil may result in lesser root growth, lower water uptake, an increase in potentially toxic ions (Na^+ and Cl^-), nutritional imbalance and lower yield, but these effects depend on the level of salinity (Grewal 2010a) as well as on the degree of tolerance of the plant (Grewal 2010b; Nuttall et al. 2010). In the case of rainfed agriculture, these negative impacts may be more intense in drought years when the evaporative demand is higher. These conditions result in severe water deficit in the soil surface layer, forcing the roots to further explore subsoil where they encounter salinity and sodicity constraints (Sadras et al. 2003).

Another very common field situation in salt-affected areas is the occurrence of transient waterlogging in the topsoil and subsoil, both in irrigated and non-irrigated areas (Barrett-Lennard 2003; Saqib et al. 2005; Qureshi et al. 2008; Bakker et al. 2010). In transient salinity, transient waterlogging in the subsoil will commonly coincide with wetting of the sodic horizons. The interaction between these two abiotic factors causes simultaneous increase in Na^+ and/or Cl^- concentrations and decreases in K^+ concentrations in shoots, which can have adverse effects on plant growth and survival (Barrett-Lennard and Shabala 2013). In fact, salinity along with excess water causes damage to plant growth much more than that caused only by salinity. However, these impacts are greater the longer the duration of waterlogging and are less expressive in the species capable of producing adventitious roots and aerenchyma (Barrett-Lennard 2003; Saqib et al. 2005). The latter are root structures that favour the aeration of the root system under conditions of excess water in the soil. This justifies, at least in part, the success of rice cultivation under flooding in saline areas in Brazil, this species being used commonly during the processes of reclamation of salt-affected soils (Gomes et al. 2000).

The interaction between salinity and excess water is quite complex, since it depends on the degree of tolerance of the plants to the osmotic, ionic and oxygen deficiency stresses, which may be present at different intensities in the root environment. For example, the beneficial effect of flooding on rice crops in saline soils (Gomes et al. 2000), a species considered sensitive to salinity, is due to the effect of dilution of salts by excess water and to the fact that this species has mechanisms to adapt to the oxygen deficiency conditions in the soil. Opposite results can be obtained with species that are more tolerant to salts, but that do not have the capacity to produce aerenchyma that favours the oxygenation of the root system (Medeiros et al. 2018). Therefore, any generalisation about this topic may not be recommended.

4.5 Effect of Exchangeable Cations (Including Sodicity) and Soluble Anions on Soil Structural Stability

It is well known that when exchangeable sodium percentage (ESP) of a soil increases above a certain level, the soil structural stability is impaired due to swelling and dispersion of clay particles in soil aggregates. As a result, soil physical parameters such as aeration porosity, hydraulic conductivity and soil strength are highly altered to levels not conducive to plant growth (So and Aylmore 1993). Since it has been well established that cations other than Na^+ have specific roles in clay swelling and dispersion, the term 'dispersive soil' is more appropriate than 'sodic soil' when soil physical phenomena in salt-affected soils are considered.

The schematic diagram (Fig. 4.3) using the data of Rengasamy et al. (1992) compares aeration porosity, mechanical resistance and relative water content for a dispersive and a non-dispersive subsoil. Aeration porosity >15% and mechanical resistance <1 MPa are considered to be conducive to good crop performance, and both are related to the soil water content. In this diagram for a dispersive soil, point A is the water content where mechanical resistance starts to decrease below 1 MPa, while point B is the water content of a dispersive soil where aeration porosity starts to increase above 15%. The water contents between A and B are defined as the 'non-limiting water range' of the dispersive soil, i.e. the water content range that does not limit crop production (see Letey 1985). In contrast when these points are determined for a non-dispersive soil (points C and D), the non-limiting water range is many times larger (Fig. 4.3).

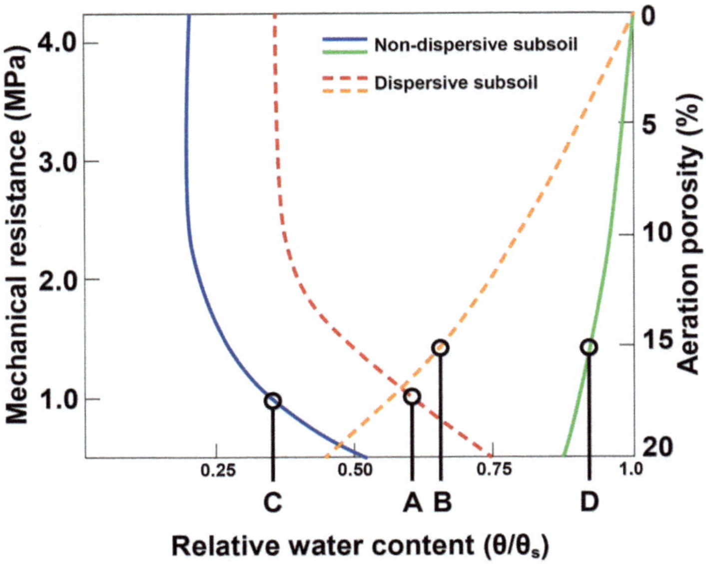

Fig. 4.3 Relationships of mechanical resistance and aeration porosity to soil relative water contents for a dispersive and a non-dispersive subsoil. Relative water content is the ratio between soil water content (θ) and the water content at saturation (θ_s). The 'non-limiting water range' can be estimated from these data based on critical values for mechanical resistance (1 MPa) and aeration porosity (15%), i.e. A to B for the dispersive subsoil and C to D for the non-dispersive subsoil. (Reproduced from Rengasamy 2016b with permission from Oxford University Press USA)

4.5.1 Role of Exchangeable Cations and Soluble Anions in Clay Swelling and Dispersion

Soil aggregates are composed of clay, silt, sand and organic matter linked by different types of chemical bonding. The combination of these materials results in an electrical charge different from the charge contributed by the individual clay mineral and organic materials. This specific 'charge' changes also with soil pH. With increasing pH, negative charge on soil particles increases, while pH values below zpc (zero point of charge) induce positive charge. This charge, when negative, is usually balanced by adsorbing cations and, when positive, by adsorbing anions. Most of the salt-affected soils have net negative charge balanced predominantly by Ca^{2+}, Mg^{2+}, K^+ and Na^+ and lead to the linking of soil particles via covalent and/or ionic bonding. Water molecules being polar react with ionic bonds (and not covalent bonds). Therefore hydration reactions depend on the ionicity index (Marchuk and Rengasamy 2011), a molecular-scale geochemical parameter which estimates the relative bonding (ionic and covalent) tendency of a cation with charged particles. Following these principles, Rengasamy (2002b) determined the dispersive powers of these cations relative to Ca. They are Ca = 1, Mg = 1.7, K = 25 and Na = 45. While, in dry soils, the exchangeable cations react with water molecules inducing clay dispersion, the soluble cations present in soil water oppose the dispersive reactions because of their flocculating powers, and relative to Na, they are Ca = 45, Mg = 27, K = 1.8 and Na = 1 (Rengasamy and Sumner 1998).

Recently, Rengasamy et al. (2016) proposed a new concept of 'net dispersive charge' which determines the amount of clay dispersed from a dispersive soil and is defined as:

$$\text{Net dispersive charge} = \text{Dispersive charge} - \text{Flocculating charge}$$

The dispersive charge of a soil is derived from the concentrations of exchangeable cations ($cmol_c\ kg^{-1}$) measured at the given soil pH, weighted by the respective dispersive power, and is defined as:

$$\text{Dispersive charge}\left(cmol_c\ kg^{-1}\right) = (Ca) + 1.7(Mg) + 25(K) + 45(Na)$$

Similarly, the flocculating charge is derived from the concentrations of the cations ($cmol_c\ kg^{-1}$ on soil basis) in the dispersed soil-water suspension, weighted by the respective flocculating power, and is defined as:

$$\text{Flocculating charge}\left(cmol_c\ kg^{-1}\right) = 45(Ca) + 27(Mg) + 1.8(K) + (Na)$$

When the dispersive charge is above a threshold value, clay particles dissociate and form a diffuse double layer. The flocculating effects of cations operate only when a diffuse double layer is developed. At low dispersive charge, soil aggregates slake, as macro aggregates break into micro aggregates; this process is not affected by the electrolyte concentration (i.e. salinity) of the soil solution. The point of zero dispersion (Fig. 4.2) occurs when flocculating charge equals the dispersive charge. If the flocculating charge contributed by the cations in a saline soil is equal to or greater than the dispersive charge, the soil structural effects of salinity will be minimal, and crop growth is mainly affected by osmotic and ion toxicity effects. When the flocculating charge is lower than the dispersive charge, degradation of soil physical properties will affect crop growth in addition to salt effects on crops. When the electrolyte concentration is low, cations do not reach toxic concentrations in the soil solution. For example, non-saline soils with high exchangeable sodium and low EC do not have toxic levels of sodium in soil solutions. However, seasonal changes in salt concentration due to soil wetting and drying in the field will determine whether soil structural or salinity effects influence crop performance.

Because of the difficulties in the estimation of exchangeable cations, concentrations estimated in soil solution have been used to derive values of sodium absorption ratio (SAR) to predict ESP and associated soil physical conditions. However, K is not included in the derivation of SAR, and Mg is considered as equal to Ca in the SAR model. In order to include the different roles played by K^+ and Mg^{2+} in dispersion-flocculation processes, Rengasamy and Marchuk (2011) proposed the 'cation ratio of soil structural stability (CROSS)' parameter which is defined as follows:

$$\text{CROSS} = \left(\text{Na} + 0.56\,\text{K}\right) / \left(\text{Ca} + 0.6\,\text{Mg}\right)^{0.5}$$

where the concentrations of cations are expressed in mmol L^{-1}. Unlike SAR, CROSS includes the dispersive effects of K and the flocculating effect of Mg and Ca. Values for CROSS were effective in predicting clay dispersion and soil physical behaviour, particularly when K and Mg are present in significant amounts in soils (Rengasamy and Marchuk 2011; Oster et al. 2016).

In soils, dominated by iron and aluminium oxides and kaolinite, such as Oxisols, when soil pH is lower than the 'zero point of charge' (zpc), the net charge is positive, and conditions can be dispersive. Exchangeable anions and soluble anions such as Cl^- and SO_4^{2-} will play a vital role in clay swelling and dispersion when soils have a net positive charge. In addition, the anions, HCO_3^- and CO_3^{2-}, induce high pH, higher than zpc, and, hence, increased negative charge in soils leading to clay dispersion.

This new concept of net dispersive charge of dispersive soils resolves the controversies over the definition of sodic soils on the basis of arbitrary ESP values which vary with many soil factors such as mineralogy, organic matter, exchangeable cations, electrolyte concentration of soil solution, soil texture and pH. The long-standing debate on the roles of K and Mg in clay dispersion is settled by introducing appropriate weighting for their dispersive powers. Establishing the quantitative relationship between net dispersive charge and the amount of dispersed clay and also threshold values for clay dispersion and their effect on soil physical conditions in the field will be necessary for decisions on soil management.

4.6 Alkalinity and High Soil pH

When the pH of a soil is >7, it is described as alkaline. However, soils with pH between 7 and 8 produce limited constraints for plants. Generally, soils with pH > 8 are considered to be 'alkaline soils'. As the pH increases above 8, soil constraints, like nutrient deficiency and ion toxicity, increase in severity. Alkalinity of soil solution with pH > 7 can be defined (Sposito 2008), without considering organic species, as follows:

$$\begin{aligned}\text{Alkalinity} = {} & \left[HCO_3^-\right] + 2\left[CO_3^{2-}\right] + \left[H_2PO_4^-\right] + 2\left[HPO_4^-\right] + 3\left[PO_4^{3-}\right] \\ & + \left[B(OH)_4^-\right] + \left[OH^-\right] - \left[H^+\right]\end{aligned}$$

Bicarbonate and carbonate ions dominate in alkaline soils with pH > 8 and are related to the 'carbonate alkalinity' as below:

$$\text{Carbonate alkalinity} = \left[HCO_3^-\right] + 2\left[CO_3^{2-}\right]$$

Calcareous soils contain lime (calcite, $CaCO_3$) as dispersed powder or as nodules, and irrespective of the concentration of lime, the soil pH is buffered between 8 and 8.5. Because of the low solubility of lime, soil pH levels are determined by the chemical buffering of calcite. Very alkaline environments (pH $\geq$ 9) are found in large areas of Western Asia, East Europe, Southern South America and Southern Australia and have expanded with irrigation in dry valleys such as those watered by the river Ganges in India and the Nile in Egypt (Kelly 1951; Jobbagy et al. 2017). High pH ($\geq$ 9) environments are developed when concentrations of ($HCO_3^- + CO_3^{2-}$) species in solutions are higher than ($Ca^{2+} + Mg^{2+}$). Soil solutions are also dominated by $Na^+ + K^+$. Thus, when Na salinity increases in calcareous soils, protons induced by sodium chloride react with calcite, and bicarbonate and carbonate salts of sodium are formed. This also leads to the increase in exchangeable sodium. But, it is important to recognise that high alkalinity (or high pH) is caused by bicarbonate and carbonate concentrations, and not by high ESP or excessive Na^+ in soil solutions (Tavakkoli et al. 2015). Jobbagy et al. (2017) cite the examples where seawater having high sodium [SAR >58 (mmol $L^{-1})^{0.5}$] and low carbonate species causes the formation of saline soils and not alkaline-sodic soils. Irrigation with river waters containing low sodium [SAR <15 (mmol $L^{-1})^{0.5}$] with a strong dominance of carbonate species leads to the development of alkaline-sodic soils. Generally, in Australian sodic soils, topsoils are less alkaline (pH < 8) than subsoils (Northcote and Skene 1972). Physical conditions of sodic subsoils lead to water stagnation and anoxic environment where concentrations of CO_2 increase leading to an increase of $HCO_3^- + CO_3^{2-} > Ca^{2+} + Mg^{2+}$ in soil solution and hence high pH. Jobbagy et al. (2017) estimated that the extent of soils with pH > 9 in the top metre are 2.7% globally but 18% in areas with low slope (<0.05%) and semi-arid-subhumid climate. Australia has the largest proportion of very alkaline soils compared to other continents, many of which are located under more arid climates and in areas with higher slopes (De Caritat et al. 2011). This may be due to dispersive subsoils with transient waterlogging and anoxic conditions leading to higher concentration of carbonate species.

Accumulation of chloride and sulphate salts in alkaline subsoils reduces the soil pH. Generally, in salt-affected soils, a negative relationship between soil pH and salinity (EC) has been observed (e.g. Al-Busaidi and Cookson 2003). In our observations, measurements of pH of high alkaline soils (pH in water >9) using 0.01 M $CaCl_2$ always result in a pH value ≈ 8.2 reflecting that $Ca^{2+} + Mg^{2+} > HCO_3^- + CO_3^{2-}$ in the solutions, because of the precipitation of carbonate species by $CaCl_2$. However, when increasing concentrations of NaCl (EC of 0.5 to 2.6 dS m^{-1}) were added to an alkaline soil, pH gradually decreased from 9.7 to 8.9 (Rhiannon Schilling, Personal communication), in spite of the fact NaCl is not involved in either carbonate precipitation or the increase of $Ca^{2+} + Mg^{2+}$ in solution. The role of NaCl in decreasing soil pH remains to be explored.

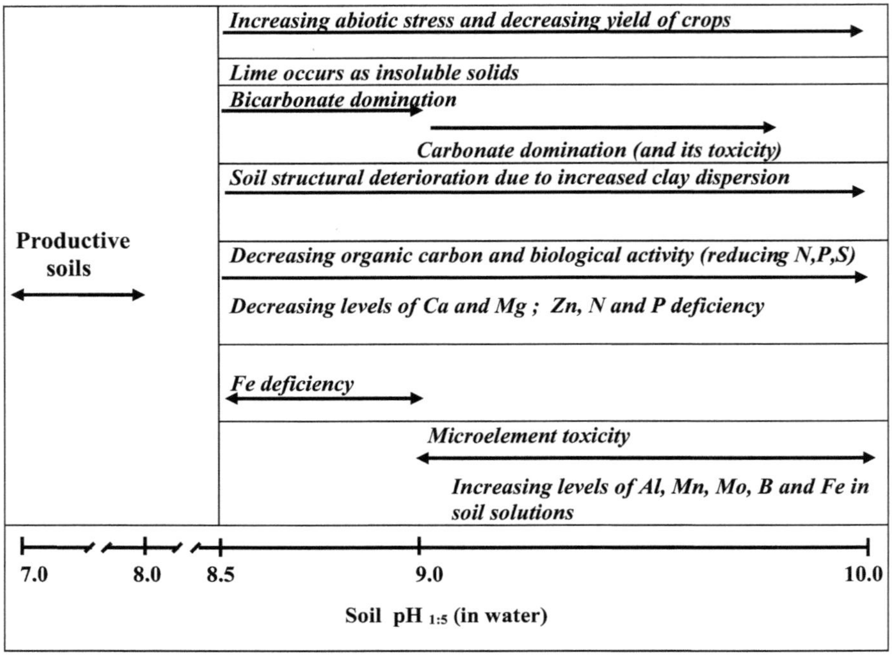

Fig. 4.4 Abiotic stress in alkaline soils in the pH range between 8.5 and 10. (Adapted from Rengasamy 2016b with permission from Oxford University Press USA)

Specific constraints in alkaline soils (pH in water between 8.5 and 10) depend on soil pH (Fig. 4.4). For example, bicarbonate toxicity may prevail between pH values of 8 and 9, while carbonate toxicity dominates above pH 9. Iron deficiency may be a problem when pH is between 8 and 9, while Fe toxicity can occur above pH 9 (Setter et al. 2009). Similarly, boron deficiency and toxicity are dependent on soil pH. As the pH increases above 9, $B(OH)_4^-$ species increase and are adsorbed by soil particles (Sposito 2008), perhaps leading to B deficiency. Recent studies have shown that Al toxicity can be a problem when soil pH increases above 9 due to the formation of soluble anionic aluminate species (Brautigan et al. 2012). With increasing pH, monovalent ions, particularly Na^+, increase in soil solution, while concentrations of divalent Ca^{2+} and Mg^{2+} decreased, leading to high clay dispersivity and soil structural degradation. This also renders subsoil with transient waterlogging and anoxic conditions, where pH and pE interactions induce microelements (such as Mn, Fe, Al and B) toxicities due to their increasing concentrations in soil solutions and/or reduced energy of plant roots to exclude these ions (Setter et al. 2009). When alkaline soils are salinised by chloride and sulphate salts of sodium, high soil pH is considerably reduced, and toxicity due to carbonates and elements (Mn, Al and Fe) is diminished, and osmotic effects become dominant. Alkaline soils sequester less organic carbon, high pH being responsible for the increase in dissolved organic carbon (McDonald et al. 2017) and reduced microbial activity.

4.7 Categories of Salt-Affected Soils

The classification of salt-affected soils has mainly been based on the chemical characteristics of the soil, such as electrical conductivity (EC_{se}) and sodium adsorption ratio (SAR_{se}) measured in a soil saturation extract, pH of a saturated paste (pH_{sp}) and exchangeable sodium percentage (ESP). These characteristics allow the identification of problems of salinity, sodicity and alkalinity in the soil.

The most commonly employed classification was initially proposed by the US Salinity Laboratory (USSL/ARS/USDA) (Richards 1954) which recognised three categories of salt-affected soils: saline, saline-alkaline or saline-sodic soils and alkali or sodic soils. Saline soils are those with EC_{se} higher than 4 dS m^{-1} at 25 °C, ESP less than 15 and pH_{sp} generally lower than 8.5. Saline soils are often recognised by the presence of white crusts of salts on the surface and thus received the name of white alkali soils (Hilgard 1906). In the Old Russian classification, such soils were denominated as Solonchaks. The presence of high salt concentrations and low ESP values results in well-structured soils, and salinity problems can be corrected by leaching using a good drainage system.

Saline-alkaline or saline-sodic soils are the result of the processes of salinisation and alkalisation or sodification. These soils have EC_{se} higher than 4 dS m^{-1} at 25 °C and ESP greater than 15. The pH_{sp} of these soils is high but usually not higher than 8.5. The management of these soils is more complex in comparison to saline soils, since simple leaching, while removing excess salts, has no effect on ESP which remains above 15 and increases pH_{sp}. The net effect of leaching is damage to the soil structure. The reclamation of this type of soil requires a chemical treatment in advance, especially when the gypsum content in soil is very low.

Sodic soils as defined by Richards (1954) are those that have an ESP greater than 15, but the EC_{se} is less than 4.0 dS m^{-1}, at 25 °C. In alkaline soils, the pH_{sp} can reach values between 8.5 and 10. These high values of pH_{sp} can cause dispersion and dissolution of the organic matter, producing a dark colour thin layer (or film) on the surface, hence the name black alkali soils (Hilgard 1906). In the Russian classification, these soils were called Solonetz. High exchangeable sodium in these soils results in the dispersion of clay and promotes the formation of compacted layer at certain depth due to deposition of clay that considerably reduces water permeability and root development, so that few plant species are able to grow under these conditions.

Despite its wide diffusion and application throughout the world, the classification proposed by Richards (1954) has also generated controversy. The first one concerns the EC_{se} value of 4.0 dS m^{-1} as the limit for saline soils. According to the Terminology Committee of the Soil Science Society of America (SSSA 1976), the value of 2.0 dS m^{-1} would be more appropriate, since many agricultural crops especially the fruit crops are seriously affected in EC_{se} values between 2.0 and 4.0 dS m^{-1}. The Committee also recommended the use of SAR of the saturation extract in place of ESP because of simplicity in its determination and suggested a value of 13 $(mmol\ L^{-1})^{0.5}$ to distinguish between sodic and non-sodic soils.

Even greater divergence of opinion is related to the ESP values. Australian and Spanish researchers have defined the limit value of 6% for sodic soils, considering that this value is enough to cause damage to the soil structure (Northcote and Skene 1972; Pizarro 1978; Isbell 2002). Earlier experiments on hydraulic conductivity of sodic soils to define critical ESP levels for sodicity in the USA and Australia used waters containing different amounts of electrolytes. Generally, increased levels of electrolytes in water reduce the harmful effects of sodicity on hydraulic conductivity. Tap water containing salts were used in US Salinity Laboratory, whereas pure water was used in Australian experiments. This resulted in the different critical ESP values adopted in the USA, Australia and Spain (Sumner 1993; Ribeiro et al. 2016).

Other authors and organisations have preferred to increase the number of categories of salt-affected soil or adopted hierarchical levels within a soil classification system, such as the Brazilian Soil Classification System (Ribeiro et al. 2016) and WRB (World Reference Base for Soil Resources; FAO 2015). The classification of the WRB allows, for example, for salic ($EC_{se} \geq 8.0$ dS m^{-1}), hypersalic ($EC_{se} \geq 30.0$ dS m^{-1}) and hyposalic soil layers ($EC_{se} \geq 4.0$ dS m^{-1}). In a way, this classification of WRB accepts the values of 4.0 dS m^{-1} and 15%, as limits for salinity and sodicity, respectively, in agreement with the USSL classification (Richards 1954). The term Solonchak indicates the existence of salic horizon, and Solonetz indicates a natric horizon; these terms are currently used in the first classification level of WRB.

Another more detailed classification has been presented by Rengasamy (2016a). It proposed a set of categories of salt-affected soils, which uses the electrical conductivity and SAR measured in the soil saturation extract (EC_{se} and SAR_{se}) and the pH measured in 1:5 (soil:water) extract. The critical value of SAR_{se} is based on the Australian criteria of sodicity, which is ESP > 6 (Isbell 2002). The 12 categories recognise that the problems of salinity and sodicity vary with soil acidity and alkalinity (Table 4.2).

It is important to note that the classification shown in Table 4.2 is dynamic and can be altered depending on soil management and other processes. In addition, toxicity, deficiency or imbalance due to various ions will depend on the ionic composition of soil solution (Rengasamy 2016a). Moreover, the actual effects of these 12 categories of salt-affected soils on crop growth also depend on crop tolerance.

4.8 Managing Subsoil Constraints Due to Salinity, Sodicity (Dispersivity) and Alkalinity

Inefficient use of 'green water' stored in subsoil is considered as a major factor for the low yield of crops. Roots struggle to grow through subsoil with constraints, caused by salinity, sodicity and alkalinity, and are unable to make full use of the stored subsoil water. In part, plants are also unable to use stored water because of

Table 4.2 Categories of salt-affected soils based on EC_{se} (dS m^{-1}), SAR_{se} (mmol $L^{-1})^{0.5}$ and $pH_{1:5}$ of soil solutions and the possible effects on plants

Category of saline soil	Criteria	Possible impact on plants
Acidic-saline soil	$EC_{se} > 4$; $SAR_{se} < 6$; $pH_{1:5} < 6$	Osmotic effect; microelement (Fe, Al, Mn, etc.) toxicity; SO_4^{2-} toxicity in very low pH
Neutral saline soil	$EC_{se} > 4$; $SAR_{se} < 6$; $pH_{1:5}$ 6–8	Osmotic effect; toxicity of dominant anion or cation other than Na^+
Alkaline-saline soil	ECse >4; $SAR_{se} < 6$; $pH_{1:5}$ 8–9	Osmotic effect; HCO_3^- and CO_3^{2-} toxicity
Highly alkaline-saline soil	$EC_{se} > 4$; $SAR_{se} < 6$; $pH_{1:5} > 9$	Osmotic effect; HCO_3^- and CO_3^{2-} toxicity; microelement (Fe, Al, Mn, etc.) toxicity
Acidic-saline-sodic soil	$EC_{se} > 4$; $SAR_{se} > 6$; $pH_{1:5} < 6$	Osmotic effect; Na^+ and microelement (Fe, Al, Mn, etc.) toxicity
Neutral saline-sodic soil	$EC_{se} > 4$; $SAR_{se} > 6$; $pH_{1:5}$ 6–8	Osmotic effect; Na^+ toxicity; toxicity of dominant anion (Cl^- or SO_4^{2-}
Alkaline-saline-sodic soil	$EC_{se} > 4$; $SAR_{se} > 6$; $pH_{1:5}$ 8–9	Osmotic effect; Na^+ toxicity; HCO_3^- and CO_3^{2-} toxicity
Highly alkaline-saline-sodic soil	$EC_{se} > 4$; $SAR_{se} > 6$; $pH_{1:5} > 9$	Osmotic effect; Na^+ toxicity; HCO_3^- and CO_3^{2-} toxicity; microelement (Fe, Al, Mn, etc.) toxicity
Acidic-sodic soil	$EC_{se} < 4$; $SAR_{se} > 6$; $pH_{1:5} < 6$	Indirect effect due to soil structural problems; seasonal waterlogging can induce microelement (Fe, Al, Mn, etc.) toxicity
Neutral sodic soil	$EC_{se} < 4$; $SAR_{se} > 6$; $pH_{1:5}$ 6–8	Indirect effect due to soil structural problems; seasonal waterlogging; Na^+ toxicity at high SAR_{se}
Alkaline-sodic soil	ECse <4; $SAR_{se} > 6$; $pH_{1:5}$ 8–9	Indirect effect due to soil structural problems; seasonal waterlogging; Na^+ toxicity at high SAR_{se}; HCO_3^- and CO_3^{2-} toxicity
Highly alkaline-sodic soil	$EC_{se} < 4$; $SAR_{se} > 6$; $pH_{1:5} > 9$	Indirect effect due to soil structural problems; seasonal waterlogging; Na^+ toxicity at high SAR_{se}; HCO_3^- and CO_3^{2-} toxicity; microelement (Fe, Al, Mn, etc.) toxicity

After Rengasamy (2016a)

decrease in osmotic potential of soil solution due to increase in soil salinity, so the available water to plant is reduced.

Ameliorating subsoil constraints by changing crop and soil management is often difficult because of the depth at which these constraints occur and the high cost of correcting them. Genetic improvement of crops to tolerate these constraints and to increase yield is another option.

Table 4.3 Problems encountered and principles of management of saline, sodic and alkaline soils

Category	Problems encountered	Principles of management
Saline	The major impediment is the salt concentration inducing osmotic stress. Depending on the ionic composition, specific ion effects may arise	Leaching of salts to the level tolerated by crop species Leaching may lead to soil dispersivity (sodicity) Utilisation of salt-tolerant crops
Sodic (dispersive)	When the salinity (EC) is low, swelling and clay dispersion affect soil physical condition (including hard-setting of soil layers and waterlogging)	Reduction of dispersive charge and increasing flocculation by cations (ca in gypsum can reduce the dispersive charge by exchanging adsorbed monovalent cations and also increase the flocculating charge in soil solutions)
Alkalinity (pH > 8)	As shown in Fig. 4.3, depending on soil pH, ion toxicity or deficiency affects crops. Soil structural stability can also be a problem	Reduction of pH (<8.2) by phytoremediation or chemical amendments

Under field conditions, it is common to observe vertical variations of soil constraints within a given profile at a site, in addition to the horizontal variations across a farm. Subsoil constraints cannot be managed in isolation, without considering the problems in topsoil. Similarly, several subsoil constraints can occur together. Problems due to salinity, sodicity (dispersivity) and alkalinity can occur simultaneously. There is a tendency to evaluate the effects of these constraints separately, and little information is available on the relative importance of different subsoil constraints (McDonald et al. 2012). High alkalinity is usually associated with sodicity. Commonly, salinity and sodicity are interrelated when sodium salts are prevalent. Thus, several categories of salt-affected soils are possible (Table 4.2). Management of subsoil constraints requires strategies specifically suitable to a given category or a site. Principles involved in the management of different categories of saline, sodic and alkaline soils are given in Table 4.3.

4.8.1 Reclamation of Saline Subsoil

When salinity alone is the constraint, soils can be reclaimed by applying good quality water, causing leaching of excess salts to the deeper layers. This process usually requires a large volume of water and generally promotes the reclamation of saline land, especially when the soil has good natural drainage and there is no impeding layer in the subsoil, or when the water table is deep (beyond 3–4 m). When there are problems of soil permeability, it may be necessary to install subsurface drainage system and to adopt some cultivation techniques. These techniques can ensure the removal of salts from the topsoil, making it again productive, but in this case, the cost of the reclamation increases considerably (Sousa et al. 2014).

When the salinity problem occurs in the subsoil, the issue becomes more complex, since water and salt movement occur simultaneously between the upper and lower layers of the soil, depending on the inflow and outflow of water in the profile. Thus, depth, thickness of layer, degree of salinisation, physical characteristics of the subsoil, depth of the groundwater and salt tolerance of crops to be grown are decisive aspects in the reclamation of salinity in the subsoil. For example, when crops exploit only the surface layer and when it is possible to minimise capillary rise, subsoil reclamation may not be necessary. However, for crops that exploit water and nutrients from the subsoil, particularly under rainfed farming, leaching could be an option. However, when the subsoil has restricted drainage, which is not unusual, reclamation by leaching should be preceded by the installation of a subsurface drainage system and the adoption of techniques such as subsoil modification or deep ploughing.

Various models (SWAP, DrainMod-S, UnSatChem and Hydrus), based on Richard's differential equation for the movement of water in unsaturated soil, are available to predict water and solute transport in the soil. These models require input of soil characteristics like the unsaturated soil moisture content, water tension, water retention curve, unsaturated hydraulic conductivity, dispersivity and diffusivity. These relations vary to a great extent from place to place and from time to time and are not easy to measure. Further, the models are difficult to calibrate under farmer's field conditions because of the spatial variability of soil salinity.

Simpler models, like SaltMod (Oosterbaan 2000), based on monthly or seasonal water and salt balances and an empirical capillary rise function, are also available. They are useful for long-term salinity control and predictions in relation to irrigation and drainage practices.

LeachMod (www.waterlog.info/leachmod.htm), using the SaltMod principles, helps in analysing leaching experiments in which the soil salinity is monitored in various root zone layers. The model optimises the value of the leaching efficiency of each layer so that a best fit is obtained between observed and simulated soil salinity values.

4.8.2 Amelioration of Subsoil Sodicity and Soil Dispersivity

Dispersive (sodic) soils present a variety of abiotic constraints to plant growth and function, including waterlogging and anoxia and Na^+ toxicity, as well as present a physical barrier to root growth resulting from high soil strength. Measures undertaken to amend sodicity by application of Ca compounds such as gypsum (or lime in acidic-sodic soils) aim to reduce exchangeable sodium. Soluble Ca^{2+} from these compounds exchanges with Na^+ and also other cations (K^+ and Mg^{2+}), thereby decreasing dispersive charge. The Ca^{2+} in soil solution also increases the flocculating charge. As a result, clay dispersion is prevented, and soil physical condition is improved.

Lime ($CaCO_3$) is insoluble when pH is above 8, and its low solubility under acidic pH values makes it suitable for amending acidic-sodic soils with pH < 6 (Bennett et al. 2014). Lime also corrects soil acidity. Gypsum ($CaSO_4.2H_2O$) is sparingly soluble and has been effectively used for all types of sodic soils (Oster and Jayawardane 1998; Zoca and Penn 2017), particularly in 'gypsum-responsive soils' identified by clay dispersion tests rather than by relying on ESP values (Loveday 1974). Addition of organic matter, including organic and green manures, to subsoil can alleviate the dispersivity (Gill et al. 2009) but depends on the use of organic compounds which covalently bond to clays and reduce the dispersive charge. Hydrophobic organic compounds can prevent water interactions with soil particles. Synthetic organic polymers such as polyvinyl alcohol (PAV) or polyacrylamide (PAM) have been effectively used in surface soils (Shainberg and Levy 1994; Mamedov et al. 2010). Subsoil application of these polymers is rare in cropping lands.

Dense, sodic, clay subsoils, particularly under dryland conditions, pose great problems in the reclamation of dispersivity because of the difficulty in the application of amendments to the deeper soil layers, in addition to the cost involved. Deep ripping, deep ploughing and profile inversion in conjunction with the amendments such as gypsum or manures are the common techniques used. Positive and negative results have been reported with these techniques (e.g. Hamza and Anderson 2003; Nuttall et al. 2005). Farmers have to choose the options on the basis of economic analysis including cost-benefit ratio.

4.8.3 Correction of Subsoil Alkalinity

High soil pH has been shown to be a constraint to crop productivity in South Australia by Cooper (2004) and the Central Soil Salinity Research Institute (CSSRI 2007) in India. The remediation of phytotoxicity in alkaline soils may not necessitate lowering soil pH to neutral, but to <9 (Brautigan et al. 2014). Protons needed to reduce pH in soils with a pH range of 8.5 to >9 may be comparatively low (because of the absence of calcite buffering in this pH range) compared to soils with pH near 8.5 since at the lower pH, calcite dissolves faster and buffers the pH change. Thus, the condition where $HCO_3^- + CO_3^{2-} > Ca^{2+} + Mg^{2+}$ in soil solution should be the criterion for the correction of alkalinity. Management of soil alkalinity should aim in reducing the concentrations of $HCO_3^- + CO_3^{2-}$ in soil solutions and maintaining soil pH below 8.5.

Application of sulphuric acid to reduce soil pH has been discussed but may be hazardous if not managed appropriately. Use of elemental sulphur to reduce soil pH will be successful only when soil contains sulphur-oxidising microorganisms. Chemolithotrophic bacteria such as *Thiobacillus* spp. are efficient S oxidisers, but they prefer low pH (<5) soil. Heterotrophic bacteria and fungi can oxidise S but will require biologically available carbon in alkaline soils (Lawrence and Germida 1988; Zhao et al. 2015). Pyrite (iron sulphide) available as an industrial waste has been used to reclaim alkaline sodic soils. While lime can increase soil pH in acid soils, gypsum application reduces soil pH by about 0.5 unit. The Ca^{2+} from gypsum can also precipitate $HCO_3^- + CO_3^{2-}$ found in alkaline soil solutions, thereby reducing soil pH.

Plant roots can secrete protons in the rhizosphere and reduce alkalinity. Thus, growing of legumes in crop rotation has been shown to cause soil acidification (e.g. Xu et al. 2002). Addition of organic materials may increase the population of acid-secreting microbes in the soil, thereby lowering soil pH (Jones 1998; Tang and Yu 1999; Walker et al. 2004). Brautigan et al. (2014) found that the reduction of pH achieved with biological amendments was temporary, with pH returning to pre-amendment levels over a short period of time. However, they found that when gypsum was added in conjunction with organic matter, the pH reduction was sustained. Phytoremediation occurs where protons produced by roots increase the dissolution of calcite and release Ca^{2+} which will be useful in reclamation of calcareous sodic soils (Qadir et al. 2007). As mentioned earlier, the successful use of these amendments for treatment of subsoil alkalinity will depend on the economics of applying them in subsoils and the benefits obtained in terms of yield.

References

Al-Busaidi AS, Cookson B. Salinity-pH relationships in calcareous soils. Agric Mar Sci. 2003;8:41–6. https://doi.org/10.24200/jams.vol8iss1pp41-46.

Arienzo M, Christen EW, Jayawardane NS, et al. The relative effects of Na and K on soil hydraulic conductivity and implications for winery wastewater management. Geoderma. 2012;173–174:303–10. https://doi.org/10.1016/j.geoderma.2011.12.012.

Bakker DM, Hamilton GJ, Hetherington R, et al. Salinity dynamics and the potential for improvement of waterlogged and saline land in a Mediterranean climate using permanent raised beds. Soil Tillage Res. 2010;110:8–24. https://doi.org/10.1016/j.still.2010.06.004.

Barrett-Lennard EG. The interaction between waterlogging and salinity in higher plants: causes, consequences and implications. Plant Soil. 2003;253:35–54. https://doi.org/10.1023/A:1024574622669.

Barrett-Lennard EG, Shabala SN. The waterlogging/salinity interaction in higher plants revisited – focusing on the hypoxia-induced disturbance to K^+ homeostasis. Funct Plant Biol. 2013;40(9):872–82. https://doi.org/10.1071/FP12235.

Barrett-Lennard EG, Anderson GC, Holmes KW. High soil sodicity and alkalinity cause transient salinity in South-Western Australia. Soil Res. 2016;54:407–17. https://doi.org/10.1071/SR15052.

Beltrán JM. Integrated approach to address salinity problems in irrigated agriculture. In: Gheyi HR, Dias NS, Lacerda CF, et al., editors. Manejo da salinidade na agricultura: Estudos básicos e aplicados. 2nd ed. Fortaleza: INCTSal; 2016. p. 3–6.

Bennett JM, Greene RSB, Murphy BW, et al. Influence of lime and gypsum on long-term rehabilitation of a Red Sodosol, in a semi-arid environment of New South Wales. Soil Res. 2014;52:120–8. https://doi.org/10.1071/SR13118.

Brautigan DJ, Rengasamy P, Chittleborough DJ. Aluminium speciation and phytotoxicity in alkaline soils. Plant Soil. 2012;360:187–96. https://doi.org/10.1007/s11104-012-1232-5.

Brautigan DJ, Rengasamy P, Chittleborough DJ. Amelioration of alkaline phytotoxicity by lowering soil pH. Crop Pasture Sci. 2014;65:1278–87. https://doi.org/10.1071/CP13435.

Bui EN, Krogh L, Lavado RS, et al. Distribution of sodic soils: the world scene. In: Sumner ME, Naidu R, editors. Sodic soils: distribution, properties, management and environmental consequences. New York: Oxford University Press; 1998. p. 19–33.

Cooper DS. Genetics and agronomy of transient salinity in *Triticum durum and T. aestivum*. PhD thesis, The University of Adelaide; 2004.

CSSRI. Central Soil Salinity Research Institute annual report. Karnal: CSSRI; 2007.

De Caritat P, Cooper M, Wilford J. The pH of Australian soils: field results from a national survey. Soil Res. 2011;49:173–82.

Dixit PN, Chen D. Modification of a spatially referenced crop model to simulate the effect of spatial pattern of subsoil salinity. Comput Electron Agric. 2010;74(2):313–20. https://doi.org/10.1016/j.compag.2010.09.006.

FAO – Food and Agriculture Organization of the United Nations. World reference base for soil resources. International soil classification system for naming soils and creating legends for soil maps. World Soil Resources Report 106, Rome. 203p; 2015.

Gill JS, Sale PWG, Peries RR, et al. Changes in soil physical properties and crop root growth in dense sodic subsoil following incorporation of organic amendments. Field Crops Res. 2009;114:137–46. https://doi.org/10.1016/j.fcr.2009.07.018.

Gomes EM, Gheyi HR, Silva EFF. Melhorias nas propriedades químicas de um solo salino sódico e rendimento de arroz, sob diferentes tratamentos. Rev Bras Eng Agríc Ambient. 2000;4:355–61. https://doi.org/10.1590/S1415-43662000000300009.

Grewal HS. Response of wheat to subsoil salinity and temporary water stress at different stages of the reproductive phase. Plant Soil. 2010a;330:103–13. https://doi.org/10.1007/s11104-009-0179-7.

Grewal HS. Water uptake, water use efficiency, plant growth and ionic balance of wheat, barley, canola and chickpea plants on a sodic Vertosol with variable subsoil NaCl salinity. Agric Water Manag. 2010b;97:148–56. https://doi.org/10.1016/j.agwat.2009.09.002.

Hamza MA, Anderson WK. Responses of soil properties and grain yields to deep ripping and gypsum application in a compacted loamy sand soil contrasted with a sandy clay loam soil in Western Australia. Aust J Agric Res. 2003;54:273–82. https://doi.org/10.1071/AR02102.

He B, Cai Y, Rana W, et al. Spatial and seasonal variations of soil salinity following vegetation restoration in coastal saline land in eastern China. Catena. 2014;118:147–53. https://doi.org/10.1016/j.catena.2014.02.007.

Hilgard EW. Soils, their formation, properties, compositions and relations to climate and plant growth. New York/London: The Macmilan Company; 1906. https://doi.org/10.5962/bhl.title.24461.

Isbell RF. The Australian soil classification, Australian soil and land survey handbook series 4. Melbourne: CSIRO Publishing; 2002.

Jobbagy EG, Toth T, Nosetto MD, et al. On the fundamental causes of high environmental alkalinity (pH $\geq$ 9): an assessment of its drivers and global distribution. Land Degrad Dev. 2017; https://doi.org/10.1002/ldr.2718.

Jones LJ. Organic acids in the rhizosphere – a critical review. Plant Soil. 1998;205:25–44. https://doi.org/10.1023/A:1004356007312.

Kelly WP. Alkali soils: their formation, properties, and reclamation. New York: Reinhold; 1951. 176p.

Lacerda CF, Assis Júnior JO, Lemos Filho LCA, et al. Morpho-physiological responses of cowpea leaves to salt stress. Braz J Plant Physiol. 2006;18:455–65. https://doi.org/10.1590/S1677-04202006000400003.

Lacerda CF, Ferreira JFS, Liu X, et al. Evapotranspiration as a criterion to estimate nitrogen requirement of maize under salt stress. J Agron Crop Sci. 2016;202:192–202. https://doi.org/10.1111/jac.12145.

Lawrence JR, Germida JJ. Relationship between microbial biomass and elemental sulfur oxidation in agricultural soils. Soil Sci Soc Am J. 1988;52:672–7. https://doi.org/10.2136/sssaj1988.03615995005200030014x.

LeachMod for simulating leaching of saline irrigated soil. Available at www.waterlog.info/leachmod.htm
Letey J. Relationship between soil physical properties and crop production. Adv Soil Sci. 1985;1:277–94. https://doi.org/10.1007/978-1-4612-5046-3_8.
Loveday J. Recognition of gypsum-responsive soils. Aust J Soil Res. 1974;25:87–96.
Mamedov AI, Wagner LE, Huang C, et al. Polyacrylamide effects on aggregate and structure stability of soils with different clay mineralogy. Soil Sci Soc Am J. 2010;74:1720–32. https://doi.org/10.2136/sssaj2009.0279.
Marchuk A, Rengasamy P. Clay behaviour in suspension is related to the iconicity of clay-cation bonds. Appl Clay Sci. 2011;53:754–9. https://doi.org/10.1016/j.clay.2011.05.019.
McDonald GK, Taylor JD, Verbyla A, et al. Assessing the importance of subsoil constraints to yield of wheat and its implications for yield improvement. Crop Pasture Sci. 2012;63:1043–65. https://doi.org/10.1071/CP12244.
McDonald GK, Tavakkoli E, Cozzolino D, et al. A survey of total and dissolved organic carbon in alkaline soils of southern Australia. Soil Res. 2017;55:617–29. https://doi.org/10.1071/SR16237.
Medeiros WJF, Oliveira FIF, Lacerda CF, et al. Isolated and combined effects of soil salinity and waterlogging in seedlings of 'Green Dwarf' coconut. Semina: Ciências Agrárias. 2018;39:1459–68. https://doi.org/10.5433/1679-0359.2018v39n4p1459.
Munns R, Tester M. Mechanisms of salinity tolerance. Ann Rev Plant Biol. 2008;59:651–81. https://doi.org/10.1146/annurev.arplant.59.032607.092911.
Naidu R, Rengasamy P. Ion interactions and constraints to plant nutrition in Australian sodic soils. Soil Res. 1993;31:801–19. https://doi.org/10.1071/SR9930801.
Northcote JH, Skene JKM. Australian soils with saline and sodic properties, Publication 27. Melbourne: CSIRO Publishing; 1972.
Nuttall JG, Armstrong RD. Impact of subsoil physicochemical constraints on crops grown in the Wimmera and Mallee is reduced during dry seasonal conditions. Soil Res. 2010;48:125–39. https://doi.org/10.1071/SR09075.
Nuttall JG, Armstrong RD, Connor DJ. The effect of boron tolerance, deep ripping with gypsum and water supply on subsoil water extraction of cereals on alkaline soil. Aust J Agric Res. 2005;56:113–22. https://doi.org/10.1071/AR04183.
Nuttall JG, Hobson KB, Materne M, et al. Use of genetic tolerance in grain crops to overcome subsoil constraints in alkaline cropping soils. Soil Res. 2010;48:188–99.
Oosterbaan RJ. SALTMOD description of principles, user manual, and examples of application. ILRI: Wageningen; 2000.
Oster JD, Jayawardane NS. Agricultural management of sodic soils. In: Sumner ME, Naidu R, editors. Sodic soils: distribution, properties, management and environmental consequences. New York: Oxford University Press; 1998. p. 125–47.
Oster JD, Sposito G, Smith CJ. Accounting for potassium and magnesium in irrigation water quality assessment. Calif Agric. 2016;70(2):71–6. https://doi.org/10.3733/ca.v070n02p71.
Pizarro F. Drenaje agrícola y recuperacion de suelos salinos. Madrid: Editorial Agrícola Española; 1978.
Qadir M, Oster JD, Schubert S, et al. Phytoremediation of sodic and saline-sodic soils. Adv Agron. 2007;96:197–247. https://doi.org/10.1016/S0065-2113(07)96006-X.
Qureshi AS, Mccornick PG, Qadir M, et al. Managing salinity and waterlogging in the Indus Basin of Pakistan. Agric Water Manag. 2008;95:1–10. https://doi.org/10.1016/j.agwat.2007.09.014.
Rengasamy P. Transient salinity and subsoil constraints to dryland farming in Australian sodic soils: an overview. Aust J Exp Agric. 2002a;42:351–61. https://doi.org/10.1071/EA01111.
Rengasamy P. Clay dispersion. In: McKenzie J, et al., editors. Soil physical measurement and interpretation for land evaluation. Collingwood: CSIRO Publishing; 2002b. p. 200–10.
Rengasamy P. World salinization with emphasis on Australia. J Exp Bot. 2006;57:1017–23. https://doi.org/10.1093/jxb/erj108.

Rengasamy P. Soil processes affecting crop production in salt-affected soils. Funct Plant Biol. 2010;37:613–20.
Rengasamy P. Soil chemistry factors confounding crop salinity tolerance – a review. Agronomy. 2016a;6(4):53. https://doi.org/10.3390/agronomy6040053.
Rengasamy P. Soil salinization. In: Oxford research encyclopedia, environmental science; 2016b. https://doi.org/10.1093/acrefore/9780199389414.013.65.
Rengasamy P, Marchuk A. Cation ratio of soil structural stability. Soil Res. 2011;49:280–5. https://doi.org/10.1071/SR10105.
Rengasamy P, Sumner ME. Processes involved in sodic behaviour. In: Sumner ME, Naidu R, editors. Sodic soils: distribution, properties, management and environmental consequences. New York: Oxford University Press; 1998. p. 35–50.
Rengasamy P, Olsson KA, Kirby JM. Physical constraints to root growth in the subsoil. In: Proceedings, national workshop on subsoil constraints to root growth and high soil water and nutrient use by plants, Tanunda, South Australia; 1992. p. 84–91.
Rengasamy P, Chittleborough D, Helyar K. Root-zone constraints and plant-based solutions for dryland salinity. Plant Soil. 2003;257:249–60. https://doi.org/10.1023/A:1027326424022.
Rengasamy P, Tavakkoli E, McDonald GK. Exchangeable cations and clay dispersion: net dispersive charge, a new concept for dispersive soil. Eur J Soil Sci. 2016;67:659–65. https://doi.org/10.1111/ejss.12369.
Ribeiro MR, Ribeiro Filho MR, Jakomine PKT. Origem e classificação de solos afetados por sais. In: Gheyi HR, Dias NS, Lacerda CF, et al., editors. Manejo da salinidade na agricultura: Estudos básicos e aplicados. 2nd ed. Fortaleza: INCTSal; 2016. p. 9–16.
Richards LA. Diagnosis and improvement of saline and alkali soils, USDA agriculture handbook, 60. Riverside: United States Salinity Laboratory; 1954.
Roux P, Preez CC, Strydom MG, et al. Effect of irrigation on soil salinity profiles along the Lower Vaal River, South Africa. Water SA. 2007;33 https://doi.org/10.4314/wsa.v33i4.52942.
Sadras V, Baldocka J, Rogeta D, et al. Measuring and modelling yield and water budget components of wheat crops in coarse-textured soils with chemical constraints. Field Crops Res. 2003;84:241–60. https://doi.org/10.1016/S0378-4290(03)00093-5.
Saqib M, Akhtar J, Qureshi RH. Na^{+} exclusion and salt resistance of wheat (*Triticum aestivum*) in saline-waterlogged conditions are improved by the development of adventitious nodal roots and cortical root aerenchyma. Plant Sci. 2005;169:125–30. https://doi.org/10.1016/j.plantsci.2005.03.003.
Setter TL, Waters I, Sharma SK, et al. Review of wheat improvement for waterlogging tolerance in Australia and India: the importance of anaerobiosis and element toxicities associated with different soils. Ann Bot. 2009;103:221–35. https://doi.org/10.1093/aob/mcn137.
Shainberg I, Levy GJ. Organic polymers and soil sealing in cultivated soils. Soil Sci. 1994;158:267–73.
Sharma DK, Singh A. Salinity research in India – achievements, challenges and future prospects. Water Energy Int Water Resour Sect. 2015;58:35–45.
So HB, Aylmore LAG. How do sodic soils behave – the effects of sodicity on soil physical behaviour. Soil Res. 1993;31:761–77. https://doi.org/10.1071/SR9930761.
Sousa CHC, Lacerda CF, Silva FLB, et al. Yield of cotton/cowpea and sunflower/cowpea crop rotation systems during the reclamation process of a saline-sodic soil. Eng Agríc. 2014;34:867–76. https://doi.org/10.1590/S0100-69162014000500006.
Sposito G. The chemistry of soils. 2nd ed. New York: Oxford University Press; 2008.
Sposito G. Sustaining "the genius of soils". In: Churchman GJ, Landa ER, editors. The soil underfoot: infinite possibilities for a finite resource. Florida: CRC Press; 2014. p. 395–408.
SSSA – Soil Science Society of America. Glossary of soil science terms. Madison: Soil Science Society of America; 1976.
Sumner ME. Sodic soils – new perspectives. Soil Res. 1993;31:683–750. https://doi.org/10.1071/SR9930683.

Tang C, Yu Q. Impact of chemical composition of legume residues and initial soil pH on pH change of a soil after residue incorporation. Plant Soil. 1999;215:29–38. https://doi.org/10.1023/A:1004704018912.

Tavakkoli E, Rengasamy P, Smith E, et al. The effect of cation-anion interactions on soil pH and solubility of organic carbon. Eur J Soil Sci. 2015;66:1054–62. https://doi.org/10.1111/ejss.12294.

Walker DJ, Clemente M, Bernal P. Contrasting effects of manure and compost on soil pH, heavy metal availability and growth of *Chenopodium album* L. in a soil contaminated by pyritic mine waste. Chemosphere. 2004;57:215–24. https://doi.org/10.1016/j.chemosphere.2004.05.020.

Wicke B, Smeets E, Dornburg V, et al. The global technical and economic potential of bioenergy from salt-affected soils. Energy Environ Sci. 2011;4:2669–81. https://doi.org/10.1039/C1EE01029H.

Xu RK, Coventry DR, Farhoodi A, et al. Soil acidification as influenced by crop rotations, stubble management, and application of nitrogenous fertilizer, Tarlee, South Australia. Soil Res. 2002;40:483–96. https://doi.org/10.1071/SR00104.

Zhao C, Degryse F, Gupta V, et al. Elemental sulfur oxidation in Australian cropping. Soil Sci Soc Am J. 2015;79:89–96. https://doi.org/10.2136/sssaj2014.08.0314.

Zoca SM, Penn C. An important tool with no instruction manual: a review of gypsum use in agriculture. Adv Agron. 2017;144:1–44.

Chapter 5
Pyritic Subsoils in Acid Sulfate Soils and Similar Problems in Mined Areas with Sulfidic Rocks

José João Lelis Leal de Souza and Walter Antônio Pereira Abrahão

Abstract Acid drainage is one of the most dangerous and costly environmental problems that reduce soil and water quality. Acid drainage begins when pyrite and other sulfides are exposed and, in the presence of oxygen and water, undergo oxidation to form hydrated sulfates. This can occur in tailing piles, sterile or other materials moved by mining activities, cuts of pyrite materials on roads, tunnels, etc., where the oxidized compounds appear as white and yellowish crusts on the exposed surface of rocks and weathered sediment. Sulfide oxidation products, besides being highly soluble, have a strongly acidic reaction, so that they are easily dissolved in the liquid phase, acidifying the water bodies into which they discharge. Thus, the waters, when dissolving the salts produced by oxidation, become acidic and with high concentrations of sulfate and iron. Due to the low pH values (which may drop below 2.0), other elements such as aluminum, manganese, copper, zinc, lead, mercury, cadmium, etc., if present in the medium, are solubilized and mobilized in the drainage waters and may become toxic. By reaching lakes and rivers, acid drainage can compromise the ecological balance of water resources, with the potential to incorporate heavy metals in the food chain. Remediation of acid drainage involves (*i*) admixture of sufficient alkaline material to neutralize acidity and (*ii*) reducing sulfide contact with the atmosphere through a water layer or chemical or biological precipitation of sulfates and oxides. Adoption of preventive and corrective strategies should be based on balance between acidity and alkalinity generation, as well as the reaction kinetics.

Keywords Acidic soils · Heavy metals · Jarosite · Pyrite

J. J. L. L. de Souza (✉) · W. A. P. Abrahão
Soils Department, Federal University of Viçosa, Viçosa, MG, Brazil
e-mail: jjlelis@ufv.br; wabrahao@ufv.br

T. S. d. Oliveira, R. W. Bell (eds.), *Subsoil Constraints for Crop Production*,
https://doi.org/10.1007/978-3-031-00317-2_5

5.1 Introduction

The phenomenon of acid drainage from sulfide oxidation was first reported in ancient times by Theophrastus (315 B.C.) and Pliny (23–79 A.C.) (Agricolla 1950). Acid drainage is a process of widespread occurrence in tropical and temperate environments but also in periglacial landscapes. In this process, sulfide minerals exposed to oxygen and water are oxidized, yielding strong acidity and high concentrations of soluble salts in water. Additionally, the acidity favors the dissolution of other minerals and increases the activity of metal(loid)s in solution. This process can occur due to natural exposure of sulfides in potential acid sulfate soils or from sulfidic rocks (when the discharge is called acid rock drainage) or through exposure of sulfides by anthropogenic activities, usually mining (when it is named acid mine drainage) (Simate and Ndlovu 2014). The existence of sulfidic materials in the subsoil has implications for crop production, but most of the implications are for water resources and for off-site impacts. Acid drainage threatens crop production and the dwindling supply of fresh water. The United Nations recently labeled acid drainage as the second biggest problem facing the world after global climatic changes. Hence the present chapter focuses on the implications of the sulfidic subsoils (naturally occurring) and subsurface sulfidic wastes (in mine sites).

Acid drainage is the result of acidity-producing reactions from the oxidation of sulfides and ferrous ions but is influenced by dissolution reactions of some minerals, mainly carbonates, capable of producing alkalinity. Acid drainage, as well as the solubilization and precipitation of metal minerals, has been the focus of geochemical investigations in the last 60 years (Sato 1960; Nordstrom 1977; Evangelou 1995; Hemingway et al. 2002; Dold 2010; Kefeni et al. 2017). It is necessary to understand the parameters that control the formation of acid drainage and to develop effective prevention methods.

The reaction rate differs depending on the type of sulfide being oxidized by Fe^{3+}. Experiments on weathering kinetics indicate the importance of trace elements on the individual stability of sulfides. When different sulfides are present, electrochemical processes that influence the reactivity of individual sulfides are likely to occur (Aghamirian 1997).

The initial reaction of acid drainage is the oxidation of sulfide, represented by pyrite, by atmospheric O_2, producing sulfate, a metallic cation and H^+ ions (International Network for Acid Prevention 2009) (Eq. 5.1[1]). In sequence the Fe^{2+} is oxidized to Fe^{3+}, which, in turn, undergoes hydrolysis producing ferric hydroxide and more acidity (Eqs. 5.2 and 5.3). The oxidation of pyrite, followed by hydrolysis of Fe^{3+}, can generate four moles of H^+ per mole of pyrite (Eq. 5.4).

$$FeS_2 + 7/2O_2 + H_2O \leftrightarrow Fe^{2+} + 2SO_4^{2-} + 2H^+ \; \log K° = 206.91 \qquad (5.1)$$

[1] The equilibrium constants were calculated based on the free energy of formation of the reactants (ΔGr^o) and the products, according to data from Faure (1991).

$$Fe^{2+} + 1/4O_2 + H^+ \leftrightarrow Fe^{3+} + 1/2H_2O \log K^\circ = 7.67 \tag{5.2}$$

$$Fe^{3+} + 3H_2O \leftrightarrow Fe(OH)_3 + 3H^+ \log K^\circ = -3.54 \tag{5.3}$$

$$4FeS_2 + 15O_2 + 10H_2O \leftrightarrow 4FeOOH + 8H_2SO_4 \log K^\circ = 858.69 \tag{5.4}$$

The Fe^{3+} hydrolysis reaction usually occurs readily in the initial stages of pyrite oxidation, limiting the free Fe^{3+} activity in the solution. However, as acidity builds up in the medium and the pH falls below 3.5, the hydrolysis reaction is limited, significantly increasing the Fe^{3+} concentration in the solution. In these conditions, Fe^{3+} starts to act as an electron acceptor on the surface of the pyrite crystals (Singer and Stumm 1970), becoming the main mechanism of sulfide oxidation and acid production (Eq. 5.5).

$$FeS_2 + 14Fe^{3+} + 8H_2O \leftrightarrow 15Fe^{2+} + 2SO_4^{2-} + 16H^+ \log K^\circ = 98.49 \tag{5.5}$$

As the pH of the medium is sufficiently acidic, the Fe^{3+} becomes the main oxidant of the pyrite, being reduced to Fe^{2+}, and O_2 has an indirect role in the re-oxidation of Fe^{2+}, regenerating Fe^{3+} (Eq. 5.2). Under acid conditions corresponding to pH values below 3, the oxidation of pyrite by Fe^{3+} is about 10–100 times faster than by O_2 (Kefeni et al. 2017). The process becomes self-propagating, constituting a continuous cycle capable of generating large amounts of H^+ ions.

In natural environments, the presence of some species of bacteria, such as *Thiobacillus ferrooxidans*, can increase sulfide oxidation (Blowes et al. 2003). The initial step of pyrite oxidation does not require an elaborate sequence of different geochemical reactions for each pH range. *Thiobacillus* sp. attach onto the mineral particle surfaces and directly oxidize iron and sulfur (Eq. 5.6). The bacteria also can accelerate the oxidation reaction of Fe^{2+} (Eq. 5.7) and consequently catalyze sulfide oxidation (Eq. 5.5). *Thiobacillus* sp. also forms nano-environments surrounding sulfides, excreting polymers on the mineral surface (Simate 2009). Subsequently, the bacteria oxidize Fe^{2+} to Fe^{3+}, which is utilized to oxidize sulfide (Eq. 5.5).

$$FeS_2 + 7/2O_2 + H_2O \xrightarrow{\text{Bacteria}} Fe^{2+} + 2SO_4^{2-} + 2H^+ \tag{5.6}$$

$$2Fe^{2+} + 0.5O_2 + 2H^+ \xrightarrow{\text{Bacteria}} 2Fe^{3+} + H_2O \tag{5.7}$$

These bacteria occur naturally in aquatic and soil systems but appear to be more active in acid media, with pH values close to 3.0 (Rowe et al. 2007). For this reason, iron oxidation is considered the limiting step of the acid drainage process and proceeds slowly under sterile conditions.

When acidic water, rich in Fe^{2+}, from underground layers reaches the surface, it will undergo oxidation, and, by hydrolyzing, it can precipitate ferrihydrite, schwertmannite, goethite, or jarosite depending on the pH-Eh conditions, and availability of other necessary elements, such as potassium and sulfur. Jarosite, schwertmannite, and ferrihydrite are meta-stable minerals concerning goethite (Equeenuddin et al. 2010; Hemingway et al. 2002; Schwertmann 1991). The hydrolysis and precipitation of ferric hydroxides at pH above 3.5 is the largest acidity generator in this process.

Pyrite (α-FeS_2) is the main sulfide mineral capable of producing acid drainage, although other, less common sulfides may also contribute to the process. In general, iron sulfides with metal/sulfur molar ratios lower than one and sulfosalts (e.g., enargite) generate acid when reacting with oxygen and water. On the other hand, sulfides with metal/sulfur molar ratios equal to one (e.g., sphalerite, galena, chalcopyrite) tend not to produce acidity when oxygen is the oxidant. However, when aqueous $Fe^{3+}$is the oxidant, all sulfides can generate acidity. Therefore, the quantity and reactivity of iron sulfide present in potential acid sulfate soils, an ore deposit, or mine waste play a crucial role in determining the mine drainage characteristics.

5.2 Thionic Soils

The influence of sulfide materials in pedogenesis is indicated by (*i*) pH in water which can reach values lower than 4.0; (*ii*) the presence of mottles or coatings with accumulations of iron or aluminum sulfate; or (*iii*) hydroxysulfate minerals and water-soluble sulfate content higher than 0.05%. These characteristics are used to define the "thionichorizon" (IUSS Working Group WRB 2015). Similar horizons are present in Soil Taxonomy (Soil Survey Staff 2006), the Brazilian System of Soil Classification (Santos et al. 2013), the Australian Soil Classification (The National Committee on Soil and Terrain 2016), and the Canadian System of Soil Classification (Soil Classification Working Group 1998).

Thionic soils are present in all continents. In general, thionic soils are developed under tidal oscillation and occur in mangroves and actual or paleo marine fans. Sediments transported and deposited by rivers and seawater bring together Fe oxides from water basins and SO_4^{2-}, from seawater. In these areas, subsoils are typically waterlogged (hypoxic), which contributes to the accumulation of organic matter in anaerobic conditions (paludization). The low partial pressure of O_2 favors the activity of chemotrophs, and the relative natural abundance of iron and sulfur means Fe^{3+} and SO_4^{2-} is by far the most abundant terminal electron acceptors in estuarine soils (Kristensen et al. 2008). Consequently, H_2S and ferrous ion are formed and react to form at first FeS and S^0. Pyrite crystals either form from saturated solution of FeS (Wada and Seisuwan 1986), although no micromorphological evidence seems to be available for this issue (Mees and Stoops 2010). Pyrite is more abundant in deeper horizons (i.e., subsoils) and often occurs in spherical or framboid clusters of microcrystalline particles (Wilkin and Barnes 1997). Pyrite framboids take 3 hours and 3 years to form in sediments (Rickard 2019) and are one of the most reactive sulfides (Evangelou 1995).

Histosols and Gleysols are the main soil groups affected by thionic materials in coastal lowlands, while these characteristics are less common in Vertisols, Planosols, Solonchaks, Stagnosols, and Umbrisols (IUSS Working Group WRB 2015).

Thionic soils also are observed far from marine influence, in soils developed from sulfide-bearing rocks or where mining-related materials are exposed to weathering (Borba et al. 2003; Francelino et al. 2011). Although this process occurs naturally, anthropogenic activities, especially mining, can induce a thionic horizon by increasing the number of sulfides exposed (Dold 2014; Li et al. 2009). In these cases, Technosols and Inceptisols are commonly affected by thionic materials.

The colors of thionic soils reflect various stages in the process of sulfurization (Fig. 5.1). Oxidation of sulfides in surface horizons may produce a variety of colors, which range from 10YR through 2.5Y and 5GY together with low to high chroma and values (Fanning et al. 1993). By contrast, sulfidic mineral materials associated with potential acid sulfate conditions have low chroma and value, typically equal to or below 1 and 4, respectively.

The input of Fe and other metals by sulfide oxidation is essential for phytoplankton production (Dold et al. 2013; Hodson et al. 2017). However, in general, oxidation of sulfides affects the balance between macro and micronutrients in the soil (Halcomb and Fare 2002) and reduces the diversity of aquatic life to a small number of species better adapted to conditions of low pH and high pE (He et al. 2015; Rowe et al. 2007; Zhi et al. 2015). Even when the input of Fe and other metals is limited

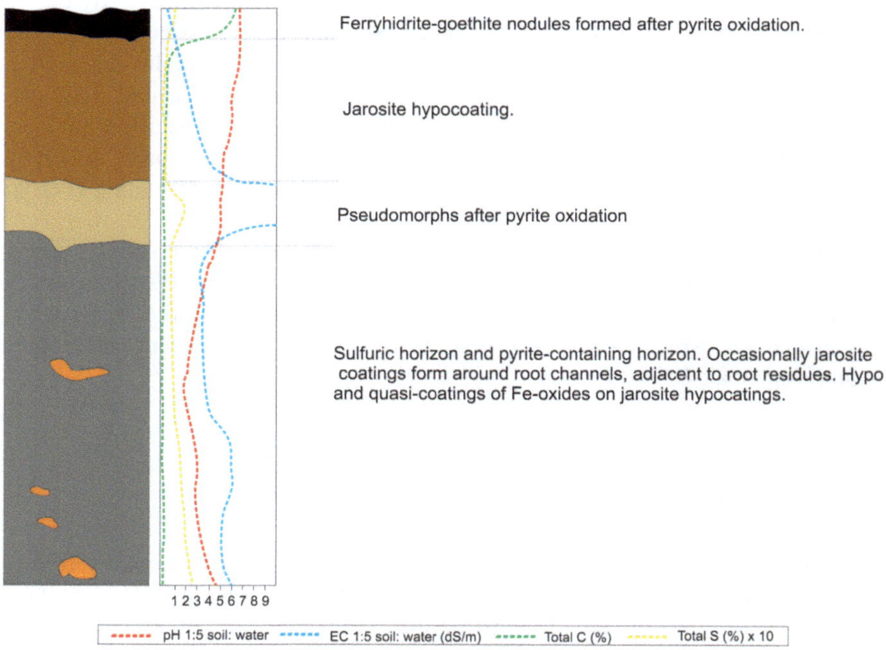

Fig. 5.1 Schematic thionic soil profile and their variability with depth

to groundwater, there may be impacts on terrestrial plants and animals depending on the substrate and rocks through which the acidic water passes, the extent of dilution or neutralization caused by contact with other water bodies, and with the hydrological regime (Lei et al. 2010).

Heavy metal contamination derived from sulfide oxidation is highly variable because it depends on the mineralogy and amount of sulfide mineral oxidized and the gangue minerals present in the rock (Durães et al. 2014; Equeenuddin et al. 2010; Li et al. 2009). Besides the ingestion of other toxins in polluted water, the ingestion of excess sulfates is harmful. Consumption of water with more than 600 mg sulfate L^{-1} by humans will lead to vomiting and diarrhea in most individuals (Dawson et al. 2015). Except for a few species, plants are sensitive to excessive dissolved ions in the soil solution. The plants growing along the riparian zone of streams containing acid water discharge are exposed to high levels of sulfates and may be replaced by the sulfate-tolerant species over time.

Even after mining ceases, the heavy metal pollution in soil and water continues to endanger the health of plants and animals (Jennings et al. 2008; Liao et al. 2016). Once organisms absorb heavy metals, they accumulate in vital organs and glands and inhibit the absorption of nutrients, thereby hindering biological functions (Singh et al. 2011). Because heavy metals accumulate in the trophic chain, they are generally related, even at low concentrations, to necrosis, tumors, cancer, and general impairment of the digestive system, cardiovascular system, and urogenital system in vertebrates (Chen et al. 2010; Li et al. 2012). Heavy metals cause cellular damage and disturbance of cellular ionic homeostasis of plants (Yadav 2010).

Acid mine drainage is associated with several lithologies (Dürr et al. 2005) and numerous mining sites worldwide (Fig. 5.2). Water and soil samples taken from these sites have concentrations above the legislation limits for several toxic elements, namely, fluoride, uranium, and aluminum and heavy metals, from coal (Campaner et al. 2014), gold (Andrade et al. 2008), copper (Larsson et al. 2018), zinc (Macías et al. 2012), silver (Sainz et al. 2003), and uranium mining (Ferrari et al. 2017).

Although the exchangeable fraction of heavy metals in thionic soils is higher than in other polluted soils (Candeias et al. 2014), approximately 70% of the heavy metals in thionic soils are in the crystal structures of goethite, hematite, jarosite, and schwertmannite, which are relatively resistant to leaching and transfer to the trophic chain (Zhao et al. 2011).

5.3 Prediction of Acid Drainage

There are two main approaches for predicting the drainage water quality from a given substrate regardless of whether it is topsoil or subsoil material. Chemical analyses of the substrate are standard but often supplemented by kinetic tests of leaching or simulated weathering. Chemical analysis of sulfides and neutralizing materials present in the substrate allows calculation of the acid generation potential

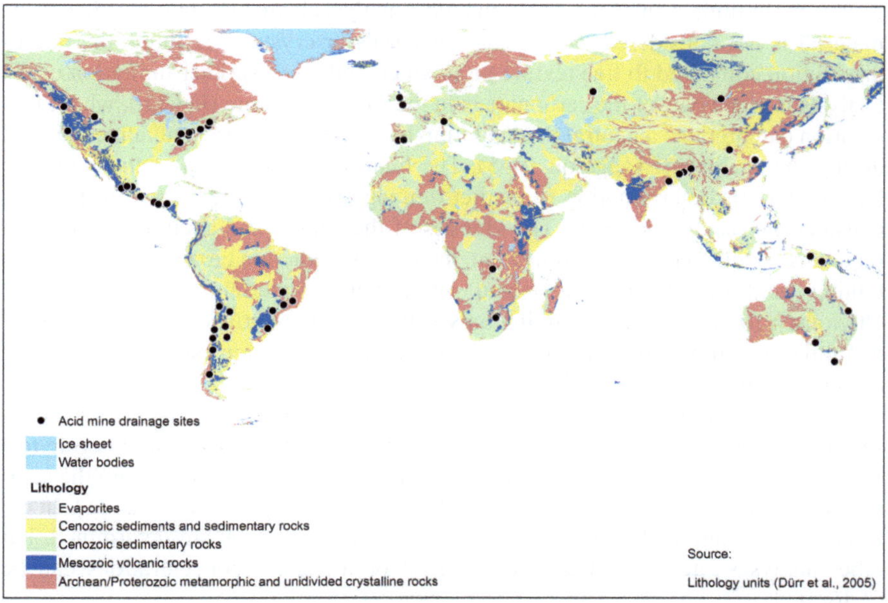

Fig. 5.2 Acid mine drainage sites in the world according to lithology

(acidity potential) and the alkali generation potential (neutralization potential), respectively. Based on the results of these analyses and the stoichiometry of the reactions discussed above, an acid-base balance can be derived (Skousen et al. 2001, 2002). The substrate is considered a possible acid drainage generator if the acidity potential is higher than the neutralization potential. Otherwise, acidic drainage from this substrate is not expected unless the rate of acid formation exceeds the rate of neutralization.

The acidity potential is determined from the total sulfide content in the sample. In this analysis, the sample is washed with acid to remove the sulfates and non-pyritic Fe present (Evangelou 1995). Subsequently, the sample is digested, and the S or Fe content in the extract is determined. The direct determination of sulfur has the advantage of providing a more exact measure of the abundance of sulfides present in the substrate. However, if any of the sulfide minerals present cannot generate acid when oxidized, the acidity potential will be overestimated. Determining the iron content to predict iron sulfide content minimizes the possibility of overestimation. However, the determination of Fe relates only to pyrite and does not consider other sulfides present that can generate acidity. Also, Fe is often present in other minerals, which are not removed in the initial acidic washing of the sample, which would overestimate the pyritic iron. The primary method used for the oxidative digestion of sulfides is hydrogen peroxide (H_2O_2) at 30% (Evangelou 1995). However, other oxidants may be used.

The main limitation of the static chemical analysis is that the particle size of the material is not considered: very coarse materials or crystal size limits the rate of dissolution compared to fine texture or small crystal size. Another limitation of the analysis is when part of the carbonate is in the form of siderite ($FeCO_3$), whose dissolution can produce a neutral final solution but is computed as an alkaline material during the analysis (Evangelou 1995). The final major limitation of the acid-base balance (ABB) determination to predict drainage water quality is that it does not consider the reaction kinetics. The kinetics of the acidity and alkalinity producing reactions are not necessarily equivalent, leading to prediction errors in samples containing acid-forming (sulfides) and neutralizing materials (e.g., limestone). Obviously, for substrates containing only sulfides, with relatively small amounts of neutralizing compounds, ABB-based acid drainage generation estimates tend to be closer to reality.

To avoid the limitations and errors of interpretation of the static substrate analyses, an alternative technique is leaching tests that simulate substrate weathering to a certain extent (Schaffie and Hosseini 2014). These tests consist of placing the sample into leach cells or columns, to which distilled water is periodically added. In the leachate solution, acidity and sulfate and iron concentrations are determined, but other analyses such as pH and toxic metal or metalloid concentrations can be performed.

The main advantage of simulated weathering tests is that data are obtained as a function of time, enabling a prediction of the oxidation kinetics of the sulfides. However, kinetics may still deviate from field weathering since the dissolution kinetics of carbonates vary according to field rainfall and the interval between leaching events. In this case, leaching studies can adjust the quantities of water and the interval between leaching to match the precipitation conditions of the target environment. The main disadvantage of this leaching technique is the time required to obtain enough data to understand the kinetics. To shorten this time, hydrogen peroxide (H_2O_2) at low concentrations as a leaching solution can be used (Evangelou 1995).

5.4 Prevention and Soil Remediation Opportunities for Acid Sulfate Materials

Prevention of acid drainage generation can be achieved by a mixture of materials containing sufficient alkaline material to neutralize acidity. Examples of admixtures include mine tailings and soil, quarried and crushed aggregate, basic sand materials, and cement (Benzaazoua et al. 2002). Placement of a capping layer above the substrate and subaqueous disposal is a possible strategy to prevent water and air, respectively, from reaching the sulfide-bearing rocks or tailings (Villain et al. 2013). Clay subsurface horizons of soils, oxide wastes, fly ashes, organic wastes, and geofabrics are commonly used to seal the sulfidic wastes (Lu et al. 2014; Olds et al. 2012).

Dissolution of silicate minerals can neutralize acidity. However, silicates generally decompose more slowly than carbonates and have lower neutralization capacity (Masindi and Gitari 2016). Consequently, addition of limestone ($CaCO_3$), slaked lime ($Ca(OH)_2$), soda ash ($NaCO_3$), caustic soda (NaOH), ammonium hydroxide (NH_4OH), magnesium hydroxide ($Mg(OH)_2$), and calcium oxide (CaO) is the most generally applied treatment of thionic soils to increase the pH and precipitate dissolved metals (Watten et al. 2005).

Limestone ($CaCO_3$) consumes two protons (H^+), according to Eq. 5.8.

$$CaCO_3 + 2H^+ \leftrightarrow Ca^{2+} + CO_2 + H_2O \qquad \log K^\circ = 9.74 \tag{5.8}$$

Equation 5.8 assumes maximum neutralization, that is, when the reaction proceeds at high acidity and the carbonate equilibrium is shifted to H_2CO_3. At lower acidity, when the carbonate buffers the pH to around 6.5–7.0, the equilibrium shifts to HCO_3^-, and the dissolution of 1-mole calcite consumes only 1 mole of protons. Most of the ABB theory assumes the condition of maximum acidity, in which the H_2CO_3 species is the most stable. This may lead to an overestimation of the neutralization potential at pH close to neutrality.

Treatment of thionic soils can be exceedingly challenging and costly. First of all, the control of water tables is needed to ensure the thionic or sulfidic materials remain submerged. In the exposition of sulfide material, the amount of lime required to neutralize soil acidity depends on its purity and pH buffer capacity of the thionic soil. Soil pH buffer capacity increases with increasing organic matter, clay minerals, Al/Fe (hydr)oxides, and carbonate contents (Kissel et al. 2012; Vasques et al. 2018; Wang et al. 2015; Zhou et al. 2007). Applications of low-cost alkaline waste products, such as cement waste (Kastyuchik et al. 2016), bentonite (Masindi and Gitari 2016), fly ashes and concrete (Jones and Cetin 2017), and biochar (Kim et al. 2014), are less effective than neutralization of acidity with limestone and require higher rates and more time for effective remediation of acid drainage (Iakovleva et al. 2015).

5.5 Treatment of Acid Discharge Water

Because the application of alkaline compounds to reduce the acidity of thionic soils is palliative and expensive (Mello et al. 2006), quite commonly, thionic soils are left untreated. Instead, the strategy applied is to treat, recover, and reuse resources from acid drainage with technologies designed to reduce the volume of waste and make financial returns. Treatments include artificial wetlands (Kuyucak 2002), sulfate-reducing bioreactors (Bwapwa et al. 2017; Wang et al. 2017; Yadav and Jamal 2015), and permeable reactive barriers (Holmes et al. 2017; Shabalala et al. 2017; Wang et al. 2016). Wetlands treated with organic materials to induce bacterial sulfate reduction can neutralize acidity and precipitate dissolved metals in acid drainage waters (Ayora et al. 2016), while similar neutralization can be achieved with passive compost bioreactors in drains containing acid discharge waters (Biermann et al. 2014). Permeable reactive barriers utilize zero-valent iron, zeolite, activated carbon, and other materials to sorb or precipitate contaminants in acid discharge water, while limestone barriers can neutralize the acidity in the water (Henderson and Demond 2007). Bioreactors use salts ($CaCl_2$, K_2HPO_4, NH_4Cl, $MgSO_4.7H_2O$, $CaCO_3$, and $(NH_4)_2Fe(SO_4)_2$) as nutrient sources and to reduce the adverse effect of toxic elements. Organic components increase media permeability and are carbon sources for microbiological activity (Muhammad et al. 2017; Neculita et al. 2007; Zagury et al. 2007). Bioreactors based on mixtures of salts and organic sources provide more effective heavy metal removal than limestone reactors from acid drainage (Kiran et al. 2017; Nancucheo et al. 2017; Zhang and Wang 2016).

Chemical precipitation, ion exchange, adsorption, membrane filtration, coagulation-flocculation, flotation, and electrochemical methods were extensively tested to recover metals from acid drainage (Fu and Wang 2011; Schaffie and Hosseini 2014). Uranium (Howard et al. 2009), rare earth elements (Ayora et al. 2016), metals (Chockalingam and Subramanian 2006; Park et al. 2015), and pigments (Michalková et al. 2013) are some of the resources efficiently recovered from contaminated waters.

5.6 Conclusions

Mineralogy, the surface area of sulfide exposed, temperature, degree of saturation with water, water movement, oxygen diffusion in soil and water, pH, and microbial activity affect sulfide oxidation. Consequently, the chemistry of sulfide exposed sites is highly variable from site to site, and remediation cannot be achieved by a single intervention practice. The treatment chosen is determined by the toxic elements present, their ionic species and concentration, land use and waste produced, sulfide-bearing rock, pH, and economic factors.

References

Aghamirian M. Reactivity of sulfide minerals and its effect on gold dissolution and its electrochemical behaviour in cyanide solution. Queen's University, Kingston, Ontario, Canada; 1997.

Agricolla G. Georgius Agricola De Re Metallica. New York: Dover Publications; 1950.

Andrade RP, Santana Filho SS, Mello JWV, et al. Arsenic mobilization from sulfic materials from gold mines in Minas Gerais State. Quim Nova. 2008;31:1127–30. https://doi.org/10.1590/S0100-40422008000500037.

Ayora C, Macías F, Torres E, et al. Recovery of rare earth elements and yttrium from passive-remediation systems of acid mine drainage. Environ Sci Technol. 2016;50:8255–62. https://doi.org/10.1021/acs.est.6b02084.

Benzaazoua M, Belem T, Bussière B. Chemical factors that influence the performance of mine sulphidic paste backfill. Cem Concr Res. 2002;32:1133–44. https://doi.org/10.1016/S0008-8846(02)00752-4.

Biermann V, Lillicrap AM, Magana C, et al. Low cost passive treatment of extremely acidic and saline drainage using municipal waste organics in a compost bioreactor. Water Res. 2014;55:83–94. https://doi.org/10.1016/j.watres.2014.02.019.

Blowes DW, Ptacek CJ, Jambor JL, et al. The geochemistry of acid mine drainage. In: Holland HD, Turekian KK, editors. Treatise on geochemistry. Elsevier; 2003. p. 149–204. https://doi.org/10.1016/B0-08-043751-6/09137-4.

Borba RP, Figueiredo BR, Matschullat J. Geochemical distribution of arsenic in waters, sediments and weathered gold mineralized rocks from Iron Quadrangle, Brazil. Environ Geol. 2003;44:39–52. https://doi.org/10.1007/s00254-003-0766-5.

Bwapwa JK, Jaiyeola AT, Chetty R. Bioremediation of acid mine drainage using algae strains: a review. S Afr J Chem Eng. 2017;24:62–70. https://doi.org/10.1016/j.sajce.2017.06.005.

Campaner VP, Luiz-Silva W, Machado W. Geochemistry of acid mine drainage from a coal mining area and processes controlling metal attenuation in stream waters, southern Brazil. An Acad Bras Cienc. 2014;86:539–54. https://doi.org/10.1590/0001-37652014113712.

Candeias C, Melo R, Ávila PF, et al. Heavy metal pollution in mine-soil-plant system in S. Francisco de Assis – Panasqueira mine (Portugal). Appl Geochem. 2014;44:12–26. https://doi.org/10.1016/j.apgeochem.2013.07.009.

Chen A, Dietrich KN, Huo X, et al. Developmental neurotoxicants in e-waste: an emerging health concern. Environ Health Perspect. 2010;119:431–8. https://doi.org/10.1289/ehp.1002452.

Chockalingam E, Subramanian S. Studies on removal of metal ions and sulphate reduction using rice husk and Desulfotomaculum nigrificans with reference to remediation of acid mine drainage. Chemosphere. 2006;62:699–708. https://doi.org/10.1016/j.chemosphere.2005.05.013.

Dawson PA, Elliott A, Bowling FG. Sulphate in pregnancy. Nutrients. 2015;7:1594–606. https://doi.org/10.3390/nu7031594.

Dold B. Basic concepts in environmental geochemistry of sulfidic mine-waste management. In: Kumar ES, editor. Waste management; 2010. p. 172–98. https://doi.org/10.5772/8458.

Dold B. Evolution of acid mine drainage formation in sulphidic mine tailings. Fortschr Mineral. 2014;4:621–41. https://doi.org/10.3390/min4030621.

Dold B, Aguilera A, Cisternas ME, et al. Sources for iron cycling in the southern ocean. Environ Sci Technol. 2013;47:6129–36. https://doi.org/10.1021/es305141b.

Durães N, Silva EF, Bobos I, et al. Rare earth elements fractionation in native vegetation from the Moncorvo iron mines, NE Portugal. Procedia Earth Planet Sci. 2014;10:376–82. https://doi.org/10.1016/j.proeps.2014.08.064.

Dürr HH, Meybeck M, Dürr SH. Lithologic composition of the Earth's continental surfaces derived from a new digital map emphasizing riverine material transfer. Global Biogeochem Cycles. 2005;19:n/a. https://doi.org/10.1029/2005GB002515.

Equeenuddin SM, Tripathy S, Sahoo PK, et al. Hydrogeochemical characteristics of acid mine drainage and water pollution at Makum Coalfield. India J Geochem Explor. 2010;105:75–82. https://doi.org/10.1016/j.gexplo.2010.04.006.

Evangelou VP. Pyrite oxidation and its control. Boca Raton: CRC Press; 1995.

Fanning DS, Bigham MC, Rabenhorst JM. Colors of acid sulfate soils. In: Bigham JM, Ciolkosz EJ, editors. Soil color. Madison: Soil Science Society of America; 1993. p. 91–108. https://doi.org/10.2136/sssaspecpub31.c6.

Faure G. Principles and applications of inorganic geochemistry. New York: Macmillan Publishing Company; 1991.

Ferrari CR, de Azevedo H, Rodgher S, et al. Effects of the discharge of uranium mining effluents on the water quality of the reservoir: an integrative chemical and ecotoxicological assessment. Sci Rep. 2017;7:1–10. https://doi.org/10.1038/s41598-017-14100-w.

Francelino MR, Schaefer CEGR, Simas FNB, et al. Geomorphology and soils distribution under paraglacial conditions in an ice-free area of Admiralty Bay, King George Island, Antarctica. Catena. 2011;85:194–204. https://doi.org/10.1016/j.catena.2010.12.007.

Fu F, Wang Q. Removal of heavy metal ions from wastewaters: a review. J Environ Manag. 2011;92:407–18. https://doi.org/10.1016/j.jenvman.2010.11.011.

Halcomb M, Fare D Soil pH explained. Agricultural Extension Service, The University of Tennessee; 2002. Disponível em: http://www.utextension.utk.

He F, Jiang W, Tang T, et al. Assessing impact of acid mine drainage on benthic macroinvertebrates: can functional diversity metrics be used as indicators? J Freshw Ecol. 2015;30:513–24. https://doi.org/10.1080/02705060.2014.998730.

Hemingway BS, Seal RR, Chou IM. Thermodynamic data for modeling acid mine drainage problems: compilation and estimation of data for selected soluble iron-sulfate minerals, USGS numbered series; 2002. https://doi.org/10.3133/ofr02161.

Henderson AD, Demond AH. Long-term performance of zero-valent iron permeable reactive barriers: a critical review. Environ Eng Sci. 2007;24:401–23. https://doi.org/10.1089/ees.2006.0071.

Hodson A, Nowak A, Sabacka M, et al. Climatically sensitive transfer of iron to maritime Antarctic ecosystems by surface runoff. Nat Commun. 2017;8:1–7. https://doi.org/10.1038/ncomms14499.

Holmes RR, Hart ML, Kevern JT. Heavy metal removal capacity of individual components of permeable reactive concrete. J Contam Hydrol. 2017;196:52–61. https://doi.org/10.1016/j.jconhyd.2016.12.005.

Howard D, Grobler C, Robinson REG et al. Sustainable purification of mine water using ion exchange technology. In: International mine water conference. Pretoria, South Africa; 2009. p. 447–53.

Iakovleva E, Mäkilä E, Salonen J, et al. Acid mine drainage (AMD) treatment: neutralization and toxic elements removal with unmodified and modified limestone. Ecol Eng. 2015;81:30–40. https://doi.org/10.1016/j.ecoleng.2015.04.046.

International Network for Acid Prevention. Global acid rock drainage guide [WWW document]. 2009. http://www.gardguide.com/index.php?title=Main_Page.

IUSS Working Group WRB. World reference base for soil resources 2014, update 2015 international soil classification system for naming soils and creating legends for soil maps, world soil resources reports. Rome: FAO; 2015. https://doi.org/10.1017/S0014479706394902.

Jennings SR, Blicker S, Neuman DR, et al. Acid mine drainage and effects on fish health and ecology: a review. Reclam Res Gr. 2008;1:1–26.

Jones SN, Cetin B. Evaluation of waste materials for acid mine drainage remediation. Fuel. 2017;188:294–309. https://doi.org/10.1016/j.fuel.2016.10.018.

Kastyuchik A, Karam A, Aïder M. Effectiveness of alkaline amendments in acid mine drainage remediation. Environ Technol Innov. 2016;6:49–59. https://doi.org/10.1016/j.eti.2016.06.001.

Kefeni KK, Msagati TAM, Mamba BB. Acid mine drainage: prevention, treatment options, and resource recovery: a review. J Clean Prod. 2017;151:475–93. https://doi.org/10.1016/j.jclepro.2017.03.082.

Kim MS, Min HG, Koo N, et al. The effectiveness of spent coffee grounds and its biochar on the amelioration of heavy metals-contaminated water and soil using chemical and biological assessments. J Environ Manage. 2014;146:124–30. https://doi.org/10.1016/j.jenvman.2014.07.001.

Kiran MG, Pakshirajan K, Das G. Heavy metal removal from multicomponent system by sulfate reducing bacteria: mechanism and cell surface characterization. J Hazard Mater. 2017;324:62–70. https://doi.org/10.1016/j.jhazmat.2015.12.042.

Kissel DE, Sonon LS, Cabrera ML. Rapid measurement of soil pH buffering capacity. Soil Sci Soc Am J. 2012;76:694. https://doi.org/10.2136/sssaj2011.0091.

Kristensen E, Bouillon S, Dittmar T, et al. Organic carbon dynamics in mangrove ecosystems: a review. Aquat Bot. 2008;89:201–19. https://doi.org/10.1016/j.aquabot.2007.12.005.

Kuyucak N. Acid mine drainage prevetion and control options. Can Inst Metall Bull. 2002;95:96–102.

Larsson M, Nosrati A, Kaur S, et al. Copper removal from acid mine drainage-polluted water using glutaraldehyde-polyethyleneimine modified diatomaceous earth particles. Heliyon. 2018;4:e00520. https://doi.org/10.1016/j.heliyon.2018.e00520.

Lei L, Song C, Xie X, et al. Acid mine drainage and heavy metal contamination in groundwater of metal sulfide mine at arid territory (BS mine, Western Australia). Trans Nonferrous Met Soc China. 2010;20:1488–93. https://doi.org/10.1016/S1003-6326(09)60326-5.

Li YT, Becquer T, Dai J, et al. Ion activity and distribution of heavy metals in acid mine drainage polluted subtropical soils. Environ Pollut. 2009;157:1249–57. https://doi.org/10.1016/j.envpol.2008.11.050.

Li QS, Chen Y, Fu HB, et al. Health risk of heavy metals in food crops grown on reclaimed tidal flat soil in the Pearl River Estuary. China J Hazard Mater. 2012;227–228:148–54. https://doi.org/10.1016/j.jhazmat.2012.05.023.

Liao J, Wen Z, Ru X, et al. Distribution and migration of heavy metals in soil and crops affected by acid mine drainage: public health implications in Guangdong Province, China. Ecotoxicol Environ Saf. 2016;124:460–9. https://doi.org/10.1016/j.ecoenv.2015.11.023.

Lu J, Alakangas L, Wanhainen C. Metal mobilization under alkaline conditions in ash-covered tailings. J Environ Manage. 2014;139:38–49. https://doi.org/10.1016/j.jenvman.2013.12.036.

Macías F, Caraballo MA, Rötting TS, et al. From highly polluted Zn-rich acid mine drainage to non-metallic waters: implementation of a multi-step alkaline passive treatment system to remediate metal pollution. Sci Total Environ. 2012;433:323–30. https://doi.org/10.1016/j.scitotenv.2012.06.084.

Masindi V, Gitari WM. Simultaneous removal of metal species from acidic aqueous solutions using cryptocrystalline magnesite/bentonite clay composite: an experimental and modelling approach. J Clean Prod. 2016;112:1077–85. https://doi.org/10.1016/j.jclepro.2015.07.128.

Mees F, Stoops G. Sulphidic and sulphuric materials. In: Stoops G, Marcelino V, Mees F, editors. Interpretation of micromorphological features of soils and regoliths. Amsterdam: Elsevier; 2010. p. 543–68. https://doi.org/10.1016/B978-0-444-53156-8.00024-6.

Mello JWV, Dias LE, Daniel AM, et al. Preliminary evaluation of acid mine drainage in Minas Gerais State, Brazil. Rev Bras Ciência do Solo. 2006;30:365–75. https://doi.org/10.1590/S0100-06832006000200016.

Michalková E, Schwarz M, Pulišová P, et al. Metals recovery from acid mine drainage and possibilities for their utilization. Polish J Environ Stud. 2013;22:1111–8.

Muhammad SN, Kusin FM, Zahar M, et al. Passive bioremediation technology incorporating lignocellulosic spent mushroom compost and limestone for metal- and sulfate-rich acid mine drainage. Environ Technol. 2017;38:2003–12. https://doi.org/10.1080/09593330.2016.1244568.

Nancucheo I, Bitencourt JAP, Sahoo PK, et al. Recent developments for remediating acidic mine waters using sulfidogenic bacteria. Biomed Res Int. 2017:2017. https://doi.org/10.1155/2017/7256582.

Neculita CM, Zagury GJ, Bussière B. Passive treatment of acid mine drainage in bioreactors using sulfate-reducing bacteria. J Environ Qual. 2007;36:1. https://doi.org/10.2134/jeq2006.0066.

Nordstrom DK. Hydrogeochemical and microbiological factors affecting the heavy metal chemistry of an acid mine drainage system. Standford University; 1977.

Olds W, Weber P, Pizey M. Alkalinity producing covers for minimisation of acid mine drainage generation in waste rock dumps. In: Proceedings of the Eighth Australian Workshop on Acid and Metalliferous Drainage; 2012. p. 253–62.

Park SM, Yoo JC, Ji SW, et al. Selective recovery of dissolved Fe, Al, Cu, and Zn in acid mine drainage based on modeling to predict precipitation pH. Environ. Sci Pollut Res. 2015;22:3013–22. https://doi.org/10.1007/s11356-014-3536-x.

Rickard D. How long does it take a pyrite framboid to form ? Earth Planet Sci Lett. 2019;513:64–8. https://doi.org/10.1016/j.epsl.2019.02.019.

Rowe OF, Sánchez-España J, Hallberg KB, et al. Microbial communities and geochemical dynamics in an extremely acidic, metal-rich stream at an abandoned sulfide mine (Huelva, Spain) underpinned by two functional primary production systems. Environ Microbiol. 2007;9:1761–71. https://doi.org/10.1111/j.1462-2920.2007.01294.x.

Sainz A, Grande J, de la Torre M. Odiel River, acid mine drainage and current characterisation by means of univariate analysis. Environ Int. 2003;29:51–9. https://doi.org/10.1016/S0160-4120(03)00006-0.

Santos HGS, Jacomine PK, Anjos LHC, et al. Sistema brasileiro de classificação de solos. Rio de Janeiro: Embrapa; 2013. ISBN 978-85-7035-198-2.

Sato M. Oxidation of sulfide orebodies: oxidation mechanisms of sulfide minerals at 25 °C. Econ Geol. 1960;55:1202–31.

Schaffie M, Hosseini MR. Biological process for synthesis of semiconductor copper sulfide nanoparticle from mine wastewaters. J Environ Chem Eng. 2014;2:386–91. https://doi.org/10.1016/j.jece.2014.01.006.

Schwertmann U. Solubility and dissolution of iron oxides. Plant Soil. 1991;130:1–25.

Shabalala AN, Ekolu SO, Diop S, et al. Pervious concrete reactive barrier for removal of heavy metals from acid mine drainage – column study. J Hazard Mater. 2017;323:641–53. https://doi.org/10.1016/j.jhazmat.2016.10.027.

Simate GS. Bacterial leaching of nickel laterites using chemolithotrophic microorganisms. University of the Witwatersrand; 2009. https://doi.org/10.1016/j.hydromet.2009.05.007.

Simate GS, Ndlovu S. Acid mine drainage: challenges and opportunities. J Environ Chem Eng. 2014;2:1785–803. https://doi.org/10.1016/j.jece.2014.07.021.

Singer PC, Stumm W. Acidic mine drainage: the rate-determinig step. Science. 1970;167:1121–3.

Singh R, Gautam N, Mishra A, et al. Heavy metals and living systems: an overview. Indian J Pharmacol. 2011;43:246. https://doi.org/10.4103/0253-7613.81505.

Skousen J, Simmons J, Ziemkiewicz P. The use of acid-base accounting to predict post-mining drainage quality on West Virginia surface mines. J Am Soc Min Reclam. 2001;2001:437–47. https://doi.org/10.21000/JASMR01010437.

Skousen J, Simmons J, McDonald LM, et al. Acid–base accounting to predict post-mining drainage quality on surface mines. J Environ Qual. 2002;31:2034. https://doi.org/10.2134/jeq2002.2034.

Soil Classification Working Group. The Canadian system of soil classification. 3rd ed. Ottawa: NRC Research Press; 1998.

Soil Survey Staff. Keys to soil taxonomy. 9th ed. Washington, DC: US Department of Agriculture, NRCS; 2006. p. 281.

The National Committee on Soil and Terrain. The Australian soil classification. 2nd ed. Clayton: CSIRO Publishing; 2016.

Vasques ICF, Mello JWV, Veloso RW, et al. Arsenite removal from contaminated water by precipitation of aluminum, ferrous and ferric (hydr)oxides. Environ Sci Pollut Res. 2018;25:12967–80. https://doi.org/10.1007/s11356-018-1458-8.

Villain L, Alakangas L, Öhlander B. The effects of backfilling and sealing the waste rock on water quality at the Kimheden open-pit mine, northern Sweden. J Geochem Explor. 2013;134:99–110. https://doi.org/10.1016/j.gexplo.2013.08.003.

Wada H, Seisuwan B. The process of pyrite formation in mangrove soils. In: Dorst H, editor. Selected papers of the Dakar symposium on acid sulphate soils. ILRI: Wageningen; 1986. p. 24–37.

Wang X, Tang C, Mahony S, et al. Factors affecting the measurement of soil pH buffer capacity: approaches to optimize the methods. Eur J Soil Sci. 2015;66:53–64. https://doi.org/10.1111/ejss.12195.

Wang Y, Pleasant S, Jain P, et al. Calcium carbonate-based permeable reactive barriers for iron and manganese groundwater remediation at landfills. Waste Manag. 2016;53:128–35. https://doi.org/10.1016/j.wasman.2016.02.018.

Wang L, Ji B, Hu Y, et al. A review on in situ phytoremediation of mine tailings. Chemosphere. 2017;184:594–600. https://doi.org/10.1016/j.chemosphere.2017.06.025.

Watten BJ, Sibrell PL, Schwartz MF. Acid neutralization within limestone sand reactors receiving coal mine drainage. Environ Pollut. 2005;137:295–304. https://doi.org/10.1016/j.envpol.2005.01.026.

Wilkin RT, Barnes HL. Formation processes of framboidal pyrite. Geochim Cosmochim Acta. 1997;61:323–39.
Yadav SK. Heavy metals toxicity in plants: an overview on the role of glutathione and phytochelatins in heavy metal stress tolerance of plants. S Afr J Bot. 2010;76:167–79. https://doi.org/10.1016/j.sajb.2009.10.007.
Yadav HL, Jamal A. Removal of heavy metals from acid mine drainage: a review. Int J New Technol Sci Eng. 2015;2:77–84. https://doi.org/10.1016/j.jhazmat.2007.11.123.
Zagury GJ, Neculita C, Management MW. Passive treatment of acid mine drainage in bioreactors: short review, applications, and research needs. In: Applications, and research needs. Ottawa Geo; 2007, p. 1439–1446. https://doi.org/10.2134/jeq2006.0066.
Zhang M, Wang H. Preparation of immobilized sulfate reducing bacteria (SRB) granules for effective bioremediation of acid mine drainage and bacterial community analysis. Miner Eng. 2016;92:63–71. https://doi.org/10.1016/j.mineng.2016.02.008.
Zhao K, Liu X, Zhang W, et al. Spatial dependence and bioavailability of metal fractions in paddy fields on metal concentrations in rice grain at a regional scale. J Soils Sediments. 2011;11:1165–77. https://doi.org/10.1007/s11368-011-0408-6.
Zhi E, Song Y, Duan L, et al. Spatial distribution and diversity of microbial community in large-scale constructed wetland of the Liao River Conservation Area. Environ Earth Sci. 2015;73:5085–94. https://doi.org/10.1007/s12665-015-4021-7.
Zhou JM, Dang Z, Cai M-F, et al. Soil heavy metal pollution around the Dabaoshan Mine, Guangdong Province, China. Pedosphere. 2007;17:588–94. https://doi.org/10.1016/S1002-0160(07)60069-1.

Chapter 6
Physical Subsoil Constraints of Agricultural and Forestry Land

Teogenes Senna de Oliveira and Raphael Bragança Alves Fernandes

Abstract Physical subsoil constraints can originate from natural or anthropic processes, and some of them can be reduced, avoided, corrected and/or predicted by the land use practices applied. While the solutions may not be economically viable in cases of hardened layers (e.g. duripan, fragipan), shallow rock layers or water tables, the elimination of compacted layers originating from soil management practices in areas of agricultural potential is generally feasible. Soil compaction is the most reported physical subsoil constraint worldwide and can affect plant productivity by reducing the root system growth and the soil volume explored for the absorption of water and nutrients, as well as inducing soil surface erosion depending on the local physical environment. This chapter explores the compaction processes and their genesis, factors affecting susceptibility to compaction, response of compaction to land use and the prediction, prevention and correction of compaction.

Keywords Compaction · Diagnosis · Ecosystem services · Mechanization · No-tillage · Soil management · Susceptibility · Tillage

6.1 Introduction

The soil has multiple functions that play a fundamental role in the environmental balance of the planet, but more attention is paid to the topsoil rather than the horizons below (subsoil). Topsoil and subsoil are a single entity, allowing the soils to perform their ecosystem functions, especially under agricultural land use. The subsoil constraints can be due to physical, chemical and biological factors and originate from natural or anthropic actions. All these factors have the potential to compromise land use and plant development, as well as all other soil functions depending on the contrast between the topsoil and the subsoil. Specifically, subsoil physical

T. S. de Oliveira (✉) · R. B. A. Fernandes
Soils Department, Federal University of Viçosa, Viçosa, MG, Brazil
e-mail: teo@ufv.br; raphael@ufv.br

T. S. d. Oliveira, R. W. Bell (eds.), *Subsoil Constraints for Crop Production*,
https://doi.org/10.1007/978-3-031-00317-2_6

constraints can have a direct impact on water and gases dynamics and the development of plant roots in the soil profile.

The most common natural subsoil physical constraints are associated with the presence of rocky material and/or gravel and hardened horizons (fragipan, duripan or cohesive horizons). Clay-textured horizons (textural B) can also be a natural subsoil constraint and is a common characteristic in soil classes of Acrisols, Luvisols and Leptosols in Brazil and in Sodosol and Chromosol soils in Australia (MacEwan et al. 2010; Davies et al. 2019; Isbell and National Committee and Soil and Terrain 2021). Schaefer et al. (2022) outline how subsoil constraints are often associated with deeply weathered and deep saprolite-derived soils, and these are particularly prevalent in tropical climates or in regions that experienced prolonged chemical weathering in earlier geological periods. Their chemical, physical and mineralogical properties reflect prolonged weathering and chemical denudation, and the nature of saprolites has a definite impact on soil constituents, irrespective of bioclimatic conditions, including subsoil physical constraints. On the other hand, land use practice in crop production, livestock raising and forestry can also cause physical restrictions in the subsoil due to machine traffic for intensive soil tillage in conventional planting, soil preparation and cultivation/harvesting operations at unsuitable soil water content, overgrazing and limited adoption of practices that maintain or increase soil organic matter.

Soil compaction is the most reported physical subsoil constraint worldwide, and its origin is frequently associated with unsuitable land use practices and soil management. The greater use of heavier machinery in annual and perennial crop production, worldwide, is a matter of concern (Lipiec et al. 2003; Wahlströma et al. 2020), and its direct consequences on soil functions and crop productivity were reviewed by Keller et al. (2019). Indirect effects of subsoil compaction include waterlogging which could happen in soils with natural physical limitations (clay enrichment at depth as in duplex soils in Australia or Inceptisols in Brazil and Australia) or induced as subsoil compaction by machinery use (Manik et al. 2019).

Although subsoil compaction is an important constraint for agricultural production, it is often overlooked and was referred to by Frisby and Pfost (2020) as "the silent thief". This chapter will address the susceptibility to subsoil compaction under different soil conditions, the effect and limitations caused by soil compaction under agriculture (conventional and no-till) and forestry activities, the methods to evaluate compaction and existing practices to manage and prevent compacted soil layers forming.

6.2 Soil Compaction and Land Use

Conceptually, induced soil compaction can be considered as the compression process (increase in soil mass per unit volume) caused by soil use and management practices, through the application of external forces that cause the reduction of pore space, a denser rearrangement of solid particles and, consequently, an increase in soil bulk density (Håkansson and Lipiec 2000). The physically compacted layers restrict root penetration, gas exchange and infiltration of water, in addition to increasing the susceptibility of soil to erosion (Raper 2005).

The breakdown of soil aggregates during conventional tillage facilitates the rearrangement of particles/aggregates into smaller spaces, promoting a decrease in volume, which is intensified by the compression exerted by the weight of the machinery and the exposure of the soil to the direct impact of rain drops (Ampoorter et al. 2007). These effects are less evident in the topsoil since the layer is commonly tilled annually, but compaction stands out in a subsurface layer in the form of dense layers coinciding with the transition zone between the disturbed and undisturbed part of the soil. Management, harvesting and transportation operations in agricultural and forest areas also cause compaction, due to uncontrolled heavy machinery traffic.

The spatial distribution of soil compaction and with depth down the profile is quite variable. In conventional agricultural systems, there is a tendency for compaction to occur superficially in the profile and homogeneously in a field, due to uniform trafficking and to ploughing in the first 30 cm of soil (Reichert et al. 2007). On the other hand, in forest soils, compaction is heterogeneous spatially (Greacen and Sands 1980; Horn et al. 2004; Reichert et al. 2007), and its intensity depends, among other factors, on the type of machine used, localized traffic, soil moisture and root growth (Arvidsson et al. 2001).

Although more limited, the effect of roots on soil compression cannot be disregarded. The growth of tree roots in forest and agricultural areas (e.g. fruit cultivation) promotes a natural increase in soil density in the rhizosphere, decreasing exponentially with the distance from the surface of the roots (Dexter 1987). Increases in soil density can be significant as a result of root elongation only, with the absolute values comparable to the compaction induced by machine traffic (Aravena et al. 2014). Increases in soil density and reduced porosity were verified in electron microscope scanning images of a root section of the corn crop that showed a 22–24% reduction in soil porosity in the layer close to the soil-root interface (0–180 μm) and an increase in soil density from 1.56 kg dm^{-3} between 1440 and 1500 μm from the root to 1.79 kg dm^{-3} at the interface (0–60 μm) (Bruand et al. 1996). However, the root compacting effect at the soil-root interface is outweighed by the beneficial effects of roots on soils. The roots cause cracks in soils with their expansion, promoting the formation of biopores (Gaiser et al. 2013), incorporate organic matter in the profile and release exudates which are important for the formation and stability of aggregates, improving the subsoil physical conditions.

Compaction compromises soil ecosystem services by reducing soil porosity and permeability, increasing mechanical resistance to root penetration and by degradation of soil structure (van den Akker and Soane 2004). These impacts affect drainage and water retention in the soil, gas exchange, absorption of nutrients and the extension of roots, compromising or limiting the growth and production of plants (Ma et al. 2022; Vance and Milroy 2022). Soil conservation itself is also affected by compaction (Thorsøe et al. 2019). Therefore, soil management practices should be designed based on the premise of avoidance of compaction (van den Akker and Schjønning 2004). Hamza and Anderson (2005) listed practical techniques which have emerged on how to avoid, delay or prevent soil compaction: (*i*) reducing pressure on soil by either decreasing axle load or increasing the contact area of wheels with the soil; (*ii*) working soil and allowing grazing at optimal soil moisture; (*iii*) reducing the number of passes by farm machinery and the intensity and frequency of grazing; (*iv*) confining traffic to certain areas of the field (controlled traffic); (*v*) increasing soil organic matter through retention of crop and pasture residues; (*vi*) removing soil compaction by deep ripping in the presence of an aggregating agent; (*vii*) crop rotations that include plants with deep, strong taproots; and (*viii*) maintenance of an appropriate base saturation ratio and complete nutrition to meet crop requirements to help the soil/crop system to resist harmful external stresses. In the following sections, we review the origins of soil compaction, its effects and amelioration techniques.

6.2.1 Conventional and No-Tillage Farming

Worldwide, agricultural activities are reported to promote the formation of compacted soil layers for both annual and perennial crops. The compacted layers are often present below the arable layer with conventional tillage practices. In the no-till systems, the compacted layer will be more superficial due to the limited soil disturbance, while maintenance of the crop residue on the surface increases soil organic matter content close to the soil surface. The increase in soil density in tilled soils is associated with the rearrangement and compression of particles and soil aggregates due to the pressure exerted by machines and implements, both on the surface (upturned layer) and in the subsoil (interface between the unturned and upturned layer).

The alteration of electrochemical soil properties with fertilizers and soil ameliorants must also be considered in the process of soil compaction, especially in dominant oxidic soils in a tropical environment. The increase in soil pH after liming increases surface negative charges due to the loss of protons from functional groups on the surface of organic matter and the clay fraction (silicates and oxides of Fe and Al) or by the replacement of trivalent cations (Al^{3+}) with bivalent cations (Ca^{2+} and Mg^{2+}). The increase in negative changes can favour the dispersion of particles. On the other hand, at higher lime rates, Ca^{2+} concentrations and ionic strength in soil solution increase causing compression of the electrical double layer and renewed flocculation. Both lime and hydroxy-Al polymers, formed by precipitation of exchangeable Al, can act as cementing agents bonding soil particles together and improving soil structure (Haynes and Naidu 1998). So, the liming effect on soil aggregation is unclear, and the intensity of this effect on soil compaction at field level is not quantified yet. The dispersion of particles in tropical soils in Brazil was observed with higher liming rates favouring higher soil density, clay dispersion and tortuosity of pores, and total porosity, saturated hydraulic conductivity and fractal dimension for 1–2 mm aggregates decreased, while larger pores were replaced by smaller ones (in 1–2 mm aggregates) (Spera et al. 2008; Ferreira et al. 2019).

The level of soil compaction increases with time and with intensity of production, even in areas with less intensive uses, such as pastures and under no-tillage. The diagnosis and the correction of soil compaction should become a routine practice, especially in more intensive land use systems such as frequent short-cycle crops under irrigation and where tillage is part of the production process (production of roots, tubers, bulbs, etc.).

Agricultural intensification is exacerbating soil compaction since mechanization in agriculture and forestry is expanding due to the efficiency gains in farm enterprises. The intensification of mechanization in agricultural and forestry use has invariably been associated with an increase in the size and mass of machines and implements, to enhance gains in productivity and profitability, but without due concern for the potential effects on the soil physical quality. In the last four decades, there was a fourfold increase in applied loads and in the frequency of machinery moving on the ground (Horn 2015).

Even in the cultivation of perennial plants, where mechanized operations are concentrated between the planting lines, compaction is a growing concern due to limitations to the development of the root system. Petry et al. (2016) found that wheel trafficking by tractors and implements (especially sprayers) caused soil compaction between the lines, limiting the lateral distribution of peach roots by up to 60%. This limitation was most associated with the central strip between the lines (1.7 m from each line of trees), where the movement of machines occurs and where resistance to penetration exceeded 2 MPa which limits root elongation of most plants (Gregory et al. 2006): root restrictions were most pronounced in the 5–30 cm layer. Furthermore, the compaction occurred across virtually the entire central portion of the orchard inter-rows since tractor traffic coupled with a range of implements with varied wheel-to-wheel distance results in trafficking of most of the inter-row space. Similar results were reported by van Dijck and van Asch (2002) who assessed the extent of soil compaction of mature orchards and vineyards. However, in other studies, despite the mechanical traffic between the lines, the soil physical quality was maintained, highlighting the differences between cultivation systems, crops and soil classes. Indeed, Sandoval et al. (2020) observed an improvement in stability and an increase in the size of soil aggregates in the inter-rows of coffee growing areas after 32 years.

The increase in the frequency of machine traffic in crops increases the risks of compaction and the possibility of reaching critical limits for plant growth (Ampoorter et al. 2007; Jaafari et al. 2014). The indiscriminate trafficking by machines and implements even in areas of no-till highlights the urgent need to decrease and control the traffic of machines and implements as much as possible. While controlled traffic is increasingly adopted in Australian cropping systems, it is not common in Brazil (https://www.actfa.net/actfa-conferences/).

The implementation of no-till systems has been advocated to improve soil physical quality by reducing mechanical disturbance of soils and by increasing the retention of crop residue on the soil surface. However, even in no-till areas that are widely adopted globally, especially in Brazil and Australia, studies have indicated the frequent presence of compacted layers and with restriction to root growth (Reinert et al. 2008). Soil compaction has been considered one of the most important barriers to soil management under no-tillage systems in the world (Spera et al. 2018). One of the reasons for soil compaction in no-till areas in Brazil is that minimum soil disturbance is the dominant practice, but crop rotation, the use of winter crops and the maintenance of crop residue cover throughout the year are not always practiced due to costs or climatic limitations, or even low acceptance of these practices by farmers.

In no-till areas in Brazil, the effects of compacted soils are more serious in non-irrigated areas during years with less rain, when the mechanical resistance of the soil to root penetration is increased. In the long term, no-tillage has been associated with gains in aggregate stability but often also with compromised soil physical quality (Martínez et al. 2008). Alternatives to tackle these problems in areas of continuous adoption of no-till include crops with different root systems and the occasional practice of scarification or even ploughing (Peixoto et al. 2019b). The use of seeders equipped with double discs reaching a depth of 7 cm or double disks + short ripper reaching a depth of 13 cm was also evaluated to decrease soil compaction in long-term no-till systems (Drescher et al. 2012). However, the effects of applying these practices, which can be costly, have not provided permanent reduction in soil compaction and crop yield gains, as they have only short-term effects.

Where soil compaction persists despite the adoption of measures to mitigate trafficking, deep subsoiling or more superficial action implements (harrows, knife rollers, scarifiers and cultivators) may be used, but these may simply transfer the compaction to deeper layers beyond the range of these implements or to the interface between the disturbed and undisturbed soil layers. The presence of compacted layers at depth may not be easily perceived by farmers. This is more problematic in areas of intensive and irrigated production, where there is no expression of restricted root development in the form of crop water stress. Negative effects of the deep compaction on nutrient uptake may go unrecognized. However, there is limited direct evidence yet in the literature about this issue.

Irrigation alone can be related to loss of structural soil quality, especially when poor quality irrigation water causes colloidal dispersion and destabilization of soil structure (Rengasamy 2018). For example, in the semi-arid region of Brazil after 23 years of continuous use of irrigation, the repetitive wetting and drying cycles provided by irrigation have improved the stability of soil aggregates but caused an increase in microporosity and a decrease in the air permeability of soil, leading to the formation of a less well-connected pore system (Costa et al. 2014). The concern with compaction in irrigated areas is greater in climatic regions which allow cropping system intensification and the production of year-round cropping.

6.2.2 Livestock/Pasture

Pasture covers 45% of the Brazilian territory (IBGE 2017) and 52% of Australia (Bell et al. 2014), and in both countries, livestock production makes up about 40% of the agricultural sector's gross domestic product (GDP) (Australian Bureau of Statistics 2012). In pasture areas, compaction is more prevalent in the first millimetres of soil (~50 mm), mainly due to the pressure exerted by the animals' hooves (Bilotta et al. 2007) at high stocking rates. The decrease of soil physical conditions such as less porosity and water infiltration is associated with reduced productivity and vegetation cover (Bell et al. 2014), and trampling by stock at rates above the recommended levels intensifies soil surface erosion and its degradation. The effect of machines and implements is also reported for pastures, although pasture is considered a minimum trafficking and soil disturbance land use compared to cropping. The reduction of superficial compaction can be achieved by shallow subsoiling or soil scarification to eliminate the physical surface restrictions in association with the application of fertilizers and correction of acidity. Alternatively, these measures and the adjustment of stocking rates can be combined with the planting of trees in pastures (Wolfe 2020). Bilotta et al. (2007) discuss a set of mitigation and remediation measures for pastures that can promote 33 ecosystem services, with the most frequent being C sequestration, forage production and water erosion control (Zhao et al. 2020). Such measures are particularly needed in the pastures of the Brazilian Atlantic Forest biome to reverse the state of degradation. The inadequate past management of pastures on undulating land together with excessive trampling and the nutritional poverty of deep weathered soils led to drastic consequences for soil quality in this environment. In addition, for decades burning the grasses in the dry period to provide forage of better nutritional quality was a common practice before the rainy season. Intensive erosive resulted from the combination of heavy rainfall and concentration of water in the paths formed by the animals and led to the formation of rills or even gullies. Soil erosion in pastures has led to soil degradation, subsoil exposure and negative effects on water quality at the basin level. Although soil conservation practices (e.g. introducing trees in pastures) are necessary to interrupt the degrading practices, improving chemical and physical conditions such as nutrient replacement, reducing acidity and reducing compaction, the creation and adoption of public policies that enable integrated actions to recover these pastures are also important.

6.2.3 Planted Forests

Plantation forestry in Brazil has been marked by great advances in productivity and innovation in recent decades. Part of these productivity gains in Brazil and the world is associated with shorter rotations and the increased use of mechanization in the field.

In the forestry sector, the heavy machinery traffic starts at the soil preparation stage and continues until the harvest of the trees and their transport outside the area, with numerous machine operations in the field in between. The use of mechanized forest harvesting systems can cause a series of disturbances in the soil physical quality (Ares et al. 2005). Harvesting is considered the most critical stage, when there is greater pressure exerted by machinery, in addition to tipping, tree dragging and storage in the field, which increases the risks of soil compaction (Marchi et al. 2016). Among the disturbances, soil compaction is considered in the forestry sector a serious trigger of disturbances already mentioned for agricultural areas, affecting the quality, the biodiversity, the emission of greenhouse gases and the soil carbon stocks, as well as forest growth and regeneration (Cambi et al. 2015). Soil compaction in forest areas is easily observed visually at the time of field operations and includes the formation of deep furrows by the machinery wheels, superficial soil disturbance, the formation of water puddles and the local deposition of plant residues (Agherkakli et al. 2010).

The effects of forest harvesting on soil are dependent on the type of harvest, which determines the presence or absence of residues on the soil surface and the soil moisture at the time of the operation. Mechanical forest harvesting systems can be divided into three categories: (*i*) short log system (cut-to-length), in which the complete processing of logs is done in the stand and the resulting residues are left on the soil surface; (*ii*) the system of long logs (tree-length), in which the processing of logs is partly done in the stand and part outside, after dragging to remove the trees; and (*iii*) the full-tree system, when the complete removal of the tree is done for full processing outside the stand (Rodrigues 2018). Depending on the system applied, heavier machines have more negative consequences for soil compaction, especially when transporting the wood out of the field by complete dragging of the log (full-tree).

Among the adopted harvest procedures, the cut-to-length has been shown to have less impact on the physical quality of the soil, by leaving residues on the surface that minimize the effects of compaction. The attenuating effect of organic waste on compaction is due to its low density and high susceptibility to elastic deformations, forming a protective layer that reduces the contact and helps to dissipate the pressure applied by the machinery, reducing the risks of compaction (Reichert et al. 2007). Therefore, the use of the harvester + forwarder module of the cut-to-length system has advantages over the tree-length feller + skidder module, as it promotes greater accumulation of residues when performing all the processing of the logs within the field.

The maintenance of forest residues may affect the re-planting by reducing the depth of action of the implements during soil preparation (Oliveira 2018). However, the gains with the residues left in the soil outweigh the disadvantages, and this practice is widespread. Nevertheless, this practice is facing a new challenge due to the demand for this organic material as a bioenergy source. The maintenance of residues in the field, especially from the bark of the trees, has economic and environmental benefits, which attenuate the reduction in productivity in mechanically harvested forest plantations (Jesus et al. 2015).

The effects of subsoil compaction are persistent, if not permanent on forestry (van den Akker and Schjønning 2004; Cambi et al. 2017). During forest harvest, only two passes of harvester + forwarder can cause soil compaction, an effect that lasted for more than 441 days in a Ferralsol in Brazil (Silva et al. 2008). Longer lasting effects of compaction on forest soils have also been reported by Hattori et al. (2013) in Malaysia. These authors evaluated the growth of seedlings planted in a logging area more than 20 years after the forest harvest and found an increase in the mortality rate in the initial phase of seedling development (0–12 months) and a reduction in elongation (81 months) of roots with increased soil resistance to penetration.

The intensity of damage to the structure depends on the external and internal factors that govern compressibility in unsaturated soils (Ampoorter et al. 2007). The external factors are related to the characteristics of the applied force and the topography of the area, while the internal factors are initial soil density, texture, structure, contents of organic matter, mineralogy and current soil water content (Bygdén et al. 2003; Dexter 2004). It is important to emphasize that soil compressibility is also directly associated with the ability of soil aggregates to withstand pressure without breaking (Han et al. 2006). Therefore, the cementing agents within aggregates, such as organic matter and Fe and Al oxides, directly influence compaction. Soil organic matter can be considered as the main cementing agent in soils in temperate regions, mainly in the superficial layers (Abdollahi et al. 2014). However, in tropical soils, Fe and Al oxides assume greater importance in cementing aggregates, helping to form aggregates with high resistance to compaction (Ferreira et al. 2007). Fine-textured soils are more susceptible to compaction when compared to coarse-textured soils (Wästerlund 1985), especially as they retain a higher volume of water, which acts as a lubricant between particles, facilitating the denser resettlement of soil.

The water content is decisive in compaction since the dry soil has a high degree of cohesion, and this increases the soil resistance to deformation. Higher moisture levels imply an increase in soil compressibility (Han et al. 2006; McDonald and Seixas 1997) and the intensity of damage to the soil structure, especially during forest harvesting, which is enhanced during rainy seasons. It is not always possible to harvest trees during ideal soil moisture, but failure to do so can have a negative impact on the growth of future trees (Wronski and Humphreys 1994) with the greatest amplitude of the compaction effects in a subsoil layer. Machine operations on soils with inadequate soil moisture may cause compaction of layers below 50 cm in depth if the load-bearing capacity of the subsoil is exceeded (Alakukk et al. 2003). The pressure applied to the ground by the machinery's wheels increases with its size and load capacity. In logging systems, the impacts caused by the forwarder on soil were greater than those caused by the harvester. Under higher soil moisture conditions, the deformations caused by forwarder traffic have become even more severe, with potential effects on soil physical conditions and tree growth (Wronski and Humphreys 1994).

The terrain slope can also accentuate compaction in forest soils since in steep terrain, the plots experience higher machine traffic compared to areas of gentle slopes to achieve the same efficiency. The increase in the number of machinery passes increases the risk of additional compaction to a threshold that exceeds the critical limits for plant growth (Ampoorter et al. 2007; Jaafari et al. 2014; Schaefer and Sohns 1993; Williamson and Neilsen 2000). Therefore, mechanized cutting operations in areas with more than 20% slope should be avoided, whenever possible, to reduce the risks of soil compaction (Jaafari et al. 2014; Naghdi et al. 2016).

The number of machine passes also affects the spatial distribution of compaction in forest systems. Localized traffic in harvest and transport operations increases the spatial variability of soil compaction due to differences in the frequency with which a given area is trafficked (Arvidsson et al. 2001). Tracking the movement of a skidder used in the forest harvest, it was found that the resistance to penetration in the first 0.15 m in the post-harvest traffic area increased by 81–272% (Zenner and Berger 2008).

There is a tendency for soil compaction to decrease with increased depth due to the dissipation of the force applied by the weight of the machinery. However, high moisture content at the time of operations in conjunction with the stresses induced by machine traffic above the load-bearing capacity of the subsoil can cause compaction of layers below 0.50 m in depth (Alakukk et al. 2003). Then, the heterogeneity and depth of compaction in forested areas make the control and correction measures even more complex than in agricultural areas or pastures (Reichert et al. 2007; Soane et al. 1980). In general, the following recommendations for minimizing compaction in forest harvest involve (Cambi et al. 2017) reducing machine movement; limiting the area trafficked by machines; and promoting the use of brush mat on the skid trail to reduce the pressure exerted on the ground.

6.3 Compaction and Impacts on Soil Functions

Compaction affects several soil functions and can compromise the provision of essential ecosystem services. The main impacts of compaction are on soil physical properties and include the increase in soil density and the mechanical resistance of the soil to root growth, in addition to the reduction in the volume of macropores, the rate of infiltration and the movement of water and gas (Cambi et al. 2017; Schjønning et al. 2015; van den Akker and Schjønning 2004).

The damage caused to the pore continuity affects the transport pathways and the mineralization rates of soil organic matter, changing the transport and the amount of nutrients in solution (Berisso et al. 2012). In a mechanized forestry system, an average increase in soil density by 60% in the traffic areas caused decreases in the soil levels of organic C (−38%) and concentration of N (−57%), P (−25%) and K (−31%) (Naghdi et al. 2016) compared with undisturbed areas. Biological processes driven by soil macro and microfauna, such as the formation of macropores and the humification of organic matter, are also affected by soil compaction (Jaafari et al. 2014). There is an interdependence between the soil biological processes and soil compaction, since the latter affects biodiversity, while the activity of soil organisms can alleviate compaction (Beylich et al. 2010).

The preservation of soil functions or at least the mitigation of the effects of compaction on these functions should guide soil management practices. However, most studies about soil compaction are concerned with only one or a few soil properties, and not the multiplicity of functions that soils can offer. Seeking to overcome this challenge, Pöhlitz et al. (2020) defined critical values of stress caused by compaction on soil physical, morphological and biological attributes as well as in the development of plants in soils of different textures. Considering the optimum soil density for soil to perform all its functions at close to field capacity, the following critical ranges were found: 45–71 kPa for mechanical pre-compression stresses; 12–155 kPa for the minimum desirable macroporosity (0.08 $m^3 m^{-3}$); 100–200 kPa for pore connectivity; 11–80 kPa for biopores built by earthworms; 7–56 kPa for grain production; and 30–50 kPa for straw production.

Concerns about compaction affecting soil functions are usually restricted to its effect on soil structure. Rabot et al. (2018) review the relationship between different methods of assessing soil structure with soil functions such as biomass production, storage and filtering of water in the soil, dynamics and cycling of nutrients, soil carbon storage, habitat for biological activity and the role of soil in stability and physical support. Among the indicators of soil structure, porosity, macroporosity, pore distances and pore connectivity are the most relevant for the soil functions evaluated.

Agricultural adaptation practices associated with reducing erosion and increasing soil carbon stocks can maintain soil functions and ecosystem services which may be compromised by climate change. However, a study that evaluated 20 agricultural adaptation case studies in Europe to address threats to the world's soils, within the context of the Sustainable Development Goals (Hamidov et al. 2018), concluded that reducing soil compaction is still a major challenge to overcome, not only because of the direct but also indirect repercussions (e.g. soil erosion).

Also concerning climate change, soil compaction can affect the flow of greenhouse gases. While there is no consensus on this effect and its persistence, some studies associate compaction with increased CH_4 (Cardoso et al. 2019) and N_2O emissions (Bessou et al. 2010; Ruser et al. 2006) but a reduction in CO_2 emission (Mordhorst et al. 2014). The effects of compaction on the dynamics of soil gases are directly related to porosity, diffusivity of gases and water-air relationship in the pores and indirectly to the activity of the roots and microbial communities. Compaction reduces macroporosity, the key determinant of water drainage and gas exchange (Ampoorter et al. 2007), which occurs as both decreased pore space and pore connectivity. With increasing the degree of soil compaction (Suzuki et al. 2007; Reichert et al. 2009; Millan et al. 2014), macroporosity decreased by 53–67% with the application of 0.9 MPa pressure (Silva et al. 2006) or from 0.29 $m^3 m^{-3}$ to 0.04 $m^3 m^{-3}$ when soils were artificially compacted from 1.0 to 1.6 kg dm^{-3} (Stone et al. 2002). By contrast, increases in macroporosity of compacted soils can lead to greater root development and plant production (Colombi et al. 2017). The minimum value of soil macroporosity suggested by some authors is 0.10 $m^3 m^{-3}$ and is also considered adequate by Reichert et al. (2009) for clay soils, although the same value may be low for sandy soils, since they have less connectivity among macropores.

This same limit of 0.10 $m^3\ m^{-3}$ is considered to be the critical total porosity for root development (Lipiec and Hatano 2003; Schjønning et al. 2003).

The increase in microporosity is another possible effect of compaction (Ampoorter et al. 2007). While compacted soils may have higher water retention, it does not necessarily imply more available water, given that very small pores can retain moisture at high matric potential, limiting the absorption by plants (Van Der Weert 1974). In addition, the reduction of macroporosity slows soil drainage, causing increased run-off, which can trigger erosive processes (Zimmermann and Elsenbeer 2009). Compaction also affects the continuity of pores, impairing air and water flows. These effects can be observed by the reduction of infiltration rate, hydraulic conductivity and availability of water for plants in compacted soils (Berisso et al. 2012; Frey et al. 2009; Nadezhdina et al. 2012). It is important to emphasize that the water flow in soils is essential for the transport of nutrients to roots, and hence compaction may compromise the nutrition of the plants. The continuity of soil voids can compromise the mineralization rates of organic matter in the soil, changing the mobility and the amount of nutrients in solution (Berisso et al. 2012).

Changes in pore size distribution by compaction affect biological activity, as the effects on soil water and air flow and storage can restrict the activity and diversity of microorganisms and compromise their functions in soil and in carbon and nutrient cycling (Beylich et al. 2010). Increases in soil density are associated with changes in microbial biomass, soil respiration, phospholipid fatty acid profile (PLFA) (Schnurr-Pütz et al. 2006; Busse et al. 2006), soil enzymes (Li et al. 2002) and the density and diversity of soil macro and mesofauna (Battigelli et al. 2004) such as earthworms (Lees et al. 2016).

The effects of soil compaction on plant development can be expressed as smaller plant size, yellowed leaves, shallow and horizontal root systems and tortuous or malformed roots (Dias Júnior 2000). The root system is the first component to reflect the effects of compaction, with reduced growth, limiting the volume of soil explored by the roots. Then, the possibility of water and nutritional deficits increases leading to reduction of leaf photosynthesis rates and, consequently, of crop productivity (Gaertig et al. 2002).

The reduction of soil porosity caused by compaction can also cause changes in the hydrological cycle of river basins, affecting the flood regime. The reduction of water infiltration and soil hydraulic conductivity with a consequent increase in run-off is the source of this relationship between compaction and flooding. Alaoui et al. (2018) reported a relationship between increased areas of compacted soils and changes in land use with discharge peaks in rivers, mainly in plots and at hill slope scale. At the catchment scale, this relationship is stronger in small basins with shorter response times.

In contrast to the above concerns about soil compaction, in a few situations, it can be beneficial. This is the case when soil compaction improves the contact zone between roots and soil particles on coarse soils. Slight soil compaction can increase water retention by increasing soil bulk density and reducing macroporosity. A moderate level of soil compaction could increase soil water retention and be beneficial for soils where rains are scarce (Agrawal 1991). However, these effects can be limited to short time periods and may cease to be beneficial over time. In addition to negative effects and no effects, Armando Gomez et al. (2002) noted positive effects of soil compaction on cumulative stem volume for a *Pinus* species on sandy loam soil. Powers et al. (2005) observed positive effects of severe compaction on the productivity of sites with sandy soil textural classes after a long-term (10 years) soil productivity study. Gains of more than 40% are indicated and associated with improvements in available water holding capacity on sands after compaction.

6.4 Assessment Methods of Soil Compaction

Subsoil compaction can be detected by visual qualitative approaches (Shepherd 2009; Ball et al. 2007). Decreases in productivity, premature haying off, surface concentration of the root system, toppling of plants, surface erosion and waterlogging are some of the signs that may indicate the presence of compacted layers at depth. These signs can also be associated with the natural soil density occurring in the subsoil due the high clay concentration (textural B) or even another depth limiter such as the presence of rock or gravel layers. The history of land use of an agricultural area also can help to rule out one or another of these possible causes.

The compacted subsurface layers can be identified by the greater hardness and density than the layers just above and below, what is often observed by the massive structure in contrast to the adjacent layers. Also, it is possible to observe platy structure neo-formed by soil management, especially in weakly structured silty soils, as observed by Sasal et al. (2017) in soils under no-tillage. The visual evaluation can be comparatively effective based on prior knowledge and technical experience with the soils of the region and provided the assessment depth is no greater than 60 cm in depth.

Soil physical attributes that are frequently measured as indicators of compaction include soil bulk density, soil resistance to penetration, porosity and soil structure (Dexter 2004). However, the use of these indicators can be limited by the lack of sensitivity for the direct and short-term response to changes in the soil physical condition imposed by the management or to support decision-making in the agricultural systems.

Rabot et al. (2018) discuss the advantages and limitations of different soil structure assessment methods to assess compaction. The soil structure is assessed by field (visual assessment of topsoil morphology and the entire soil profile) and laboratory methods (soil bulk density, aggregate size distribution and stability and estimated indexes of stability expression), by direct characterization of soil pores (imaging techniques) and by indirect measures (mercury porosimetry, soil water retention curve methods and derived indicators and gas adsorption). Imaging techniques are modern tools and include computer-assisted tomography (CAT) and nuclear magnetic resonance (Lipiec and Hatano 2003) but are mostly restricted to research use.

Soil bulk density, defined by the ratio of mass of solids to total soil volume, is one of the parameters commonly used to express the effects of soil compaction, as it represents the degree of mechanical impediment that restricts the growth of root system (Håkansson and Lipiec 2000). Critical values of soil bulk density associated with the restriction of root development in Brazilian soils ranged between 1.21 and 1.84 kg dm^{-3} (average of 1.60 kg dm^{-3}) for soils with clay contents between 9% and 75% (Reichert et al. 2009). Any impact on soil bulk density reflects changes in soil porosity. Nevertheless, the pore network is also dependent on the structure, texture, content of organic matter and soil mineralogy. Thus, although soil bulk density is relevant, it should not be used as the only indicator of soil compaction (Horn et al. 2004; Naderi-Boldaji et al. 2016). An alternative is the use of relative soil density (or degree of compaction, expressed as a percentage), which relates soil density to a maximum reference density (Lipiec and Hatano 2003; Naderi-Boldaji et al. 2016) obtained at a given soil moisture condition and standardized tension (Proctor 1933). The use of relative density aims to normalize the current density to allow its comparison with other situations and soils (Naderi-Boldaji et al. 2016). As an indicator for plant development, Oliveira et al. (2016) reported the relative density or degree of compaction equivalent to 98% was a critical value for soybean, regardless of soil texture. However, Silva et al. (2014) indicate that the degree of critical compaction of a crop will depend on which attribute is being evaluated and found that for soybean critical values varied from 82% (plant height) to 75% (root dry matter production), 87% (shoot dry matter production) and 93% (evapotranspiration).

Another parameter commonly used to assess soil compaction status is the soil mechanical resistance to penetration. This evaluation aims to simulate the resistance that the roots face when growing vertically down the soil profile. The penetration test can be performed in the laboratory or in the field, and the results are expressed in units of pressure (MPa) (Moraes et al. 2014b). Although measurements of soil resistance to penetration are simple to perform, the results can be affected by soil moisture, texture and density (Vaz et al. 2011). In general, for the same soil density, the resistance decreases with the increase of soil moisture, but the resistance increases with the increase of the soil density at a particular soil moisture level. Under dry conditions, clayey soils are more resistant to penetration than sandy soils. Therefore, assessments of resistance to penetration should also report soil moisture and/or soil bulk density values (Busscher 1990; Vaz et al. 2011).

The effect of moisture on soil resistance to penetration can cause misinterpretations about the magnitude of management effects on compaction of the soil. The most common recommendation is to assess penetration resistance at close to field capacity (Arshad et al. 1996). However, this recommendation is not unanimous, and Peixoto et al. (2019a) proposed for no-till areas in Brazil that the ideal range of soil water content for diagnosis of compaction (IRDC) is −0.03 and −0.50 MPa soil water potential. At the field level, the basic recommendation is to determine the soil resistance to penetration across soil types or management zones at a fixed time when differences in soil moisture are minimal.

The critical value most associated with soil resistance to root penetration is 2 MPa. While many authors attribute this value to Taylor et al. (1966), they reported that 2.5 MPa was the limit for the development of cotton roots, from the seed, for the four soils of the semi-arid and western United States, with sand contents varying from 44% to 83%. The value of 2 MPa also coincides with that proposed in the Soil Survey Manual (USDA 2019) as the lower limit of soil resistance to penetration to use when describing soil profiles. However, Moraes et al. (2014a) indicate 3 MPa as the critical value for soils under minimum tillage with chiselling and 3.5 MPa for soils in no-tillage. Miller et al. (2004) reviewed other studies and found values varying from 2 to 4.2 MPa as limits for the beginning of severe restriction or cessation of the growth of pine and eucalyptus tree roots. Ribeiro (2010) proposed the following classes of soil resistance to penetration (RP) and degrees of limitation to the growth of roots: <2 MPa as low RP and without impediment to plant growth; from 2 to 4 MPa as moderate RP when restrictions begin to exist; 4–6 MPa as high RP, indicating that there are restrictions on root development; and >6 MPa, as very high RP. Oliveira et al. (2016) propose that the critical limit of resistance to penetration for soybean roots should also consider the soil texture as follows: 2.8 (sandy loam soil), 5.6 (sand clay loam soil), 3.5 (clayey soil) and 5.2 MPa (very clayey soil).

Recently, a series of soil compaction sensors have been evaluated for the diagnosis of this phenomenon (Lee et al. 2010). Sensors have acquired importance for rapid assessment of compaction for precision agriculture (Lui et al. 1996). Some sensors can assess soil compaction through electrical resistivity (Islam et al. 2012), using a piezoceramic-based active sensor (Yang et al. 2018) or a strength profile sensor (Agüera et al. 2013), and others are based on electromagnetic conductivity (Hoefer et al. 2010). When linked to GPS equipment, they are capable of mapping the spatial variability of soil compaction in rural properties (Hemmat and Adamchuk 2008).

6.5 Soil Compaction Management

Soil compaction management should focus on minimization (McPhee et al. 2020), using a range of approaches: reduction of machine and implement traffic, which avoid trafficking during excess soil moisture; adoption of crop rotation, green manure, cover crops, etc.; vary the depth of soil preparation; select machines and implements that exert less pressure on soil; and control tyre pressure and load distribution on machines. For remediation, soil compaction can be reduced with the use of deep ploughing, chiselling or subsoiling according to the depth of occurrence of the compacted layer, followed by the use of cover plants with vigorous roots. In remediated soils, trafficking should be controlled, restricting the compaction to permanent wheel tracks that cover less than 20% of the land area (Davies et al. 2019).

Conservative soil management systems seek to safeguard the soil against the negative effects of compaction, and they can be organized (Lebert et al. 2007) considering four principles: (*i*) practices that aim to reduce stress on the soil (larger tyres and avoid mechanical disturbance of soils with higher moisture); (*ii*) improved soil support capacity (no-till, minimum cultivation, conservation practices);

(*iii*) development of new solutions (low-pressure tyres and pressure regulation systems); and (*iv*) crop adaptation methods, such as adapting the width of the planting lines to receive larger tyres.

The spatial identification of areas susceptible to compaction can aid in its future management; however, such predictions are scarce and limited by methodological gaps. Low-cost methodologies are also needed for long-term monitoring of soil compaction at field scale.

In addition to general recommendations, some rules have been suggested to minimize soil compaction. The 50–50 rule, for example, proposed by Schjønning et al. (2012) for sustainable traffic on soils stated that: "at water contents around field capacity, traffic on agricultural soil should not exert vertical stresses in excess of 50 kPa at depths >50 cm". To predict the depth (d50) at which the vertical stress of 50 kPa is reached, the same authors suggest an equation that can be used by farmers that considers the wheel load (WL, in tonnes) and tyre pressure (tyre in bar): d50 = 30 + 8WL + 8log2 (tyre). The 8–8 rule states that "the depth of the 50-kPa stress isobar increases by 8 cm for each additional tonne increase in wheel load and by 8 cm for each doubling of the tyre inflation pressure".

In addition to the tyre pressure and contact area, attention should be paid to the implements used in agricultural management, including those used to correct soil compaction such as the subsoiler. The pressure exerted at depth and its localized action may be contributing to the formation of a new compacted layer at the interface between the disturbed/subsoiled layer and the undisturbed layer. This condition is well exemplified with the soil preparation in areas of melon cultivation produced for export in Northeastern Brazil (Mossoró-RN). In soils with a textural B horizon (Red Dystrophic Argisol), the conventional preparation uses implements for forming ridges for planting melons causing the accumulation of topsoil in ridges and exposure of lower layers between the ridges. The densified soil caused by the pressure of the implements is evident from the spatial distribution of pores, showing the localized action of the implements compacted the subsoil layers (white dots indicate the presence of pores) (Fig. 6.1). While the negative effects on soil water availability and root volume of soil are minimized due to irrigation, more pronounced effects are likely for the productivity in rainfed systems, and these may be exacerbated by climate change.

The plough harrow is one of the most used implements in conventional tillage in Brazil that performs two operations in one pass: ploughing and harrowing. Its mass varies between 2000 and 4000 kg, distributed over a very reduced contact area with the ground, which leads to high pressure on the trafficked soil layer. This implement has been associated with the formation of compacted layers in the subsurface.

Selection of crop species or cultivars tolerant to soil compaction seems to be an efficient alternative for compacted soils and may even reduce compaction in subsurface layers. More aggressive root systems, with some capacity to overcome or even break compacted soil layers (Meyer et al. 2014), should be considered in rotations. This practice has been called biodrilling or biological drilling (Cresswell and Kirkegaard 1995) or "biological tillage" (Chen and Weil 2010). Williams and Weil (2004) present a very illustrative example when photographing the development of

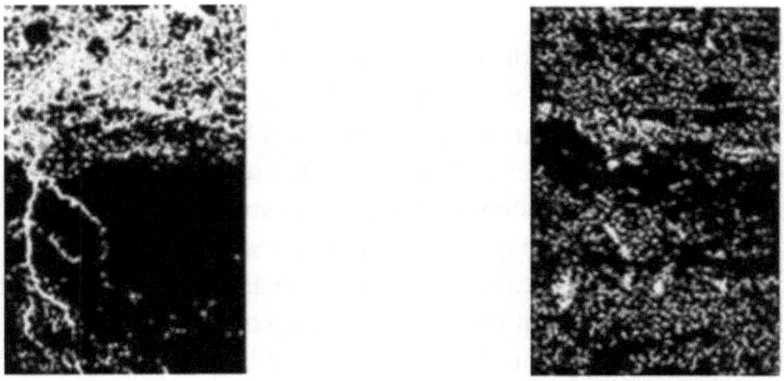

Cultivated area
34% porosity
Horizon Ap (3-13 cm)

Uncultivated area
32% porosity
Horizon Ap (5-15 cm)

Fig. 6.1 Soil porosity quantified by image analysis of the superficial horizon of cultivated and non-cultivated profiles of a dystrophic Red Argisol, in the Angico cultivation area located at Fazenda MAISA, Mossoró, Rio Grande do Norte, Brazil. The black dots refer to the soil matrix and the white patches to soil pores (Alencar 2000)

soybean plant roots in channels left previously by canola roots. Other plants are reported in the literature, both in field or laboratory studies, as capable of penetrating compacted layers, namely: pasture bush (*Senna occidentalis* L.) (Alvarenga et al. 1996), alfalfa (*Medicago sativa*) (Löfkvist et al. 2005), guar (*Cyamopsis tetragonoloba* L.) (Bassegio et al. 2018), forage radish (*Raphanus sativus*) (Chen and Weil 2010, 2011), rapeseed (*Brassica napus*) (Chen and Weil 2011) and Tifton 85 grass (*Cynodon* spp.) (Magalhães et al. 2009). The combination of these plant species capable of biodrilling, in rotation with mechanical practices, can give interesting results. These are the cases reported for no-till areas when subsurface compaction was attenuated with the use of scarification in combination with ruzi grass (*Brachiaria ruziziensis*) and castor bean (*Ricinus communis*) (Rosolem and Pivetta 2017) or with brachiaria grass (*Urochloa* spp.) (Spera et al. 2018).

The most common mechanical practices for coping with soil compaction are deep ploughing, chiselling and subsoiling. Deep ploughing consists of the same traditional soil preparation operation but with a greater depth than the compacted layer. For the greatest efficiency of this practice, the soil must be in a friable condition, and the disc plough or mouldboard can be used, and even the chisel plough. Then, in some cases, it has been common to use a light levelling and grinding-type harrow before planting.

The chisel plough has been recommended lately, including in no-till areas with compaction problems (Camara and Klein 2005) and in association with biodrilling plants. The implement cuts the soil without inversion of soil and maintains the cultural residues on the soil surface, minimizing soil and water losses. The advantages of using the chisel plough in reducing erosion have been shown for maize crops (Levien et al. 1990). The reduction in soil losses when the chisel plough replaced the conventional tillage was from 16.7 t ha^{-1} to 0.79 t ha^{-1} in the corn/wheat succession, while the total percentage of rainfall run-off decreased from 44.6% to 29.6%. However, chiselling is not so effective as subsoiling to eliminate compaction in no-till areas at depth greater than 0.20 m (Peixoto et al. 2019b), having a short-term effect when performed as an isolated practice (Drescher et al. 2016). The lower efficiency of scarification associated with no-till areas has been attributed to the failure to adopt a crop rotation scheme that favours soil protection in conditions that allow the production of straw to be maintained on the soil surface.

Subsoiling is a corrective practice directed towards more serious situations of soil compaction. The implement tines can reach up to 1.0 m in depth and break up dense and/or compacted layers in the subsurface. This practice has been considered efficient to increase the soil aeration, favour deep root growth and increase the production of agricultural crops, especially where there are no chemical restrictions in the subsurface (Davies et al. 2019; Hall et al. 2020). The high energy consumption determines the costs involved in carrying out this practice, which requires high-powered machines, including four-wheel tractors or crawler tractors, both operating at reduced speed, which causes greater fuel consumption, the main operational cost of subsoiling (Simões et al. 2011). Also, the greater the depth, the greater the energy expenditure and the lower the crop yield response.

Subsoiling efficiency is dependent on the operation's depth and soil moisture. In more humid soils, the tine can contribute to compaction at the interface with underlying soil. Subsoiling is more recommended in drier soils, when the breaking of compacted layers is more effective, and the effect extends over greater distances in the soil profile. The ideal soil moisture for subsoiling varies among soils, as observed by Sasaki et al. (2007) when they evaluated three Ferralsols in Brazil with different textures. The ideal subsoiling water content was 0.07–0.13 $m^3\ m^{-3}$, 0.12–0.19 $m^3\ m^{-3}$ and from 0.14 to 0.27 $m^3\ m^{-3}$, respectively, for soils with 24%, 46% and 62% clay.

The potential benefits of subsoiling depend on the land use (Sasaki et al. 2007). In the forestry sector, soil preparation is often characterized by the subsoiler only in the seedling planting line to guarantee the full development of the root system. This practice is also being adopted in other perennial crops, such as coffee (Oliveira et al. 2019), where the deep soil preparation (80 cm) with the subsoiler is associated with the application of liming, ensuring gains in aeration and available water in the soil (Barbosa et al. 2020). Subsoiling is becoming a frequent practice also in irrigated areas, but the recommendation to apply this practice at low soil moisture conditions is often not taken into consideration. Finally, in recovery programs of mined areas, subsoiling up to 1 m depth is practiced since soils are deeply compacted by the intense trafficking by heavy machinery during mining.

Although controlled traffic has been evaluated in many countries, its adoption is still restricted. The practice consists of always using the same traffic lines, thus sacrificing <20% of the land affected by the wheel tracks and preserving most of the cultivation area (Raper and Kirby 2006). In this case, two zones are created, one cultivated and not trafficked and the other with machine and implement traffic, both optimized in their functions (Chamen 2015). The advent of global position system (GPS) technology and the development of precision agriculture can greatly increase the use of traffic plans on farms. However, its adoption is limited by the lack of standardization of equipment specifications, since all machines and implements, from preparation to harvest, must have similar width of the wheel base. Even with the adoption of controlled traffic, other practices to prevent or minimize compression, as mentioned above, should be applied.

In modern agriculture, which is increasingly mechanized and technological, the gain in profit is often associated with larger tractors and implements that are also heavier. It is also common to acquire machinery that exceeds needs, which implies the use of excessive loads over the soil. The gain in load increases the depth of compaction layers worldwide, although the rate of increase in depth has slowed with the advent of radial tyres. However, the efficiency gains from radial tyres have been offset due to the increase in load over time (Chamen 2015).

In addition to the weight, Miller et al. (2004) indicate that the effect of machinery on soil compaction depends on soil characteristics (depth to water table, moisture movement, depth of A horizon, texture), the season, presence of crop residues on the soil surface, static weight and dynamic load of the equipment and knowledge and skill of the operator. Nevertheless, choosing tractors and implements with no greater capacity than is necessary and following the manufacturers' recommendations for use would go a long way towards mitigation of compaction risks.

The selection of tyres and their operation can help to mitigate the effects of compaction. Tyre size, type and pressure directly affect the pressure distribution on the ground. Radial tyres and lower pressure increase the tyre-to-ground contact area, providing a better distribution of the machines' weight, reducing the effects of the pressure exerted. However, simply using tyres with a larger diameter, a smaller cross section, less rigid walls and operated at low inflation pressure (Richart et al. 2005) has not been sufficient to prevent soil compaction. Other options are to increase the number of axles or wheels or switch to tracks, develop lighter machines (Schjønning et al. 2015) or use smaller and autonomous vehicles (McPhee et al. 2020).

The susceptibility to soil compaction increases with soil water content, but there seems to be a consensus that the maximum soil compaction occurs when soil moisture is at or near the field capacity. Trafficking when the soil is in a plastic or sticky condition is harmful to the soil structure and favourable to soil compaction. By contrast, soil water content close to the state of friability will minimize the forces of adhesion and cohesion which result in soil compaction. With soil water content close to the state of friability, there is less propensity for deformation, destruction and reorganization of soil particles. Trafficking on friable soil will reduce machinery wear and fuel expense.

Knowing the soil physical properties and their response to soil water levels can help in choosing the best time for mechanical operations while avoiding soil water levels that will increase soil compaction. Grouping of soils based on the number of days required to achieve the friable moisture content after rains is a practical example (Milde et al. 2010) of managing machine trafficking to minimize compaction. However, in practice the timing of machinery movement, tillage and cultivation operations often fails to prioritize concerns with soil quality. The pressures of market demands, harvest time, the time of the year (rainy) and the needs for yield and maintenance of post-harvest quality will determine the time for machine operation, which will not always be associated with the ideal soil condition for its execution. Controlled traffic systems provide much greater operational flexibility in the timing of machinery operations, without the consequences of greater soil compaction.

6.6 Compaction Susceptibility

The soil susceptibility to compaction depends on external and internal factors that govern its compressibility (Ampoorter et al. 2007). The external factors are related to the characteristics of the applied pressure and the topography of the area, while the internal factors are associated with the history of disturbance experienced by the soil, the initial soil bulk density, the texture, the structure and its compressibility, the content of organic matter, mineralogy and current soil moisture (Bygdén et al. 2003; Dexter 2004; Han et al. 2006).

Lebert et al. (2007) indicate that measures of soil stability and soil strength, soil water content and soil water potential, vehicle data and soil stress distribution are essential for a full risk assessment of soil compaction. This study proposed a four-step method to assess the risk of soil compaction and ways to avoid it and suggests two concepts that can be used to assess the effects and susceptibility of soils to compaction: pre-compression stress and loading ratio. In the first step, these two indicators are used to define the soil compaction status. Once the compaction is diagnosed, the next step considers the use of good agricultural practices to reduce the loads applied on soil. The third step is actually the repetition of the first, seeking to evaluate the effectiveness of the adopted practices, and, if the compaction persists, the fourth step proposes more in-depth physical analyses to identify suitable measures to eliminate soil compaction.

It is also possible to predict the susceptibility to compaction through information existing in soil surveys (Schneider and Don 2019). These authors assessed the extent, cause and effect of root-restricting layers in agricultural soils in Germany and found that compactness was the most common cause of root restriction, affecting 51% of cropland and 32% of grasslands. The German study is noteworthy by showing that it is possible to produce subsoil constraints mapping at farm scale and to map subsoil constraints that arise from land management practices. The restrictions to root elongation due to compactness was evaluated by the packing density (PD) – a parameter well correlated with other indices describing the soil compactness, such as least limiting water range (da Silva and Kay 1997; Kaufmann et al. 2010), S-index (Dexter 2004; Kaufmann et al. 2010) and degree of compactness (Naderi-Boldaji et al. 2016). Packing density (PD) is obtained by the dry bulk density of the <2 mm soil fraction plus clay (%) and silt (%) (Renger et al. 2014). Then, it was assumed that the compactness in a cropland (PD_{crop}) can be described considering (*i*) the PD reference (PD_{ref}) which is the theoretical, site-specific value without anthropogenic influence; (*ii*) the land use-induced change in compactness due to SOC losses after conversion to cropland (use); and (*iii*) the management-induced change in compactness due to trafficking/tillage (man). In the German study, grassland was considered the PD_{ref} as it is not ploughed (no plough pan) and typically receives a lower trafficking intensity than cropland. To quantify man, a random forest model was trained only on data from permanent grassland (0–100 cm with soil depth as the predictor variable) and used to predict the PD of cropland as a function of depth ($\widehat{PD}_{crop}$). If $PD_{ref,i}$ of a specific site, $i > 1.75$ g cm^{-3}, severe soil compactness was assumed to be of pedogenic or geogenic origin. If $PD_{ref,i} < 1.75$ g cm^{-3} and $PD_{ref,i}$ + (man) >1.75 g cm^{-3}, severe soil compactness was assumed to be due to tillage/trafficking. If $PD_{ref,i} < 1.75$ g cm^{-3} and $PD_{ref,I} + \overline{use} > 1.75$ g cm^{-3}, severe soil compactness was assumed to be land use derived (due to SOC loss). The authors defined "compacted" as the soil layers with PD above the critical limit of 1.75 g cm^{-3} and "compaction" as the compression of soil from initially $\leq$1.75 g cm^{-3} to >1.75 g cm^{-3}.

Finer textured soils are generally more susceptible to compaction than coarse textured soils (Wästerlund 1985). Finer textured soils retain a higher volume of water compared to sands, and water acts as a lubricant between particles, facilitating the denser rearrangement of soil particles. Even in soils considered more resistant and with a stronger structure, such as the Ferralsols of the Brazilian Central Plateau, the clay effect is evident. Horn et al. (1995) propose three texture-based groups of soils in terms of susceptibility to compaction: sandy, silty soils with low clay content, medium and fine textured loam and clay soils. The first group represents soils that are slightly susceptible to soil compaction which, even if it reaches a high degree of compaction, does not cause a significant deterioration in the soil physical quality. The second represents soils with a weaker structure that are easily compacted. The third group comprises soils that are more resistant to mechanical pressure when dry, but that are very susceptible to compaction when moist or wet.

Notwithstanding the above classification, subsoil compaction can become a significant subsoil constraint on deep sands subject to heavy machinery trafficking (Davies et al. 2019) as in Western Australia. Average yield responses to deep ripping of about 0.5 t ha^{-1} have been widely measured (Moodie et al. 2020), although these yield gains may be short lived (e.g. 2–3 years). Hence, subsoil compaction cannot be associated simply with the particle diameter classes, as proposed by Horn et al. (1995), but with their propensity for particle arrangement and high packing density (Hall et al. 2020). Packing models need to consider all soil particle classes (finer and coarse) and their distribution within the soil profile, even in sandy texture soils classes. The reorganization of soil particles could increase density in specific soil layers, and the crop response to these changes can be a good indicator of the effectiveness of amendments adopted in the tillage.

The susceptibility to compaction can be predicted by the degree of pedality (the proportion of soil in aggregates/peds and their strength), by clay activity expressed by cation exchange capacity (CEC) and also by soil consistency (the differential between cohesion within aggregates and adhesion between aggregates) that is associated with the aggregate's resistance to destruction by external forces. When the soils are wet (field capacity), the structure/aggregates breakdown will be higher in soils with high clay activity and less in soils with low clay activity. Generally, in soils with higher clay activity, the grade of structure/aggregates is weak/moderate (Cambisols/Luvisols) when wet, while with low activity clays, strong degree of structure/aggregation is maintained (Ferralsol). Table 6.1 shows the differences of soil bulk density between weak (Cambisols) and strong (Ferralsol) grade of structure/aggregates at depth of 0–5 cm (the layer of direct effect of the pressure exerted by machines and implements) under different land use and natural vegetation. The compaction will be higher in weakly structured/aggregated Cambisol than in the Ferralsols, under natural vegetation and/or even under intensive agricultural use.

The pre-consolidation pressure (σ_p) of a weakly structured Red-Yellow Ferralsol declined with an increase of soil moisture to a greater extent than that in a moderately to strongly structured Cambisol (Fig. 6.2). The maximum pressure that should be applied to soil in order to avoid soil compaction was lower in soils of a weak grade of structure/aggregate (Cambisols). In addition, the increase in soil bulk density values under land use was higher compared to that under natural vegetation, highlighting the proximity of the maximum (Bd_{max}), especially in the irrigated areas of Cambisols.

In apedal soils (massive or single grain), compaction can occur through the reorganization of the particles (single grain), naturally or under pressure, while massive apedal soils are theoretically already in a condition of maximum compaction. Sandy soils naturally tend to have higher pre-consolidation pressure than soils with some degree of pedality. The pre-consolidation pressure can increase by the reorganization of the soil particles with tillage and/or cultural crop practices. Table 6.1 shows several examples of bulk density changes between the plant rows (*R*) and inter-rows (*I*), where the machine movement on the ground is minimum and maximum, respectively. The differences between soil bulk densities (ΔBd) of inter-rows are lower for both situations than that obtained in the rows, and the pre-consolidation pressure could be similar in the inter-rows to that obtained in the rows (6 and 10 years of cherry Indian). The exception occurs when the application of palm straw as mulch (a common management practice in the situation studied) is done in the first years of crop development, which increases the pre-consolidation pressure required to induce compaction. The results (Table 6.1) show that sandy soils (apedal) become susceptible to increasing their density with land use, due to the rearrangement of particles and not to their susceptibility to aggregate breakdown.

The higher susceptibility of soils with a low degree of structure development to the disintegration and reorganization of particles and aggregates caused by soil compression soils can be characterized at a microscale. Cambisols originating from limestone rocks in the Chapada do Apodi (Ceará state, Brazil) and intensively used for crop production under pivot irrigation systems become denser due to pore filling with particles and microaggregates (Fig. 6.3). There were different types and composition of pore fillings in all the evaluated land use systems, from the most intensive with successive corn and bean crops (2 per year) under central pivot to soils under natural vegetation and guava and banana (perennials) with localized irrigation, in addition to pasture areas under central pivot (Oliveira et al. in press). The pore fillings can be composed of primary particles of Si (Si mapped) or Fe (Fe mapped) compatible with quartz, Fe oxides (Fe mapped) or microaggregates from

Table 6.1 Effects of grade of structure/aggregates[1], cation exchange capacity (CEC)[2] and consistency[3] on bulk density (Bd) and pre-consolidation pressure (σ_p) of rows (*R*) and inter-rows (*I*) of Cambisols, Arenosols and Ferralsols under different agricultural land uses at 0–5 cm depth

Soil class	Land use	Bd (kg dm^{-3})	Bd$_{max}$ (kg dm^{-3})	ΔBd	σ_p (kPa)
Weak moderate structure, high CEC, very firm to very strong consistence (Cambisol)	Pasture (central pivot)[a]	1.52	1.56	0.04	118
	Banana$_I$ (15 years – micro sprinkler)[a]	1.68	1.72	0.04	131
	Corn$_I$ (central pivot)[a]	1.62	1.73	0.11	83
	Natural vegetation[a]	1.50	1.79	0.29	128
Primary particles, low CEC, loose consistence (Sandy soil)	Cherry Indian$_I$ 1 year (central pivot)[b]	1.69	1.84	0.15	199
	Cherry Indian$_I$ 6 years (irrigation with micro sprinkler)[b]	1.70	1.75	0.05	124
	Cherry Indian (*I*) 10 years (inter-rows and micro sprinkler irrigation)[b]	1.44	1.64	0.20	125
	Cherry Indian (*R*) 1, 6 and 10 years[b]	1.44	1.69	0.25	170
Strong/very strong granular structure, low CEC, weak to firm consistence (Ferralsol)	No-tillage (soybean)[c]	1.21	–	–	–
	Annual crops 31 years (central pivot irrigation)[d]	1.21b	–	–	303
	Eucalyptus harvesting and trafficked eight times by forwarder[e]	1.00	–	–	349
	Eucalyptus harvesting and trafficked eight times by forwarder[e]	0.99	–	–	–
	Natural vegetation				
		1.06[f]	–	–	168
		0.96[f]	–	–	227
		1.30[f]	–	–	153
		1.22[g]	–	–	–
		0.99[g]	–	–	–
		0.83[g]	–	–	–
		0.96			156
		1.06			–
		1.34			227

All observations were made in inter-rows unless otherwise indicated for the effects of mechanization of the different managements

Bd bulk density, *Bd$_{max}$* maximum density, *ΔBd* Bd$_{max}$ – Bd, *σp* pre-consolidation pressure (matric potential of −10 kPa), *I* inter-rows, *R* rows, *Pivot* irrigation by central pivot, CEC

[a]Watanabe et al. (2017)
[b]Watanabe (2013)
[c]Souza et al. (2010)
[d]Klein and Libardi (2002)
[e]Silva et al. (2010)
[f]Ayayi et al. (2009)
[g]Severiano et al. (2011)

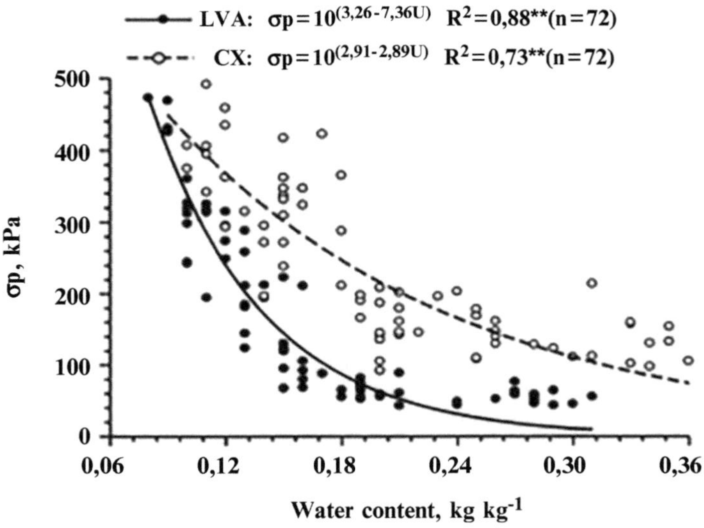

Fig. 6.2 Variation of pre-consolidation pressure (σ_p) as a function of water content of a Red-Yellow Ferralsol (LVA weak structure) and Cambisol Háplico (CX moderate to strong structure) (Severiano et al. 2008)

the soil matrix (Al mapped), among others. In this case, the pore filling is the primary evidence that explains the increase in the soil bulk density and compaction, the obstruction of pores and the reduction of the porous space, not only due to direct effects of tillage but also due to the pressure exerted by machines and equipment.

Although the structure of Ferralsols is more resilient to the impact of intensive land use, these soils are still susceptible to the damaging effects of soil management practices. In Ferralsols, the granular microstructure is favoured by the electrochemical flocculation condition and by the activity of cementing agents, especially Fe and Al oxides, and shows high resistance to the impacts of soil physical degradation (Ferreira et al. 2007). A study was conducted with very clayey, oxidized Red Ferralsols from Alto Paranaíba (Minas Gerais state, Brazil) characterized by strong to very strong structure that was quite resistant to disruption. However, even these well-structured soils change over time with intensive land use. Pereira (2020) reported a decline in the stability of aggregates with land use, especially in the 0–20 cm layer of well-structured Ferralsols after 1, 17, 22, 28, 32 and 43 years of growing vegetables. Change was reflected in the reduction of the weighted and geometric and macroaggregate average diameters and the increase of meso and microaggregates at a rate of 1.7% per year. The physical changes were greatest in the first year, when cracks are observed in aggregates examined in soil thin sections and by electron microscope. The first year coincides with the first deep tillage, pH changes with liming, the addition of ions by fertilizers and the reduction of organic matter contents. The soil under natural vegetation (Cerrado vegetation) presents a well-developed rounded to sub-rounded granular microstructure separated by compound

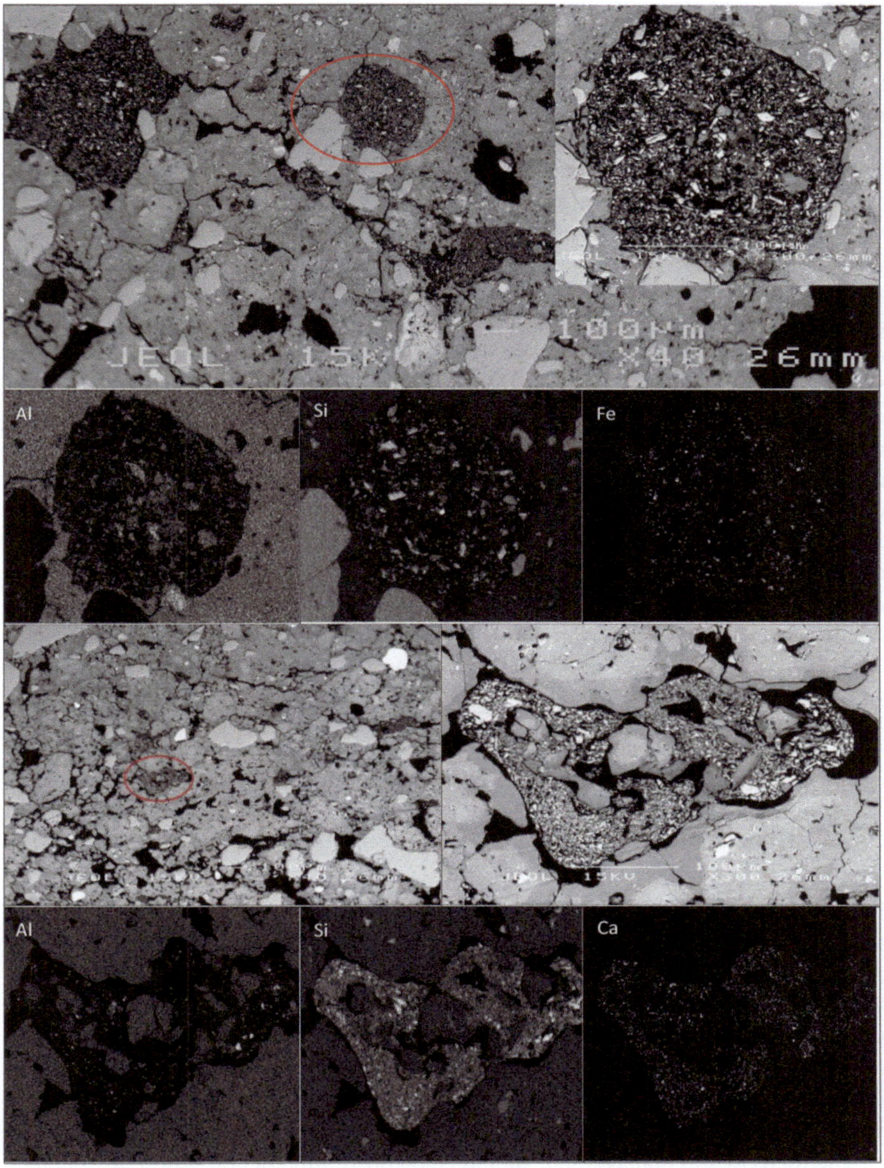

Fig. 6.3 Backscattered electron micrographs of infillings consisting of Al-, Fe-, Si- and Ca-enriched particles or clods for samples taken under irrigated crops in the Jaguaribe-Apodi Plateau, Ceará state, Brazil (Oliveira et al. in press)

packing voids and subangular blocks internally constituted by coalesced granules. Contrastingly, in the soils under cultivation for 17, 22, 28, 32 and 43 years, the granular microstructure declined in size, and there was closer contact between the granule surfaces, forming curved planar voids. The total intra-aggregate porosity (connected and isolated pore) in Cerrado vegetation (22%) was similar to the 1-year treatment (22%) but declined after 17 years of cultivation (14%) and then increased after 43 years of cultivation (32%). On the other hand, there is a reduction in the number of voids with the time of cultivation, ranging from 39 thousand isolated voids under Cerrado vegetation to 21, 14 and 12 thousand voids in soils after 1, 17 and 43 years of cultivation, respectively. Although the number of voids in the aggregates decreased with time of land use, an increase in the volume of isolated voids was found with an average of 167.6 μm^3 after 17 and 43 years compared with 52.9 μm^3 under Cerrado vegetation. With increasing time under this land use, larger voids formed, but pore network connectivity decreased. The tortuosity data show that the inter-aggregate connected voids (main pore) presented more tortuous paths until 17 years of cultivation and at 43 years more straight paths when compared with Cerrado vegetation. According to Peth et al. (2008), more aligned voids can be associated with a better interconnected network of more continuous flow channels. However, as observed for the soil after 43 years of land use, the better alignment in connected voids (less tortuosity) was not reflected in greater connectivity (greater number of Euler). Changes in the shape of the intra-aggregate voids were also observed. For instance, the sphericity reduced in areas under cultivation compared with Cerrado vegetation, corroborating to the findings of Carducci et al. (2014) and Tippkötter et al. (2009).

6.7 Final Comments

Subsoil physical restrictions may be natural properties or originate from management practices that increase soil density. Restrictions on air and water circulation can affect the development of cultivated plants and their productivity. Soil and subsoil biological restrictions are intensified by the physical changes imposed by management, especially in the interactions between the root system and microorganisms. Alleviation of subsoil physical constraints will favour the development of the root system and increase nutrient absorption and water acquisition by plants so long there are no other physical, chemical or biological constraints in the subsoil.

The focus of soil management should be on prevention or correction of soil physical constraints, but it is necessary to improve our understanding of the changes caused by management, since some effects will persist after amelioration. Even when aggregates are stable, they can be modified on a microscopic scale, or spatially in the soil profile, even when these changes are not detected macroscopically or by visual observations. Better characterization of residual physical limitations, even after amelioration, is needed to design management practices that avoid degradation of soil physical properties.

Increases of soil organic matter promote soil resilience. The chemical, physical and biological advantages associated with organic matter can be achieved not only by its presence on the surface with the deposition of crop and pasture residues but also by land use systems in which the soil volume explored by root systems favours the increase of organic matter at depth. In addition to the direct incorporation of organic matter, root systems can alter physical conditions at depth when biopores are created by the penetration of roots and the subsequent degradation of tissues. Understanding the role of organic matter, either by its direct addition or by its indirect effects in increasing porosity, is necessary to minimize the effects of physical constraints in soils managed for agricultural production.

The rapid and direct identification of subsoil physical constraints, as well as the development of models and methods for prediction, requires investment in training more students, field technicians and farmers to identify and record the problems. Diagnostic methods to identify soil physical limitations in field conditions are essential for the development of integrated land management practices that maintain soil quality and plant productivity. In addition, the advances in machine learning techniques allow the predictive assessment of soil physical constraints and their susceptibility to change at spatial scale based on local or national surveys sampling. These spatially explicit models could better orient the management of land use systems, avoiding the generation or intensification of soil restrictions by mechanical practices. Therefore, more priority should be given to approaches that predict possible physical subsoil restrictions according to soil susceptibility and specific management practices rather than a sole focus on intervention and correction techniques.

Regardless of a soil's susceptibility to compaction, a focus on avoidance of subsoil compaction should have the highest priority. Soil compaction is a silent, widespread and hidden phenomenon, which can compromise the quality and sustainability of agroecosystems, affecting essential soil functions and soil ecosystem functions. The diagnosis of the problem at early stages can be decisive for the most effective and least costly management strategy for its mitigation.

References

Abdollahi L, Schjønning P, Elmholt S, et al. The effects of organic matter application and intensive tillage and traffic on soil structure formation and stability. Soil Tillage Res. 2014;136:28–37. https://doi.org/10.1016/j.still.2013.09.011.

Agherkakli B, Najafi A, Sadeghi SH. Ground based operation effects on soil disturbance by steel tracked skidder in a steep slope of forest. J For Sci. 2010;56(6):278–84. https://doi.org/10.17221/93/2009-JFS.

Agrawal RP. Water and nutrient management in sandy soils by compaction. Soil Tillage Res. 1991;19(121–130):1991. https://doi.org/10.1016/0167-1987(91)90081-8.

Agüera J, Carballido J, Gil J, et al. Design of a soil cutting resistance sensor for application in site-specific tillage. Sensors. 2013;13:5945–57. https://doi.org/10.3390/s130505945.

Alakukk L, Weisskopf P, Chamen WCT, et al. Prevention strategies for field traffic-induced subsoil compaction: a review: part 1. Machine/soil interactions. Soil Tillage Res. 2003;73(1–2):145–60. https://doi.org/10.1016/S0167-1987(03)00107-7.

Alaoui A, Rogger M, Peth S, et al. Does soil compaction increase floods? A review. J Hydrol. 2018;557:631–42. https://doi.org/10.1016/j.jhydrol.2017.12.052.
Alencar GV Compactação e/ou adensamento de solos cultivados com melão em litosequência na região da chapada do Apodi-RN. Thesis, Universidade Federal do Ceará; 2000.
Alvarenga RC, Costa LM, Moura Filho W, et al. Crescimento de raízes de leguminosas em camadas de solo compactadas artificialmente. Rev Bras Cienc Solo. 1996;20:319–26.
Ampoorter E, Goris R, Cornelis WM, et al. Impact of mechanized logging on compaction status of sandy forest soils. For Ecol Manag. 2007;241(1–3):162–74. https://doi.org/10.1016/j.foreco.2007.01.019.
Aravena JE, Berli M, Ruiz S, et al. Quantifying coupled deformation and water flow in the rhizosphere using X-ray microtomography and numerical simulations. Plant Soil. 2014;376(1):95–110. https://doi.org/10.1007/s11104-013-1946-z.
Ares A, Terry TA, Miller RE, et al. Ground-based forest harvesting effects on soil physical properties and Douglas-fir growth. Soil Sci Soc Am J. 2005;69(6):1822–32. https://doi.org/10.2136/sssaj2004.0331.
Armando Gomez G, Singer MJ, Powers RF, et al. Soil compaction effects on water status of ponderosa pine assessed through 13C/12C composition. Tree Physiol. 2002;22:459–67. https://doi.org/10.1093/treephys/22.7.459.
Arshad MA, Lowery B, Grossman B. Physical tests for monitoring soil quality. In: Doran JW, Jones AJ, editors. Methods for assessing soil quality, Special publication 49. Madison: SSSA; 1996. p. 123–41.
Arvidsson J, Trautner A, Van den Akker JJH, et al. Subsoil compaction caused by heavy sugarbeet harvesters in southern Sweden: II. Soil displacement during wheeling and model computations of compaction. Soil Tillage Res. 2001;60(1–2):79–89. https://doi.org/10.1016/S0167-1987(01)00168-4.
Australian Bureau of Statistics (ABS). Year book Australia—2012. Canberra: Australian Bureau of Statistics; 2012.
Ayayi AE, Dias Junior MS, Curi N, et al. Relation of strength and mineralogical attributes in Brazilian latosols. Soil Till Res. 2009;102:14–8. https://doi.org/10.1016/j.still.2008.05.013.
Ball BC, Batey T, Munkholm LJ. Field assessment of soil structural quality - a development of the Peerlkamp test. Soil Use Manag. 2007;23:329–37. https://doi.org/10.1111/j.1475-2743.2007.00102.x.
Barbosa SM, Silva BM, Oliveira GC, et al. Deep furrow and additional liming for coffee cultivation under first year in a naturally dense Inceptisol. Geoderma. 2020;357:113934. https://doi.org/10.1016/j.geoderma.2019.113934.
Bassegio D, Sarto MVM, Rosolem CA, et al. Guar root and shoot growth as affected by soil compaction. Pesqui Agropecu Trop. 2018;48:163–9. https://doi.org/10.1590/1983-40632018v4852189.
Battigelli JP, Spence JR, Langor DW, et al. Short-term impact of forest soil compaction and organic matter removal on soil mesofauna density and oribatid mite diversity. Can J For Res. 2004;34:1136–49. https://doi.org/10.1139/x03-267.
Bell LW, Hayes RC, Pembleton KG, et al. Opportunities and challenges in Australian grasslands: pathways to achieve future sustainability and productivity imperatives. Crop Pasture Sci. 2014;65:489–507. https://doi.org/10.1071/CP13420.
Berisso FE, Schjønning P, Keller T, et al. Persistent effects of subsoil compaction on pore size distribution and gas transport in a loamy soil. Soil Tillage Res. 2012;122:42–51. https://doi.org/10.1016/j.still.2012.02.005.
Bessou C, Mary B, Léonard MJ, et al. Modelling soil compaction impacts on nitrous oxide emissions in arable fields. Eur J Soil Sci. 2010;61:348–63. https://doi.org/10.1111/j.1365-2389.2010.01243.x.
Beylich A, Oberholzer HR, Schrader S, et al. Evaluation of soil compaction effects on soil biota and soil biological processes in soils. Soil Tillage Res. 2010;109(2):133–43. https://doi.org/10.1016/j.still.2010.05.010.

Bilotta GS, Brazier RE, Haygarth PM. The impacts of grazing animals on the quality of soils, vegetation, and surface waters in intensively managed grasslands. Adv Agron. 2007;94:237–80. https://doi.org/10.1016/S0065-2113(06)94006-1.

Bruand A, Cousin I, Nicoullaud B, et al. Backscattered electron scanning images of soil porosity for analyzing soil compaction around roots. Soil Sci Soc Am J. 1996;60:895–901. https://doi.org/10.2136/sssaj1996.03615995006000030031x.

Busscher WJ. Adjustment of flat-tipped penetrometer resistance data to a common water content. Trans ASAE. 1990;33(2):519–24. https://doi.org/10.13031/2013.31360.

Busse MD, Beatle SE, Powers RF, et al. Microbial community responses in forest mineral soil to compaction, organic matter removal, and vegetation control. Can J For Res. 2006;36:577–88. https://doi.org/10.1139/x05-294.

Bygdén G, Eliasson L, Wästerlund I. Rut depth, soil compaction and rolling resistance when using bogie tracks. J Terramech. 2003;40(3):179–90. https://doi.org/10.1016/j.jterra.2003.12.001.

Camara RK, Klein VA. Escarificação em plantio direto como técnica de conservação do solo e da água. Rev Bras Cienc Solo. 2005;29:789–96. https://doi.org/10.1590/S0100-06832005000500014.

Cambi M, Certini G, Neri F, et al. The impact of heavy traffic on forest soils: a review. For Ecol Manag. 2015;338:124–38. https://doi.org/10.1016/j.foreco.2014.11.022.

Cambi M, Hoshika Y, Mariotti B, et al. Compaction by a forest machine affects soil quality and Quercus robur L. seedling performance in an experimental field. For Ecol Manag. 2017;384:406–14. https://doi.org/10.1016/j.foreco.2016.10.045.

Cardoso AS, Quintana BG, Janusckiewicz ER, et al. How do methane rates vary with soil moisture and compaction, N compound and rate, and dung addition in a tropical soil? Int J Biometeorol. 2019;63:1533–40. https://doi.org/10.1007/s00484-018-1641-0.

Carducci CE, Oliveira GC, Curi N, et al. Scaling of pores in 3D images of Latosols (Oxisols) with contrasting mineralogy under a conservation management system. Soil Res. 2014;52:231. https://doi.org/10.1071/SR13238.

Chamen T. Controlled traffic farming – from worldwide research to adoption in Europe and its future prospects. Acta Technol Agric. 2015;3:64–73. https://doi.org/10.1515/ata-2015-0014.

Chen G, Weil RR. Penetration of cover crop roots through compacted soils. Plant Soil. 2010;331:31–43. https://doi.org/10.1007/s11104-009-0223-7.

Chen G, Weil RR. Root growth and yield of maize as affected by soil compaction and cover crops. Soil Tillage Res. 2011;117:17–27. https://doi.org/10.1016/j.still.2011.08.001.

Colombi T, Braun S, Keller T, et al. Artificial macropores attract crop roots and enhance plant productivity on compacted soils. Sci Total Environ. 2017;574:1283–93. https://doi.org/10.1016/j.scitotenv.2016.07.194.

Costa PA, Mota JCA, Romero RE, et al. Changes in soil pore network in response to twenty-three years of irrigation in a tropical semiarid pasture from northeast Brazil. Soil Tillage Res. 2014;137:23–32. https://doi.org/10.1016/j.still.2013.11.004.

Cresswell HP, Kirkegaard JA. Subsoil amelioration by plant roots – the process and the evidence. Aust J Soil Res. 1995;33:221–39. https://doi.org/10.1071/SR9950221.

da Silva AP, Kay B. Estimating the least limiting water range of soils from properties and management. Soil Sci Soc Am J. 1997;61:877–83. https://doi.org/10.2136/sssaj1997.03615995006100030023x.

Davies SL, Armstrong R, Macdonald L, et al. Soil constraints: a role for strategic deep tillage. In: Pratley J, Kirkegaard J, editors. Australian agriculture in 2020: from conservation to automation. Wagga Wagga: Agronomy Australia and Charles Sturt University; 2019. p. 117–35.

Dexter AR. Compression of soil around roots. Plant Soil. 1987;97:401–6. https://doi.org/10.1007/BF02383230.

Dexter AR. Soil physical quality Part I. Theory, effects of soil texture, density, and organic mailer, and effects on root growth. Geoderma. 2004;120:201–14. https://doi.org/10.1016/j.geoderma.2003.09.004.

Dias Júnior MDES. Compactação do solo. In: Novais RF, Alvarez VH, Schaefer CEGR, editors. Tópicos em Ciência do Solo. 3rd ed. Sociedade Brasileira de Ciência do Solo: Viçosa; 2000. p. 55–94.

Drescher MS, Eltz FLF, Denardin JE, et al. Resistência a penetração e rendimento da soja após intervenção mecânica em Latossolo Vermelho sob plantio direto. R Bras CI Solo 2012; 38: 1836–1844. https://doi.org/10.1590/S0100-06832012000600018.

Drescher MS, Reinert DJ, Denardin JE, et al. Duração das alterações em propriedades físico-hídricas de Latossolo argiloso decorrentes da escarificação mecânica. Pesq Agropec Bras. 2016;51(2):159–68. https://doi.org/10.1590/S0100-204X2016000200008.

Ferreira FP, Azevedo ACD, Dalmolin RSD, et al. Carbono orgânico, óxidos de ferro e distribuição de agregados em dois solos derivados de basalto no Rio Grande do Sul-Brasil. Cienc Rural. 2007;37(2):381–8. https://doi.org/10.1590/S0103-84782007000200013.

Ferreira TR, Pires LF, Wildenschild D, et al. Lime application effects on soil aggregate properties: use of the mean weight diameter and synchrotron-based X-ray μCT techniques. Geoderma. 2019;338:585–96. https://doi.org/10.1016/j.geoderma.2018.10.035.

Frey B, Kremer J, Rüdt A, et al. Compaction of forest soils with heavy logging machinery affects soil bacterial community structure. Eur J Soil Biol. 2009;45(4):312–20. https://doi.org/10.1016/j.ejsobi.2009.05.006.

Frisby JC, Pfost DL. Soil compaction: the silent thief. University of Missouri Extension; 2020. Available at: https://extension2.missouri.edu/g1630.

Gaertig T, Schack-Kirchner H, Hildebrand EE, et al. The impact of soil aeration on oak decline in southwestern Germany. For Ecol Manag. 2002;159(1–2):15–25. https://doi.org/10.1016/S0378-1127(01)00706-X.

Gaiser T, Perkons U, Küpper PM. Modeling biopore effects on root growth and biomass production on soils with pronounced sub-soil clay accumulation. Ecol Model. 2013;256:6–15. https://doi.org/10.1016/j.ecolmodel.2013.02.016.

Greacen EL, Sands R. Compaction of forest soils. A review. Aust J Soil Res. 1980;18(2):163–89. https://doi.org/10.1071/SR9800163.

Gregory AS, Whalley WR, Watts CW, et al. Whit more calculation of the compression index and precompression stress from soil compression test data. Soil Tillage Res. 2006;89:45–57. https://doi.org/10.1016/j.still.2005.06.012.

Håkansson I, Lipiec J. A review of the usefulness of relative bulk density values in studies of soil structure and compaction. Soil Tillage Res. 2000;53(2):71–85. https://doi.org/10.1016/S0167-1987(99)00095-1.

Hall DJM, Davies SL, Bell RW, et al. Soil management systems to overcome multiple constraints for dryland crops on deep sands in the Mediterranean climate of the south coast of Western Australia. Agronomy. 2020;10(12):1881.

Hamidov A, Helming K, Bellocchi G, et al. Impacts of climate change adaptation options on soil functions: a review of European case-studies. Land Degrad Dev. 2018;29:2378–89. https://doi.org/10.1002/ldr.3006.

Hamza MA, Anderson WK. Soil compaction in cropping systems: a review of the nature, causes and possible solutions. Soil Tillage Res. 2005;82:121–45. https://doi.org/10.1016/j.still.2004.08.009.

Han HS, Page-Dumroese D, Han SK, et al. Effects of slash, machine passes, and soil moisture on penetration resistance in a cut-to-length harvesting. Int J For Eng. 2006;17(2):11–24. https://doi.org/10.1080/14942119.2006.10702532.

Hattori D, Kenzo T, Irino KO. Effects of soil compaction on the growth and mortality of planted dipterocarp seedlings in a logged-over tropical rainforest in Sarawak, Malaysia. For Ecol Manag. 2013;310:770–6. https://doi.org/10.1016/j.foreco.2013.09.023.

Haynes R, Naidu R. Influence of lime, fertilizer and manure applications on soil organic matter content and soil physical conditions: a review. Nut Cycl Agroecosyst. 1998;51:123–37. https://doi-org.ez35.periodicos.capes.gov.br/10.1023/A:1009738307837.

Hemmat A, Adamchuk VI. Sensor systems for measuring soil compaction: review and analysis. Comput Electron Agric. 2008;63:89–103. https://doi.org/10.1016/j.compag.2008.03.001.

Hoefer G, Bachmann J, Hartge KH. Can the EM38 probe detect spatial patterns of subsoil compaction? In: Viscarra Rossel RA, et al., editors. Proximal soil sensing, Progress in soil science 1. Springer; 2010. p. 265–73. https://doi.org/10.1007/978-90-481-8859-8_22.

Horn R. Soil compaction and consequences of soil deformation on changes in soil functions. In: Northcliff S, editor. Task force – solutions under foot. Catena Verlag; 2015. p. 28–32.

Horn R, Domżżał H, Słowińska-Jurkiewicz A, et al. Soil compaction processes and their effects on the structure of arable soils and the environment. Soil Tillage Res. 1995;35:23–36. https://doi.org/10.1016/0167-1987(95)00479-C.

Horn R, Vossbrink J, Becker S. Modern forestry vehicles and their impacts on soil physical properties. Soil Tillage Res. 2004;79(2):207–19. https://doi.org/10.1016/j.still.2004.07.009.

IBGE – Instituto Brasileiro de Geografia e Estatística. Censo agropecuário 2017. Brasília-DF. 2017. https://censoagro2017.ibge.gov.br/templates/censo_agro/resultado/estabelecimentos.html.

Isbell RF, National Committee on Soil and Terrain. The Australian soil classification. Australian soil and land survey handbooks series. 3rd ed. Melbourne: CSIRO Publishing; 2021.

Islam T, Chik Z, Mustafa MM, et al. Modeling of electrical resistivity and maximum dry density in soil compaction measurement. Environ Earth Sci. 2012;67:1299–305. https://doi.org/10.1007/s12665-012-1573-7.

Jaafari A, Najafi A, Zenner EK. Ground-based skidder traffic changes chemical soil properties in a mountainous Oriental beech (*Fagus orientalis* Lipsky) forest in Iran. J Terramech. 2014;55:39–46. https://doi.org/10.1016/j.jterra.2014.06.001.

Jesus GL, Silva IR, Almeida LFJ, et al. Produtividade do eucalipto, atributos físicos do solo e frações da matéria orgânica influenciadas pela intensidade de tráfego e resíduos de colheita. Rev Bras Cienc Solo. 2015;39:1190–203. https://doi.org/10.1590/01000683rbcs20140494.

Kaufmann M, Tobias S, Schulin R. Comparison of critical limits for crop plant growth based on different indicators for the state of soil compaction. J Plant Nutr Soil Sci. 2010;173:573–83. https://doi.org/10.1002/jpln.200900129.

Keller T, Sandin M, Colombi T, Horn R, Or D. Historical increase in agricultural machinery weights enhanced soil stress levels and adversely affected soil functioning. Soil Tillage Res. 2019;194:10493. https://doi.org/10.1016/j.still.2019.104293.

Klein VA, Libardi PL. Densidade e distribuição do diâmetro dos poros de um latossolo vermelho, sob diferentes sistemas de uso e manejo. Rev Bras Ci Solo. 2002;26:857–67. https://doi.org/10.1590/S0100-06832002000400003.

Lebert M, Böken H, Glante F. Soil compaction - indicators for the assessment of harmful changes to the soil in the context of the German Federal Soil Protection Act. J Environ Manag. 2007;82:388–97. https://doi.org/10.1016/j.jenvman.2005.11.022.

Lee WS, Alchanatis V, Yang C, et al. Sensing technologies for precision specialty crop production. Comput Electron Agric. 2010;74:2–33. https://doi.org/10.1016/j.compag.2010.08.005.

Lees KJ, McKenzie AJ, Newell-Price JP, et al. The effects of soil compaction mitigation on belowground fauna: how earthworms respond to mechanical loosening and power harrow cultivation. Agric Ecosyst Environ. 2016;232:273–82. https://doi.org/10.1016/j.agee.2016.07.026.

Levien R, Cogo NP, Rockenbach CA. Erosão na cultura do milho em diferentes sistemas de cultivo anterior e métodos de preparo do solo. Rev Bras Cienc Solo. 1990;14:73–80.

Li CH, Ma BL, Zhang TQ. Soil bulk density effects on soil microbial populations and enzyme activities during the growth of maize (*Zea mays* L.) planted in large pots under field exposure. Can J Soil Sci. 2002;82:147–54. https://doi.org/10.4141/S01-026.

Lipiec J, Hatano R. Quantification of compaction effects on soil physical properties and crop growth. Geoderma. 2003;116(1–2):107–36. https://doi.org/10.1016/S0016-7061(03)00097-1.

Lipiec L, Medvedev VV, Birkas M, et al. Effect of soil compaction on root growth and crop yield in Central and Eastern Europe. Int Agrophys. 2003;17:61–9.

Löfkvist J, Whalley R, Clark LJ. A rapid screening method for good root-penetration ability: comparison of species with very different root morphology. Acta Agric Scand B Soil Plant Sci. 2005;55:120–4. https://doi.org/10.1080/09064710510008504.

Lui W, Upadhyaya SK, Kataoka T, et al. Development of a texture/soil compaction sensor. In: Robert PC, Rust RH, Larson WE, editors. Precision agriculture. Proceedings of the 3rd international conference on precision agriculture. ASA, CSSA, SSSA; 1996. p. 617–30. https://doi.org/10.1071/SR09076.

Ma Q, Bell RW, Mattiello E. Nutrient acquisition with particular reference to subsoil constraints. In: Oliveira TS and Bell RW, editors. Subsoil Constraints for Crop Production, 1st edn. LOCAL: Springer; 2022. p. 289–321.

MacEwan RJ, Crawford DM, Newton PJ, et al. High clay contents, dense soils, and spatial variability are the principal subsoil constraints to cropping the higher rainfall land in south-eastern Australia. Aust J Soil Res. 2010;48:150–66. https://doi.org/10.1071/SR09076.

Magalhães EM, Oliveira GC, Severiano EC, et al. Recuperação estrutural e produção de capim-tifton 85 em um Argissolo Vermelho-Amarelo compactado. Cienc Anim Bras. 2009;10:68–76. https://doi.org/10.5216/cab.v10i1.2211.

Manik SMN, Pengilley G, Dean G, et al. Soil and crop management practices to minimize the impact of waterlogging on crop productivity. Front Plant Sci. 2019;10:140. https://doi.org/10.3389/fpls.2019.00140.

Marchi E, Picchio R, Mederski PS. Impact of silvicultural treatment and forest operation on soil and regeneration in Mediterranean Turkey oak (*Quercus cerris* L.) coppice with standards. Ecol Eng. 2016;95:475–84. https://doi.org/10.1016/j.ecoleng.2016.06.084.

Martínez E, Fuentes JP, Silva P, et al. Soil physical properties and wheat root growth as affected by no-tillage and conventional tillage systems in a Mediterranean environment of Chile. Soil Tillage Res. 2008;99:232–44. https://doi.org/10.1016/j.still.2008.02.001.

McDonald TP, Seixas F. Effect of slash on forwarder soil compaction. J For Eng. 1997;8(2):15–26.

McPhee JE, Antille DL, Tullberg JN, et al. Managing soil compaction – a choice of low-mass autonomous vehicles or controlled traffic? Biosyst Eng. 2020;195:227–41. https://doi.org/10.1016/j.biosystemseng.2020.05.006.

Meyer C, Lüscher P, Schulin R. Recovery of forest soil from compaction in skid tracks planted with black alder (*Alnus glutinosa* (L.) Gaertn.). Soil Tillage Res. 2014;143:7–16. https://doi.org/10.1016/j.still.2014.05.006.

Milde GA, Dedecek RA, Gava JL. Unidades de colheita: estratégia para evitar a compactação dos solos florestais. Rev Arvore. 2010;34:1075–83. https://doi.org/10.1590/S0100-67622010000600013.

Millan E, Ruiz HA, Fernandes RBA, et al. Condutividade hidráulica, porosidade, resistência mecânica e intervalo hídrico ótimo em Latossolos artificialmente compactados. Rev Bras de Eng Agricola e Ambient. 2014;18:1003–9. https://doi.org/10.1590/1807-1929/agriambi.v18n10p1003-1009.

Miller RE, Colbert SR, Morris LA. Effects of heavy equipment on physical properties of soils and on long-term productivity: a review of literature and current research, Technical bulletin no 887. Research Triangle Park: National Council for Air and Stream Improvement, Inc (NCASI); 2004. p. 76. https://www.fs.fed.us/pnw/olympia/silv/publications/opt/499_MillerEtal2004.pdf.

Moodie M, Saunders C, Ucgul M, et al. The how's and whys for ripping deep sandy soils of the low rainfall Mallee. Grains Research & Development Corporation – GRDC; 2020. https://grdc.com.au/resources-and-publications/grdc-update-papers/tab-content/grdc-update-papers/2020/02/the-hows-and-whys-for-ripping-deep-sandy-soils-of-the-low-rainfall-mallee.

Moraes MT, Debiasi H, Carlesso R, et al. Critical limits of soil penetration resistance in a Rhodic Eutrudox. Rev Bras Cienc Solo. 2014a;38:288–98. https://doi.org/10.1590/S0100-06832014000100029.

Moraes MT, Silva VR, Zwirtes AL, et al. Use of penetrometers in agriculture: a review. Eng Agricola. 2014b;34(1):179–93. https://doi.org/10.1590/S0100-69162014000100019.

Mordhorst A, Peth S, Horn R. Influence of mechanical loading on static and dynamic CO2 efflux on differently textured and managed Luvisols. Geoderma. 2014;219-220:1–13. https://doi.org/10.1016/j.geoderma.2013.12.020.

Naderi-Boldaji M, Weisskopf P, Stettler M, et al. Predicting the relative density from on-the-go horizontal penetrometer measurements at some arable top soils in Northern Switzerland. Soil Tillage Res. 2016;159:23–32. https://doi.org/10.1016/j.still.2015.12.002.

Nadezhdina N, Prax A, Čermák J. Spruce roots under heavy machinery loading in two different soil types. For Ecol Manag. 2012;282:46–52. https://doi.org/10.1016/j.foreco.2012.06.054.

Naghdi R, Solgi A, Ilstedt U. Soil chemical and physical properties after skidding by rubber-tired skidder in Hyrcanian forest, Iran. Geoderma. 2016;265:12–8. https://doi.org/10.1016/j.geoderma.2015.11.009.

Oliveira JM. Impactos do manejo florestal mecanizado sobre as propriedades físicas de Latossolos argilo-arenosos. Dissertation, Universidade Federal de Viçosa; 2018.

Oliveira PD, Sato MK, Lima HV, et al. Critical limits of the degree of compactness and soil penetration resistance for the soybean crop in N Brazil. J Plant Nutr Soil Sci. 2016;179:78–87. https://doi.org/10.1002/jpln.201400315.

Oliveira GC, Silva BM, Carducci CE. Melhoria físico-hídrica do ambiente radicular do cafeeiro em condições de sequeiro e implicações no uso da água. In: Severiano EC, Moraes MF, Paula AM, editors. Tópicos em Ciência do Solo, vol. 10. 1st ed. Viçosa: Sociedade Brasileira de Ciência do Solo; 2019. p. 70–140.

Oliveira TS, Gomes LC, Lacerda NB, et al. Soil void infillings: a potential source of information to better understand the effects of agricultural practices on soil structure. Rev Bras Cienc Solo (Submitted); in press.

Peixoto DS, Silva BM, Oliveira GC, et al. A soil compaction diagnosis method for occasional tillage recommendation under continuous no tillage system in Brazil. Soil Tillage Res. 2019a;194:104307. https://doi.org/10.1016/j.still.2019.104307.

Peixoto DS, Silva BM, Silva SHG, et al. Diagnosing, ameliorating, and monitoring soil compaction in no-till Brazilian soils. Agrosyst Geosci Environ. 2019b;2:180035. https://doi.org/10.2134/age2018.09.0035.

Pereira MTJ (2020) Impactos do cultivo intensivo de olerícolas em longo prazo na mineralogia, química e estrutura de Latossolo oxídico. Thesis, Universidade Federal de Viçosa.

Petry HB, Mazurana M, Marodin GAB, et al. Root distribution of peach rootstocks affected by soil compaction and acidity. Rev Bras Cienc Solo. 2016;40:e0150135. https://doi.org/10.1590/18069657rbcs20150135.

Peth S, Horn R, Beckmann F et al. Threedimensional quantification of intra-aggregate porespace features using synchrotron-radiation-based microtomography. Soil Sci Soc Am J. 2008;72: 897–907. http://dx.doi.org/10.2136/sssaj2007.0130.

Pöhlitz J, Rücknagel J, Schlüter S, et al. Estimation of critical stress ranges to preserve soil functions for differently textured soils. Soil Tillage Res. 2020;200:104637. https://doi.org/10.1016/j.still.2020.104637.

Powers RF, Scott DA, Sanchez FG, Stone DM, et al. The North American long-term soil productivity experiment: findings from the first decade of research. For Ecol Manag. 2005;220(1–3):31–50. https://doi.org/10.1016/j.foreco.2005.08.003.

Proctor RR. Fundamental principles of soil compaction. ENR. 1933;111(9):245–8.

Rabot E, Wiesmeier M, Schlüter S, et al. Soil structure as an indicator of soil functions: a review. Geoderma. 2018;314:122–37. https://doi.org/10.1016/j.geoderma.2017.11.009.

Raper RL. Agricultural traffic impacts on soil. J Terramech. 2005;42(3–4):259–80. https://doi.org/10.1016/j.jterra.2004.10.010.

Raper RL, Kirby JM. Soil compaction: how to do it, undo it, or avoid doing it. ASAE distinguished lecture, n. 30. Agricultural equipment technology conference, Louisville, KY, USA, p 14. ASABE publication number 913C0106; 2006.

Renger M, Bohne K, Wessolek G. Bestimmung und Aussagemöglichkeiten der effektiven Lagerungsdichte. In: Wessolek G, Kaupenjohann M, Renge M, editors. Bodenphysikalische Kennwerte und Berechnungsverfahren für die Praxis, Teil II. Berlin: TU Berlin; 2014;27–41.

Reichert JM, Suzuki LEAS, Reinert DJ. Compactação do solo em sistemas agropecuários e florestais: identificação, efeitos, limites críticos e mitigação. In: Cerreta CA, Silva LS, Reichert JM, editors. Tópicos em Ciência do Solo. 5th ed. Viçosa: Sociedade Brasileira de Cíência do Solo.2007;49–134.

Reichert JM, Suzuki LEAS, Reinert DJ, et al. Reference bulk density and critical degree-of-compactness for no-till crop production in subtropical highly weathered soils. Soil Tillage Res. 2009;102:242–54. https://doi.org/10.1016/j.still.2008.07.002.

Reinert DJ, Albuquerque JA, Reichert JM, et al. Limites críticos de densidade do solo para o crescimento de raízes de plantas de cobertura em Argissolo Vermelho. Rev Bras Cienc Solo. 2008;32:1805–16. https://doi.org/10.1590/S0100-06832008000500002.

Rengasamy P. Irrigation water quality and soil structural stability: a perspective with some new insights. Agronomy. 2018;8(5):72. https://doi.org/10.3390/agronomy8050072.

Ribeiro CA. Variabilidade espacial da resistência mecânica do solo à penetração em áreas mecanizadas de cana-de-açúcar (*Saccharum officinarum*). Dissertation, Jaboticabal: FCAV/UNESP; 2010.

Richart A, Tavares Filho J, Brito OR, et al. Compactação do solo: causas e efeitos. Semin Cienc Agrar. 2005;26:321–44.

Rodrigues CK. Colheita e transporte florestal. Curitiba: Carla Krulikowski Rodrigues; 2018. p. 68.

Rosolem CA, Pivetta LA. Mechanical and biological approaches to alleviate soil compaction in tropical soils: assessed by root growth and activity (Rb uptake) of soybean and maize grown in rotation with cover crops. Soil Use Manag. 2017;33:141–52. https://doi.org/10.1111/sum.12313.

Ruser R, Flessa H, Russow R, et al. Emission of N_2O, N_2 and CO_2 from soil fertilized with nitrate: effect of compaction, soil moisture and rewetting. Soil Biol Biochem. 2006;38:263–74. https://doi.org/10.1016/j.soilbio.2005.05.005.

Sandoval FHB, Souza ZM, Lima ES, et al. Structural quality and load-bearing capacity of an Ultisol (Argissolo Vermelho amarelo) in mechanized coffee areas with different deployment times. Rev Bras Cienc Solo. 2020;44:e0190009. https://doi.org/10.36783/18069657rbcs20190009.

Sasaki CM, Gonçalves JLM, Silva AP. Ideal subsoiling moisture content of Latosols used in forest plantations. For Ecol Manag. 2007;243:75–82. https://doi.org/10.1016/j.foreco.2007.02.024.

Sasal MC, Léonard J, Andriulo A, Boizard H. A contribution to understanding the origin of platy structure in silty soils under no tillage. Soil Tillage Res. 2017;173:42–8. https://doi.org/10.1016/j.still.2016.08.017.

Schaefer T, Sohns D. Reduction of soil compaction by covering soil with brushwood. AFZ-Der Wald. 1993:452–5.

Schaefer CEGR, Souza JJLL, Oliveira FS, et al. The geological, geomorphological, climatic and hydrological background of tropical regoliths and hostile subsoils: the Brazilian landmass. In: Oliveira TS, Bell RW, editors. Subsoil constraints for crop production. 1st ed. Cham: Springer; 2022, p. 11–52.

Schjønning P, Thomsen IK, Moldrup P, et al. Linking soil microbial activity to water-and air-phase contents and diffusivities. Soil Sci Soc Am J. 2003;67(1):156–65. https://doi.org/10.2136/sssaj2003.1560.

Schjønning P, Lamandé M, Keller T, et al. Rules of thumb for minimizing subsoil compaction. Soil Use Manag. 2012;28:378–93. https://doi.org/10.1111/j.1475-2743.2012.00411.x.

Schjønning P, van den Akker JJ, Keller T, et al. Driver-pressure-state-impact-response (DPSIR) analysis and risk assessment for soil compaction – a European perspective. Adv Agron. 2015;133:183–237. https://doi.org/10.1016/bs.agron.2015.06.001.

Schneider F, Don A. Root-restricting layers in German agricultural soils. Part I: extent and cause. Plant Soil. 2019;442:433–51. https://doi.org/10.1007/s11104-019-04185-9.

Schnurr-Pütz S, Bååth E, Guggenberger G, et al. Compaction of forest soil by logging machinery favours occurrence of prokaryotes. FEMS Microbiol Ecol. 2006;58:503–16. https://doi.org/10.1111/j.1574-6941.2006.00175.x.

Severiano EC, Oliveira GC, Dias Júnior MS, et al. Pressão de preconsolidação e intervalo hídrico ótimo como indicadores de alterações estruturais de um latossolo e de um Cambissolo sob cana-de-açúcar. Rev Bras Cienc Solo. 2008;32(4):1419–27. https://doi.org/10.1590/S0100-06832008000400006.

Severiano EC, Oliveira GC, Dias Júnior MS, et al. Structural changes in latosols of the Cerrado region: i – relationships between soil physical properties and least limiting water range. R Bras Ci Solo. 2011;35:773–82. https://doi.org/10.1590/S0100-06832011000300013.

Shepherd TG. Visual soil assessment. Volume 1. Field guide for pastoral grazing and cropping on flat to rolling country. 2nd ed. Palmerston North: Horizons Regional Council; 2009. p. 106.

Silva SR, Barros NF, Costa LM. Atributos físicos de dois Latossolos afetados pela compactação do solo. Rev Bras de Eng Agricola e Ambient. 2006;10:842–7. https://doi.org/10.1590/S1415-43662006000400009.

Silva SR, Barros NF, Costa LM, et al. Soil compaction and eucalyptus growth in response to forwarder traffic intensity and load. Rev Bras Cienc Solo. 2008;32:921–32. https://doi.org/10.1590/S0100-06832008000300002.

Silva ARS, Dias Junior MS, Leite FP. Propriedades físicas e mecânicas de latossolos em diferentes manejos florestais. Ciênc Agrotec. 2010;34(6):1483–91. https://doi.org/10.1590/S1413-70542010000600019.

Silva FR, Albuquerque JA, Costa A. Crescimento inicial da cultura da soja em Latossolo Bruno com diferentes graus de compactação. Rev Bras Ciênc Solo. 2014;38:1731–9. https://doi.org/10.1590/S0100-06832014000600008.

Simões D, Silva MR, Fenner PT. Desempenho operacional e custos da operação de subsolagem em área de implantação de eucalipto. Biosci J. 2011;27:692–700.

Soane BD, Blackwell PS, Dickson JW, et al. Compaction by agricultural vehicles: a review I. Soil and wheel characteristics. Soil Tillage Res. 1980;1:207–37. https://doi.org/10.1016/0167-1987(80)90026-4.

Souza FR, Rosa Junior EJ, Fieltz CR, et al. Atributos físicos e desempenho agronômico da cultura da soja em um latossolo vermelho distroférrico submetido a dois sistemas de manejos. Ciênc Agrotec. 2010;34(6):1357–64. https://doi.org/10.1590/S1413-70542010000600001.

Spera ST, Denardin JE, Escosteguy PAV, et al. Clay dispersion in microaggregates of soil incubated with limestone. Rev Bras Ciênc Solo. 2008;32(spe) https://doi.org/10.1590/S0100-06832008000700002.

Spera ST, Magalhães CAS, Denardin JE, et al. Estratificação química e física em solos manejados com sistema plantio direto em Mato Grosso – entraves à produção das culturas. Sinop: Embrapa; 2018. p. 34.

Stone LF, Guimarães CM, Moreira JAA. Compactação do solo na cultura do feijoeiro. I: efeitos nas propriedades físico-hídricas do solo. Rev Bras de Eng Agricola e Ambient. 2002;6:207–12. https://doi.org/10.1590/S1415-43662002000200004.

Suzuki LEAS, Reichert JM, Reinert DJ, Lima CLR, et al. Grau de compactação, propriedades físicas e rendimento de culturas em Latossolo e Argissolo. Pesqui Agropecu. 2007;42:1159–67.

Taylor HM, Roberson GM, Parker Junior JJ. Soil strength-root penetration relations for medium- to coarse-textured soil materials. Soil Sci. 1966;102:18–22.

Thorsøe MH, Noe EB, Lamandé M, et al. Sustainable soil management – farmers' perspectives on subsoil compaction and the opportunities and barriers for intervention. Land Use Policy. 2019;86:427–37. https://doi.org/10.1016/j.landusepol.2019.05.017.

Tippkötter R, Eickhorst T, Taubner H, et al. Detection of soil water in macropores of undisturbed soil using microfocus X-ray tube computerized tomography (µCT). Soil Tillage Res. 2009;105:12–20. https://doi.org/10.1016/j.still.2009.05.001.

U.S. Department of Agriculture. Soil Survey Manual (U.S. Department of Agriculture Handbook No. 18). Paperback. 2019;638.

Van den Akker JJH, Schjønning P. Soil compaction and ways to prevent it. In: Schjønning P, Elmholt S, Christensen BT, editors. Managing soil quality: challenges in modern agriculture. Wallingford: CAB International; 2004. p. 163–84.
Van den Akker JJH, Soane B. Compaction. In: Hillel D, editor. Encyclopedia of soils in the environment. Amsterdam: Academic Press; 2004. p. 285–93.
Van Der Weert R. Influence of mechanical forest clearing on soil conditions and the resulting effects on root growth. Trop Agric. 1974;51(2):325–31.
Van Dijck SJE, Van Asch TWJ. Compaction of loamy soils due to tractor traffic in vineyards and orchards and its effect on infiltration in southern France. Soil Tillage Res. 2002;63:141–53. https://doi.org/10.1016/S0167-1987(01)00237-9.
Vance WH, Milroy SP. Water acquisition by roots from the subsoil: impact of physical constraints on the dynamics of water capture. In: Oliveira TS, Bell RW, editors. Subsoil constraints for crop production. 1st ed. Cham: Springer; 2022, p. 323–45.
Vaz CM, Manieri JM, De Maria IC, et al. Modeling and correction of soil penetration resistance for varying soil water content. Geoderma. 2011;166(1):92–101. https://doi.org/10.1016/j.geoderma.2011.07.016.
Wahlströma EM, Kristensenb HL, Thomsen IK, et al. Subsoil compaction effect on spatio-temporal root growth, reuse of biopores and crop yield of spring barley. Eur J Agron. 2020;123:126225. https://doi.org/10.1016/j.eja.2020.126225.
Wästerlund I. Compaction of till soils and growth tests with Norway spruce and scots pine. For Ecol Manag. 1985;11(3):171–89. https://doi.org/10.1016/0378-1127(85)90025-8.
Watanabe R. Susceptibilidade a compactação de solos em agroecossistemas no semiárido cearense. Dissertation, Universidade Federal do Ceará; 2013.
Watanabe R, Figueiredo GC, Silva AP, et al. Soil compressibility under irrigated perennial and annual crops in a semi-arid environment. Rev Bras Cienc Solo. 2017;41:e0160206. https://doi.org/10.1590/18069657rbcs20160206.
Williams SM, Weil RR. Crop cover root channels may alleviate soil compaction effects on soybean crop. Soil Sci Soc Am J. 2004;68:1403–9. https://doi.org/10.2136/sssaj2004.1403.
Williamson JR, Neilsen WA. The influence of forest site on rate and extent of soil compaction and profile disturbance of skid trails during ground-based harvesting. Can J For Res. 2000;30(8):1196–205. https://doi.org/10.1139/x00-041.
Wolfe EC (2020) Country pasture profile Australia. Retrieved from Graham Centre website: https://www.csu.edu.au/research/grahamcentre/resources.
Wronski EB, Humphreys NA. Method for evaluating the cumulative impact of ground- based logging systems on soils. J For Eng. 1994;5(2):9–20. https://doi.org/10.1080/08435243.1994.10702656.
Yang W, Kong Q, Ho SCM, et al. Real-time monitoring of soil compaction using piezoceramic-based embeddable transducers and wavelet packet analysis. IEEE Access. 2018;6:5208–14. https://doi.org/10.1109/ACCESS.2018.2790902.
Zenner EK, Berger AL. Influence of skidder traffic and canopy removal intensities on the ground flora in a clearcut-with-reserves northern hardwood stand in Minnesota, USA. For Ecol Manag. 2008;256(10):1785–94. https://doi.org/10.1016/j.foreco.2008.05.030.
Zhao Y, Liu Z, Wu J. Grassland ecosystem services: a systematic review of research advances and future directions. Landsc Ecol. 2020;35:793–814. https://doi.org/10.1007/s10980-020-00980-3 (0123456789.
Zimmermann B, Elsenbeer H. The near-surface hydrological consequences of disturbance and recovery: a simulation study. J Hydrol. 2009;364(1–2):115–27. https://doi.org/10.1016/j.jhydrol.2008.10.016.

Chapter 7
Subsoil and Surface Soil Constraints of Mined Land and Tailings

Hwat Bing So, Ivo Ribeiro da Silva, Lucas Carvalho Gomes, and Teogenes Senna de Oliveira

Abstract Mining causes extreme forms of soil degradation and creates new surfaces and topography. The new landforms have increased average slopes with implications for increased run-off and erosion, while mine pits and tailing dams may alter groundwater as well as surface water. Soil biota is removed or otherwise hampered, and in addition, nutrient cycling, biodiversity and resilience of vegetation are all altered by mine site disturbance. The process of rehabilitation involves the reduction of excessively steep slopes and requires significant earth movement using large machinery, which in turn causes subsoil compaction. Deep ripping to improve soil aeration and water infiltration is usually required to create suitable conditions for root growth and development in the subsurface layers. Capping the area with the original surface and/or subsoil, which supplies organic matter, microorganisms and a seedbank from the original ecosystem, will greatly support root growth and plant development. The subsoil conditions control the growth and development of the root system, which is specifically affected by high bulk density, pans, hardsetting, compaction layers, extreme texture (sandy, silty or clayey) and adverse chemical properties (high sodicity, alkalinity, salinity, acidity and other possible toxic ions and materials). The limitations of the subsoil of mined land and tailings are discussed, and case studies of rehabilitation after bauxite and iron ore mining are described to demonstrate the main practices applied and results. Overall, the types and intensity of soil constraints are site specific, and consequently the practices adopted to restore or rehabilitate each area must consider the local conditions to achieve the best outcomes. .

Keywords Acidity · Alkalinity · Compaction · Mined lands · Rehabilitation · Salinity · Textural limitations

---

H. B. So (✉)
Australian Rivers Institute, Griffith University, Nathan, QLD, Australia
e-mail: h.so@grifith.edu.au

I. R. da Silva · L. C. Gomes · T. S. de Oliveira
Soils Department, Federal University of Viçosa, Viçosa, MG, Brazil
e-mail: ivosilva@ufv.br; teo@ufv.br

T. S. d. Oliveira, R. W. Bell (eds.), *Subsoil Constraints for Crop Production*,
https://doi.org/10.1007/978-3-031-00317-2_7

7.1 Introduction

Mining causes extreme forms of land degradation and creates new surfaces and new topography. In some instances, the pre-existing land use and ecosystem can be restored to the land after mining, but in many cases restoration requires the creation of new ecosystems because the substrate and landform are no longer suitable for the pre-existing land use, species or plant communities. Most modern open-cut mining requires the stripping and storage of the original surface soil and subsoil for restoration activities if their properties are suitable for later use as root zone capping during the final stages of post-mining land restoration.

Land disturbance due to mining has many forms, and soils are compacted, eroded, buried, mixed, stored, dug up and relocated or altered in their chemical properties. New landforms are created by waste rock dumps, tailing dams and pits. Except for the tailings, the new landforms commonly have increased average slopes compared to the pre-existing landforms with implications for increased run-off and erosion. The new landforms, particularly mine pits and tailings dams, may alter groundwater as well as surface water hydrology. In addition, nutrient cycling, soil biota, biodiversity and resilience of vegetation may be altered by mine site disturbance. These disturbances, including those in the subsoil, should be reversed by acceptable post-mining land use to minimise further damage and pollution to the surrounding environment.

In Australia and most developed countries, legislation and public opinion require that these highly disturbed post-mining landscapes be satisfactorily rehabilitated into an approved post-mining land use. This involves the reduction of excessively steep slopes, which require significant earth movement using large machinery, which in turn causes topsoil and subsoil compaction. Acceptable post-mining land use generally involves developing a sustainable ecosystem or cropping system on this landscape, and that requires the creation of a suitable medium for root growth and development including the subsurface soil.

Soils in mature undisturbed ecosystems would generally have at least three horizons underneath the litter layer. However, in a profile on a mine waste dump or one that is otherwise disturbed by mining, the substrate may be uniformly or heterogeneously mixed, or layered, but generally does not conform to the horizons of a natural soil profile. Hence the characteristics and functions of each horizon of the disturbed soil, which vary with depth, are rather different and may not be suitable for a productive soil. In a natural soil profile, the surface soil plays an important role in the supply of nutrients, water infiltration and air supply into the subsoil, while the subsoil (B and C horizons) plays an important role in the storage and supply of water, oxygen and nutrients. On the other hand, it can present a major limitation to plant growth if toxic chemicals are present in the soil. The soil profile controls the growth and development of the root system which is responsible for water and nutrient acquisition in such deeper soil layers. In a mature native ecosystem that is in equilibrium with the prevailing climate, the subsoil (B and C horizons) is able to supply adequate water and nutrients during the drier months for the vegetation to survive until the next wet season. This is an important criterion that will determine the success or failure of rehabilitated ecosystems on mine sites located in areas where the climate is characterised by distinct wet and dry seasons.

In contrast to undisturbed soils, rehabilitated mined land and tailings generally consist of one (if they are not capped) or two uniform or heterogeneous mixtures of material. In the latter case, the first layer (topsoil) is of limited depth, and the second layer (subsoil) can be of considerable depth. The first layer is commonly a capping made up of the stripped and stored surface horizons of the original soil or a similar suitable material, and the quantity of this material determines the depth of the topsoil of the newly constructed soil profile (e.g. Ni et al. 2014). The second layer consists of regolith or the broken-up waste rock overburden material (at various stages of weathering) or tailing material in the case of the tailing dam. To develop a sustainable ecosystem or cropping system on this newly constructed soil, the first layer provides an important medium for the initial establishment of the plant community, and it can be the source of the necessary seedbank, organic matter, microorganisms as well as nutrients. The second layer or new subsoil is important for plant survival and sustainability and must be able to (*i*) store and supply adequate water to the vegetation and to keep it alive during the dry season; (*ii*) supply adequate air and oxygen; (*iii*) supply adequate nutrients; (*iv*) be free from materials that are toxic to the plant or plant roots; and (*v*) display low mechanical resistance to allow adequate root growth to explore and access the above resources.

These processes will be affected by adverse subsoil structure (e.g. high bulk density and strength in pans or hardsetting and compaction layers), texture (extremely sandy, silty or clayey) and chemical properties (e.g. high sodicity, alkalinity (from bicarbonate, carbonate or hydroxyl), salinity, acidity (inherent or triggered by oxidation of exposed sulphidic materials), and other possible toxic elements or materials). Any limitation imposed by these properties will affect the desired post-mining land use. If the land is to be returned to productive seasonal cropping, limitations are important during the cropping season, generally the rainy season and/or dry season where irrigation is available. If the land is to be returned to permanent pastures, forestry or native ecosystem, subsoil water and nutrient supply should be adequate to support the vegetation during the dry season and to survive into the next wet season.

This chapter will outline the limitations of the subsoil of mined land and tailings and possible ways to ameliorate them.

7.2 Compaction and High Soil Strength

In natural soil profiles or soil profiles used for agricultural/forestry production, the presence of compacted layers or pans will reduce the depth of soil that can support root growth and the storage of plant-available water and nutrients (Duiker 2004). Hakansson and Reeder (1994) reviewed the results of an international series of 25 experiments on the immediate and residual effects of compaction of surface and subsoil compaction on the yield of wheat. They summarised the results and proposed an approximate model that describes the average residual effect following a single compaction event from agricultural vehicles with a high axle load of 10 Mg on the yield responses of wheat (Fig. 7.1). They were able to show the contribution of surface (0–25 cm), upper subsoil (25–40 cm) and deeper subsoil (>40 cm depth) compaction on the average relative yield of wheat. Figure 7.1 shows that shallow compaction (0–40 cm) is likely to be ameliorated naturally by processes such as wetting and drying, and freezing and thawing, but that this process will require time depending on the depth of compaction (5 years for the surface and 10 years for the upper subsoil), while the effect of deeper subsoil compaction appears to be permanent. Although these results were derived from agricultural soils, this pattern of response is relevant to mined lands, as vehicles of much higher axle loads are used in developed countries (100 to 200 Mg) with multiple passes during the rehabilitation process irrespective of the soil water contents. The effects will most probably be greatly magnified. To establish plants on these post-mining soils will require positive action to ameliorate subsurface compaction.

Compaction increases soil strength, which depends on soil water content, with the effects being lowest at higher water contents and increasing as the soil dries out. Although there are differences among species, roots have reduced growth rates as soil strength increases above a threshold range and decrease to zero when soil

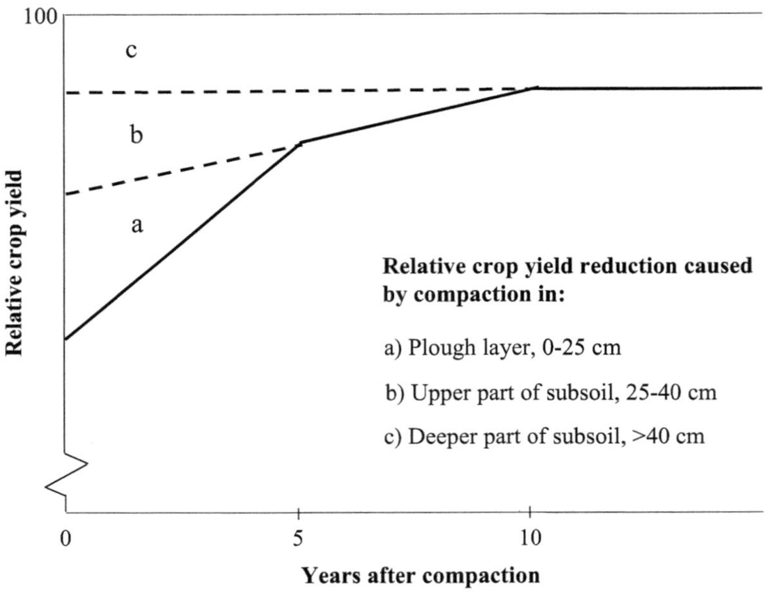

Fig. 7.1 Schematic diagram showing the contributions to subsequent relative crop yield reductions from compaction in different soil layers caused by high axle load traffic. The magnitude and persistence of the individual components vary considerably among soils. The diagram illustrates the situation for a clay loam soil. (After Hakansson and Reeder 1994)

strength (measured as the resistance to a cone penetrometer) reaches about 2 MPa (Cornish et al. 1984; Duiker 2004; Daquiado 1998).

Modern mining operations use large machinery and trucks with high axle loads to efficiently reshape the post-mining landscape and to apply topsoil to the disturbed mine site resulting in significant soil compaction to depth. This has been reported as a major problem with coal mining reclamation in the Appalachian region (Fields-Johnson et al. 2014; Sweigard et al. 2007) and lignite mining in Germany (Sagel 2015), as well as on Australian bauxite mine reclamation (Gardner and Bell 2007).

Haigh and Sansom (1999) reported that soil bulk densities on reclaimed coal lands in the United Kingdom are low in the surface 10 cm and increase rapidly to levels of 1.8 g cm^{-3} at depths of 50 cm or more. These levels of soil bulk density are close to critical densities that suppress root extension. In addition, these high densities result in low hydraulic conductivity and consequently waterlogged conditions when the prevailing rainfall is high. They found that a major contributor to high bulk densities is the breakdown of unstable primary particles in the mine spoils which releases large quantities of fine particles that accumulate in the pore spaces of the soil. Disruption due to heavy machinery trafficking added to the increase in bulk density.

In most Australian open-cut coal mining, rehabilitation of the damaged landscape is compulsory and requires the reduction of slope gradient and slope lengths to stabilise the landscapes against geotechnical failures (landslips) and erosion (So et al. 2018). This necessitates moving large quantities of spoil and soil to achieve the desired landscape using heavy machinery resulting in significant compaction particularly when operations are conducted under high soil/spoil water contents during the wet season. Where the overburden is highly saline and sodic, the soil becomes less permeable and water contents remain high and the spoil prone to compaction. In addition, dispersion of clay-sized materials associated with high sodicity could also fill the pore spaces leading to increased bulk densities.

Rehabilitation to native ecosystems requires adequate growth of trees and shrubs, and these in turn require loose soil/spoil to a depth of at least 120 cm to grow optimally (Sweigard et al. 2007). Root growth and development of an establishing ecosystem will be restricted in the presence of a compacted layer with its associated high strength that restricts the ability of the root system to explore the reconstructed soil profile.

The best management strategy for dealing with compaction is to avoid it in the first place, using techniques that can reduce soil/spoil bulk density at lower cost than correction of soil compaction (Sweigard et al. 2007). In the Appalachian region, coal operators are using such techniques successfully to establish commercial forests (Burger et al. 2005). Techniques consist of loosely dumping surface materials combined with minimal grading necessary to shape the landscape creating loose soils and rough surfaces, thus increasing rainfall infiltration and increasing the survival and growth of trees. Garcia and Stearns (2015) described a successful rehabilitation operation of open-cut coal mining in Colorado where the topsoil is stripped and used immediately to topsoil nearby rehabilitated areas to preserve the soil fertility and seedbank, followed by deep ripping using tined implements to reduce compaction prior to tree planting.

This technique of preventing compaction may not apply to all areas, and in that case, compacted areas should be deep ripped on the contour to a depth of 80–120 cm to increase water infiltration and storage capacity. This has been shown to significantly improve tree growth and development (Fields-Johnson et al. 2014). This is standard practice on open-cut coal mines in Queensland, Australia, and bauxite mining in southwest Australia (Lardner and Tibett 2013).

7.3 Soil Texture: Soil Water and Chemical Constraints

Profiles formed from waste rock dumps, from dredging and from tailings commonly have a narrow range of particle sizes creating extremes of soil texture that limit water and nutrient availability to plants. For example, in tailings which are developed as by-products from the processing of mineral rocks, waste is often deposited as a slurry into ponded structures. In the process, it is separated into different sized particles with the coarser sand particles deposited near the discharge outlet or spigots and the finer particles further downstream in the tailing dam or other dams. Typical examples of this are sand and silt tailings from open-cut tin mining. The most common method of mining depositional tin in Australia, Malaysia and Thailand is the gravel pump method where the sedimentary ore deposits are dislodged and washed into a sump using high-pressure waterjets (Tanavud 1992; Yap 2007). The gravel pump is then used to pump the ore-sand-soil slurry from the sump onto a sloping giant sluice or palong with baffles across the flow direction that separate and trap the heavy (high density) ore particles from the lighter sand and silt tailing slurry. The tailings are directed into a large pond or through a series of ponds resulting in each pond having progressively finer particles (Tanavud 1992; Lau 1999). As the soil has a sedimentary origin, there are more coarse/sandy materials than the silt and clay in the tailings. Sandy tailings have low fertility and a low water holding capacity but allow high infiltration rates. Figure 7.2 shows coarse tailing ponds in Thailand and in the background a stockpile of the sandy tailings which will remain unvegetated despite the high prevailing rainfall conditions. To reclaim the sandy tailings, they are either capped with topsoil or where adequate topsoil is not available, the finer tailings or other available organic materials are mixed into the surface of the coarser tailings before revegetation can be attempted. As many tin mining operations are located in the wet tropics, water is generally not a limitation, but soil fertility is and needs appropriate management until the desired ecosystem is established. Where the dry season is pronounced, the water holding capacity of the sandy subsoil is critical to the survival of the vegetation on these reclaimed tailings. On the other hand, where the subsoil consists of predominantly silty or clayey materials, waterlogging may become the limiting factor to vegetation growth under the

Fig. 7.2 Tin tailing dump in Thailand – tailings are stored in impounded dams (after Yap 2007) and in the background is a coarse sandy tailing stockpile where very little vegetation will grow due to nutrient limitations despite the high prevailing tropical rainfall conditions

high tropical rainfall conditions, and the provision of adequate drainage facilities (agricultural drains or open ditch drains) is necessary for the success of rehabilitation of these heavy-textured tailing dumps. By contrast, with the above examples, soil texture was generally not a limitation in the subsoil of an open-cut coal mine in Australia that usually consists of pulverised and degraded rocks (mudstones or sandstones) derived from sedimentary rocks (Hannan and Gordon 1996; Roe et al. 1996).

Water and nutrient stress are also likely to be the limiting factors in the establishment and growth of plants on the fine sandy tailings associated with gold mining in a monsoonal climate with distinct wet and dry seasons, e.g. the tailings from Kidston Gold Mine in North Queensland. A water spillway was incorporated in the design of the tailing impoundment wall that can maintain high water tables across the dam with a seasonal fluctuation of approximately 1.5 m (Williams and Currey 2002). Vegetation was established directly on the tailings without soil capping using a combination of fertilisation, supplemental irrigation, seeding and tube stock planting (Edraki et al. 2017). It was anticipated that the deep roots from trees and shrubs will reach the water table and contribute to the sustainability of the new ecosystem.

7.4 Chemical Limitations of the Subsoil

In contrast to the relatively benign tin mine tailings, the processing of bauxite into alumina ore produced sandy and silty-clayey residues (red mud) that are highly saline (EC >30 dS m^{-1}) with a high pH (pH >10 associated with the use of NaOH in the Bayer process for extracting the alumina) and high in exchangeable sodium (>70%) (Wehr et al. 2006). The insoluble materials (residue sand and mud) from the Bayer process are washed, sometimes partially neutralised (with seawater, CO_2 or acids) to reduce its pH and salinity and deposited in impoundments or tailing dams (after transport from the processing plant) as a wet slurry (15–30% solids) or thickened slurry (50–65% solids) (Jones and Haynes 2011). Thick slurries are deposited by dry stacking as shown in Fig. 7.3 for highly saline and alkaline thick slurry from the bauxite refinery at Gove, Northern Territory, Australia. These toxic and hostile conditions in combination with low concentrations of plant-available nutrients make it difficult to establish vegetation directly on these tailings, particularly on the red mud. They require capping with an adequate depth of surface soil or surface soil-like material that allows vegetation to grow and then sufficient time to leach out the salt and alkaline substances.

Another option includes capping the red mud with residue sand and then topping it with soil, or mixing residue sand with the red mud to make it more permeable (Anderson et al. 2011). The low water holding capacity of the residue sand is

Fig. 7.3 Dry stacking of thick slurries from bauxite processing at Gove, Northern Territory, Australia

increased by mixing with the finer red mud (silt and clay) (Anderson et al. 2011), but mixtures that give minimum void ratios and high densities (70% sand with 30% mud) should be avoided as they result in hardsetting mixtures with high soil strength (Buchanan et al. 2010) which is not conducive to healthy root development.

The physical, chemical and nutritional limitations of these residues can be improved using large quantities of amendments such as gypsum and lime, manure/compost and sewage sludge (Wehr et al. 2006; Jones and Haynes 2011). Seawater neutralisation also improves some nutritional limitations by increasing the concentrations of plant-available Ca and Mg and reducing the plant-available sodium and pH of the residue. When this is followed with freshwater flushing (easier on the sandy compared to the fine tailings), the salinity of the residue is further reduced and renders it more suitable for plant growth. However, some alumina processing residues contain high concentrations of reactive silica which forms sodium alumino-silicates and continues to release sodium and hydroxyl ions over time resulting in increased sodicity and alkalinity that will hamper plant growth in the longer term (Wehr et al. 2006).

Revegetation experience at the Alcan refinery at Gove, Northern Australia, executed progressively from 1978 to 1982, shows that after 10 years the establishment of the vegetation cover was highly variable (Wehr et al. 2006). The red mud had a pH of 10.5, while the sandy residues had a pH of 9.7. Both were highly saline (EC >30 dS m^{-1}). As these are hostile conditions for plant growth, they were capped with 75–150 cm depth of a clayey subsoil material and topped with 15 cm of sandy topsoil. The soil was ripped to 60 cm depth, adequately fertilised and sown with a mixture of salt- and alkali-tolerant exotic pasture and native species (grass, shrub and trees). The climate at Gove is monsoonal with an average rainfall of 1330 mm, but 90% falls during the wet season between December and April. In 1990, an aerial survey showed that 52% of the area was predominantly grass, 39.5% was predominantly trees and 8.5% was bare. It was observed that areas with shallow soil cover (0.6 ± 0.3 m) supported mostly grass cover, while the deep soil cover (1.4 ± 0.4 m) supported a dense vegetation cover consisting of grass, shrubs and trees. The bare areas had sodic surface soil with an elevated salinity (pH >8.5, $EC_{1:5}$ >1 dS m^{-1}) associated with waterlogging during the wet season and upward capillary rise of the salt and NaOH from the underlying red mud. The incorporation of a capillary break above the residue mud, as part of the capping, would prevent capillary rise, reduce the variability of the vegetation and eliminate the bare areas.

Both overburden and tailings generally have low plant-available nutrient contents and are biologically hampered, hence the necessity for capping with topsoil which supplies soil nutrients, organic matter, microbial populations and a seedbank of native species. In addition, fertilisers will be required for the establishment of suitable vegetation and until nutrient cycling has stocks and fluxes that are in balance with the established ecosystem.

Another important limitation on post-mining landscapes is the presence of acids derived from the oxidation of pyrite (FeS_2), commonly present in the overburden material above coal seams. The action of mining brings the sulphide-bearing minerals into contact with water and oxygen which leads to the production of sulphuric acid as shown by the following reactions (McQuade and Riley 1996):

$$FeS_2 + \frac{1}{2}O_2 + 10H_2O \rightarrow FeSO_4 + H_2SO_4$$

$$2FeSO_4 + H_2SO_2 + \frac{1}{2}O_2 \rightarrow Fe_2(SO_4)_3 + H_2O$$

$$FeS_2 + Fe_2(SO_4)_3 + 2H_2O + 3O_2 \rightarrow 3FeSO_4 + 2H_2SO_4$$

$$MS + Fe_2(SO_4)_3 + ½O_2 + H_2O \rightarrow MSO_4 + 2FeSO_4 + H_2SO_4$$

where M represents a metallic cation present in the soil.

If sulphide-bearing minerals are present in the subsoil or root zone, it will impose a severe limitation on plant growth. It is important that sulphide-bearing minerals are buried sufficiently deep to minimise their chemical transformation into sulphuric acid and to keep them away from plant roots to ensure that sustainable ecosystems can be developed on the post-mining landscapes. It is also important to prevent or minimise leaching of the acids into the surrounding environment.

7.5 Overcoming Multiple Soil Constraints for Brazilian Bauxite Rehabilitation in High Rainfall Environments

The biggest challenge of mining activities is the restoration or rehabilitation of the area after mining, when the subsoil and tailings present physical and chemical limitations for plant growth and development. For this, it is essential to address the limitations of the soil surface and subsurface in an integrated way, especially when the objective is to rehabilitate the area for long-term productive agricultural use. However, given the variety of natural and agroecosystems where mining occurs, the restoration and rehabilitation alternatives must be evaluated and applied considering principles of rehabilitation discussed above according to the local conditions and possibilities. Mining is one of the principal economic activities in Brazil (52% of the Brazilian trade balance in 2019), and results from an experiment on a bauxite mine under rehabilitation will be presented as a case study.

In a post-mined bauxite area in the Zona da Mata, Minas Gerais state, researchers from the Universidade Federal de Viçosa set up, in 2011, a series of experiments (Fig. 7.4) aimed at rehabilitating the area for coffee, eucalyptus and pastures that are the main land uses and sources of income in the strongly undulating relief in this region. The climate is predominantly Cwa (Köppen), with hot and rainy summers (October–February) and a well-defined dry season (April–August). The annual precipitation is 1287 mm and temperature average is 20.3 °C (INMET, 2016). During the mining process, the topsoil of the typic dystrophic Red-Yellow Latosol (an

Fig. 7.4 Replacement of the topsoil (**a**), reconfiguration of the mining area (**b**), installation of the experiments (**c** and **d**) and general view of the mining area under recovery with the installed experiments: tree species, including monocropped eucalyptus and a mixture of native species (**e**); coffee (**f**); and monocropped and intercropped *Urochloa* and *Stylosanthes* (**g**)

Oxisol) was removed and saved for later use in the rehabilitation. After mining and the topography reconfiguration to decrease hillslopes, the bottom spoil of the mine pit was deep ripped to 60 cm depth, then a layer (average 40 cm) of the saved topsoil was placed over the mine spoil and a second deep ripping to 60 cm depth was done. Additionally, conservation practices, such as vegetated terraces, were implemented and suitable chemical fertilisation and conditioning applied to the surface soil layer. All experiments were conducted with the soil physical conditions as described above.

In one experiment, pasture (with a mixture of *Urochloa brizantha*, *Stylosanthes capitata* and *Stylosanthes macrocephala*) was compared to coffee (*Coffea arabica* var. Catuai red IAC 144) with grasses as cover plants between the rows of coffee trees.

In another experiment different tree species were compared: (*i*) short rotation of clonal eucalyptus (AEC144; a hybrid between *Eucalyptus urophylla* and *E. grandis*) (*ii*) a mixed plantation including 16 native tree species (pioneer+non-pioneer) from the region; and (*iii*) a plantation of the nitrogen-fixing tree, *Anadenanthera peregrina.*

In a third experiment, chemical fertiliser practices normally adopted by the mining company were compared to the supply of readily available poultry litter in the region.

A post-mined area was maintained without vegetation cover and an adjacent area covered by a fragment of native Atlantic forest vegetation defined as semi-deciduous seasonal montane forest were selected as the reference for the recovery process of these experiments.

The landscape reshaping and soil ripping together with chemical/organic fertilisation alleviated the soil physical and chemical limitations after the bauxite mining and led to successful rehabilitation for important land uses in the region. Several studies explored the plant growth in these experiments and the effects on soil attributes, with specific studies about pasture (Borges 2013; Oliveira et al. 2016, 2017), coffee (Cavalcante et al. 2019), forest species (Valente et al. 2019) and integrative studies detailed by Borges et al. (2019) and Teixeira et al. (2019). Overall, the results indicate that the use of organic fertiliser, either separately or combined with the chemical fertiliser, improved the biomass of *Urochloa* and *Stylosanthes* grasses (Borges 2013), eucalyptus and the native pioneer+non-pioneer species mixture after 56 months (Valente et al. 2019). Benefits for soil carbon and nitrogen stocks, soil aggregation and carbon management index (CMI) (Blair et al. 1995), an indicator of recovery of impacted areas, were also observed in these experiments with higher values in pasture, followed by tree species and then coffee (Oliveira et al. 2016; Cavalcante et al. 2019). These results highlight the effective role of rapid-growing plants, particularly grasses, in speeding up the post-mining rehabilitation of the disturbed soil and its ecosystem, particularly when adequate amounts of nutrients are applied via mineral and organic fertilisers at the very beginning of the rehabilitation phase.

The lack of surface cover (ground cover) and soil development (aggregation) in combination with intense rainfall events leads to severe water erosion on post-mined areas. Grasses and trees with rapid initial growth are suitable as pioneer vegetation to reduce erosion and water run-off (Table 7.1). The relative reduction in soil and water loss varied between 39 and 78% for pasture and from 68 to 80% for coffee compared with post-mined areas without vegetation (Borges 2013). Nevertheless,

Table 7.1 Mean values of accumulated run-off water and soil loss in an area under rehabilitation after bauxite mining, for agricultural use as pasture (December 2010 to March 2012) or coffee with *Urochloa* as a ground cover between the rows (March 2010 to August 2012), using chemical and organic fertiliser (poultry litter)

Land use	Type of fertiliser	Ground cover plant	Total run-off (% of the total accumulated precipitation)	Relative reduction of run-off with *Urochloa*	Total soil loss (t ha^{-1})	Relative reduction of soil loss with *Urochloa*
Pasture[a]	Organic	*Urochloa*	0.5	77	16.0	78
		No plants	2.4		74.5	
	Chemical	*Urochloa*	1.8	39	42.0	52
		No plants	3.0		87.2	
Coffee[b]	Organic	*Urochloa*	0.3	80	2.3	81
		No plants	1.6		12.3	
	Chemical	*Urochloa*	1.4	71	9.4	68
		No plants	4.7		29.4	

[a]Accumulated precipitation was 2787 mm between December 2010 and March 2012; [b]Accumulated precipitation was 2391 mm between March 2011 and August 2012

the smaller losses of soil and water under these conditions may also reflect the effect of physical practices introduced for recovering the mined area. For instance, the deep ripping conducted after topsoil application reduced soil strength and improved water infiltration and rooting conditions.

The improvements in soil chemical, physical and biological quality arising from the practices adopted and reported above are probably restricted to the depth of deep ripping that reached around 1 meter depth. These experiments did not evaluate the effect of different subsoil deep ripping or other topsoil physical practices, but the results suggest that deeper subsoil ripping may have improved the eucalyptus development (total height, diameter at breast height and tree biomass). Even though genetically uniform clonal material was planted, Valente et al. (2019) found that the eucalyptus tree size was variable (large, medium and small). A possible explanation for such variation could be variable penetration resistance at depth, due to a heterogeneous (horizontally and vertically) subsoiling throughout the mining area or that seedlings were not being planted exactly above the ripline of the subsoiler. Therefore, it is possible that some eucalyptus plants encountered resistance to their full root growth and development, which is reflected in the trunk volume.

7.6 Recovery After Iron Ore Mining and Tailing Dam Collapse in Brazil

In certain cases, the soil conditions in mining areas are severely limited by physical, chemical and biological conditions such that the rehabilitation to the preferred post-mining land use is not economic. Therefore, a return to a native ecosystem seems preferable. The Brazilian Quadrilátero Ferrífero is one of the biggest iron reserves in the world, and it is associated with the Campo Rupestre vegetation, an endemic plant community characterised mainly by shrubs on rocky outcrops. This vegetation is adapted to the natural soil limitations in the region, such as poor nutrient availability, shallow soils and low plant-available water retention. The main strategies adopted for recovery of the mined areas are the use of topsoils from other areas and the dispersion of seeds from the Campo Rupestre vegetation. For instance, an experiment evaluating the application of different topsoil thickness and re-introduction of plants from nearby ecosystems identified that a layer of 40 cm, compared with 20 cm, increased the vegetation cover, especially when there was no fertiliser application (Rezende 2013). The challenge of this practice is the availability of seeds of Campo Rupestre vegetation, and studies are currently identifying suitable species that can be used in the restoration processes for this region (Garcia et al. 2009).

Failure of tailing storage facilities is risky, as highlighted by the collapse of tailing dams of the Fundão (2015) and Brumadinho (2019) mines that apart from loss of human life covered agricultural areas and forests with mined waste. The collapse of the Fundão dam added more than 34 million cubic meters of iron ore tailings to the rivers and nearby agricultural areas. This disaster created several socio-ecological problems, affecting the water quality of rivers and covering agricultural soils along the river with the deposition of a thick layer of tailings. The iron mine waste covered important areas used by family farms that cultivate vegetables and dairy cows with up to 3 meters of sediment over soil. The material was rich in SiO_2, Al_2O_3 and Fe_2O_3, and physical structure was composed of particle sizes between 0.075 and 2.38 mm (62%) and smaller than 0.075 mm (38%) (Couto et al. 2021), with high soil and particle density and low porosity (Silva et al. 2016). To overcome the constraints imposed by the tailing material, several experiments were set up to rehabilitate these areas and also to recover the riparian forests that were destroyed. The rehabilitation of some areas was done using topsoil material from the mountains nearby, allowing the farmers to grow grasses again and some vegetables, as well as fruit trees. The recovery of riparian forest was done using phytoremediation with native species and physico-chemical remediation by incorporating organic matter (OM) into the sediment (Scotti et al. 2020).

7.7 Final Remarks

The reconstruction of the root zone is essential to overcome the subsoil and surface soil constraints of mined land and tailings and to create a profile that is physically and hydro-geochemically stable. Recovery of mined or tailing areas has their specific challenges, firstly concerning the appropriated physical conditions for root development and secondly the need for hydro-geochemically stable conditions for the establishment of plant and soil biological communities since unstable geochemistry and toxic chemical conditions may be present. Huang et al. (2012) indicate that the reconstruction of a stable and sustainable root zone is essential to support plant communities and for phytostabilisation in the long term. These authors propose a conceptual model for the reconstruction of root zones in mine tailings, which also could be applied in mined sites. Two main stages should be considered: first, the hydro-geochemical stabilisation through remediation measures to mitigate physico-chemical constraints (high bulk density/root penetration resistance and/or low pH and nutrient availability) and, second, rehabilitation of soil biological capacity and ecological linkages between the reconstructed root zones and plant communities. Specifically, for the remediation strategies aiming to have long-term impacts on the root zone, studies should investigate the geochemistry of the tailing materials and its physical and chemical interaction with remediation practices and materials. In addition, the local climatic conditions, especially annual precipitation and its variability, should be also taken into consideration when planning remediation activities in tailings or mined areas. Depending on whether the location has a humid or semi-arid climate, the hydrological response will be different for the same mined or tailing remediation strategy.

Although some experiments are describing the recovery of mined or tailing areas around the world, more long-term studies focusing on the reconstruction (Huang et al. 2012) are needed to overcome the soil limitations after mining activities, especially in Brazil after the two tailing dam collapses. For this, it is essential to establish cooperation among mining companies, local farmers and scientists to plan experiments that help to design and validate recovery practices. These experiments should cover the best vegetation practices, such as plant species, but also different subsoil conditions and methods of physical amelioration. With well-researched practices, future recovery and rehabilitation processes may occur faster and mitigate the socio-ecological effects of mining activities.

References

Anderson JA, Bell RW, Phillips IR. Bauxite residue fines added residue sands enhance plant growth potential – a glasshouse study. J Soils Sediments. 2011;11:889–902. https://doi.org/10.1007/s11368-011-0379-7.

Blair GJ, Lefroy RD, Lisle L. Soil carbon fractions based on their degree of oxidation, and the development of a carbon management index for agricultural systems. Aust J Agric Res. 1995;46:1459–66. https://doi.org/10.1071/AR9951459.

Borges SR. Qualidade do solo em áreas em recuperação com forrageiras e cafeeiro pós-mineração de bauxita. Minas Gerais, Brazil: PhD Thesis, Federal University of Viçosa; 2013.

Borges SR, Santos RS, Oliveira DMS, et al. Practices for rehabilitating bauxite-mined areas and an integrative approach to monitor soil quality. Land Degrad Dev. 2019;30(7):866–77. https://doi.org/10.1002/ldr.3273.

Buchanan SJ, So HB, Kopittke PM, Menzies NW. Influence of texture in bauxite residues on void ratio, water holding characteristics, and penetration resistance. Geoderma. 2010;158:421–6. https://doi.org/10.1016/j.geoderma.2010.06.016.

Burger J, Graves D, Angel P, Davis V, Zipper C. The forestry reclamation approach. Forest reclamation advisory no. 2. Appalachian regional reforestation initiative, U.S. Office of Surface Mining. 2005:4.

Cavalcante DM, Castro MF, Chaves MTL, et al. Effects of rehabilitation strategies on soil aggregation, C and N distribution and carbon management index in coffee cultivation in mined soil. Ecol Indic. 2019;107:105668. https://doi.org/10.1016/j.ecolind.2019.105668.
Cornish PS, So HB, McWilliam JR. Effects of soil bulk density and water regime on root growth and uptake of phosphorus by ryegrass. Aust J Agric Res. 1984;35(5):631–44.
Couto FR, de Mello FA, Pontes PP, Marques AR. Physical, chemical and microbiological characterization of the soils contaminated by iron ore tailing mud after Fundão dam disaster in Brazil. Appl Soil Ecol. 2021;158:103811. https://doi.org/10.1016/j.apsoil.2020.103811.
Daquiado NP. Assessing the limitations on soil productivity imposed by hardsetting properties. PhD thesis, University of Queensland; 1998.
Duiker SW. Effects of soil compaction. college of agricultural sciences. The Pennsylvania State University, University Park, PA 16802. (http://pubs.cas.psu.edu/freepubs/pdfs/uc188.pdf), diakses 2 April 2015; 2004.
Edraki M, Baumgartl T, Mulligan D, et al. Geochemical characteristics of rehabilitated tailings and associated seepages at Kidston gold mine, Queensland, Australia. Int J Mining Reclam Environ. 2017;33:133–47. https://doi.org/10.1080/17480930.2017.1362542.
Fields-Johnson CW, Burger JA, Evans DM, Zipper CE. Ripping improves tree survival and growth on unused reclaimed mined lands. Environ Manag. 2014;53:1059–65. https://doi.org/10.1007/s00267-014-0261-7.
Garcia LC, Barros FV, Lemos-Filho JP. Fructification phenology as an important tool in the recovery of iron mining areas in Minas Gerais, Brazil. Braz J Biol. 2009;69(3):887–93.
Garcia J, Stearns M. The colowyo mine: a case for successful mine reclamation. Cornerstone. 2015;4(3):31–5. Winter 2015.
Gardner JH, Bell DT. Bauxite mining restoration by Alcoa world alumina Australia in Western Australia: social, political, historical, and environmental contexts. Restor Ecol. 2007;15(4):3–10. https://doi.org/10.1111/j.1526-100X.2007.00287.x.
Haigh M, Sansom B. Soil compaction, runoff and erosion on reclaimed coal-lands (UK). Int J Mining Reclam Environ. 1999;13:135–46. https://doi.org/10.1080/09208119908944239.
Hakansson I, Reeder RC. Subsoil compaction by vehicles with high axle load – extent, persistence and crop responses. Soil Tillage Res. 1994;29:277–304. https://doi.org/10.1016/0167-1987(94)90065-5.
Hannan JC, Gordon RM. Environmental management of coal mines in the Hunter Valley, New South Wales. In: Mulligan DR, editor. Environmental management in the Australian mining and energy industries: principles and practices. UNSW Press; 1996. p. 265–316.
Huang L, Baumgartl T, Mulligan D. Is rhizosphere remediation sufficient for sustainable revegetation of mine tailings? Ann Bot. 2012;110(2):223–38. https://doi.org/10.1093/aob/mcs115.
Jones BEH, Haynes RJ. Bauxite processing residue: a critical review of its formation, properties, storage and revegetation. Crit Rev Environ Sci Technol. 2011;41:271–315. https://doi.org/10.1080/10643380902800000.
Lardner TD, Tibett M. Deep ripping after topsoil return affects root proliferation and floristic diversity in a restored biodiverse forest after bauxite mining. In: Tibett M, Fourie A, Digby C, editors. Mine closure. Australian Centre for Geomechanics; 2013. p. 363–76.
Lau WCR. TED case studies: Tin mining in Malaysia present and future. https://web.archive.org/web/20161130185311/http://www1.american.edu/TED/tin.htm; 1999.
McQuade CV, Riley SJ. Water management. In: Mulligan D, editor. Environmental management in the Australian minerals and energy industries: principles and practices. Sydney: UNSW Press; 1996. p. 189–226.
Ni CY, Bell RW, McGrath W, Jasper DA, McNeill P. Amelioration of soil chemical constraints for revegetation on gold oxide processing residues. Ecol Eng. 2014;64:66–76. https://doi.org/10.1016/j.ecoleng.2013.12.039.
Oliveira D, Mayrink G, Barreto M, et al. Soil quality and grass yield in a mined area under poultry litter application in Minas Gerais state, Brazil. Annals of Agricultural & Crop Sciences; 2016. p. 1–6.

Oliveira DMS, Silva IR, Mendes GO, et al. Carbon fluxes from different pools in a mined area under reclamation in Minas Gerais state, Brazil. Land Degrad Dev. 2017;28:507–14. https://doi.org/10.1002/ldr.2601.

Rezende LAL. Restauração ecológica de campos rupestres ferruginosos. PhD thesis, Federal University of Viçosa, Minas Gerais, Brazil. https://locus.ufv.br//handle/123456789/6814; 2013.

Roe PA, Mulligan DR, Bell LC. Environmental management of coal mines in the Bowen Basin, Central Queensland. In: Mulligan DR, editor. Environmental management in the Australian Mining and energy industries: principles and practices. UNSW Press; 1996. p. 317–82.

Sagel R. Ecology & reclamation: life after mining! – Rhineland (Germany). Europe Solidaire Sans Frontières. http://www.europe-solidaire.org/spip.php?auteur14584; 2015.

Scotti MR, Gomes AR, Lacerda TJ, et al. Remediation of a riparian site in the Brazilian Atlantic Forest reached by contaminated tailings from the collapsed Fundão dam with native woody species. Integr Environ Assess Manag. 2020;16:669–75. https://doi.org/10.1002/ieam.4272.

Silva AC, Cavalcante LCD, Fabris JD, et al. Características químicas, mineralógicas e físicas do material acumulado em terraços fluviais, originado do fluxo de lama proveniente do rompimento de barragem de rejeitos de mineração de ferro em Bento Rodrigues, Minas Gerais. Brasil Rev Espinhaço UFVJM. 2016:44–53.

So HB, Khalifa AM, Yu B, et al. MINErosion 3: using measurements on a tilting flume-rainfall simulator facility to predict erosion rates from post-mining landscapes in Central Queensland, Australia. PLoS One. 2018;13(3):1–21. https://doi.org/10.1371/journal.pone.0194230.

Sweigard R, Graves D, Barton C, et al. Loosening compacted soils on mined sites. In: The Appalachian regional reforestation initiate: Forest reclamation advisory, Vol. 4; 2007. p. 1–4.

Tanavud C. Reclamation of land disturbed by tin mining. PhD Thesis, University of Queensland; 1992.

Teixeira RS, Souza IF, Santos RS, et al. Impact of fertilization on cover crops and microbial community on a bauxite-mined soil undergoing reclamation. J Plant Nutr Soil Sci. 2019;182:515–23. https://doi.org/10.1002/jpln.201800586.

Valente FDA, Castro MF, Torres CMME, et al. Growth, biomass and carbon stocks in forest cover planted in an area of bauxite mining in rehabilitation. Rev Bras Cienc do Solo. 2019;43:1–15. https://doi.org/10.1590/18069657rbcs20180212.

Wehr JB, Fulton I, Menzies NW. Revegetation strategies for bauxite refinery residue: a case study of Alcan Gove in Northern Territory. Australia Environ Manage. 2006;37(3):297–306. https://doi.org/10.1007/s00267-004-0385-2.

Williams DJ, Currey NA. Engineering closure of an open pit gold operation in a semi-arid climate. Int J Mining Reclam Environ. 2002;16:270–88. https://doi.org/10.1076/ijsm.16.4.270.8632.

Yap KM. Tin mining in Malaysia- is there any revival? http://dspace.unimap.edu.my/xmlui/bitstream/handle/123456789/15965/feature%20tin%20m.ning%205pp.pdf?sequence=1; 2007.

Chapter 8
Sand and Gravel Subsoils

Craig A. Scanlan, Karen W. Holmes, and Richard Willian Bell

Abstract Approximately 5 and 29% of soils used for crop production globally have a sand or gravelly subsoil. The proportion of cropping soils with sand subsoil is greatest in Africa and Australia and Oceania. The countries with the greatest area of soil with gravel subsoil used for cropping are India and China. Sand and gravel subsoils have a limited capacity to supply soil and water to crops. Sand subsoils by definition have low clay content and, as a result, low water and nutrient storage capacity. The capacity of gravel subsoils to store water and nutrients decreases as gravel content increases. Although crop roots can access water and nutrients from these subsoils, the depth of these resources and physical constraints to root growth limit the efficiency of their use. Sand and gravel subsoils can constrain root growth although the mechanisms differ. Root growth is constrained in sand subsoils by constraints that can develop under crop production: compaction and aluminium toxicity. The impact of gravel subsoils on crop growth depends upon the penetrability of the gravel layer by crop roots. For impenetrable gravel layers, the properties of the topsoil will have the greatest influence on crop growth. For penetrable layers, root depth or length decreases as gravel content increases. There is potential to adapt agronomic management to maximise production on soils with sand or gravel subsoils. Split applications of nutrients can minimise leaching risk. There is evidence that the constraints that develop on sand subsoils due to crop production can be ameliorated profitably.

C. A. Scanlan (✉)
Department of Primary Industries and Regional Development, Northam, WA, Australia
e-mail: craig.scanlan@dpird.wa.gov.au

K. W. Holmes
Department of Primary Industries and Regional Development, South Perth, WA, Australia
e-mail: karen.holmes@dpird.wa.gov.au

R. W. Bell
Centre for Sustainable Farming Systems, Food Futures Institute, Murdoch University, Murdoch, WA, Australia
e-mail: r.bell@murdoch.edu.au

T. S. d. Oliveira, R. W. Bell (eds.), *Subsoil Constraints for Crop Production*,
https://doi.org/10.1007/978-3-031-00317-2_8

Keywords Crop growth · Gravel subsoil · Sand subsoil · Subsoil water · Subsoil nutrients

8.1 Introduction

Soils used for crop production that have sand texture or gravel in the subsoil have one common feature; they have a limited capacity to supply resources to the crop. Subsoil with a sand texture (sand and loamy sand) have less than 15% clay (IUSS Working Group WRB 2015) and, as a result, low water holding capacity, high permeability when saturated or near saturation and low cation exchange capacity. Gravelly subsoil, which we define as greater than 15% of material by volume with a diameter greater than 2 mm, based on the criteria used for topsoil in the Australian Soil Classification (Isbell 2002), can vary significantly in soil matrix texture, from sand to clay, and in the properties of the gravel particles. However, in general the capacity of the matrix-gravel mixture to store and supply water and nutrients is lower than non-gravel soils and decreases as gravel content increases.

The capacity of subsoils to supply resources to the crop is an important factor for determining grain yield, particularly in dryland cropping. The supply of nitrogen (N), phosphorus (P) and potassium (K) from the subsoil ranges from 8 to 75, 3 to 85 and < 3 to 70% of total uptake, respectively (Kautz et al. 2013). Also, the supply of soil water to the crop from the subsoil can be important for grain yield in water-limited environments (Kirkegaard et al. 2007). However, the influence of the limited storage capacity of sand and gravelly subsoils on the subsoil supply of soil water and nutrients has not been explored.

In this chapter we review the global distribution of sand and gravelly subsoils in use for crop production and their effect on root growth and function, plant-soil water relations and nutrient acquisition from subsoils. We also review options available for ameliorating constraints on these soils and the economic gains from these.

8.2 Definitions

This chapter focuses on soils used for crop production which have a sand or gravelly subsoil (below the cultivation layer, typically >10 cm depth). Here, we define soils with a sand-textured subsoil as those with a sand or loamy sand texture (> 65% sand and < 15% clay) between 15 and 100 cm depth, which includes Arenosols, many Podzols and other soils with an arenic qualifier (IUSS Working Group WRB 2015; Jahn et al. 2006). We define soils with gravelly subsoil as those with more than 15% of soil volume as gravel (> 2 mm) between 15 and 100 cm depth in the profile, which includes, but is not limited to, Plinthosols and Durisols, or soils with a calcic, duric, ferric, pisoplinthic or plinthic horizon.

8.3 Geological Origin

The geological origin of sand particles is diverse and influences the physical properties of the sand matrix. 'Sand' refers to particle size (approximately 0.05–2 mm), but composition and grain shape vary according to local rock types and processes of landscape formation. Sand-sized particles can be formed by biogenic processes, or the physical and chemical weathering of igneous, metamorphic or sedimentary rock (Yang et al. 2016). Most sand particles are quartz, derived from quartz crystals in granitic rocks by weathering, or grains released from the weathering of sandstones. The size and shape of the quartz sand grain that is released are influenced by the shape of the original crystal and forces acting on this (Smalley 1966). Post release, the shape of the sand particles is determined by a combination of exposure time, mineralogy and the transportation, depositional and weathering conditions encountered. Typically, roundness increases with age because there is a greater likelihood of exposure to chemical effects and abrasion during transport (Santamarina and Cho 2004). For example, quartz grains that have not been transported far are angular or sub-angular, but those that have been eroded and deposited are well-rounded to sub-rounded (Pye and Tsoar 2014). Sands derived from calcareous or mafic parent material (e.g. sand associated with limestone, coral-dominated coastlines and mafic bedrock) are generally not quartz dominated; these sands are less likely to be as well sorted or as geochemically stable over long time periods as quartz sand and account for a small proportion of sand subsoils under crop production.

The pedogenesis of soils with sand subsoil varies, and this can affect their suitability for crop production due to chemical and physical differences of the grains and arrangement in the soil profile. The sand subsoils used for crop production in south-west Western Australia were most likely formed in situ by prolonged weathering with localised redistribution (Newsome 2000). Any clay remaining in these profiles is low activity, and high levels of free iron and aluminium are common, which affect suitability for crop production. Deep sands or arenosols are widespread in most regions of Brazil particularly in the Cerrado (savannah) region, the northeast (caatingas), the Parana Basin and some parts of the Amazon basin, where the predominant soils are sandy Podzols (campinas e campinaras) (IUSS Working Group WRB 2015). Most of these sands are derived from sandstone and are commonly formed on deeply weathered regoliths (Amado et al. 1999; da Silva et al. 2013; Costa et al. 2013). By contrast, large areas of mobile sand dunes occur in northeast Brazil, but the majority of these dunes are in the Lençóis Maranhenses National Park and not used for agriculture. Termite activity can be significant for modifying the topsoil properties of sands; in sands of southern Bahia state, formed on sandstone, termite activity contributed to the enrichment of topsoils with clay, organic matter and nutrients (Sarcinelli et al. 2013). The role of termites in bioturbation in sands and other soil profile types is discussed by Schaefer and Oliveira (2022) in this volume.

The geological origin of gravelly subsoils is also diverse. For example, the relatively young soils of central Slovenia have developed on acidic igneous, metamorphic and sedimentary rocks and have a gravelly, or stony, B horizon (Vrščaj et al. 2017). Rendzinas (Leptosol, calcaric), also young soils, such as those found in Croatia (Bašić 2016), are formed from calcareous parent rock. The hillside skeletal soils of the Ratchaburi province, Thailand, are residuum of metamorphic rocks (Khetdan et al. 2017). In contrast, the gravelly subsoils of the wheat-growing region of Western Australia are formed in highly weathered landscapes, by the deposition of sand over ironstone gravels which are remnants of lateritic profiles (Mulcahy 1960).

8.4 Geographical Distribution

We estimated the geographical distribution of sand and gravelly subsoils by overlaying maps of global cropland distribution with models of sand and gravelly subsoil extent. The cropland map, a raster dataset at approximately 1-km pixel resolution (30 arcseconds), was used to estimate area of cropland per country and continent and was intersected with maps of soil information to derive subsoil areas under cropping. The cropland dataset is one of a series of basic land cover type maps produced by the Food and Agriculture Organization (FAO 2014) by way of classification of remote sensing imagery covering the six inhabited continents. Each pixel's value represents the percentage of a particular land cover, in this case cropland. Global gridded soil information was used to map sand and gravelly subsoils (https://soilgrids.org). These SoilGrids were generated using a compilation of soil data from around the world, which was harmonised to the same measurement depths, laboratory methods and soil description categories (e.g. texture class definitions) (Hengl et al. 2017). The SoilGrids are predicted soil property surfaces for fixed depth layers (e.g. percent sand in 0–5 cm). Sand and gravel percentage grids for the 15–30, 30–60 and 60–100 cm depth increments were combined using depth-weighted averaging to produce a single raster of subsoil texture percentage for the 15–100 cm, representing the subsoil. Because the raw soil measurements were harmonised in a database prior to spatial modelling, the SoilGrids do not depend on interpretation of soil classification systems to infer subsoil texture. The accuracy of area estimates of cropland, sand and gravelly subsoils is influenced by the 1-km pixel size of the maps, potential bias in soil profile distribution and density and possible under-representation in areas with low investment in soil research or clear authority for soil information management. The results for Australia were carefully evaluated against multiple national soil mapping products. Local sand and gravelly subsoil patterns did not always match those in more detailed mapping, but the continental estimates were very similar, suggesting the maps at the global scale provide a reasonable estimate of cropland and subsoil extent.

The area of land with sand or gravelly subsoils used for crop production in each continent is shown in Figs. 8.1 and 8.2 and is summarised in Table 8.1. Approximately 4.7 M km^2 or 29% of soils used for crop production globally have gravelly subsoil.

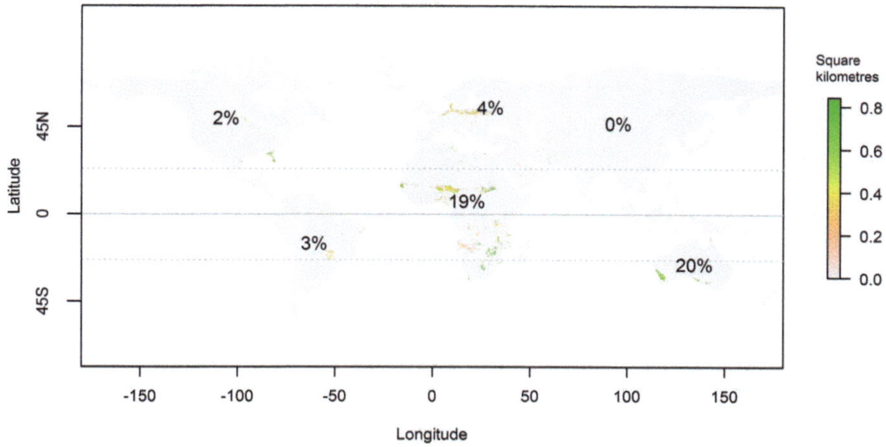

Fig. 8.1 Cropland x sand subsoils. The square kilometres of cropland raster were multiplied by the binary (1/0) sand subsoil map, so those pixels with sand subsoil are assigned square kilometres of sand cropland, and all others are set to zero. Percentages written on map refer to the percent of cropland on each continent that has sand subsoils

The countries with the greatest percentage of soils used for crop production with gravelly subsoil are India (1.1 M km^2), China (0.53 M km^2), Nigeria (0.22 M km^2), Australia (0.21 M km^2), Spain (0.19 M km^2) and Mexico (0.53 M km^2). A much smaller area of land used for crop production globally has sand subsoil. Approximately 0.9 M km^2 or 5% of soils used for crop production globally have sand subsoil, and the countries with the highest percentage of this soil type are Zimbabwe (0.68 M km^2), Poland (0.56 M km^2), Australia (0.14 M km^2), Nigeria (0.083 M km^2), Niger (0.072 M km^2) and Sudan (0.072 M km^2). While a much smaller proportion of sand subsoils are cropped globally, they are important in Africa and Australia + Oceania, accounting for 19% and 20% of soils used for crop production in those continents, respectively.

8.5 Effect of Sand and Gravel Subsoils on Root Density and Function

The shape of the sand grains in sand subsoil may be important for root growth. Root growth of wheat seedlings was more restricted in a rough (angular) compared to a smooth (rounded) sand; root length was less and root diameter was greater in the rough sand (Lipiec et al. 2016). The differences in root growth were attributed to a greater resistance to mechanical displacement of sand particles by the growing root tip in the rough sand, due to interlocking of the angular sand particles. Other studies indicate that soil strength increases as the surface roughness of sand particles

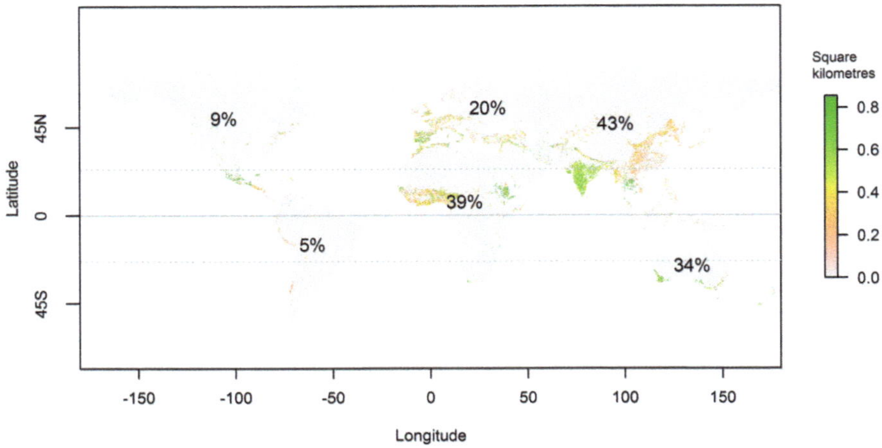

Fig. 8.2 Cropland x gravelly subsoils. The square kilometres of cropland raster were multiplied by the binary (1/0) gravelly subsoil map, so those pixels with gravelly subsoil are assigned square kilometres of gravelly cropland. Percentages written on map refer to the percent of cropland on each continent that has gravelly subsoils

increases; the angular grains with rough surfaces interlock and greater mechanical force is required to displace them (Panayiotopoulos 1989). Thus sand grains derived from in situ weathering of granite that releases angular grains are likely to result in subsoils with greater penetration resistance for roots than subsoils weathered from sediments such as sandstone.

The shape of sand particles in sand subsoil may also have implications for the level of compaction caused by agricultural machinery. Round sand particles are more susceptible to compression than rough particles, while a well-graded sand will compress more easily than a sand with a narrow particle size distribution (Panayiotopoulos 1989). For example, the maximum bulk density of sandy subsoils from some cropping soils in Western Australia was highest for the soils with the broader particle size distribution (Henderson et al. 1988). Also, Cruse et al. (1980) showed that for the same vibration energy, a soil with smooth particle surfaces had a higher packing density than a soil with rough particle surfaces. They suggested that soils with smooth particle surfaces would develop compaction layers more rapidly and require tillage to ameliorate compaction more frequently than soils with rough particle surfaces.

In comparison to subsoils with loam or clay texture, sand subsoils are more susceptible to compaction by machinery, which can lead to restricted root growth. Spoor et al. (2003) developed a risk matrix for subsoil compaction based on texture and packing density, and the susceptibility of coarse and medium (< 18% clay) textured subsoils was very high where packing density was less than <1.4 g cm^{-3} and moderate where packing density was >1.75 g cm^{-3}. These subsoils are extremely vulnerable to subsoil compaction when packing density was <1.4 g cm^{-3} and where the subsoil is rated as wet or moist. The compaction of sandy subsoils has

Table 8.1 Summary of global estimates, by continent, for cropland area, and cropland with sand or gravel subsoils. Data shown here are based on the modelling shown in Figs. 8.1 and 8.2

	Landmass area		Soils with sand subsoil		Soils with gravelly subsoil		Cropland		Cropland with sand subsoil		Cropland with gravelly subsoil	
	sq km	%	*sq km*	%	*sq km*	%	*sq km*	%	*sq km*	%	*sq km*	%
World	134,836,183	100	21,230,002	16	48,319,635	36	16,542,224	12	891,303	5	4,731,733	29
Asia	44,905,869	33	2,600,758	6	22,337,347	50	5,829,962	13	14,568	0	2,501,624	***43***
North America	24,266,661	18	700,357	3	9,206,540	38	3,090,037	13	54,604	2	281,338	9
Europe	9,939,514	7	495,204	5	3,019,848	30	2,987,704	30	128,070	4	604,592	***20***
South America	17,703,512	13	538,996	3	3,421,492	19	1,269,532	7	36,859	3	64,111	5
Africa	29,904,832	22	13,545,808	45	8,082,313	27	2,682,224	9	520,342	***19***	1,050,128	***39***
Australia and Oceania	8,115,795	6	3,348,879	41	2,252,093	28	682,765	8	136,861	***20***	229,940	***34***

implications for root growth; root growth is negatively related to soil strength, and soil strength increases as soil bulk density increases and matric potential increases (Unger and Kaspar 1994). An increase in subsoil bulk density by about 0.13 g cm^{-3} in the 10–30 and 30–45 cm layers, due to a compaction treatment using heavy machinery, caused a 30% yield loss in corn (Voorhees et al. 1986, 1989). However, the implications of subsoil compaction and the likelihood of crop response to alleviation of compaction will vary with rainfall and soil strength; a yield benefit from deep tillage is most likely when there is a restriction to root growth in the subsoil and the crop experiences water stress (Schneider et al. 2017).

The susceptibility of sandy subsoils to compaction has implications for nutrient acquisition from these subsoils. A review by Lipiec and Stpniewski (1995) revealed that compaction reduced the availability of the less mobile nutrients P and K more than the mobile N, which was attributed to the negative effect of compaction on root elongation. However, the restriction of root growth and access to nutrients in the subsoil may be compensated by increased uptake in other layers; Shierlaw and Alston (1984) showed that the proportion of P taken up from the surface layer (0–10 cm) increased from approximately 30–100% as the bulk density of a subsoil layer (10–17 or 10–20 cm depending on bulk density) increased from 1.2 to 1.75 g cm^{-3}. The ability of the crop to compensate for restricted root growth, and therefore nutrient uptake, from subsoils will be dependent on the level of soil nutrients in the topsoil layer and their availability.

The acidity of sand subsoils or their susceptibility to acidification is also an important factor for root function. Sand subsoils will develop an acidity constraint more rapidly than other soil textures under the same environmental and management conditions because they have a lower pH buffering capacity (pHBC). pHBC is a measure of the rate of change in soil pH after an addition of alkalinity or acidity and is related to soil organic carbon (SOC) concentration in combination with, depending upon the soil type, clay content and mineralogy, exchangeable aluminium (Al), effective cation exchange capacity and pH (Weaver et al. 2004). For example, the pHBC of subsoils (10–20 cm) of cropping soils in Western Australia was positively related to organic carbon and exchangeable Al (Moore et al. 1998). pHBC typically shows a positive linear relationship with SOC and in sandy soils is usually highest in the surface and lower at depth (Dolling and Porter 1994). The development of subsoil acidity is significant for crop production because of the relationship between soil pH and soluble Al; as pH falls below 4.8 ($CaCl_2$), the activity of Al^{3+} in soil solution increases to levels that are toxic to roots, causing a reduction in root elongation rate and the development of root hairs (George et al. 2012).

The development of Al phytotoxicity in sand subsoils is significant because it reduces access of the crop to water and nutrients. Carr et al. (1991) compared the grain yield of wheat to soil extractable Al (0.005 M $CaCl_2$) in depth increments from 0 to 5 cm down to 75–100 cm and found that soil-extractable Al at 15–25 cm accounted for 95% of the variation in grain yield and attributed this to decreased root growth in this layer and a reduction in the availability of water from the subsoil. However, the amount of yield lost due to subsurface Al phytotoxicity is also influenced by seasonal rainfall and the location of the toxic Al in the soil profile. An empirical relationship between soil extractable Al and root growth was used in a simulation study to investigate the interaction between subsoil extractable Al concentration and seasonal rainfall on grain yield of wheat. This study showed that grain yield losses were greater at the high (392 mm average growing season rainfall (GSRF)) compared to the low (265 mm GSRF) rainfall site and that yields were lower where soil extractable Al increased linearly with depth compared to soils where soil extractable Al peaked at about 25 cm depth, even though the Al profiles at 0 to 25 cm were similar (Tang et al. 2003). In addition to constraining soil water availability, subsoil Al toxicity can reduce the availability of subsoil P leading to the need for an increase in P fertiliser level to maximise profit (Scanlan et al. 2017).

Root growth in gravelly subsoil is likely to decrease as its gravel content increases. Vine et al. (1981) found that rooting depth decreased as soil gravel content increased and that rooting depth was negatively related to penetrometer resistance, suggesting that the mechanism by which gravel content affected root growth was mechanical impedance. Babalola and Lal (1977) also observed that root depth decreased as soil gravel content increased and observed that roots grown in gravelly subsoil had a greater root diameter than those grown in a gravel-free topsoil, as well as symptoms of mechanical impedance. In addition to gravel content, the physical properties of the soil matrix are also an important factor for the influence of gravelly soil on root growth. In a sand and loamy sand, there was a slight increase in root depth as gravel content increased from 0 to 20–30% and a decrease in root depth as gravel content increased to 80%, i.e. there was a stimulatory effect at low gravel content. By contrast, on a clay soil, root depth decreased almost linearly as gravel content increased from 0 to 80% (Babalola and Lal 1977). The impact of gravel content on root growth is important for grain yield; the decrease in grain yield as soil gravel content increased was attributed to a reduced soil water and nutrient supply (Ercoli et al. 2006; Grewal et al. 1984).

8.6 Effect of Sand and Gravel Subsoils on Plant-Soil Water Relations

Subsoil water supply can make an important contribution to crop yield in water-limited environments. Kirkegaard et al. (2007) found that an additional 10 mm of soil water extracted from the subsoil (1.35–1.85 m) of a Kandosol (Profondic Lixisol (WRB classification), McKenzie et al. 2004) by wheat led to a grain yield increase of 0.6 t ha^{-1} compared to the control treatment. The yield increase was attributed to an increased assimilation rate post-anthesis leading to an increase in grain size. There are no studies that report how water uptake from the subsoil varies in response to the texture of the subsoil; however, the evidence available suggests there are a number of constraints that would limit crop access to subsoil water in sand and gravelly subsoils.

The most significant constraint to crop water uptake from sand and gravelly subsoils appears to be the limited capacity of these soils to store water. The cumulative plant-available water capacity (PAWC), defined as the difference between field capacity and wilting point, with depth for four representative soils from south-west Western Australia with sand and gravelly subsoils is shown in Fig. 8.3. This figure illustrates the importance of rooting depth for access to subsoil water; PAWC for the soil with a sand subsoil at Irwin for 0–135 cm depth was 59 mm; however, if root depth was constrained to 35 cm by soil compaction or Al toxicity, total PAWC was decreased by 40 mm. The amount of water that can be stored by these soils is small compared to other soils used for crop production, for example, Kandosols which stored 82–254 mm PAWC and Vertosols and Dermosols which stored 70–245 mm PAWC (Oliver and Robertson 2009).

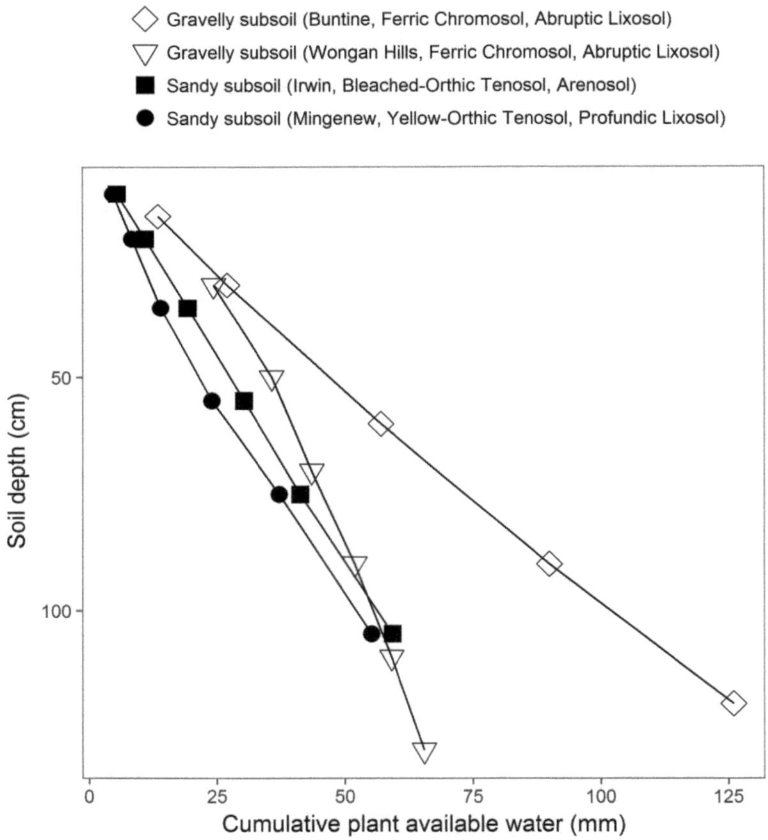

Fig. 8.3 Cumulative plant-available soil water with increasing depth for four soils in Western Australia. Soil profile data from the APSoil database (www.apsim.info). Text in parentheses in the legend is location, Australian Soil Classification (Isbell 2002) and World Reference Base for Soil Resources classification (IUSS Working Group WRB 2015)

The impact of limited PAWC in sand and gravelly subsoils on grain yield is likely to depend on the distribution and amount of rainfall during the growing season. An analysis of the relationship between grain yield of wheat and PAWC from a range of soil types and seasons in south-west Western Australia showed that when PAWC was less than 65 mm, grain yield showed a positive relationship with PAWC, ranging from 17 to 58 kg^{-1} ha^{-1} mm^{-1} PAWC. However, when PAWC was greater than 65 mm, the relationship between grain yield and PAWC depended upon rainfall. Grain yield showed a positive relationship with PAWC in seasons where growing season rainfall exceeded 220 mm and plant-available water was high at anthesis. No relationship or a negative relationship between grain yield and PAWC was observed when soil water storage at anthesis was less than half PAWC, i.e. there was inadequate soil water stored to meet crop demand (Lawes et al. 2009). In this study, PAWC was varied by using soil profiles with different Australian Soil Classifications (e.g. Tenosols, Kandosols and Chromosols) and levels of PAWC and while it did not explicitly address the influence of subsoil water capacity, it does provide an insight into how the PAWC of subsoils and rainfall are likely to interact. For gravelly subsoils that have a fine matrix texture and a high PAWC, these soils are only likely to provide a grain yield benefit when there is adequate water storage to meet crop demand and little or no rainfall late in the growing season. For sandy or gravelly soils with a similar PAWC profile as the examples in Fig. 8.3, ameliorating constraints to root growth such as subsoil compaction or Al toxicity to allow roots to access soil water at 60–100 cm is likely to have a positive effect on grain yield. However, the actual plant-available water to the crop is unlikely to increase linearly as depicted in Fig. 8.2 due to inefficiency in subsoil water extraction by crops (Kirkegaard et al. 2007).

The efficiency of water extraction from subsoils is lower than soil from near or at the surface. Kirkegaard et al. (2007) attributed the incomplete water use from subsoil by wheat to the short residence time of roots in that layer, low root density and clumpy root distribution. Also, a soil layer which restricts root growth can reduce water extraction from the soil below because root growth and branching have been restricted (Tardieu 1988). In addition, hydraulic cut-off occurs at higher matric potential in sands than in finer textured soils; it occurs between −0.3 and − 0.8 MPa for soils with less than 15% clay and between −0.8 and − 1.25 MPa for soils with 15–25% clay (Czyż and Dexter 2013), which could further reduce the efficacy of extraction of water from sandy subsoils as they dry.

The influence of the properties of gravelly subsoils on water supply to crops is likely to depend on the penetrability of the gravelly horizon by roots. For gravelly subsoils that are impenetrable by roots, the impact of these subsoils on grain yield may be driven by the properties of the topsoil associated with the gravelly subsoil. For example, in a field where the depth of a sandy topsoil to a cemented gravel subsoil varied between 30 and 90 cm, PAWC was positively correlated to the depth to the gravel layer (Wong et al. 2009). For gravelly subsoils that are penetrable by roots, gravel content and particle size distribution will be important factors for subsoil water supply.

Attempts to establish a relationship between the hydraulic properties of soils and the gravel content and size have been inconclusive. For example, simulation modelling showed that saturated hydraulic conductivity (Ks) decreased linearly as rock fragment volume increased (Hlaváčiková et al. 2016), whereas laboratory measurements showed an increase in Ks as gravel content increased from 0 to 20% (Beckers et al. 2016). *In situ* measurements of Ks were not significantly affected by rock content of soil (Khetdan et al. 2017). However, the impact of gravel-induced changes to Ks of the subsoil on root growth is questionable; it will only affect crop growth where infiltration rate exceeds drainage rate. The influence of gravel content on unsaturated hydraulic conductivity is likely to be more important. Unsaturated hydraulic conductivity with 20% gravel at 540 cm tension was about half that of the gravel-free soil (Beckers et al. 2016). The combination of restricted root growth described above, lower unsaturated hydraulic conductivity and lower extraction efficiency of water from subsoil by crop plants will constrain water uptake from gravelly subsoils, and the impact of this on grain yield will be greatest in water-limited cropping regions. However, there has been no definitive study of the relationship between soil gravel content and crop water use.

In addition to affecting hydraulic conductivity, gravel particles can also modify the water retention of gravelly soils. The water content of an ironstone gravel soil at −1.5 MPa ranged from 0.17 to 0.24 $m^3\ m^{-3}$, showing an increasing trend with depth. The water retained by the gravel changed PAWC (Brouwer and Anderson 2000), but this is often overlooked in PAWC estimates for gravelly subsoils.

The hydrological behaviour of soil profiles with gravelly or stony subsoils is poorly understood (Zhang et al. 2016). While we can draw on knowledge of how gravel changes soil hydraulic properties at the soil core scale, its influence at the soil profile scale in the field, the scale most relevant to soil water-plant-climate interactions, has not received the same attention.

8.7 Effect of Sand and Gravel Subsoils on Nutrient Acquisition

Nutrient uptake by crops from subsoils is influenced by the concentration of nutrients in the topsoil and subsoil, the vertical distribution of roots and the moisture conditions in the topsoil and subsoil. Nye and Foster (1961) attributed the greater uptake of P from the subsoil (13–25 cm depth) by perennial grasses compared to annual species to a greater concentration of P in the subsoil where the perennial ryegrass was grown. In a similar study, the proportion of total uptake of P by perennial ryegrass from the subsoil (18–43 cm depth) varied over a growing season and was greatest when the soil surface was dry (Newbould et al. 1971). The temporal component of root distribution with depth is also important; a simulation study showed that the availability of subsoil K relative to the topsoil (0–10 cm) declined exponentially as soil depth increased, which was attributed to the decrease in root residence time as soil depth increased (Scanlan et al. 2015).

A study on crop growth response to increasing subsoil nutrient levels on a range of soil types provides some insight into how subsoil texture, and the fertility associated with it, influences nutrient uptake (Graham and Ascher 1993). On a deep calcareous sand with low soil nutrient levels, enrichment of the subsoil (10–100 cm) with 707 kg N ha^{-1} and 782 kg P ha^{-1} increased the grain yield of barley by 180–270% in the first, third, fifth and seventh year of the experiment. There was no yield benefit to the application of the same nutrient treatment on a more fertile sandy earth. These results suggest that subsoil nutrient supply is more likely to have an effect on crop growth when nutrient supply from the whole root zone is not adequate. This response is highly relevant to sand and gravel subsoils where the capacity to store nutrients is constrained, suggesting that the constrained ability of these subsoils to store nutrients will only impact on crop growth when nutrient supplies from the topsoil and fertiliser do not meet crop demand.

The specific effects of sand and gravel subsoils on nutrient acquisition have not been reported previously; however, they are most likely to reflect an interaction between climatic conditions and the constraints associated with these subsoils. Sand subsoils are likely to make a contribution to nutrient uptake when the surface is dry and there are no restrictions to root growth, e.g. compaction and Al phytotoxicity. For gravelly soils, the decrease in root depth as soil gravel content increases will decrease the availability of subsoil nutrients. Gravelly subsoils may only make a significant contribution to nutrient uptake where the gravel content is below a threshold (20–30%) that restricts root growth.

There are potential benefits to nutrient acquisition for crops grown on sand and gravel subsoils. Long-term surplus application of P to agricultural soils in Denmark led to an accumulation of subsoil (25–50 cm depth) P, and the increase was greatest in soils with <15% clay and < 40% fine sand, mostly Podzols and Arenosols, while there was little or no increase in subsoil P in the finer textured Luvisols and Cambisols (Rubæk et al. 2013). Phosphorus is most likely to leach in a sandy subsoil because of the low capacity to adsorb P and high infiltration rates (Lewis et al. 1981). Based on the discussion above, this accumulation of subsoil P may be advantageous for crop growth in environments where the topsoil is frequently dry because the subsoil will provide a plant-available supply; however, the potential benefit to crops from accumulated subsoil P will depend upon the presence of constraints to root growth.

The soil properties that lead to an accumulation of subsoil P can also lead to increased N losses by leaching of nitrate beyond the root zone. Nitrate leaching occurs to a greater extent in sandy soils than in other soils due to a greater infiltration rate and depth (Lehmann and Schroth 2003). A simulation study of nitrate leaching on deep sand showed that the amount on N leached below the root zone (150 cm) ranged from 0 to 115 kg ha^{-1}. Leaching losses were greatest for simulations that were initialised with high levels of soil moisture and soil nitrate concentration and lowest for low initial values for soil water and soil nitrate (Asseng et al. 1998).

8.8 Agronomic Management of Sand and Gravelly Subsoils

The review above has highlighted subsoil constraints on soils with sand or gravel subsoils; however, profitable amelioration of these has been reported. Deep ripping to 500 mm led to an increase in net present value (NPV) of about \$100 ha^{-1} (AUD) after 3 years where soil strength below 20 cm depth ranged from 2 to 3.5 MPa (Hall et al. 2010). Parker et al. (2017) showed that the return on investment for deep ripping to 550 mm on compacted deep sands ranged from 2 to 7 \$ for each \$ (AUD) invested after 2 years. There is some evidence that an acidic layer in sand subsoil can be ameliorated economically by incorporation of lime with inversion tillage; the net margin after one growing season for an untreated control was not significantly different to the treatments where 3 t ha^{-1} lime had been incorporated with a rotary spader where no fertiliser was applied. In this case, the yield benefit from the inversion tillage, and additional income, was sufficient to cover the cost of the lime and tillage operation (Scanlan et al. 2014). In a long-term experiment, incorporation of 2 t ha^{-1} lime with a mouldboard plough into an acidic, sandy subsoil led to an increase in NPV of \$395 after 7 years of annual crop production (Davies et al. 2015). By contrast with the effectiveness of deep incorporation of lime for alleviation of subsoil acidity in sands, surface (0–10 cm) incorporation was ineffective in raising subsoil pH on sands in dryland agriculture in a Mediterranean climate of south-west Western Australia (Whitten et al. 2000). Under the monsoonal climate of southeast Cambodia, Hin (2018) measured pH increase in subsoils below the depth of lime incorporation on deep sands.

The incorporation of biochar in sand subsoils has potential to ameliorate two constraints associated with these subsoils: low water holding capacity and nutrient leaching. The addition of biochar to 75 cm in subsoil (equivalent to 100 and 200 t ha^{-1}) increased the water content of a coarse sandy subsoil after drainage; it was 8.0, 11.0 and 13.9% (v/v) for the 0, 1 and 2% biochar treatments, respectively (Bruun et al. 2014). A 4-year field experiment on a fluvial sand showed that a topsoil treated with biochar at 15 and 30 t ha^{-1} led to greater nitrate retention in the surface layer than the untreated control (Haider et al. 2017). These results suggest that biochar incorporated into sand subsoil could also reduce the loss of plant-available N due to leaching; however, the potential to reduce nitrate leaching is likely to depend on the implement used for incorporation, i.e. its spatial distribution in the subsoil. For example, amendments broadcast on the soil surface before incorporation with a mouldboard plough are likely to be distributed in seams between approximately 10 and 30 cm depth in the same direction as ploughing, at intervals similar to the distance between the plough faces (Scanlan and Davies 2019).

The economic benefit of ameliorating gravelly subsoils is not clear. Rock breakers and grinders have been used to improve the rocky soils of Puglia, Italy, for vineyards (Ferrara et al. 2012); however, the long-term profitability of this practice can be limited by increased erosion risk and a gradual decline in soil fertility (Shelef et al. 2016). The high level of investment in rock breaking and grinding may be prohibitive for soils used for grain production. Delving implements, with tines

20 cm in width approximately 1 m apart, have been used to cultivate to 60 cm and deeper in texture contrast soils to improve the wettability of the sandy surface layer (Betti et al. 2015). This technology could be used to create vertical seams of gravel-free soil in the gravelly subsoil, since the gravel-free topsoil falls into the trench behind the delving tine as it travels forward; however, the agronomic and economic viability of this practice needs to be assessed.

There is scope to vary cropping inputs within a field where the depth to the gravelly subsoil varies based upon yield potential. Proximal sensing may be used to delineate zones of sand or gravelly subsoils in a field so that different agronomic management can be applied in each zone. For example, Wong et al. (2009) showed that gamma K counts were closely related to the depth of sand topsoil overlaying cemented lateritic gravel and a moderate correlation ($r^2 = 0.5$) between PAWC and gamma K intensity. For this type of soil profile, the gamma radiometric survey could be used to delineate deep sand suitable for deep ripping (e.g. 500 mm) or for developing variable rate application maps for economically optimal nitrogen rates based on local calibrations between PAWC, yield potential and yield response to N fertiliser application.

The use of crop sequence to create and utilise biopores may provide short-term amelioration of gravelly subsoils. Hulugalle and Lal (1986) studied the interaction between tillage system and crop sequence on the growth of maize. They found that roots of maize were only present in the gravelly B horizon in the reduced tillage treatment and the pigeon pea-maize crop sequence. Grain yield of maize was higher in the pigeon pea-maize rotation than for continuous maize. The use of primer plants to create biopores (e.g. Yunusa and Newton 2003) in gravelly subsoils to improve the access of succeeding crop roots to soil water and nutrients may be constrained by the negative influence of gravel on root growth. To our knowledge, the sensitivity of different crop species to gravel content has not been established and is a critical knowledge gap for utilising crop sequences to improve root growth in gravelly subsoils.

Crop sequence may also be an approach to maximise nutrient use efficiency in soils with sand subsoil. The deep-rooted blue lupin (*Lupinus cosentinii*) led to an increase in topsoil extractable K concentration compared to continuous subclover (*Trifolium subterraneum*) and a wheat (*Triticum aestivum*)-subclover rotation over 4 years. The increase in topsoil K concentration was attributed to the recycling of subsoil K by the lupin roots (Edwards 1993). Similarly, the deep-rooted (2.4 m) forage radish was more effective at extracting nitrate than ryegrass (0.7 m) and winter rye (1.1 m) from the subsoil of a sandy loam (Kristensen and Thorup-Kristensen 2004). However, the use of deep-rooted crop species to cycle nutrients from depth to the soil surface is likely to be heavily influenced by the profitability of this sequence compared to other sequences. For example, the economically optimal area of the deep-rooted lucerne in a mixed farming enterprise in south-west Western Australia is sensitive to grain and wool prices; as grain price increased, the optimal area decreased (Bathgate and Pannell 2002). Although the study of Bathgate and Pannell (2002) did not include the potential benefits of nutrient cycling from depth, it does provide an illustration of how the mix of land use is influenced by commodity prices.

Agronomic management can be used to mitigate the risk of nutrient leaching on soils with sand subsoil. Splitting fertiliser applications can be a successful strategy to manage this risk; applying fertiliser in three applications at 0 (20%), 30 (40%) and 50 (40%) days after emergence (DAE) produced higher grain yield and N uptake compared to the same amount (28 kg N ha^{-1}, 12 kg P ha^{-1}, 23 kg K ha^{-1}) applied at planting and 30 DAE (86 kg N ha^{-1}) for maize grown on an Oxic Paleustult (loamy sand at 0–15 cm and sandy loam at 15–30 cm) (Sitthaphanit et al. 2009). Similarly, splitting N applications between seeding and 40 days after sowing led to less N loss and higher grain yield than a single application for wheat grown on a deep sand (Asseng et al. 1998). Increasing plant density with the aim of increasing root length density in the subsoil also appears to be a prospective strategy. Dai et al. (2014) found that as the plant density of winter wheat increased from 135 to 405 plants m^{-2}, root length density in the subsoil (400–1200 mm) increased and was identified as the mechanism leading to an increase in shoot N uptake and grain yield.

8.9 Conclusion

The global area of soils with sand or gravel subsoils that are used for crop production is significant. Approximately 29% and 5% of global cropland have gravelly or sand subsoil, respectively, based on conservative estimates using current global land cover and soil information.

The mechanism by which root growth is constrained differs in sand and gravel subsoils. In sand subsoils, root growth is constrained by limitations that can develop rapidly on these soils, particularly soil compaction and Al toxicity. In gravelly subsoils, the presence of gravel particles has a negative effect on root growth, although this relationship is affected by the properties of the soil matrix.

There are physical constraints to crop access to soil water in soils with sand and gravel subsoils. Although the PAWC of a soil with sand subsoil can be increased by increasing root depth, the deep soil water may not be used effectively by the crop because of short residence time of roots in that layer, low root density, clumpy root distribution and higher hydraulic cut-off within the range of plant-available water content.

Research conducted in Mediterranean cropping environments has shown an economic benefit from ameliorating subsoil acidity and compaction in profiles with subsoil sand. The economics of ameliorating constraints associated with gravelly subsoils in cropping systems has not been reported and is a significant knowledge gap given the larger global extent of these soils used for crop production.

References

Amado TJ, Reinert DJ, Reichert JM. Soil quality of very fragile sandy soils from southern Brazil. In: Sustaining the global farm international soil conservation organization meeting, 10th, West Lafayette, IN; 1999. p. 24–29.

Asseng S, Fillery IRP, Anderson GC, Dolling PJ, Dunin FX, Keating BA. Use of the APSIM wheat model to predict yield, drainage, and nitrate leaching for a deep sand. Aust J Agric Res. 1998;49 https://doi.org/10.1071/A97095.

Babalola O, Lal R. Subsoil gravel horizon and maize root growth: II. Effects of gravel size, inter-gravel texture and natural gravel horizon. Plant Soil. 1977;46(2):337–46. https://doi.org/10.1007/BF00010090.

Bašić F. The soils of Croatia. Dordrecht: Springer; 2016.

Bathgate A, Pannell DJ. Economics of deep-rooted perennials in western Australia. Agric Water Manag. 2002;53(1):117–32. https://doi.org/10.1016/S0378-3774(01)00160-3.

Beckers E, Pichault M, Pansak W, Degré A, Garré S. Characterization of stony soils' hydraulic conductivity using laboratory and numerical experiments. Soil. 2016;2(3):421–31. https://doi.org/10.5194/soil-2-421-2016.

Betti G, Grant C, Churchman G, Murray R. Increased profile wettability in texture-contrast soils from clay delving: case studies in South Australia. Soil Res. 2015;53(2):125–36.

Brouwer J, Anderson H. Water holding capacity of ironstone gravel in a typic plinthoxeralf in south-east Australia. Soil Sci Soc Am J. 2000;64(5) https://doi.org/10.2136/sssaj2000.6451603x.

Bruun EW, Petersen CT, Hansen E, Holm JK, Hauggaard-Nielsen H. Biochar amendment to coarse sandy subsoil improves root growth and increases water retention. Soil Use Manag. 2014;30(1):109–18. https://doi.org/10.1111/sum.12102.

Carr SJ, Ritchie GSP, Porter WM. A soil test for aluminium toxicity in acidic subsoils of yellow earths in Western Australia. Aust J Agric Res. 1991;42(5):875–92. https://doi.org/10.1071/AR9910875.

Costa A, Albuquerque JA, Costa A, Pértile P, Silva FR. Water retention and availability in soils of the State of Santa Catarina-Brazil: effect of textural classes, soil classes and lithology. Rev Bras Cienc Solo. 2013;37:1535–48. https://doi.org/10.1590/S0100-06832013000600010.

Cruse RM, Cassel DK, Averette FG. Effect of particle surface roughness on densification of coarse-textured soil. Soil Sci Soc Am J. 1980;44(4):692–7. https://doi.org/10.2136/sssaj1980.03615995004400040006x.

Czyż EA, Dexter AR. Plant wilting can be caused either by the plant or by the soil. Soil Res. 2013;50(8):708–13. https://doi.org/10.1071/SR12189.

da Silva PHM, Poggiani F, Libardi PL, Gonçalves AN. Fertilizer management of eucalypt plantations on sandy soil in Brazil: initial growth and nutrient cycling. For Ecol Manag. 2013;301:67–78. https://doi.org/10.1016/j.foreco.2012.10.033.

Dai X, Xiao L, Jia D, Kong H, Wang Y, Li C, Zhang Y, He M. Increased plant density of winter wheat can enhance nitrogen–uptake from deep soil. Plant Soil. 2014;384(1):141–52. https://doi.org/10.1007/s11104-014-2190-x.

Davies S, Hagan J, Reynolds C, Smee V, Lefroy W, Andrew J, Newman P. Long-term responses to soil amelioration – benefits will last more than 10 years! In: Crop Updates, Perth, Western Australia. Department of Agriculture and Food Western Australia. 2015

Dolling PJ, Porter WM. Acidification rates in the central wheatbelt of Western Australia. 1. On a deep yellow sand. Aust J Exp Agr. 1994;34(8):1155–64. https://doi.org/10.1071/EA9941155.

Edwards NK. Distribution of potassium in the soil profile of a sandplain soil under pasture species. Plant Soil. 1993;155-156(1):407–10. https://doi.org/10.1007/BF00025069.

Ercoli L, Masoni A, Mariotti M, Arduini I. Dry matter accumulation and remobilization of durum wheat as affected by soil gravel content. Cereal Res Commun. 2006;34(4):1299–306. https://doi.org/10.1556/CRC.34.2006.4.272.

FAO. Global Land Cover Share Database (2013). In: FAO Land and Water Division [online]. Rome; 2014. [Last accessed 30 September 2017]. http://www.fao.org/geonetwork/srv/en/main.home?uuid=ba4526fd-cdbf-4028-a1bd-5a559c4bff38

Ferrara G, Farrag K, Brunetti G. The effects of rock fragmentation and/or deep tillage on soil skeletal material and chemical properties in a Mediterranean climate. Soil Use Manag. 2012;28(3):394–400. https://doi.org/10.1111/j.1475-2743.2012.00423.x.
George E, Horst WJ, Neumann E. Chapter 17 – adaptation of plants to adverse chemical soil conditions. In: Marschner P, editor. Marschner's mineral nutrition of higher plants. 3rd ed. San Diego: Academic Press; 2012. p. 409–72. https://doi.org/10.1016/B978-0-12-384905-2.00017-0.
Graham RD, Ascher JS. Nutritional limitations of subsoils. In: Plant nutrition – from genetic engineering to field practice. Dordrecht: Springer; 1993. p. 739–42.
Grewal SS, Singh K, Dyal S. Soil profile gravel concentration and its effect on rainfed crop yields. Plant Soil. 1984;81(1):75–83. https://doi.org/10.1007/BF02206896.
Haider G, Steffens D, Moser G, Müller C, Kammann CI. Biochar reduced nitrate leaching and improved soil moisture content without yield improvements in a four-year field study. Agric Ecosyst Environ. 2017;237(Supplement C):80–94. https://doi.org/10.1016/j.agee.2016.12.019.
Hall DJM, Jones HR, Crabtree WL, Daniels TL. Claying and deep ripping can increase crop yields and profits on water repellent sands with marginal fertility in southern Western Australia. Aust J Soil Res. 2010;48:178–87. https://doi.org/10.1071/SR09078.
Henderson C, Levett A, Lisle D. The effects of soil water content and bulk density on the compactibility and soil penetration resistance of some Western Australian sandy soils. Soil Res. 1988;26(2):391–400. https://doi.org/10.1071/SR9880391.
Hengl T, de Jesus JM, Heuvelink GB, Gonzalez MR, Kilibarda M, Blagotić A, Shangguan W, Wright MN, Geng X, Bauer-Marschallinger B. SoilGrids250m: global gridded soil information based on machine learning. PLoS One. 2017;12(2):e0169748. https://doi.org/10.1371/journal.pone.0169748.
Hin S. Acid sands of south-eastern Cambodia: Their origin, properties and management for upland crops. PhD thesis. Murdoch University; 2018.
Hlaváčiková H, Novák V, Šimůnek J. The effects of rock fragment shapes and positions on modeled hydraulic conductivities of stony soils. Geoderma. 2016;281:39–48. https://doi.org/10.1016/j.geoderma.2016.06.034.
Hulugalle NR, Lal R. Root growth of maize in a compacted gravelly tropical alfisol as affected by rotation with a woody perennial. Field Crops Res. 1986;13:33–44. https://doi.org/10.1016/0378-4290(86)90005-5.
Isbell RF. The Australian soil classification. Collingwood: CSIRO; 2002.
IUSS Working Group WRB. World reference base for soil resources 2014, update 2015. International soil classification system for naming soils and creating legends for soil maps. World soil resources reports no. 106. Rome: FAO; 2015.
Jahn R, Blume H, Asio V, Spaargaren O, Schad P. Guidelines for soil description. Rome: FAO; 2006.
Kautz T, Amelung W, Ewert F, Gaiser T, Horn R, Jahn R, Javaux M, Kemna A, Kuzyakov Y, Munch J-C, Pätzold S, Peth S, Scherer HW, Schloter M, Schneider H, Vanderborght J, Vetterlein D, Walter A, Wiesenberg GLB, Köpke U. Nutrient acquisition from arable subsoils in temperate climates: a review. Soil Biol Biochem. 2013;57:1003–22. https://doi.org/10.1016/j.soilbio.2012.09.014.
Khetdan C, Chittamart N, Tawornpruek S, Kongkaew T, Onsamrarn W, Garré S. Influence of rock fragments on hydraulic properties of Ultisols in Ratchaburi Province, Thailand. Geoderma Reg. 2017;10(Supplement C):21–8. https://doi.org/10.1016/j.geodrs.2017.04.001.
Kirkegaard JA, Lilley JM, Howe GN, Graham JM. Impact of subsoil water use on wheat yield. Aust J Agric Res. 2007;58(4):303–15. https://doi.org/10.1071/AR06285.
Kristensen HL, Thorup-Kristensen K. Root growth and nitrate uptake of three different catch crops in deep soil layers. Soil Sci Soc Am J. 2004;68(2):529–37. https://doi.org/10.2136/sssaj2004.5290.
Lawes RA, Oliver YM, Robertson MJ. Integrating the effects of climate and plant available soil water holding capacity on wheat yield. Field Crops Res. 2009;113(3):297–305. https://doi.org/10.1016/j.fcr.2009.06.008.

Lehmann J, Schroth G. Nutrient leaching. In: Trees, crops and soil fertility. Wallingford: CABI Publishing; 2003. p. 151–66.

Lewis D, Clarke A, Hall W. Factors affecting the retention of phosphorus applied as superphosphate to the sandy soils in south-eastern South Australia. Soil Res. 1981;19(2):167–74. https://doi.org/10.1071/SR9810167.

Lipiec J, Stpniewski W. Effects of soil compaction and tillage systems on uptake and losses of nutrients. Soil Till Res. 1995;35(1):37–52. https://doi.org/10.1016/0167-1987(95)00474-7.

Lipiec J, Siczek A, Sochan A, Bieganowski A. Effect of sand grain shape on root and shoot growth of wheat seedlings. Geoderma. 2016;265(Supplement C):1–5. https://doi.org/10.1016/j.geoderma.2015.10.022.

McKenzie N, Jacquier D, Isbell R, Brown K. Australian soils and landscapes: an illustrated compendium. Collingwood: CSIRO Publishing; 2004.

Moore G, Dolling PJ, Porter WM, Leonard L. Soil acidity. In: Soil guide-a handbook for understanding agricultural soils. Perth: Department of Agriculture; 1998. p. 127–40.

Mulcahy MJ. Laterites and lateritic soils in south-western Australia. J Soil Sci. 1960;11(2):206–25. https://doi.org/10.1111/j.1365-2389.1960.tb01080.x.

Newbould P, Taylor R, Howse KR. The absorption of phosphate and calcium from different depths in soil by swards of perennial ryegrass. Grass Forage Sci. 1971;26(4):201–8. https://doi.org/10.1111/j.1365-2494.1971.tb00665.x.

Newsome D. Origin of sandplains in Western Australia: a review of the debate and some recent findings. Aust J Earth Sci. 2000;47(4):695–706. https://doi.org/10.1046/j.1440-0952.2000.00802.x.

Nye PH, Foster WNM. The relative uptake of phosphorus by crops and natural fallow from different parts of their root zone. J Agric Sci. 1961;56(3):299–306. https://doi.org/10.1017/S0021859600049741.

Oliver YM, Robertson MJ. Quantifying the benefits of accounting for yield potential in spatially and seasonally responsive nutrient management in a Mediterranean climate. Aust J Soil Res. 2009;47(1):114–26. https://doi.org/10.1071/SR08099.

Panayiotopoulos KP. Packing of sands – a review. Soil Till Res. 1989;13(2):101–21. https://doi.org/10.1016/0167-1987(89)90085-8.

Parker W, Isbister B, Hall D, McDonald G, Riethmuller G, Blackwell P. Longevity of deep ripping and topsoil inclusion in soils under traffic farming; evidence from the second season. Paper presented at the GRDC Research Updates, Perth. 2017.

Pye K, Tsoar H. Characteristics of windblown sediments. In: Aeolian sand and sand dunes. Berlin/Heidelberg: Springer; 2014.

Rubæk GH, Kristensen K, Olesen SE, Østergaard HS, Heckrath G. Phosphorus accumulation and spatial distribution in agricultural soils in Denmark. Geoderma. 2013;209:241–50. https://doi.org/10.1016/j.geoderma.2013.06.022.

Santamarina J, Cho G. Soil behaviour: the role of particle shape. In: Advances in geotechnical engineering: the skempton conference, 2004. London: Thomas Telford; 2004. p. 604–17.

Sarcinelli TS, Schaefer CEGR, Fernandes Filho EI, Mafia RG, Neri AV. Soil modification by termites in a sandy-soil vegetation in the Brazilian Atlantic rain forest. J Trop Ecol. 2013;29(5):439–48. https://doi.org/10.1017/S0266467413000497.

Scanlan CA, Davies SL. Soil mixing and redistribution by strategic deep tillage in a sandy soil. Soil Till Res. 2019;185:139–45. https://doi.org/10.1016/j.still.2018.09.008.

Scanlan C, Brennan RF, Sarre G. Shifting investment from nutrients to lime and cultivation on acid soils: is an immediate payback possible. In: Crop updates, Crown Perth. Department of Agriculture and Food Western Australia; 2014

Scanlan CA, Bell RW, Brennan RF. Simulating wheat growth response to potassium availability under field conditions in sandy soils. II. Effect of subsurface potassium on grain yield response to potassium fertiliser. Field Crops Res. 2015;178:125–34. https://doi.org/10.1016/j.fcr.2015.03.019.

Scanlan CA, Brennan RF, D'Antuono MF, Sarre GA. The interaction between soil pH and phosphorus for wheat yield and the impact of lime-induced changes to soil aluminium and potassium. Soil Res. 2017;55(4):341–53. https://doi.org/10.1071/SR16274.

Schaefer CEGR, Oliveira FS. Live subsoils: tropical regolith and biota interactions. In: Oliveira TS, Bell RW, editors. Subsoil constraints for crop production. Switzerland: Springer; 2022. p. 373–96.

Schneider F, Don A, Hennings I, Schmittmann O, Seidel SJ. The effect of deep tillage on crop yield – what do we really know? Soil Till Res. 2017;174:193–204. https://doi.org/10.1016/j.still.2017.07.005.

Shelef O, Stavi I, Zdruli P, Rachmilevitch S. Land use change, a case study from southern Italy: general implications for agricultural subsidy policies. Land Degrad Dev. 2016;27(4):868–70. https://doi.org/10.1002/ldr.2343.

Shierlaw J, Alston AM. Effect of soil compaction on root growth and uptake of phosphorus. Plant Soil. 1984;77(1):15–28. https://doi.org/10.1007/BF02182808.

Sitthaphanit S, Limpinuntana V, Toomsan B, Panchaban S, Bell RW. Fertiliser strategies for improved nutrient use efficiency on sandy soils in high rainfall regimes. Nutr Cycl Agroecosyst. 2009;85(2):123–39. https://doi.org/10.1007/s10705-009-9253-z.

Smalley IJ. Formation of quartz sand. Nature. 1966;211:476. https://doi.org/10.1038/211476a0.

Spoor G, Tijink FGJ, Weisskopf P. Subsoil compaction: risk, avoidance, identification and alleviation. Soil Till Res. 2003;73(1):175–82. https://doi.org/10.1016/S0167-1987(03)00109-0.

Tang C, Asseng S, Diatloff E, Rengel Z. Modelling yield losses of aluminium-resistant and aluminium-sensitive wheat due to subsurface soil acidity: effects of rainfall, liming and nitrogen application. Plant Soil. 2003;254(2):349–60. https://doi.org/10.1023/A:1025597905001.

Tardieu F. Analysis of the spatial variability in maize root density. Plant Soil. 1988;109(2):257–62. https://doi.org/10.1007/BF02202092.

Unger PW, Kaspar TC. Soil compaction and root growth: a review. Agron J. 1994;86(5):759–66. https://doi.org/10.2134/agronj1994.00021962008600050004x.

Vine NP, Lal NR, Payne ND. The influence of sands and gravels on root growth of maize seedlings. Soil Sci. 1981;131(2):124–9. https://doi.org/10.1097/00010694-198102000-00009.

Voorhees WB, Nelson WW, Randall GW. Extent and persistence of subsoil compaction caused by heavy axle loads. Soil Sci Soc Am J. 1986;50(2):428–33. https://doi.org/10.2136/sssaj1986.03615995005000020035x.

Voorhees WB, Johnson JF, Randall GW, Nelson WW. Corn growth and yield as affected by surface and subsoil compaction. Agron J. 1989;81(2):294–303. https://doi.org/10.2134/agronj1989.00021962008100020031x.

Vrščaj B, Repe B, Simončič P. The soils of Slovenia. Dordrecht: Springer; 2017.

Weaver A, Kissel D, Chen F, West L, Adkins W, Rickman D, Luvall J. Mapping soil pH buffering capacity of selected fields in the coastal plain. Soil Sci Soc Am J. 2004;68(2):662–8. https://doi.org/10.2136/sssaj2004.6620.

Whitten MG, Wong MTF, Rate AW. Amelioration of subsurface acidity in the south-west of WesternAustralia: downward movement and mass balance of surface incorporated lime after 2-15 years. Aust J Soil Res. 2000;38:711–728.

Wong MTF, Oliver YM, Robertson MJ. Gamma-radiometric assessment of soil depth across a landscape not measurable using electromagnetic surveys. Soil Sci Soc Am J. 2009;73(4):1261–7. https://doi.org/10.2136/sssaj2007.0429.

Yang H, Baudet BA, Yao T. Characterization of the surface roughness of sand particles using an advanced fractal approach. Proc R Soc A Math Phys. 2016;472(2194):20160524. https://doi.org/10.1098/rspa.2016.0524.

Yunusa IAM, Newton PJ. Plants for amelioration of subsoil constraints and hydrological control: the primer-plant concept. Plant Soil. 2003;257(2):261–81. https://doi.org/10.1023/A:1027381329549.

Zhang Y, Zhang M, Niu J, Li H, Xiao R, Zheng H, Bech J. Rock fragments and soil hydrological processes: significance and progress. Catena. 2016;147:153–66. https://doi.org/10.1016/j.catena.2016.07.012.

Chapter 9
Soilborne Pathogens

Andréa Bittencourt Moura, David Backhouse, Ismail Teodoro de Souza Júnior, and Cesar Bauer Gomes

Abstract The biological subsoil constraints include those caused by a heterogeneous group of microorganisms known as soilborne pathogens. They cause diseases on most important species of plants, and almost all crop plants are susceptible to one or more species of soilborne pathogens. These pathogens cause a wide range of symptoms, and the damage caused to the plant can include death of germinating seeds and seedlings, rotting of roots, blocking of xylem, soft rot, deformation and necrotic lesions on stem bases. Symptoms are usually seen as poor plant stand, stunting or slow growth and discolouration and wilting of the shoot that mimics nutrient deficiencies or drought stress. The losses caused by these biological constraints are huge, so their mitigation is essential. However, it is necessary to know the specific cause of the biological constraint of a given area to take the most appropriate control measure. This chapter deals with soil pathogens of different groups: bacteria, fungi, Oomycetes and nematodes. It describes the characteristics of the soilborne pathogens, *Fusarium*, *Macrophomina*, *Phytophthora*, *Pythium*, *Rhizoctonia*, *Sclerotium*, *Meloidogyne*, *Heterodera*, *Pratylenchus*, *Ralstonia*, *Rhizobium* (previously known as *Agrobacterium*), *Pectobacterium* and *Dickeya* (previously known as *Erwinia*), their geographical distribution and host range, favourable conditions, typical symptoms and plant damage. Control measures, including the use of chemical compounds and products based on biological agents, are reviewed. Subsoil management procedures including liming, chemical fertilisation, green fertilisation, crop rotation, tillage, solarisation, biofumigation and green

A. B. Moura (✉) · I. T. de Souza Júnior
Plant Protection Department, Federal University of Pelotas, Pelotas, RS, Brazil
e-mail: abmoura@ufpel.edu.br; agrojunior1@yahoo.com.br

D. Backhouse
School of Environmental and Rural Science, University of New England, Armidale, NSW, Australia
e-mail: dbackhou@une.edu.au

C. B. Gomes
Embrapa Temperate Agriculture, Brazilian Agricultural Research Corporation, Pelotas, RS, Brazil
e-mail: cesar.gomes@embrapa.br

T. S. d. Oliveira, R. W. Bell (eds.), *Subsoil Constraints for Crop Production*,
https://doi.org/10.1007/978-3-031-00317-2_9

manure are also discussed. Control examples associated with different soilborne pathogenic species are included. However, overall there has been limited study of subsoil biological constraints for agricultural and horticultural species.

Keywords Biofumigation · Crop rotation · Damping-off · Fertilisation · Gall · Root rot · Soilborne disease · Solarisation · Wilt

9.1 Introduction

Soil pathogens are microorganisms that survive and act in the soil for at least part of their life cycle. Many of them persist in the soil primarily as resistant structures (sclerotia, chlamydospores, oospores, etc.). All of them cause economic impacts, directly reducing productivity and indirectly increasing planting costs, affecting the environment and so on. Sometimes the losses reach 100% and may even cause the abandonment of the arable area, since there are soilborne pathogens that practically infect almost all species of economic crops.

They cause diseases on most important species of plants, and almost all crops plants are susceptible to one or more species of soilborne pathogens. These pathogens cause a wide range of symptoms, and the damage caused to the plant can include death of germinating seeds and seedlings, rotting of roots, blocking of xylem, soft rot, deformation and necrotic lesions on stem bases. Symptoms are expressed as poor plant stand, stunting or slow growth and discolouration and wilting of the shoot that resemble nutrient deficiencies or drought stress.

As a consequence of their soilborne life, they are heavily influenced by the abiotic and biotic components of the soil, as well as by soil management procedures such as liming, chemical fertilisation, green manure, crop rotation, tillage, solarisation and biofumigation, besides the use of chemical compounds and products based on biological agents.

This chapter deals with soilborne pathogens of different groups: bacteria, fungi, Oomycetes and nematodes. It describes the characteristics of each group, their geographical distribution and host range, favourable conditions, typical symptoms and damage. Control measures, especially subsoil management examples, associated with different pathogenic species, are also discussed.

9.2 Fungi and Fungus-Like Pathogens

Soilborne fungi and fungus-like organisms (Oomycetes) cause diseases on most species of plants. Soilborne fungal pathogens are typically adapted for long periods of survival in the absence of host plants, have relatively wide host ranges and can grow for at least short distances through soil to plant surfaces. Damage caused to the plant can include seedling damping-off, rotting of roots and necrotic lesions on the basal stem. Canopy symptoms are discolouration and wilting, usually associated with poor plant stand, stunting or slow growth. This section describes several of the most important genera of fungi and Oomycetes that cause soilborne diseases.

9.2.1 *Fusarium*

Fusarium is a large genus of fungi that are mostly found in soil or associated with plants as pathogens. It is very diverse, and species within it cause a wide range of diseases on most types of host plants.

Soilborne species of *Fusarium* are associated with three main types of symptom: root rots, stalk rots and vascular wilts. Root rots are typically caused by members of the *F. solani* species complex, which cause aggressive decay of the root system, often with the involvement of toxins. For example, *F. virguliforme* causes sudden death syndrome in soybean (Brar et al. 2011). Stalk rots are commonly seen in cereals. The fungi that cause them, such as *F. verticillioides* in maize, can often grow without symptoms within the host tissues but become aggressive pathogens when the plant is stressed (Murillo-Williams and Munkvold 2008). Vascular wilts are caused by strains of the *F. oxysporum* species complex. Growth of the fungus within the xylem vessels, together with plant reactions, leads to disruption of sap flow. Early symptoms are wilting of lower leaves and necrosis of leaf margins that can progress to death of the plant.

Almost all crop plants are susceptible to one or more species of *Fusarium*. *Fusarium* wilts are major diseases of crops like banana, cotton, tomato and cucurbits among many others. Many legumes, including peas, beans and soybeans, cereals and many fruit and vegetable crops suffer from *Fusarium* root rots. The host range of *Fusarium* species varies greatly. *F. culmorum* can infect small grain cereals and many pasture grasses, maize and many legumes. Strains of *F. oxysporum* can generally only infect one or a very small number of closely related host species and may be restricted to particular genotypes within those species.

Many species of *Fusarium* produce resistant survival spores known as chlamydospores that are able to remain viable in soil or plant residue for several years. The vertical distribution of chlamydospores reflects the pattern of root length density of the host (Dryden and Van Alfen 1984) and can extend into the subsoil if there is subsoil root penetration. Infestation with *Fusarium* may be considered a biotic subsoil constraint, especially where management practices like biofumigation are only effective in the surface soil.

Fusarium diseases are often regarded as being typical of drier soils. As a genus, *Fusarium* species can grow at lower water potentials than many other pathogens. They are thus tolerant of a wide range of moisture conditions, but infection is still favoured by wetter soils (Swan et al. 2000). There has not been enough work done to make generalisations about interactions with other subsoil constraints, except for compaction. Disease severity is typically worse in association with compaction (Tu 1994; Vick et al. 2003) and may be alleviated by subsoiling or deep tillage (Vick et al. 2003). The mechanisms for the effect of compaction on *Fusarium* diseases have not been established.

9.2.2 Macrophomina

There is one important soilborne pathogen in the ascomycete genus *Macrophomina*, *M. phaseolina*. This fungus has a very wide host range and has been recorded from many hundreds of plant species. It is particularly important on legumes, including soybean, green bean, peanut and chickpea, and on other warm-season crops such as cotton, sorghum, maize, sunflower and many vegetables. It was estimated to cause 2 million tonnes of yield loss in soybean in the top 10 producing countries in 1998, equivalent to 1.5% of total production (Wrather et al. 2001). However, losses in individual fields can be much higher.

Belowground infection causes root rot. Infected roots become discoloured brown or grey, and fine roots may decay completely. Tap roots may become brittle, and necrosis can extend from the roots into the stem base. Root damage causes reduced growth and wilting, which is most evident after flowering in most hosts, and may ultimately cause death of the whole plant. Seedling infection can appear as elongated lesions on the hypocotyls, which coalesce into large necrotic lesions. A characteristic feature of the lesions is the production of large numbers of pycnidia (asexual fruiting bodies) and microsclerotia (survival structures), which give a black, sooty appearance to diseased tissue and lead to the common name of charcoal rot. Microsclerotia in soil and plant residues are the main method of survival in the absence of a host. These germinate in response to root exudates to produce hyphae which infect roots and stem bases.

The microsclerotia are generally concentrated in the upper 30 cm of soil in a range of soil types and crops (Bruton and Reuveni 1985), so *M. phaseolina* can be considered as an inhabitant of the subsoil. However, subsoil conditions which affect plant stress could be important factors in disease expression.

M. phaseolina is favoured by hot dry soils. The optimum temperature for infection in controlled environments is around 30 °C (Meyer et al. 1974), which is much higher than for most other soilborne pathogens. Moisture is required for infection, but disease is more severe when soils are dry, because water stress makes host plants susceptible to aggressive colonisation by the fungus (Diourte et al. 1995). Subsoil constraints which increase the risk of water stress would be expected to increase severity of charcoal rot.

Salinity increases the susceptibility of common bean, one of the most important economic hosts, to infection by *M. phaseolina* (You et al. 2011). Experiments on the interaction between salinity and charcoal rot in this and other plants have used saline irrigation water. The effect on charcoal rot of subsoil salinity, as with other subsoil constraints, has not been tested adequately.

9.2.3 *Phytophthora*

Phytophthora is a genus of Oomycetes (not true fungi) with over 100 described species, all of which are plant pathogens and most of which are soilborne. *Phytophthora* species are among the most destructive of plant pathogens and can cause devastating losses. Up to 72% yield loss has been reported in individual crops of soybean when infection by *P. sojae* results in early death (Ryley et al. 1989). Infection by *Phytophthora* usually involves swimming spores (zoospores) and is favoured by wet soils. Waterlogging can also increase host susceptibility (Duncan and Kennedy 1989).

The classic symptom of many soilborne *Phytophthora* diseases is rotting of the fine roots, which generally occurs under wet soil conditions and may lead to a large reduction in root mass without obvious effects on growth. As soils dry, the plant is unable to take up enough water from the soil and shows signs of water stress. In trees like avocado and *Eucalyptus*, this can result in dieback of the crown. Plants can be killed either by the direct effects of water stress or by rotting of the stem base, and death may occur very rapidly after wilting symptoms first appear. If soils remain wet, above-ground symptoms may not be seen, and extensive infection may occur without detection.

Because of the role of water in epidemiology of *Phytophthora* diseases, there is strong evidence for an effect of subsoil constraints that impede drainage. Compaction has been shown to increase the incidence of disease, caused by *P. sojae* in soybean (Moots et al. 1988) and *P. cinnamomi* in chestnut seedlings (Rhoades et al. 2003), and could be expected to be important for other *Phytophthora* diseases. Subsoil compaction can increase soil water content and risk of waterlogging by impeding drainage. It may also increase disease severity by restricting the volume of soil from which infected root systems can access water as the soil dries out.

Phytophthora species are generally sensitive to antagonism by other soil microorganisms and so are favoured by conditions that reduce microbial activity, such as low organic carbon concentrations (Nesbitt et al. 1979). This effect will be most important in surface soils.

9.2.4 *Pythium*

Pythium is a genus of Oomycetes, closely related to *Phytophthora*. However, there are some significant differences. Most species tend to be weaker pathogens than *Phytophthora* and typically cause disease of juvenile tissues. *Pythium* species rarely cause spectacular epidemics and losses are difficult to quantify. However, they cause damping-off (death of seedlings) either before or immediately after emergence; and impairment of root function in almost all cropping systems, being the main restrictions to productivity. All crop, pasture and orchard plants are susceptible to infection by one or more *Pythium* species, and most species have very broad host ranges. They are particularly important in situations where host vigour is reduced by other factors such as low temperatures.

Most plant species become immune to *Pythium* infection of stem tissues within a few days after emergence. Early root infections appear as small brown lesions on the root surfaces. Heavily infected roots have a greatly reduced number of lateral roots, with brown lesions at root tips and points where lateral roots emerge. Above-ground symptoms are stunting and yellowing associated with the reduction in effective size of the root system.

As with *Phytophthora*, infection by *Pythium* species is generally favoured by wet soil conditions, which is required for zoospore production and their high optimum water potentials for growth (Stanghellini and Burr 1973). Species differ in their temperature requirements, with *P. aphanidermatum* causing most disease at 24–36 °C and *P. debaryanum* and *P. ultimum* at 16–20 °C (Thomson et al. 1971).

Waterlogging is also an important predisposing factor for *Pythium* root rots (Li et al. 2015). However, there has been no recent work done on the interaction between subsoil factors that influence soil wetness and *Pythium* root rots.

9.2.5 *Rhizoctonia*

Rhizoctonia in the broad sense is the term for a number of loosely related soil-inhabiting basidiomycete fungi with characteristic branching patterns in their mycelium. The most important pathogens are placed in the species *R. solani*. This is a complex of forms referred to as anastomosis groups (AG) some of which may represent distinct biological species.

Crop loss due to *R. solani* can take two main forms. One is a general reduction in plant stand density and early growth due to damping off and the involvement of the fungus in the root disease complex of seedlings. Secondly, *R. solani* infection causes severe stunting and death of older plants, typically in patches. Over 80% of plants of sugar beet can be lost in some rotations (Rush and Winter 1990). Bare patch in wheat due to *R. solani* AG-8 can cause death or severe stunting of plants in patches totalling up to 15% of the area of each field (Schillinger and Paulitz 2006).

Infection of germinating seeds can lead to death (damping off) before or immediately after emergence. Root infections lead to brown lesions, which girdle the root. When the plant is pulled from the ground, the roots break at the lesions leaving characteristic 'spear points' with the stele projecting from the rotting cortex. Root rots prune the root system, reducing water and nutrient uptake so that infected plants may be chlorotic and stunted. Infection of the crown of the plant may lead to death. Sclerotia are produced by most AG and may be seen on infected tissue, for example, on potato tubers infected with AG-3.

R. solani is thought to be favoured by warm, slightly dry conditions, but there is considerable variation among AG in their optimum temperatures for infection (Carling and Leiner 1990). Optimum temperatures are lower at higher water potentials because of antagonism by other soil organisms (Gill et al. 2001a). Infection by some AG depends on intact hyphal networks in the soil, and disease is often more severe in untilled soils (Gill et al. 2001b).

Root rot of wheat caused by *R. solani* AG8 is made more severe by compaction in sandy soils (Gill et al. 2004). In this case, compaction reduces the rate at which roots can grow into the subsoil and avoid inoculum of the fungus, which is concentrated near the surface. Roget et al. (1996) described a modified sowing point for wheat which disturbed the soil 0.5–5 cm below sowing depth. Use of this point reduced the severity of *Rhizoctonia* root rot in wheat, and this practice is widely adopted in Australia. The work of Gill et al. (2004) suggests that disturbance below seeding depth may affect disease by reducing subsoil constraints on the rate of root growth.

9.2.6 *Sclerotium*

Athelia rolfsii can cause major losses in individual crops, for example, up to 50% loss of yield in peanuts (Bowen et al. 1992; Fery and Dukes 2002). Unlike many other soilborne diseases, *A. rolfsii* can be controlled by fungicides (Bowen et al. 1992), but this adds to crop production cost.

A. rolfsii causes aggressive necrotic lesions, usually at the stem base or other parts of the plant in contact with soil or plant residues. There may be a web-like white mycelial growth over the lesions, and sclerotia similar in size and appearance to mustard seeds appear within a few days.

A. rolfsii is cosmopolitan in tropical, subtropical and warm temperate parts of the world where there is a warm season with moderate to high rainfall. Almost any warm season crop can be infected. The disease is commonly found in legumes, vegetables, turf, rice, maize and sugar cane.

The disease is favoured by warm to hot and wet conditions. Optimal temperature for growth is 27–30 ° C. Germination of sclerotia is stimulated by alcohols released by decaying vegetation, so the disease may be more severe in weedy crops or after green manure.

Although a sexual stage is known, basidiospores are considered unimportant in epidemiology. There are no asexual spores. Survival is as sclerotia, which can persist in significant numbers for several years in soil. Root exudates stimulate germination to produce hyphae. These grow towards the stem base, and mycelium may form on the plant surface before penetration. Sclerotia are formed abundantly on diseased tissue.

9.3 Bacterial Pathogens

Soilborne phytobacteria constitute an important group of pathogens of cultivated plants, causing losses in fruit, vegetables, ornamental plants, field crop and forest species of global economic importance, and are capable of making an entire cultivated area unviable. This group of pathogens is diverse, comprising several genera grouped by the principal symptoms they cause: deformation (hyperplasia), soft rot and wilt.

Abiotic factors such as temperature, pH, moisture and soil type, in addition to biotic factors such as presence and activity of predators, antagonists and/or competitor organisms, may influence the survival of bacteria in the soil (Van Veen et al. 1997).

9.3.1 Wilt

Ralstonia solanacearum bacterial wilt is probably the most destructive bacteriosis in the world. Due to its wide geographic distribution, broad host range and pathogenic behaviour, *R. solanacearum* is apparently genetically diverse and for over a century different classification systems were proposed. Strains were initially differentiated into five races according to host range, while in biovars this was based on the capacity for using or oxidising carbon sources. Later, the strains were grouped into phylotypes and sequevars. Safni et al. (2014)) reclassified the phylotypes of *R. solanacearum* into species, with phylotypes I and III being reclassified as *R. pseudosolanacearum*; IV, the most diversified phylotype, was subdivided into *R. syzygii* subsp. *syzygii*, *R. syzygii* subsp. *indonesiensis* and *R. syzygii* subsp. *celebensis*; phylotype II remained as *R. solanacearum*. The species belonging to the former phylotypes I, II and III mainly occur in countries of Asia, America and Africa, respectively. Phylotype IV, on the other hand, is composed mainly of strains from Indonesia with some isolates from Australia, Japan and the Philippines. However, for the purposes of this chapter, all new species in the complex were grouped as *R. solanacearum*.

Infection by these bacteria causes wilt, where the loss of leaf turgor starts in young leaves and spreads to the whole plant. The bacterium invades the plant through roots and systemically colonises shoot tissue, obstructing the xylem vessels, which are filled by the bacteria and by gums and tyloses formed as a host response to infection. The wilt progresses irreversibly until it causes plant death (Champosiseau et al. 2009).

At very high temperature and humidity, losses can reach 100%, especially in successive plantings, due to the ability of the bacteria to survive in the soil for long periods. Losses vary from 0% to 91% in the tomato, 33% to 90% in the potato, 10% to 30% in tobacco, 80% to 100% in the banana and up to 20% in the peanut. It is difficult to quantify the direct economic impact of *R. solanacearum*, but in the potato losses per year worldwide are estimated at around USD 1 billion (Elphinstone 2005).

Bacterial wilt is distributed in regions of tropical, subtropical and temperate climate. Species of *Ralstonia* have a wide range of hosts and infect at least 50 different plant families representing more than 200 species, especially Solanaceae.

Under European Union (EPPO) and North American (NAPPO 2017) legislation, *R. solanacearum* is listed as one of the major quarantine organisms requiring processes of eradication and control. Furthermore, it has been included in the list of potential bioterrorism agents in the United States due to the economic impact of bacterial wilt and its potential for dissemination in the form of a latent infection (EPPO 2017; Champosiseau et al. 2009). The genetic variability of *R. solanacearum* generates problems for disease management due to the difficulties in obtaining resistant cultivars and in the diagnosis of strains in the field.

9.3.2 Deformations

Among the bacteria that cause deformation in plants, the most important belong to the genus *Rhizobium*. Previously, *Rhizobium* was classified as *Agrobacterium*; however, it was reclassified based on comparative analysis of the 16S rRNA gene (Young et al. 2001). *R. rhizogenes* causes symptoms known as root hair in woody species; *R. vitis* and *R. rubi* cause, respectively, grapevine and sugar cane galls; but the disease with the greatest geographic distribution, frequency and importance is crown gall, caused by *R. radiobacter* (*A. tumefaciens*).

Tumours caused by *R. radiobacter* are often seen on crown and main roots. In herbaceous plants, they are soft, clear and noncorticate, while in woody plants, they are hard, dark and corticate. Tumours arise from host cells transformation: the bacterium penetrates through the wounds and transfers the Ti (tumour-inducing) plasmid (Kado 2010) that integrates into the plant cell genome. The transformed cells then start to encode enzymes involved in plant hormone biosynthesis such as cytokinins and auxins, leading to hormonal imbalance and hyperplasia, characterising the gall (Pitzschke and Hirt 2010).

Yield decrease of diseased plants is probably caused by several factors: reduced water and nutrient flux due to vascular tissue damage and/or their compression by the gall development (Aloni et al. 1998). Plants with several galls become weakened, stunted and unproductive, while the young plants die.

Crown gall affects dicotyledonous plants and some monocotyledons. Experimental inoculation has shown that at least 93 plant families are susceptible (De Cleene and Ley 1976). In the field, the disease often occurs in woody plants and in some herbaceous species. In addition, high variability is found within the species *R. radiobacter*, which forms up to nine different groups (Mougel et al. 2002).

9.3.3 Soft Rot

The phytopathogenic pectolytic bacteria include species of the genus *Pectobacterium* and *Dickeya* (previously a member of the genus *Erwinia*). These bacteria typically induce soft rot, but other symptoms include wilt, blackleg, hollow stem and seedling damping-off.

P. carotovorum subsp. *carotovorum* has a wide range of hosts including onion, carrot, potato, lettuce, tomato, cucumber and ornamental plants (Rashid et al. 2012). *P. atrosepticum* is restricted to the potato, especially in temperate regions (Ma et al. 2007).

The genus *Dickeya*, previously known as *Pectobacterium chrysanthemi*, currently comprises six species: *D. dadantii*, *D. zeae*, *D. dianthicola*, *D. chrysanthemi*, *D. paradisiaca* and *D. solani* (Toth et al. 2011), which are characterised as causal agents of dry stem rot, wilt and blackleg in the potato and in various plant species in tropical and subtropical climates (Charkowsky 2006). The variability within this genus is quite high, where the strains are divided into six species and seven biovars (Palacio-Bielsa et al. 2006).

The losses caused by these pathogens can be high and vary with crop, disease severity, subspecies, environmental and crop management. Potato losses vary from 10% to 100% in storage, while in the field, losses are around 10–40% (Pereira and Daniels 2003).

The initial soft rot symptom is small wet lesions, which increase fast and cause extensive rotting. The leaves may wilt and become yellow in advanced disease stages. Soft rot at the base of the stem is known as blackleg, a consequence of stem colonisation, with dark pigment production. In the case of hollow stem, the stem becomes literally empty and has a tubular appearance, since the bacterium displays greater pectolytic activity in the non-woody central region.

Pectolytic bacteria survive epiphytically in host leaves, as saprophytes in the soil, in infected debris, in propagative material and in the rhizosphere of cultivated plants and weeds (Goto 1992). Their survival in the soil depends on pH, temperature and moisture. The bacteria can survive in the soil from 1 week to 6 months, even in the absence of plant debris; however, this period may be greater with plant tissue presence (Czajkowski et al. 2011).

9.4 Plant-Parasitic Nematodes

Phyto-nematodes are soilborne pathogens responsible for causing serious damage, either directly by root system parasitism, by predisposing the attacked plant to other pathogens or as viral vectors. Sedentary nematodes, such *Meloidogyne* and *Heterodera*, and migratory endoparasites, such as *Pratylenchus*, are included among the world 'top ten' list (Jones et al. 2013).

9.4.1 Root-Knot Nematodes

There are around a hundred described *Meloidogyne* species, but *M. incognita*, *M. javanica* and *M. arenaria* (tropical climate), as well as *M. hapla* (temperate climate), are the most frequent (Hunt and Handoo 2009). Other species such as *M.*

chitwoodi, *M. fallax*, *M. minor*, *M. paranaensis*, *M. graminicola* and *M. enterolobii* are considered emergent and are under quarantine in some countries (Onkendi et al. 2014; Negretti et al. 2017).

The genus *Meloidogyne* is considered the most important cosmopolitan group because of its wide polyphagia, its genetic diversity and its aggressiveness in various host plants, including vegetables, fruit trees and annual crops, causing USD157 billion global annual losses (Onkendi et al. 2014).

The main symptom is root galls; however, depending on the infestation level and the plant species resistance, the plants may become weakened. Other symptoms are usually in spots: yellowing, defoliation and wilt at the hottest hours and, in severe cases, death.

The temperature directly interferes with the life cycle length of root-knot nematodes, and its effect is species dependent: optimal temperatures for *M. hapla* range from 18 to 25 °C, and for *M. javanica*, between 28 and 30 °C. Above 40 °C or below 5 °C, most species show little activity (Ferraz and Brown 2016). Another important factor is soil texture, since sandy soils favour nematode movement and multiplication (Barker and Weeks 1991). Other factors also affect the life cycle, such as soil humidity and aeration, light and host species.

The roots are infested by second-stage juveniles (J2) that move between cells. They establish permanent feeding sites and become sedentary on the periphery of the central vascular cylinder. The nematode causes hypertrophy, forming a cluster of giant cells for its nutrition and development that function as a drain, resulting in thickening of the roots, known as galls (Palomares-Rius et al. 2017). Females become obese and produce eggs in a gelatinous matrix external to the roots. The males, vermiforms, lose their parasitic ability, may fertilise the females or not and leave the root (Ferraz and Brown 2016).

Root-knot nematodes occur at high levels of genetic diversity worldwide, at both population level and species. Their diversity in part can be explained by its different reproduction modes: sexual and/or different modes of parthenogenesis, depending on the presence or absence of males (Castagnone-Sereno et al. 2013).

9.4.2 Cyst Nematodes

Cyst nematodes, the second largest plant parasitic nematodes group, associated with losses in cereals, vegetables and oilseeds, are mainly represented by the genera *Heterodera* and *Globodera* (Jones et al. 2013). *Heterodera* is the type of genus that contains the largest number of species and has an extensive range of host plant species, such as the soybean (*H. glycines*), beetroot (*H. goettigiana* and *H. schachtii*), rice (*H. oryzicola*), Brassicaceae (*H. cruciferae*), cereals (*H. avenae* and *H. fillipjevi*) as well as vegetables and forages (*H. trifolii*) (Evans and Rowe 1998).

Typical soybean symptoms can be seen in leaf spots, poorly developed plants, yellowing foliage, deteriorated roots, flower abortion and few or no pods. At highly infested sites, dead plants may be found.

The greatest losses are associated with soybean. In the United States, losses are estimated at more than USD 1.2 billion a year (Koenning and Wrather 2010). However, losses vary according to such factors as soil fertility, cultivar susceptibility and nematode population and management (Dias et al. 2009). The development of cyst nematode populations in the soybean crop is favoured in sandy soils with high pH and base saturation and associated with temperatures between 22 and 29 °C (Schmitt and Riggs 1989; Garcia and Silva 1996; Avendaño et al. 2004).

9.4.3 Root Lesion Nematodes

Pratylenchus species have a wide range of hosts, including many perennial and annual species from different botanical groups. More than 70 species of *Pratylenchus* are described. The most important is *P. penetrans*, distributed mainly in temperate areas, parasitising fruit and potato plants; *P. brachyurus*, *P. zeae*, *P. coffeae*, *P. vulnus* and *P. scribineri* are common in tropical and subtropical regions, attacking annual crops such as maize, sorghum, wheat, sugar cane, soybean, several vegetables as well as perennials and forage plants (Castillo and Vovlas 2007). Population density of *Pratylenchus* spp. on different plant species is negatively correlated with the exchangeable base cation contents and percentage of clay and organic matter of the soil (Kandji et al. 2001; Sundararaju and Jeyabaskaran 2003; Franchini et al. 2007) but positively correlated with pH (Franchini et al. 2011, 2018). Likewise, there is also correlation between the variables associated with precipitation and crop intensity (Kandel et al. 2013) and soil temperature (Zirakparvar et al. 1980) on the root lesion nematode population densities.

Crop yield losses from infection reach 85% in wheat and 30% in the banana. In the northwestern United States, root lesion nematodes cause an economic loss of USD 51 million annually (Yu et al. 2012). In Australia, economic losses occur on 5.53 million ha of wheat (Fosu-Nyarko and Jones 2016).

All motile forms are able to enter and leave the infected roots. Root lesion nematodes typically probe a number of cells before selecting one cell from which to feed. Root invasion is intracellular and is achieved by a combination of mechanical probing and enzymatic secretion that enable root lesion nematodes to migrate within the roots. The type of symptoms seen in the roots gives rise to the common name for the infection (Fosu-Nyarko and Jones 2016).

Typical necrotic lesions, initially yellow, evolve to dark brown or black. Severely infected plants have a smaller number of roots, which reduces their ability to absorb water and nutrients, resulting in yellowing leaves and lower growth, as well as reducing productivity and the quality of tubers (Jones et al. 2013).

All life cycle stages of *Pratylenchus* are infective and can cause penetration of the roots, rhizomes and tubers. In species such as *P. penetrans*, *P. vulnus* and *P. coffeae*, males are abundant, and reproduction is therefore sexual. In parthenogenetic species such as *P. brachyurus*, *P. scribneri* and *P. zeae*, females are common and males are rare. The root lesion nematode life cycle varies with the species, host plant and soil temperature. Tropical species like *P. brachyurus* have an optimum temperature in the range of 29–30 °C and complete their cycle in 3–4 weeks, while for temperate species, such as *P. loosi*, the temperature is between 18 and 20 °C, with a cycle of 5–7 weeks (Ferraz and Brown 2016).

9.5 Control Measures

Chemical compounds to reduce the soilborne pathogen population were used extensively and intensively until the end of the twentieth century. The biocide, methyl bromide, has 100% efficiency; however, it affects the ozone layer, and after the signing of the Montreal Protocol (1987), its production and use were progressively reduced. Since then, the existing alternatives are either not as efficient or too specific. For example, use of the eradicant Gallex® is recommended in the curative control of crown gall. Acylalanines (metalaxyl, furalaxyl, etc.) have been widely used for *Phytophthora* and *Pythium*, but resistance can develop (Qi et al. 2012). Moreover these chemicals are degraded rapidly in soil, especially with repeated use (Droby and Coffey 1991). Fungicides like azoxystrobin may be useful against *Sclerotium* (Koehler and Shew 2017) and *Rhizoctonia* (Kirk et al. 2008). There are also nematicides used for soil drench or fumigation (Onkendi et al. 2014) and for seed treatment (against *H. glycines*) (Vitti et al. 2014).

Following a worldwide trend towards green technologies and production of healthier and residue-free food, many biological products for the control of soilborne pathogens are now available on the global market. In Brazil, the number of biological products has been growing since the first registration in 2000, but the diversity of soilborne pathogens targeted is still narrow. The biological products show low toxicity, are ecofriendly and are less harmful to human health. These new products are composed of a single microorganism or by different combinations of *Pasteuria nishizawae*, *Pochonia chlamydosporia*, *Trichoderma harzianum*, *Purpureocillium lilacinum* and species of the *Bacillus* genus.

Biological control of soilborne pathogens is currently one of the most successful strategies. Since the beginning of the 1980s, products containing the antagonist *R. radiobacter* (*A. radiobacter*) have been used to control crown gall in different host plants in several countries. Initially the biocontrol isolate was K84 (Agrogall 30®, Dygall® and Galltrol-A®), which, after displaying resistance to agrocin (compound responsible for the pathogenic *R. radiobacter* control), was partially replaced by K1026 (Nogall™), an isolate obtained through genetic manipulation.

The use of resistant cultivars would be the best alternative to reliance on chemical treatment especially when the organisms are in the subsoil; however, there are few such cultivars available. Single gene/high-level resistance has been identified for some soilborne *Phytophthora* diseases, for example, against *P. sojae* (Dorrance and Schmitthenner 2000) where it is widely used in soybean cultivars. Resistant cultivars with high-level/monogenic resistance are also available to some races of some formae speciales of *Fusarium oxysporum*, e.g. *F. oxysporum* f. sp. *lycopersici*

(Catanzariti et al. 2015), and a few *H. glycines* and *Meloidogyne* races (Dias et al. 2009; Castagnone-Sereno et al. 2013). Partial or polygenic resistance is used for many diseases, for example, *F. oxysporum* f. sp. *vasinfectum* on cotton (Sanogo and Zhang 2016), *Macrophomina* on many crops (Coser et al. 2017) and *Pectobacterium* on potatoes (Chung et al. 2013).

The growing need to produce more food and to minimise the environmental impact of agriculture has led to increased research, as well as the registration and marketing of products containing biological control agents. The global market offers products to control most of the pathogens addressed in this chapter (Bettiol et al. 2009); however, they are generally recommended as seed treatments for a restricted host range, produced on a small scale and only marketed in one or a few countries.

Soil management procedures can also play an essential role in control of soilborne pathogens and, in some cases, are the only effective measure available. Practices include irrigation/drainage, loosening compacted soil, organic waste incorporation and solarisation, as well fertilisation and correcting soil pH.

Nitrogen occurs as a constituent of resistance-related compounds such as phytoalexins and phenols. However, plants fertilised with high doses of N present a larger amount of young and succulent tissues, prolonged vegetative stage and delayed maturation phase, which favours the attack of pathogens. Additionally, high concentrations of nitrogen can reduce the production of phenolic compounds (fungistatic compounds) and leaf lignification, reducing resistance to obligate pathogens; however, they have no action on the facultative pathogens (Marschner 1995). On the other hand, nitrogen deficiency favours the slow growth of plants, a condition that increases their susceptibility to pathogens (Zambolim and Ventura 1996).

Nitrogen excess generally makes the plant tissues more tender, favouring the development of most pathogens, especially soft rot (Czajkowski et al. 2011) and *Fusarium* diseases in cereals (Davis et al. 2009). However, in some cases the effect may be positive, such as the reduction in bacterial wilt and an increase in potato production (Lemaga et al. 2005). Increasing doses of potassium generally make the plants more tolerant to different pathogens and reduce the number of females of *H. glycines* (Barbosa et al. 2010) probably due to the interference of this nutrient in the establishment of the root nematode feeding site (Hussey and Davis 2004). Likewise, Franchini et al. (2011) demonstrated a negative correlation between *P. brachyurus* population density in the soil and potassium and sulphur levels in soybean leaf tissue levels. According to the same authors, the soil correction with gypsum incorporation can reduce the losses in soils infested by this nematode.

The organic matter content in soils can also influence the severity of diseases that are lower in organic matter-rich soils, because the high levels of organic matter are usually related to a high soil microbial population, increasing the suppressive effect against pathogens. Therefore, the addition of organic matter to the soil generally results in lower losses caused by soilborne pathogens and in the long run can lead to a progressive reduction in pathogen and disease incidence, turning conductive soil into suppressive soil (Cook and Baker 1983).

There are variations in the occurrence of diseases due to soil pH, especially those located near the roots (rhizosphere). However, usually correcting soil pH and increasing availability of calcium, which increases rigidity to the middle lamella of the host plant cells, reduces damage caused by pectolytic bacteria such as *P. carotovorum* (Czajkowski et al. 2011), as well as the wilt severity and the *R. solanacearum* population in tomato (Yamazaki et al. 2000). It also reduces *Pythium* germ tube length and sporangial germination (Ko and Kao 1989). For the soybean-*H. glycines* pathosystem, despite the antagonism between pH and nematode population levels, the incorporation of limestone has been associated with decreases in soil pathogens. According to Sfredo et al. (1994), the addition of lime to the soil contributes to the increase of pH and the increase of the availability of some micro- and macronutrients to the plants, especially calcium, favouring the increase of the resistance of the roots to the penetration by the nematode (Rocha et al. 2006).

On the other hand, pH changes caused by nitrogen fertilisation can also affect diseases caused by soilborne pathogens. In general, the damage caused by pathogens of the genera *Fusarium* and *Rhizoctonia* can be attenuated by the increase of rhizospheric pH, due to the higher absorption of nitrate (Huber and Watson 1974). Similarly, damages caused by *Pythium* diseases are correlated with ammonium absorption due to increased acidification of the rhizosphere. The maintenance of pH and soil correction are important in reducing the incidence of diseases caused by *Fusarium*. An example is the case of banana cv. Mysore, considered resistant to *Fusarium* wilt, but, when cultivated in soils with very low pH and unbalanced K: Mg and Ca: Mg ratios, becomes susceptible to the disease (Ventura and Hinz 2002).

Soil moisture has a great impact on the development of soilborne pathogens and can directly affect them, as well as their host and the interaction between them. Flowing water or rainwater plays a relevant role in the distribution and dispersal of pathogens among plants. Excess water increases the juiciness of plant tissues and decreases the plant's ability to defend itself due to reduced oxygen availability to the roots, increasing susceptibility to most pathogens. Additionally, under these circumstances, increased root exudation is likely to occur, which attracts pathogens that travel in chemotaxic response (Oomycetes, flagellate bacteria and nematodes), causing spore germination and egg hatching and, finally, penetration in the host plant, resulting in infection and disease development.

Most soil pathogens, such as *Phytophthora*, *Pythium*, *Rhizoctonia*, *Sclerotinia*, *Dickeya*, *Pectobacterium*, *Pseudomonas*, *Ralstonia* and nematodes, cause more severe symptoms when the soils area close to field capacity. On the other hand, water scarcity, which normally occurs in dryland regions, generally negatively affects the pathogen's life cycle and weakens plant defences. Few pathogens are favoured by low soil moisture, such as *F. solani*, *F. roseum* and *M. phaseolina*, developing well in dry environments, causing more severe diseases in plants under water stress. Additionally, dry soils favour the survival of the pathogen propagules and increase their tolerance to high temperatures, but inhibit their germination, growth and spread. For example, *M. phaseolina* sclerotia survive for almost 5 months in dry soil; however, those maintained in soils at field capacity had their viability reduced by 60%, so, maintaining soil moisture is a possible means for controlling this fungus.

At the same time, reports associate soil compaction with a higher incidence of many diseases. Waterlogging and compacted soils favour the development and penetration of the pathogen in the host plant roots, which, in addition to less aeration, negatively affect most plants exposed to diseases. In soybean, the damping-off occurs if heavy rains that soak the soil coincide with high temperatures. The situation is aggravated by the presence of crop residues and compacted soil. Subsoil compaction hinders the drainage of water which, under high temperatures, favours the rapid development of soilborne fungi. Generally, damping-off occurs in patches, being commonly observed in the crop borders, in the most compacted areas and in the lower part of the contours above the terraces, where the water accumulates.

Thus, excess moisture and high compaction cause stress on the root system and allow rapid dispersion and penetration of pathogens. Therefore, proper water management through drainage and irrigation as well as soil loosening can be used to control the diseases and some soilborne pathogens, or at least to reduce the severity of symptoms as plants will not be under these stresses.

Perhaps the most effective measure, and the one with the broadest action, is solarisation. The technique consists of covering moist soils with transparent plastic film, which is exposed to solar radiation for a variable period to reduce the populations of different pathogens. With heating of the soil, the microbiota is selectively reduced, since the lethal temperature of most soilborne pathogens is below that reached by solarisation. In addition, many saprophytic species, as well as the biocontrol agents such as *Bacillus*, remain at a relatively high level, avoiding a biological vacuum and consequently delaying reinfestation, so that the process has a lasting effect. Virtually all the pathogens covered in this text can be controlled through solarisation. The time taken to kill a high percentage of the population of these pathogens varies as a function of the solarisation time, the insolation and the season, as well as the pathogen species and the type of its structures present in the soil (De Vay et al. 1991). So, to ensure high process efficiency, usually at least 30–60 days of solarisation is required.

The efficiency and/or duration of the effect of solarisation can be intensified by the prior incorporation of organic substrate (also known as biofumigation or biosolarisation), or by the later treatment of soils with biological control agents, notably those containing *Trichoderma* and *Streptomyces* (Levy et al. 2015; Minuto et al. 2006), or even by using a minimum dose of pesticides and planting resistant/tolerant cultivars.

Biofumigation through the incorporation of *Brassica* species is the most commonly used. The decomposition of plant tissue rich in glucosinolates by the soil microbiota releases isothiocyanates, toxic compounds with a bactericidal, fungicidal and nematicidal action (Oliveira and Dhingra 2008). Other plant species are used to control various soilborne pathogens such as the castor bean (*Ricinus communis*) (Santos and Gomes 2011), pigeon pea (*Cajanus cajan*), sunn hemp (*Crotalaria juncea*) (Cardoso et al. 2006) and weeds such as *Geranium carolinianum* (Ooshiro et al. 2004). Animal manure can also be efficient (Baptista et al. 2006; Melero-Vara et al. 2011).

Using organic fertiliser and intercropping with antagonistic and bait plants are recommended for the control of pathogens (Ratnadass et al. 2012). Simple incorporation, without covering the soil with plastic film, is also used for control, although the efficiency varies with the pathogen and the plant species incorporated (Zasada et al. 2009; Ratnadass et al. 2012).

Finally, if control has not been achieved after one or more of the above measures, there still remains the possibility of crop rotation (Ratnadass et al. 2012). In most cases, for the rotational cultivation success, it is necessary to have knowledge of the pathogen species and race as well as its range of hosts. That said, it is possible to select the species to be used and establish a crop succession scheme and the minimum time before return to planting the pathogen's host species. However, the fact that most soilborne pathogens are polyphagous minimises the species diversity available to be planted in rotation. Additionally, pathogens that form resistant structures that remain viable for several years are not controlled by crop rotation. On the other hand, this practice is generally recommended for nematode control (Ferraz and Brown 2016). Rotation also can be a good option for *Fusarium* wilts because of the narrow host range of most formae speciales (Fang et al. 2012) but can be limited in effectiveness if chlamydospores can survive for a long time in soil, or the fungus can colonise the roots of symptomless hosts. Rotation is used successfully for *Fusarium* crown and root rots of cereals (Evans et al. 2010). Rotation with non-hosts such as cereals is possible also to control *Sclerotium* (Wokocha 1988).

Examples for different pathogen species and host plants, grouped by control measure, are shown in Table 9.1.

9.6 Perspectives

Much effort has been invested in the development of biological products, containing microorganisms as active ingredients, and extracts of plant and animals. This market has been growing annually, with soilborne pathogens as the main target. Therefore, we can anticipate an increase in the number of these products and the range of pathogens and hosts affected and, in particular, formulations with greater efficiency and shelf life.

On the other hand, considerable future impact should be provided by genetic engineering. The explosion of the omics era has allowed the identification of effectors and genes responsible for the parasitism of different pathogens, as well as those related to the plants' defence mechanisms. Therefore, the development of resistant cultivars that over-express defence genes of plants or that silence those of susceptibility will bring new hopes for the control of soilborne pathogens.

In addition, the rapid expansion in microbiome studies is generating global information about the soil microbiota. New findings on the composition and functioning of complex soil microbial communities during interaction with the plant revealed

Table 9.1 Control measures recommended for soilborne disease control

Control measure	Pathogen	Host	Effect	Reference
Liming	*Pectobacterium carotovorum*	Potato	Reduce severity	Czajkowski et al. (2011)
	Ralstonia solanacearum	Tomato	Reduce severity Reduce incidence	Yamazaki et al. (2000)
	Heterodera glycines	Soybean	Decrease populations – increasing limestone incorporation	Rocha et al. (2006)
Coarse sand, fine sand and pH	*R. solanacearum*	Pepper	Negative correlation	Felix et al. (2012)
Solarisation	*R. solanacearum*	Tomato	Reduce incidence	Vinh et al. (2005)
Green manure	*Fusarium oxysporum*	Strawberry	Reduce severity	Fang et al. (2012)
Crop rotation	*Fusarium* species		Reduce incidence	Evans et al. (2010)
	Rhizoctonia solani		Reduce incidence	Rush and Winter (1990)
	Meloidogyne spp.		Decrease nematode populations	Carneiro et al. (2000)
	H. glycines	Vegetables	Decrease nematode populations	Dias et al. (2009)
	Pratylenchus brachyurus	Soybean	Decrease nematode populations	Vedoveto et al. (2013)
	R. solanacearum	Potato	Reduced incidence	Katafiire et al. (2005)
Tillage	*R. solani*	Barley	Reduce severity	Roget et al. (1996)
	Fusarium virguliforme	Soybean	Reduce severity	Vick et al. (2003)
Fertilisation	*H. glycines*	Soybean	Decrease populations – increasing potassium levels	Barbosa et al. (2010)
	P. brachyurus	Soybean	Decrease populations – increasing potassium and Sulphur levels; gypsum incorporation	Franchini et al. (2011)
	P. carotovorum	Pepper	Reduce severity – increasing calcium phosphite and potassium phosphite	Silva et al. (2014)

(continued)

Table 9.1 (continued)

Control measure	Pathogen	Host	Effect	Reference
Biofumigation	*Meloidogyne arenaria*		Decrease nematode populations – biofumigation with radish and arugula	Aydinli and Mennan (2018)
Green manure + biological control	*Pratylenchus neglectus* *Meloidogyne chitwoodi*	Potato	Decrease populations – oil radish and/or rapeseed green and/or *Bacillus megaterium*	Al-Rehiayani et al. (1999)
Biocide (*R. radiobacter)*	*Agrobacterium tumefaciens*		Reduce incidence – eradicant	http://agbiochem.com/wp-content/uploads/2014/08/Revised-Galltrol-Gallex-Pamphlet-Oct-2012.pdf

that soil microorganisms are determinants of plant physiology and metabolism, nutrient acquisition, tolerance against abiotic stresses and protection against soilborne pathogens. Therefore, this knowledge can lead to the possibility of managing the soil microbiota. Changes in soil fertility, in the plant species grown, in the roots' exudates, in the quality and quantity of organic matter, etc. can modify the soil microbiota in order to increase beneficial microorganisms and/or decrease the pathogenic ones, making soil healthier and more productive. However, this review has identified major knowledge gaps with regard to the prevalence, impact and management of subsoil pathogens.

References

Aloni R, Wolf A, Feigenbaum P, Avni A, Klee HJ. The never ripe mutant provides evidence that tumor-induced ethylene controls the morphogenesis of *Agrobacterium tumefaciens*-induced crown galls on tomato stems. Plant Physiol. 1998;117:841–9. https://doi.org/10.1104/pp.117.3.841.

Al-Rehiayani SSL, Hafez M, Thornton M, Sundararaj P. Effects of *Pratylenchus neglectus* and *Bacillus megaterium* and oil radish or rapeseed green manure on reproductive potential of *Meloidogyne chitwoodi* on potato. Nematropica. 1999;29:37–49.

Avendaño F, Pierce FJ, Schabenberger O, Melakeberhan H. The spatial distribution of soybean cyst nematode in relation to soil texture and soil map unit. Agron J. 2004;96:181–94. https://doi.org/10.2134/agronj2004.1810.

Aydinli G, Mennan S. Biofumigation studies by using *Raphanus sativus* and *Eruca sativa* as a winter cycle crops to control root-knot nematodes. Braz Arch Biol Technol. 2018;61:e18180249. https://doi.org/10.1590/1678-4324-2018180249.

Baptista MJ, Souza RB, Pereira W, Lopes CA, Carrijo OA. Efeito da solarização e biofumigação na incidência da murcha bacteriana em tomateiro no campo. Hortic Bras. 2006;24:161–5.
Barbosa KAG, Garcia RA, Santos LC, Teixeira RA, Araujo FG, Rocha MR, Lima FSO. Avaliação da adubação potássica sobre populações de *Heterodera glycines* em cultivares de soja resistente e suscetível. Nematol Bras. 2010;34(3):150–8.
Barker KR, Weeks WW. Relationships between soil and levels of *Meloidogyne incognita* and tobacco yield and quality. J Nematol. 1991;23:82–90.
Bettiol W, Morandi MAB, Pinto ZV, Correa EB, Moura AB, Lucon CMM, Costa JCB, Bezerra JL. Bioprotetores comerciais para o controle de doenças de plantas. Revisão Anual de Patologia de Plantas. 2009;17:111–47.
Bowen KL, Hagan AK, Weeks R. Seven years of *Sclerotium rolfsii* in peanut fields: yield losses and means of minimization. Plant Dis. 1992;76(10):982–5.
Brar HK, Swaminathan S, Bhattacharyya MK. The *Fusarium virguliforme* toxin FvTox1 causes foliar sudden death syndrome-like symptoms in soybean. MPMI. 2011;24(10):1179–88. https://doi.org/10.1094/MPMI-12-10-0285.
Bruton BD, Reuveni R. Vertical distribution of microsclerotia of *Macrophomina phaseolina* under various soil types and host crops. Agric Ecosyst Environ. 1985;2(2):165–9. https://doi.org/10.1016/0167-8809(85)90078-7.
Cardoso SC, Soares ACF, Brito AS. Control of tomato bacterial wilt through the incorporation of aerial part of pigeon pea and crotalaria to soil. Summa Phytopathol. 2006;32(1):27–33. https://doi.org/10.1590/S0100-54052006000100004.
Carling DE, Leiner RH. Effect of temperature on virulence of *Rhizoctonia solani* and other *Rhizoctonia* on potato. Phytopathology. 1990;80(10):930–4. https://doi.org/10.1094/Phyto-80-930.
Carneiro RMDG, Randig O, Almeida MRA, Campos AD. Resistance of vegetable crops to *Meloidogyne* spp.: suggestion for a crop rotation. Nematol Bras. 2000;24:49–54.
Castagnone-Sereno P, Danchin EG, Perfus-Barbeoch L, Abad P. Diversity and evolution of root-knot nematodes, genus *Meloidogyne*: new insights from the genomic era. Annu Rev Phytopathol. 2013;51:203–20. https://doi.org/10.1146/annurev-phyto-082712-102300.
Castillo P, Vovlas N. *Pratylenchus* (Nematoda: Pratylenchidae): diagnosis, biology, pathogenicity and management, nematology monographs and perspectives. Boston: Brill Leiden; 2007.
Catanzariti AM, Lim GTT, Jones DA. The tomato I-3 gene: a novel gene for resistance to Fusarium wilt disease. New Phytol. 2015;207(1):106–18. https://doi.org/10.1111/nph.13348.
Champosiseau PG, Jones JB, Allen C. *Ralstonia solanacearum* race 3 biovar 2 causes tropical losses and temperate anxieties. Plant Health Progress, St. Paul. 2009. http://www.apsnet.org/publications/apsnetfeatures/Pages/Ralstonia.aspx. Accessed 02 Mar 2018. https://doi.org/10.1094/PHP-2009-0313-01-RV.
Charkowsky AO. The soft rot *Erwinia*. In: Gnanamanickam SS, editor. Plant-associated bacteria. Dordrecht: Springer; 2006. p. 423–505.
Chung YS, Goeser NJ, Cai XK, Jansky S. The effect of long term storage on bacterial soft rot resistance in potato. Am J Potato Res. 2013;90:351–6. https://doi.org/10.1007/s12230-013-9311-6.
Cook RJ, Baker KF. The nature and practice of biological control of plant pathogens. St. Paul: APS Press; 1983.
Coser SM, Chowda-Reddy RV, Zhang J, Mueller DS, Mengistu A, Wise KA, Allen TW, Singh A, Singh AK. Genetic architecture of charcoal rot (*Macrophomina phaseolina*) resistance in soybean revealed using a diverse panel. Front Plant Sci. 2017;8:1626. https://doi.org/10.3389/fpls.2017.01626.
Czajkowski R, Pérombelon MCM, van Veen JA, van der Wolf JM. Control of blackleg and tuber soft rot of potato caused by *Pectobacterium* and *Dickeya* species: a review. Plant Pathol. 2011;60:999–1013. https://doi.org/10.1111/j.1365-3059.2011.02470.x.
Davis RA, Huggins DR, Cook JR, Paulitz TC. Nitrogen and crop rotation effects on Fusarium crown rot in no-till spring wheat. Can J Plant Pathol. 2009;31(4):456–67. https://doi.org/10.1080/07060660909507620.

De Cleene M, Ley D. The host range of crown gall. J Bot Rev. 1976;42:389–446.
De Vay JE, Stapleton JJ, Elmore CL. Proceedings of the first international conference on soil solarization, Roma, 1991. FAO Plant Protection Paper 109. 1991. 396p.
Dias WP, Silva JFV, Carneiro GES, Garcia A, Arias CAA. Nematoide de cisto da soja: biologia e manejo pelo uso da resistência genética. Nematol Bras. 2009;33(1):1–16.
Diourte M, Starr JL, Jeger MJ, Stack JP, Rosenow DT. Charcoal rot (*Macrophomina phaseolina*) resistance and the effects of water stress on disease development in sorghum. Plant Pathol. 1995;44(1):196–202. https://doi.org/10.1111/j.1365-3059.1995.tb02729.x.
Dorrance A, Schmitthenner A. New sources of resistance to *Phytophthora sojae* in the soybean plant introductions. Plant Dis. 2000;84(12):1303–8. https://doi.org/10.1094/PDIS.2000.84.12.1303.
Droby S, Coffey MD. Biodegradation process and the nature of metabolism of metalaxyl in soil. Ann App Biol. 1991;118(3):543–53. https://doi.org/10.1111/j.1744-7348.1991.tb05344.x.
Dryden P, van Alfen NK. Soil moisture, root system density, and infection of roots of pinto beans by *Fusarium solani* f. sp. phaseoli under dryland conditions. Phytopathology. 1984;74:132–5. https://doi.org/10.1094/Phyto-74-132.
Duncan JM, Kennedy DM. The effect of waterlogging on Phytophthora root rot of red raspberry. Plant Pathol. 1989;38(2):161–8. https://doi.org/10.1111/j.1365-3059.1989.tb02129.x.
Elphinstone JG. The current bacterial wilt situation: a global overview. In: Allen C, Prior P, Hayward AC, editors. Bacterial wilt disease and the *Ralstonia solanacearum* species complex. St. Paul: APS Press; 2005. p. 9–28.
EPPO European and Mediterranean Plant Protection Organization. EPPO A1 and A2 lists of pests recommended for regulation as quarantine pests. 2017. http://www.eppo.org/. Accessed 01 Set 2017.
Evans K, Rowe JA. Distribution and economic importance. In: Sharma SB, editor. The cyst nematodes. Dordrecht: Kluwer Academic Publishers; 1998. p. 1–30.
Evans ML, Hollaway GJ, Dennis JI, Correll R, Wallwork H. Crop sequence as a tool for managing populations of *Fusarium pseudograminearum* and *F. culmorum* in South-Eastern Australia. Australas Plant Path. 2010;39(4):376–82. https://doi.org/10.1071/AP09092.
Fang X, You MP, Barbetti MJ. Reduced severity and impact of Fusarium wilt on strawberry by manipulation of soil pH, soil organic amendments and crop rotation. Eur J Plant Pathol. 2012;134(3):619–29. https://doi.org/10.1007/s10658-012-0042-1.
Felix KCS, Souza EB, Michereff SJ, Mariano RLR. Survival of *Ralstonia solanacearum* in infected tissues of *Capsicum annuum* and in soils of the state of Pernambuco, Brazil. Phytoparasitica. 2012;40:53–62. https://doi.org/10.1007/s12600-011-0200-6.
Ferraz LCCB, Brown DJF. Nematologia de plantas: fundamentos e importância. 2016. http://docentes.esalq.usp.br/sbn/ferbro/FerrazBrown2016v02.pdf. Accessed 23 Feb 2018.
Fery RL, Dukes PD. Southern blight (*Sclerotium rolfsii* Sacc.) of cowpea: yield-loss estimates and sources of resistance. Crop Prot. 2002;21(5):403–8. https://doi.org/10.1016/s0261-2194(01)00122-3.
Fosu-Nyarko J, Jones MGK. Advances in understanding the molecular mechanisms of root lesion nematode host interactions. Annu Rev Phytopathol. 2016;54:253–7. https://doi.org/10.1146/annurev-phyto-080615-100257.
Franchini JC, Crispino CC, Souza RA, Torres E, Hungria M. Microbiological parameters as indicators of soil quality under various soil management and crop rotation systems in southern Brazil. Soil Till Res. 2007;92:18–29. https://doi.org/10.1016/j.still.2005.12.010.
Franchini JC, Moraes MT, Debiasi H, Dias WP, Ribas LN, Silva JFV. Variabilidade espacial de atributos químicos do solo e relação com os danos pelo nematoide das lesões radiculares em soja. Paper presented at the 33th Congresso Brasileiro de Ciência do Solo, Sociedade Brasileira de Ciências do Solo, Uberlândia, Jul 31–Aug 4 2011.
Franchini JC, Debiasi H, Dias WP, Ribas LN, Silva JFV, Balbinot-Junior AA. Relationship among soil properties, root-lesion nematode population and soybean growth. Revista de Ciências Agroveterinárias. 2018;17(1):30–5. https://doi.org/10.5965/223811711712018030.

Garcia A, Silva JFV. Interação entre a população de cistos de *Heterodera glycines* e o pH do solo. Fitopatol Bras. 1996;21(Supl.): 420.
Gill JS, Sivasithamparam K, Smettem KRJ. Effect of soil moisture at different temperatures on Rhizoctonia root rot of wheat seedlings. Plant Soil. 2001a;231(1):91–6. https://doi.org/10.1023/A:101039411952222.
Gill JS, Sivasithamparam K, Smettem KRJ. Influence of depth of soil disturbance on root growth dynamics of wheat seedlings associated with *Rhizoctonia solani* AG-8 disease severity in sandy and loamy sand soils of Western Australia. Soil Till Res. 2001b;62(1):73–83. https://doi.org/10.1016/S0167-1987(01)00217-3.
Gill JS, Hunt S, Sivasithamparam K, Smettem KRJ. Root growth altered by compaction of a sandy loam soil affects severity of rhizoctonia root rot of wheat seedlings. Aust J Exp Agric. 2004;44(6):595–9. https://doi.org/10.1071/EA02093.
Goto M. Fundamentals of bacterial plant pathology. San Diego: Academic Press; 1992.
Huber DM, Watson RD. Nitrogen form and plant disease. Annu Rev Phytopathol. 1974;12:139–65. https://doi.org/10.1146/annurev.py.12.090174.001035.
Hunt DJ, Handoo ZA. Taxonomy, identification and principal species. In: Perry RN, Moens M, Starr JL, editors. Root-knot nematodes. Cambridge: CABI International; 2009. p. 1–17.
Hussey RS, Davis EL. Nematode esophageal glands and plant parasitism. In: Chen ZX, Chen SY, Dickson DW, editors. Nematology advances and perspectives, vol. 1. Walllingford: International CABI Publishing; 2004. p. 258–94.
Jones JT, Haegeman A, Danchin EGJ, Gaur HS, Helder J, Jones MGK, Kikuchi T, Manzanilla-Lopez R, Palomares-Rius JE, Wesemael WML, Perry RN. Top 10 plant-parasitic nematodes in molecular plant pathology. Mol Plant Pathol. 2013;14:946–61. https://doi.org/10.1111/mpp.12057.
Kado CI. Plant bacteriology. St Paul: APS Press; 2010.
Kandel SL, Smiley RW, Garland-Campbell K, Elling AA, Abatzoglou J, Huggins D, Rupp R, Paulitz TC. Relationship between climatic factors and distribution *Pratylenchus* spp. and their relationship with climatic factors in dryland wheat production areas of eastern Washington. Plant Dis. 2013;97:1448–56. https://doi.org/10.1094/PDIS-11-12-1060-RE.
Kandji ST, Ogol CKPO, Albrecht A. Diversity of plant-parasitic nematodes and their relationships with some soil physico-chemical characteristics in improved fallows in western Kenya. Appl Soil Ecol. 2001;18:143–57. https://doi.org/10.1016/S0929-1393(01)00157-3.
Katafiire M, Adipala E, Lemaga B, Olanya M, El-bedewy R. Management of bacterial wilt of potato using one-season rotation crops in southwestern Uganda. In: Allen C, Prior P, Hayward AC, editors. Bacterial wilt disease and the *Ralstonia solanacearum* species complex. St Paul: APS Press; 2005. p. 197–204.
Kirk WW, Wharton PS, Schafer RL, Tumbalam P, Poindexter S, Guza C, Fogg R, Schlatter T, Stewart J, Hubbell L, Ruppal D. Optimizing fungicide timing for the control of Rhizoctonia crown and root rot of sugar beet using soil temperature and plant growth stages. Plant Dis. 2008;92(7):1091–8. https://doi.org/10.1094/PDIS-92-7-1091.
Ko WH, Kao CW. Evidence for the role of calcium in reducing root disease infected by *Pythium* spp. In: Engelhard AW, editor. Soilborne plant pathogens: management of diseases with macro- and microelements. St Paul: APS Press; 1989. p. 205–17.
Koehler AM, Shew HD. Enhanced overwintering survival of stevia by QoI fungicides used for management of *Sclerotium rolfsii*. Plant Dis. 2017;101(8):1417–21. https://doi.org/10.1094/PDIS-02-17-0277-RE.
Koenning SR, Wrather JA. Suppression of soybean yield potential in the continental United States from plant diseases estimated from 2006 to 2009. Plant Health Prog. 2010;11(1) https://doi.org/10.1094/PHP-2010-1122-01-RS. Accessed 23Feb 2018.
Lemaga B, Kakuhenzine R, Kassa B, Ewell PT, Priou S. Integrated control of potato bacterial wilt in eastern Africa: the experience of African highlands initiative. In: Allen C, Prior P, Hayward AC, editors. Bacterial wilt disease and the *Ralstonia solanacearum* species complex. St Paul: APS Press; 2005. p. 145–58.

Levy NO, Harel YM, Haile ZM, Elad Y, Rav-David E, Jurkevitch E, Katan J. Induced resistance to foliar diseases by soil solarization and *Trichoderma harzianum*. Plant Pathol. 2015;64:365–74. https://doi.org/10.1111/ppa.12255.

Li YP, You MP, Colmer TD, Barbetti MJ. Effect of timing and duration of soil saturation on soil-borne Pythium diseases of common bean (*Phaseolus vulgaris*). Plant Dis. 2015;9(1):112–8. https://doi.org/10.1094/PDIS-09-13-0964-RE.

Ma B, Hibbing ME, Kim HS, Reedy RM, Yedidia I, Breuer J, Glasner JD, Perna NT, Kelman A, Charkowski AO. Host range and molecular phylogenies of the soft rot enterobacterial genera *Pectobacterium* and *Dickeya*. Phytopathology. 2007;97:1150–63. https://doi.org/10.1094/PHYTO-97-9-1150.

Marschner H. Mineral nutrition of higher plants. 2nd ed. London: Academic Press; 1995.

Melero-Vara JM, López-Herrera CJ, Prados-Ligero AM, Vela-Delgado MD, Navas-Becerra JA, Basallote-Ureba MJ. Effects of soil amendment with poultry manure on carnation Fusarium wilt in greenhouses in Southwest Spain. Crop Prot. 2011;30(8):970–6. https://doi.org/10.1016/j.cropro.2011.03.022.

Meyer WA, Sinclair JB, Khare MN. Factors affecting charcoal rot of soybean seedlings. Phytopathology. 1974;64(6):845–9. https://doi.org/10.1094/Phyto-64-845.

Minuto A, Spadaro D, Garibaldi A, Gullino ML. Control of soilborne pathogens of tomato using a commercial formulation of *Streptomyces griseoviridis* and solarization. Crop Prot. 2006;25(5):468–75. https://doi.org/10.1016/j.cropro.2005.08.001.

Montreal Protocol. https://en.wikipedia.org/wiki/Montreal_Protocol. 1987. Accessed 02 Apr 2019.

Moots CK, Nickell CD, Gray LE. Effects of soil compaction on the incidence of *Phytophthora megasperma* f. sp. *glycinea* in soybean. Plant Dis. 1988;72(10):896–900. https://doi.org/10.1094/PD-72-0896.

Mougel C, Thioulouse J, Perrière G, Nesme X. A mathematical method for determining genome divergence and species delineation using AFLP. Int J Syst Evol Microbiol. 2002;52:573–86. https://doi.org/10.1099/00207713-52-2-573.

Murillo-Williams A, Munkvold GP. Systemic infection by *Fusarium verticillioides* in maize plants grown under three temperature regimes. Plant Dis. 2008;92(12):1695–700. https://doi.org/10.1094/PDIS-92-12-1695.

NAPPO. North American Plant Protection Organization's (NAPPO): Phytosanitary alert system. http://www.pestalert.org/. Accessed 01 Set. 2017.

Negretti RRRD, Gomes CB, Mattos VS, Somavilla L, Manica-erto R, Agostinetto D, Castagnone-Sereno P, Carneiro RMDG. Characterization of a *Meloidogyne* species complex parasitizing rice in Southern Brazil. Nematology. 2017;19(4):403–12. https://doi.org/10.1163/15685411-00003056.

Nesbitt HJ, Malajczuk N, Glenn AR. Effect of organic matter on the survival of *Phytophthora cinnamomi* Rands in soil. Soil Biol Biochem. 1979;11(2):133–6. https://doi.org/10.1016/0038-0717(79)90089-0.

Oliveira RDL, Dhingra OD. Biofumigação com *Brassica* sp. para o controle de doenças de plantas. In: Venzon M, Paula Junior TJ, Pallini A, editors. Avanços no controle alternativo de pragas e doenças. Viçosa: EPAMIG; 2008. p. 237–57.

Onkendi EM, Kariuki GM, Marais M, Molleki LN. The threat of root-knot nematodes (*Meloidogyne* spp.) in Africa: a review. Plant Pathol. 2014;63(4):727–37. https://doi.org/10.1111/ppa.12202.

Ooshiro A, Takaesu K, Natsume M, Taba S, Nasu K, Uehara M, Muramoto Y. Identification and use of a wild plant with antimicrobial activity against *Ralstonia solanacearum*, the cause of bacterial wilt of potato. Weed Biol Manag. 2004;4(4):187–94. https://doi.org/10.1111/j.1445-6664.2004.00137.x.

Palacio-Bielsa A, Cambra MA, Lopez MM. Characterization of potato isolates of *Dickeya chrysanthemi* in Spain by a microtiter system for biovar determination. Ann App Biol. 2006;148(2):157–64. https://doi.org/10.1111/j.1744-7348.2006.00045.x.

Palomares-Rius JE, Escobar C, Cabrera J, Vovlas A, Castillo P. Anatomical alterations in plant tissues induced by plant-parasitic nematodes. Front Plant Sci. 2017;8:1987. https://doi.org/10.3389/fpls.2017.01987.

Pereira AS, Daniels J. O cultivo da batata na região sul do Brasil. Brasília: Embrapa Clima Temperado; 2003. 567 p.

Pitzschke A, Hirt H. New insights into an old story: *Agrobacterium*-induced tumor formation in plants by plant transformation. EMBO J. 2010;29(6):1021–103. https://doi.org/10.1038/emboj.2010.8.

Qi R, Wang T, Zhao W, Li P, Ding J, Gao Z. Activity of ten fungicides against *Phytophthora capsici* isolates resistant to metalaxyl. J Phytopathol. 2012;160(11–12):717–22. https://doi.org/10.1111/jph.12009.

Rashid A, Fahad MAB, Khan MA, Mateen A. Incidence of potato blackleg caused by *Pectobacterium atrosepticum* in district Chiniot and its management through bio-products. Afr J Agric Res. 2012;7(45):6035–48. https://doi.org/10.5897/AJAR12.1697.

Ratnadass A, Fernandes P, Avelino J, Habib R. Plant species diversity for sustainable management of crop pests and diseases in agroecosystems: a review. Agron Sustain Dev. 2012;32:273–303. https://doi.org/10.1007/s13593-011-0022-4.

Rhoades CC, Brosi SL, Dattilo AJ, Vincelli P. Effect of soil compaction and moisture on incidence of Phytophthora root rot on American chestnut (*Castanea dentata*) seedlings. For Ecol Manag. 2003;184(1–3):47–54. https://doi.org/10.1016/S0378-1127(03)00147-6.

Rocha MR, Carvalho Y, Corrêa GC, Cattini GP, Paolini G. Efeito de doses crescentes de calcário sobre a população de *Heterodera glycines* em soja. Pesqui Agropecu Trop. 2006;36(2):89–94.

Roget D, Neate S, Rovira A. Effect of sowing point design and tillage practice on the incidence of Rhizoctonia root rot, take-all and cereal cyst nematode in wheat and barley. Aust J Exp Agric. 1996;36(6):683–93. https://doi.org/10.1071/EA9960683.

Rush CM, Winter SR. Influence of previous crops on Rhizoctonia root crown rot of sugar beet. Plant Dis. 1990;74(6):421–5. https://doi.org/10.1094/PD-74-0421.

Ryley M, Mosetter H, Rose J. Yield losses of soybeans due to *Phytophthora megasperma* f.sp. *glycinea*. Aus J Agric Res. 1989;40(6):1161–9. https://doi.org/10.1071/AR9891161.

Safni I, Cleenwerck I, De Vos P, Fegan M, Sly L, Kappler U. Polyphasic taxonomic revision of the *Ralstonia solanacearum* species complex: proposal to emend the descriptions of *Ralstonia solanacearum* and *Ralstonia syzygii* and reclassify current *R. syzygii* strains as *Ralstonia syzygii* subsp. *syzygii* subsp. nov., *R. solanacearum* phylotype IV strains as *Ralstonia syzygii* subsp. *indonesiensis* subsp. nov., banana blood disease bacterium strains as *Ralstonia syzygii* subsp. *celebesensis* subsp. nov. and *R. solanacearum* phylotype I and III strains as *Ralstonia pseudosolanacearum* sp. nov. Int J Syst Evol Microbiol. 2014;64:3087–103. https://doi.org/10.1099/ijs.0.066712-0.

Sanogo S, Zhang J. Resistance sources, resistance screening techniques and disease management for Fusarium wilt in cotton. Euphytica. 2016;207(2):255–71. https://doi.org/10.1007/s10681-015-1532-y.

Santos AV, Gomes CB. Reaction of castor bean cultivars to *Meloidogyne* spp. and effect of root exudates on *Meloidogyne enterolobii* and *M. graminicola*. Nematol Bras. 2011;35(1–2):1–9.

Schillinger WF, Paulitz TC. Reduction of Rhizoctonia bare patch in wheat with barley rotations. Plant Dis. 2006;90(3):302–6. https://doi.org/10.1094/PD-90-0302.

Schmitt RD, Riggs RD. Populations dynamics of *Heterodera glycines* in the southeastern United States. In: Variability and population dynamics of root-knot and cyst nematodes in the southern region of the United States. The Texas A & M University-System; 1989. p. 1–7. (Southern Cooperative Series Bulletin, 336).

Sfredo GJC, Borkert M, Cattelan AJ, Hungria M. Adubação e calagem para soja no Brasil. Informativo Abrates. 1994;4(1):19–43.

Silva MS, Carvalho FCQ, Silva JR, Lins SRO, Oliveira SMA. Uso de antagonistas e produtos alternativos no manejo pós-colheita de podridão mole em pimentão. Rev Ciênc Agron. 2014;45(4):718–25.

Stanghellini ME, Burr TJ. Effect of soil water potential on disease incidence and oospore germination of *Pythium aphanidermatum*. Phytopathology. 1973;63(12):1496–8. https://doi.org/10.1094/Phyto-63-1496.

Sundararaju P, Jeyabaskaran KJ. Evaluation of different soil types on multiplication of *Pratylenchus coffeae* and growth of banana seedlings var. Nendran. Nematol Medit. 2003;31:151–3.

Swan LJ, Backhouse D, Burgess LW. Surface soil moisture and stubble management practice effects on the progress of infection of wheat by *Fusarium pseudograminearum*. Aust J Exp Agric. 2000;40:639–98. https://doi.org/10.1071/EA99106.

Thomson TB, Athow KL, Laviolette FA. The effect of temperature on the pathogenicity of *Pythium aphanidermatum, P. debaryanum*, and *P. ultimum* on soybean. Phytopathology. 1971;61(8):933–5. https://doi.org/10.1094/Phyto-61-933.

Toth IK, van der Wolf JM, Saddler G, Lojkowska E, Hélias V, Pirhonen M, Tsror L, Elphinstone JG. *Dickeya* species: an emerging problem for potato production in Europe. Plant Pathol. 2011;60(3):385–99. https://doi.org/10.1111/j.1365-3059.2011.02427.x.

Tu JC. Effects of soil compaction, temperature and moisture on the development of the Fusarium root rot complex of pea in southwestern Ontario. Phytoprotection. 1994;75(3):125–31. https://doi.org/10.7202/706059ar.

Van Veen JA, van Overbeek LS, van Elsas JD. Fate and activity of microorganisms introduced into soil. Microbiol Mol Biol Rev. 1997;61:121–35.

Vedoveto MVV, Dias-Arieira CR, Rodrigues DB, Arieira JO, Roldi M, Severino JJ. Adubos verdes no manejo de *Pratylenchus brachyurus* em soja. Nematropica. 2013;43(2):226–32.

Ventura JA, Hinz RH. Controle das doenças da bananeira. In: Zambolim L, Vale FXR, Monteiro AJA, et al., editors. Controle de doenças de plantas fruteiras. Viçosa: Editora UFV; 2002. p. 839–926.

Vick CM, Chong SK, Bond JP, Russin JS. Response of soybean sudden death syndrome to subsoil tillage. Plant Dis. 2003;87(6):629–32. https://doi.org/10.1094/PDIS.2003.87.6.629.

Vinh MT, Tung TT, Quang HX. Primary bacterial wilt study on tomato in vegetable areas of Ho Chi Minch city, Vietnam. In: Allen C, Prior P, Hayward AC, editors. Bacterial wilt disease and the *Ralstonia solanacearum* species complex. St. Paul: APS Press; 2005. p. 177–84.

Vitti AJ, Resende-Neto UR, Araújo FG, Santos LC, Barbosa KAG, Rocha MR. Effect of soybean seed treatment with abamectin and thiabendazole on *Heterodera glycines*. Nematropica. 2014;44(1):74–80.

Wokocha RC. Relationship between the population of viable sclerotia of *Sclerotium rolfsii* in soil to cropping sequence in the Nigerian savanna. Plant Soil. 1988;106(1):146–8. https://doi.org/10.1007/BF02371207.

Wrather JA, Anderson TR, Arsyad DM, Tan Y, Ploper LD, Porta-Puglia A, Ram HH, Yorinori JT. Soybean disease loss estimates for the top ten soybean-producing countries in 1998. Can J Plant Pathol. 2001;23(2):115–21. https://doi.org/10.1080/07060660109506918.

Yamazaki H, Kikuchi S, Hoshina T, Kimura T. Calcium uptake and resistance to bacterial wilt of mutually grafted tomato seedlings. Soil Sci Plant Nutr. 2000;46(2):529–34. https://doi.org/10.1080/00380768.2000.10408806.

You MP, Colmer TD, Barbetti MJ. Salinity drives host reaction in *Phaseolus vulgaris* (common bean) to *Macrophomina phaseolina*. Funct Plant Biol. 2011;38(12):984–92. https://doi.org/10.1071/fp11137.

Young JM, Kuykendall LD, Martinez-Romero E, Kerr A, Sawada H. A revision of *Rhizobium* Frank 1889, with an emended description of the genus, and the inclusion of all species of *Agrobacterium* Conn 1942 and *Allorhizobium undicola* de Lajudie et al. 1998 as new combinations: *Rhizobium radiobacter, R. rhizogenes, R. rubi, R. undicola* and *R. vitis*. Int J Syst Evol Microbiol. 2001;51:89–103. https://doi.org/10.1099/00207713-51-1-89.

Yu YT, Liu HL, Zhu AG, Zhang G, Zeng LB, Xue SD. A review of root lesion nematode: identification and plant resistance. Adv Microbiol. 2012;2:411–6. https://doi.org/10.4236/aim.2012.24052.

Zambolim L, Ventura JA. Resistência a doenças induzidas pela nutrição das plantas. Piracicaba: POTAFOS; 1996.

Zasada IA, Meyer SLF, Morra MJ. Brassicaceous seed meals as soil amendments to suppress the plant-parasitic nematodes *Pratylenchus penetrans* and *Meloidogyne incognita*. J Nematol. 2009;41(3):221–7.

Zirakparvar ME, Norton DC, Cox CP. Population increase of *Pratylenchus hexincisus* on corn as related to soil temperature and type. J Nematol. 1980;12(4):313–8.

Chapter 10
Root Systems of Agricultural Crops and Their Response to Physical and Chemical Subsoil Constraints

Mirian Cristina Gomes Costa and Ítalo Antônio Cotta Coutinho

Abstract Physical and chemical soil constraints in the subsoil limit crop yield by affecting root development and the uptake of both water and nutrients. Hence, improved knowledge of the characteristics of the root systems of the main cultivated species and the potential for the root system to overcome subsoil constraints is vital for crop production. Here, we review the root systems of widely cultivated species and their response to physical and chemical constraints, particularly in subsoils. For most agricultural and horticultural species studied, roots are concentrated in the top 20 cm depth, but 20–30% of the root system commonly grows below this depth. Physical restrictions caused by soil compaction reduce root biomass, density and length, besides limiting root exploitation depth. Some species such as rapeseed and black mustard may alleviate soil compaction, making the subsoil more suitable for roots of other crops. Plant root systems are negatively affected by constraints that are often present in the subsoil such as excess of aluminium, heavy metals and salts. However, deficiencies of N, P, K as well as micronutrients also impair plant root growth. Mechanized tillage systems, more tolerant genotypes, association with soil microorganisms, exudation of organic acids by roots and adequate nutrient supply are among the strategies that allow a better root development and function, enabling improved crop growth with subsoil constraints.

Keywords Chemical constraints · Nutrient deficiency · Physical constraints · Root system · Salinity

M. C. G. Costa (✉)
Soil Science Department, Federal University of Ceará, Fortaleza, CE, Brazil
e-mail: mirian.costa@ufc.br

Í. A. C. Coutinho
Biology Department, Federal University of Ceará, Fortaleza, CE, Brazil
e-mail: italo.coutinho@ufc.br

T. S. d. Oliveira, R. W. Bell (eds.), *Subsoil Constraints for Crop Production*,
https://doi.org/10.1007/978-3-031-00317-2_10

10.1 Introduction

Food production for the world population (increasing to 8.2–8.7 billion people by the year of 2100) is a growing concern (Abel et al. 2016). The land sector will also have to produce bioenergy, fibres and other products. The cropland soils often have constraints for plant development that are chemical (acidity, nutrient deficiency and salinity), physical (compaction, consolidation, poor drainage, limiting layer and stoniness) or biological (pathogens) in nature. Moreover, issues related to the improper use of agricultural lands that cause soil degradation, such as erosion, are also significant.

During the period from 1961 to 2009, the production of approximately 50 agricultural commodities expanded geographically around the world. The main cereal crops being produced are wheat (*Triticum aestivum* L.), rice (*Oryza sativa* L.), maize (*Zea mays* L.) and barley (*Hordeum vulgare* L.) as well as sugarcane (*Saccharum officinarum* L.), potatoes (*Solanum tuberosum* L.), vegetables and fruits. In the same period, the oil crops such as soybean (*Glycine max* (L.) Merr.), sunflower (*Helianthus annuus* L.), palm oil (*Elaeis guineensis* Jacq.), rapeseed (*Brassica napus* L.) and mustard (*Brassica nigra* (L.) Koch) increased the most in production and also started being cropped in a larger number of countries (Khoury et al. 2014).

In the past, improvement of crops and agricultural techniques has mainly focused on increasing shoot biomass and seed yield, particularly on high productivity soils. However, research also should focus on the development of strategies which may be applied in the field to increase crop production in soils with subsoil constraints. In this endeavour, the root system, which is critical for the uptake of nutrients and water, anchorage in the substrate as well as interaction with symbiotic organisms, cannot be overlooked anymore (Cheng et al. 2016).

In this chapter, we have synthesized key information regarding the main characteristics of the root system of the most widely cultivated species in the world as well as on modification in the root systems when developing under physical and chemical soil constraints. While we have tried to generalize the findings in the literature, it should be noted that the root systems often respond differently to the same stimulus among different species as well as different genotypes of the same species.

10.2 Overview of the Root System of the Main Cultivated Species

Plant root systems differ among crop species and among genotypes of the same species (Uga et al. 2009; Henry et al. 2011; Thorup-Kristensen et al. 2020). Although wheat, rice and maize crops have an adventitious root system (fibrous root system) (Fig. 10.1), common to monocotyledons, they have significant differences in their root systems such as depth and diameter (Table 10.1). Similarly, while soybean,

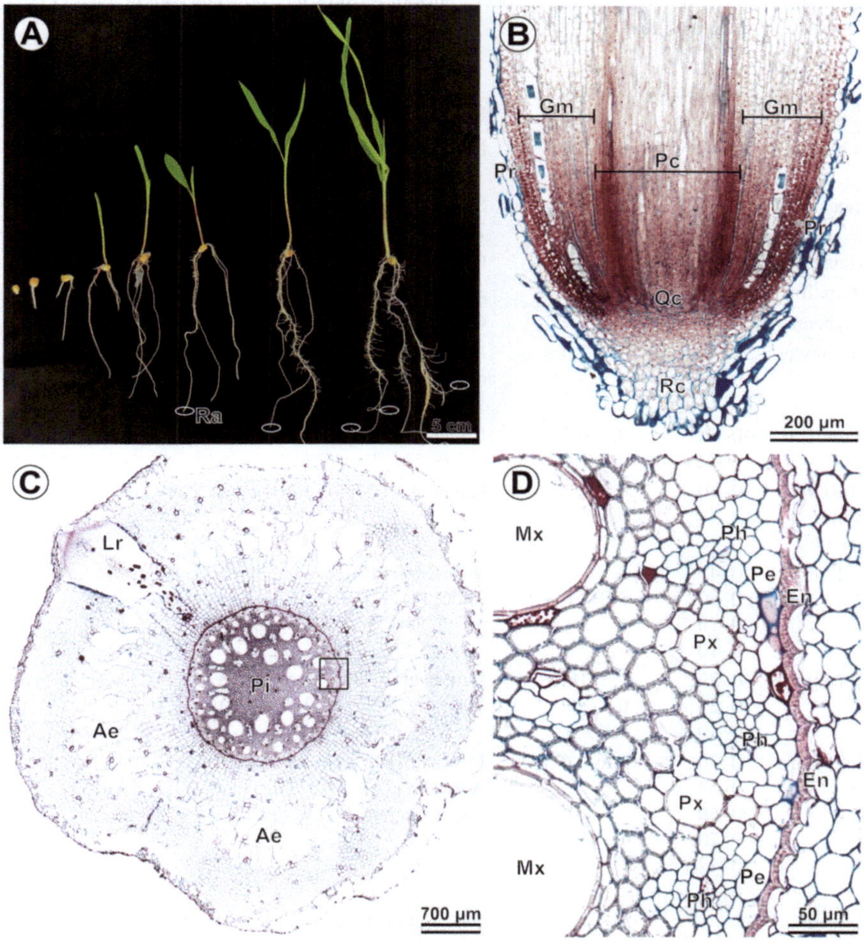

Fig. 10.1 Root architecture and anatomy in monocotyledonous species. (**a**) Root architecture of maize (*Zea mays* L.) from sprouts to seedlings showing the adventitious root system. Root architecture concerns the spatial configuration of the root system in which angles of roots (especially from lateral roots) are measured as well as root length. Please, note that only a few root apices are indicated. (**b**) Root apex of maize as observed in longitudinal section under light microscopy. (**c, d**) Root apex of banana (*Musa paradisiaca* L.) as observed in cross sections. Notice that monocotyledonous species have vascular bundles radially arranged which is known as polyarch where xylem alternates with phloem. (**d**) Detail of the square area delimited in '**c**'. (*Ae* aerenchyma, *En* endodermis, *Gm* ground meristem, *Lr* lateral root, *Mx* metaxylem, *Pc* procambium, *Ph* phloem, *Pi* pith, *Pr* protodermis, *Px* protoxylem, *Qc* quiescent centre, *Ra* root apex)

sunflower, rapeseed and mustard, along with other dicotyledons, have a tap root system (Fig. 10.2), each one has its own distinctive characteristics (Table 10.1). For temperate agricultural crops, the ratio of root mass distribution to root length distribution for monocotyledons decreases linearly at 0.1% per centimetre of depth,

Table 10.1 Range of soil depth for the occurrence of most of the roots and root diameter classes of the main cultivated crops

Crop	Soil depth (cm)	Root diameter (mm)	References
Wheat	0–40/60 0–60	0.2–0.6	Morita and Abe 1996; Muñoz-Romero et al. 2010
Rice	0–20 0–30/45	0.4–0.5	Sharma et al. 1994; Yang et al. 2004; Henry et al. 2011
Maize	0–20	0.6–0.8	Cai et al. 2014
Oil palm	0–20	1.5–4.1	Jourdan and Rey 1997; Teixeira et al. 2001
Soybean	0–68	0.6–0.7	Gao et al. 2010; Sartori et al. 2016
Sunflower	0–40 0–80 0–180	0.1–2.0	Ribeiro and Walter 2008; Manzur et al. 2014; Lisanti et al. 2013; Paul et al. 2021
Rapeseed	0–10 10–20	0.2–0.3 2.0–5.0	Liu et al. 2011a, b; Yuan et al. 2016
Mustard Black mustard	0–10	0.2–0.3 0.38–0.41	Liu et al. 2011a, b; Larson and Funk 2016
Coffee	0–40 0–50 0–20 0–10	0.07–0.8 1.0≤Ø≥3.0	Inforzato and Reis 1963; Gehrke 1962; Motta et al. 2006; Partelli et al. 2014; Ronchi et al. 2015
Eucalyptus	0–20 0–25 0–10	1.0–10.0 0.1–2.0 <2.0–5.0	Eamus et al. 2002; Bouillet et al. 2002; Moroni et al. 2003; de Mello et al. 2007; Dias et al. 2017
Sugarcane	0–30 0–54 0–60 0–20 0–470 0–40	0.2–0.4	Inforzato 1957; Smith et al. 2005; Costa 2005; de Azevedo et al. 2011; Otto et al. 2011; Castro et al. 2012; Laclau and Laclau 2009
Banana	0–45 0–30	0.2–4.3 0.7–6.6	Waser et al. 1996; Lecompte et al. 2005; Miotti et al. 2013

while for dicotyledons the rate decreases exponentially with soil depth resulting in thicker roots in topsoil for dicotyledons (Fan et al. 2016).

In order to evaluate the root system of different species and genotypes, it is important to consider the morphology, anatomy, topology, distribution and architecture of roots. Root morphology refers to features of epidermal root hairs, root diameter, undulations of the root axis, root cap, cortical senescence and so on (Lynch 1995). Topology refers to the pattern of root ramifications (root branching) not subjected to deformations which may be measured on excavated roots (Lynch 1995). Distribution refers to the root density gradient with depth in the soil. It is usually related to root dry matter or root length as a function of factors such as depth or position among surrounding plants (Lynch 1995). Architecture refers to the spatial configuration of the root system, which is relevant to the root exploration for essential resources that are unevenly available in the soil, but necessary to their development (Lynch 1995).

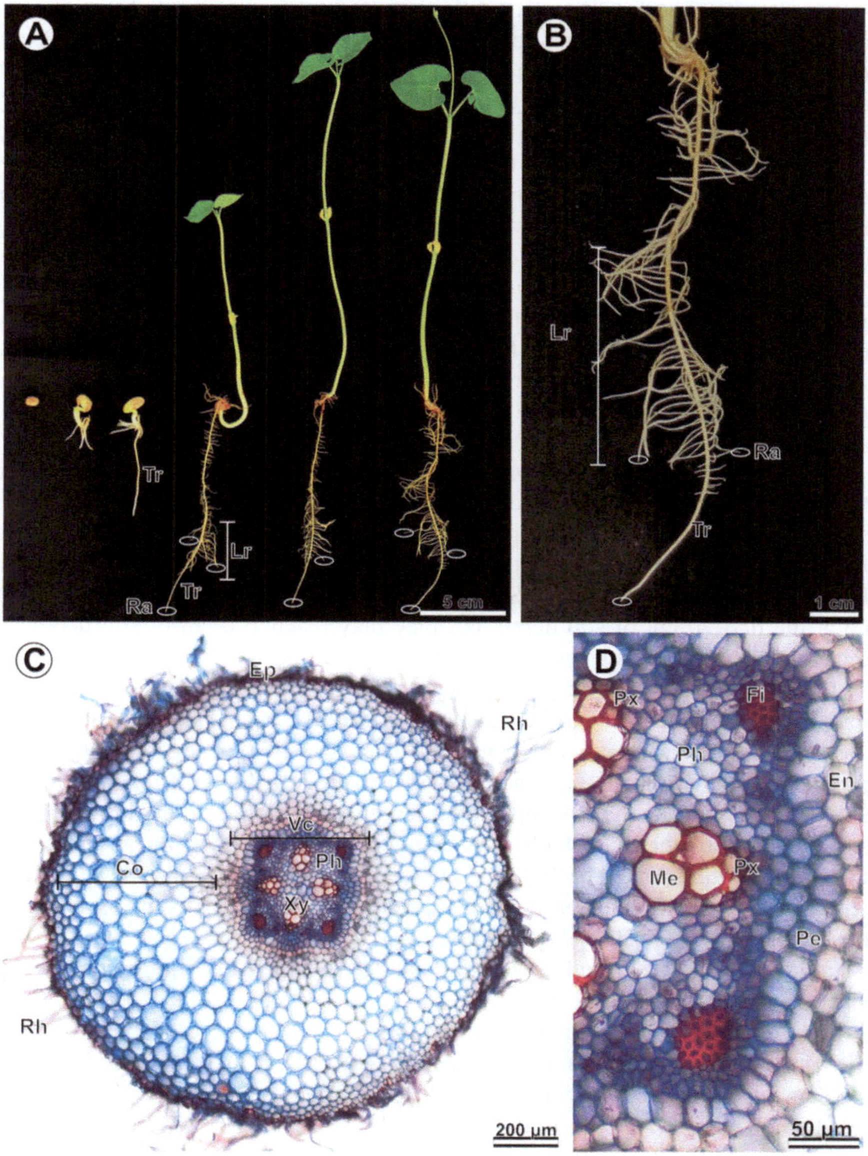

Fig. 10.2 Root architecture and anatomy in dicotyledonous species as observed in common bean (*Phaseolus vulgaris* L.). (**a**) Root architecture of common bean from sprouts to seedlings showing the tap root system. Note that only a few root apices are indicated. (**b**) Note that lateral roots form almost a right angle with the tap root. (**c, d**) Root apex as observed in cross sections. Note the absence of a pith in the central area of the root and the presence of a vascular cylinder composed of primary xylem and phloem. (**d**) Detail of the vascular cylinder showing bundles of primary xylem alternating with phloem. (*Co* cortex, *En* endodermis, *Ep* epidermis, *Fi* fibres, *Lr* lateral root, *Mx* metaxylem, *Pc* pericycle, *Ph* phloem, *Px* protoxylem, *Ra* root apex, *Rh* root hair)

A review on root distribution of different temperate crops showed that 50% of total roots accumulated from 8 to 20 cm depth (Fan et al. 2016). However, plant roots also reach the subsoil as they grow deeper. Alfalfa, for instance, reached a maximum rooting depth of 177 cm, while fescue (*Festuca arundinacea* Shreb.) reached only 78 cm depth.

For wheat, roots extend in soil to a depth of 40–60 cm, and their diameter varies from 0.2 to 0.6 mm (Muñoz-Romero et al. 2010). Both length and diameter are influenced by water availability, soil preparation (i.e. conventional tillage or no tillage) and crop development. However, Hamblin and Hamblin (1985) reported that 113 cm was the maximum rooting depth for wheat, but there was evidence of genotypic variation in depth, which implies there is genetic control to this trait. In the data compiled by Kirkegaard and Lilley (2007), maximum rooting for wheat varied between 80 and 180 cm and was influenced by soil wetting depth in such a way that root penetration was limited in the soil layers with less than ~45% of the plant available water content.

Roots of different rice cultivars show morphological variations regarding the diameter (0.647 to 1.157 mm) and number of metaxylem vessels (3.8 to 6.4) (Uga et al. 2009). Yang et al. (2004) reported that rice root diameter varies from 0.38 to 0.52 mm under different water regimes and nutrient availability. With respect to depth of root exploration in rice, Sharma et al. (1994) mentioned that 90% of roots are at the first 20 cm of soil depth, but they may reach 100 cm of soil depth for upland rice and 40 cm for lowland rice (Morita and Abe 1996). Conversely, Henry et al. (2011) reported that the greater values of root length density occur at a depth up to 30–45 cm. According to Morita and Abe (1996), seminal and nodal rice roots have different sizes. Rice root responses to water stress will depend on both its origin and size, and larger roots are more sensitive to drought than smaller ones (Morita and Abe 1996). When the development of larger roots is affected by drought, there is a compensation through a higher development of branch roots, that is, small, thin lateral roots (Morita and Abe 1996). However, thicker roots may also develop more under water stress if plants have some adaptation to drought or even if it is a non-tolerant genotype (Morita and Abe 1996). Regarding lowland rice genotypes, which are more sensitive to drought, even the smaller roots may be suppressed due to water stress (Morita and Abe 1996).

From a study of the distribution of the maize root system, Mengel and Barber (1974) reported that the maximum root density (4.1 cm cm^{-3}) occurred at a depth of 0–15 cm and, more recently, Wang et al. (2015) verified that maize roots are concentrated at a depth of 0–20 cm in the soil. The largest diameters for maize roots are observed in the 0–10 cm layer of soil, varying from 0.6 to 0.8 mm, while at greater depths the diameter reduces to ~0.4 mm (Cai et al. 2014).

The root system of oil palm forms a contrast to the annual cereal crops mentioned before because of the greater size of the tree and its perennial growth, which makes its root system more difficult to study. More than 65% of the oil palm roots are concentrated at a depth of 0–20 cm in the soil (Teixeira et al. 2001). Jourdan and Rey (1997) reported eight different morphological types of roots during the development of the oil palm: primary vertical and horizontal roots, secondary horizontal roots, upward growing secondary vertical roots and downward growing secondary vertical roots, superficial and deep tertiary roots and quaternary roots. The different types of roots were distinguished according to their development pattern and state of differentiation. As for the other monocotyledons, the radicle that emerges during the seedling stage is superseded by several adventitious roots which vary in diameter and number of xylem vessels according to the plant age.

For soybean crops, there have been many studies on roots. In the 1970s, morphological differences were observed in the root system of different genotypes of soybean. For example, the genotype Harosoy 63 root system was more extensive and had a higher surface area in comparison to genotype Aoda (Raper and Barber 1970) and roots reached a depth of 160–190 cm (Inforzato 1969; Mayaki et al. 1976). Conversely, Gao et al. (2010) demonstrated that soybean roots are able to reach only 68 cm deep in the soil in a maize/soybean intercropping system. The mean root length for soybean was 1,124–1,301 cm per three plants in 2 consecutive years, while the mean diameter for roots was 0.6–0.7 mm (Sartori et al. 2016).

Sunflower plants have a tap root system, and the greatest numbers of lateral roots are found at 0–15 cm depth. However, lateral roots may develop up to 40 cm (Ribeiro 2008) or even 180 cm depth (Lisanti et al. 2013) under field conditions. Seiler (1994) observed variations on the root length of tap roots and lateral roots, on the total length of roots and number of lateral roots among the genotypes. Sunflower plants produce supporting roots, which are important to avoid plant lodging in the field. The distribution of such supporting roots varies according to the genotype of the crop. However, almost half of the biomass of the supporting roots is concentrated at the first 5 cm depth (Manzur et al. 2014). Although thicker roots with more than 2 mm diameter are found, the root diameter of supporting roots, including the lateral ones, is ~1 mm (Manzur et al. 2014).

Rapeseed roots may develop up to a depth of 1 m, provided there are no subsoil constraints (Kjellström and Kirchmann 1992). However, Liu et al. (2011a) observed that the greatest density of the root length is found at the 0–20 cm depth. In a study to understand the dynamics of root architecture in rapeseed, it was shown that plants have roots in different diameter ranges, varying from <2 to 15 mm in diameter (Yuan et al. 2016). On the other hand, Liu et al. (2011a) report a mean diameter of roots in rapeseed at 0.3 mm.

Although the root system of the common mustard (*Brassica juncea* (L.) Czern) may reach up to 1 m depth in the soil, approximately 20% of root length and 30% of root surface area are concentrated at the first 10 cm (Liu et al. 2011a). Root length density of mustard is lower than in rapeseed (Liu et al. 2011b). Regarding root diameter, mustard plants show mean diameter between 0.3 and 0.4 mm in the superficial layer of the soil, while at deeper layers the root diameter falls to 0.3 mm (Liu et al. 2011a). Information available in the literature on the root system of black mustard (*Brassica nigra*) is scarce.

As for coffee trees (*Coffea canephora* Pierre ex A. Froehner), in irrigated plants about 35% of roots are at 0–10 cm soil depth, while in the soil without irrigation, the percentage was higher, 40% (Partelli et al. 2014; Covre et al. 2015). The mean diameter of coffee roots varies between 0.07 and 0.80 mm (Covre et al. 2015), being higher in plants propagated from seeds and at 50–60 cm soil depth (Partelli et al. 2014). For *Coffea arabica*, Ronchi et al. (2015) also found higher root length in the first 10 cm of soil depth, and superficial concentration of roots was more evident under close spacing between plants (0.4 m). According to the authors, despite the higher abundance of roots at 0–10 cm of soil depth than at 10–40 cm, roots at deeper layers were still growing in a newly coffee plantation. Regarding the mean root diameter for *Coffea arabica*, Ronchi et al. (2015) reported it was 0.3 mm, while Motta et al. (2006) found thicker roots of more than 3 mm diameter.

For hybrid eucalyptus clones (*Eucalyptus* spp.), higher root density is found at 0–25 cm soil depth, increasing with plant age, being the highest at 2 years (Bouillet et al. 2002). The fine roots (with less than 1 mm diameter) contributed more to the total root length of eucalyptus trees, being concentrated mainly at 0–10 cm soil depth (de Mello et al. 2007; Dias et al. 2017). The total length of fine roots is variable with the time of the year as well as with the tree age and following stem harvest (de Mello et al. 2007).

Sugarcane roots are concentrated at 0 to 60 cm soil depth (Inforzato 1957; Smith et al. 2005; Costa et al. 2007; Otto et al. 2011). According to Inforzato (1957), 60% of sugarcane roots are at 0–30 cm depth, with 85% of sugarcane root biomass at 0–60 cm (Smith et al. 2005), but a small amount of root biomass may still be found at 2 m depth. However, Laclau and Laclau (2009) found maximum sugarcane root depth at 425 and 470 cm in irrigated and non-irrigated crops, respectively. There are indications that the soil depth explored by sugarcane roots varies with soil texture, reaching 54 cm depth in sandy soil and 46 cm in clayey soils (Costa et al. 2007). However, besides texture, soil compaction is another factor to influence sugarcane rooting depth, with reduction in root growth if soil penetration resistance increases from 0.75 to 2.0 MPa (Otto et al. 2011).

In banana plants, root distal diameter is different according to root types. Values for banana root diameter were 2.2 to 4.3 mm for primary roots (main roots arising directly from the stem), 0.3 to 1.5 mm for secondary roots (arising from nodal roots or main roots) and 0.2 to 0.9 mm for tertiary roots (arising from secondary roots) (Lecompte et al. 2005). For plants cropped in deep soils, there is greater range of root diameter classes, due to the formation of thicker roots (2.2 to 6.6 mm) as well as thinner roots (0.68 to 0.79 mm) (Miotti et al. 2013). Soil compaction results in 48% decrease in main root elongation, as well as in visual symptoms typical of a mechanical barrier (Lecompte et al. 2005). In shallow as well as in deep soil, root growth was found up to 60 cm soil depth, but with 70% of roots concentrated at 0–30 cm soil depth. In shallow soil, the decrease in root length and surface area was pronounced at 30–40 cm depth, while in deep soil this reduction was pronounced at 40–50 cm depth (Miotti et al. 2013).

Although there is greater abundance of roots in soil surface layers, the root systems of most crops also reach the subsoil. Since it is challenging investigating deeper soil layers, the maximum root depth reported for different crops may be limited by the method used to measure root depth (Pierret et al. 2016; Thorup-Kristensen et al. 2020).

Subsoil exploitation by crop root systems is desirable because plants may uptake water from deeper soil layers as well as nutrients such as nitrogen (N-NO_3^-) that are leached downwards. Considering that actual agricultural systems do not use subsoil efficiently, the Deep Frontier project was launched to study the potential use of crops with deeper root systems to improve both plant resilience against environmental constraints and food production (Thorup-Kristensen et al. 2020). Projects of this nature are necessary to increase the knowledge on root growth in deeper layers of different soils at field conditions worldwide.

According to Thorup-Kristensen et al. (2020), there are strategies to increase soil depth exploitation by crop root systems. These strategies involve plant breeding and management measures such as crop rotation, agroforestry and intercropping systems. However, improving the knowledge on subsoil constraints (i.e. soil compaction, lower organic matter and nutrient availability, high acidity, salinity, waterlogging, toxic elements, etc.) affecting root growth and functions is necessary to technically support these strategies.

10.3 Response of the Root System of Major Agricultural Crops to Soil Physical Constraints

For wheat, the soil physical properties affect the root system (Itoh et al. 2009). In soil with a low bulk density (0.81–0.94 Mg m^{-3}), there was higher root density and better vertical distribution of roots to 36 cm deep than in soil with higher bulk density (1.31–1.44 Mg cm^{-3}) that had lower root density and vertical distribution limited to 25 cm deep. In the first year of the wheat crop, Itoh et al. (2009) found a positive relation between crop yield and development of root system. However, such a relation did not repeat in the second year. The authors concluded that, for the wheat crop, a superficial root system may result in a good yield of the crop as long as a suitable water supply is available. Under water stress conditions, reliance on the superficial root system would hamper the development of the crop.

Before discussing the root system of rice plants in relation to soil physical constraints, we need to take into account different ecosystems where rice is grown: irrigated, rainfed lowland, upland and flood-prone ecosystems (Halwart and Gupta 2004). For rice growth in uplands, physical soil constraints impair the root system of the crop (Grohmann and Queiroz Neto 1966). In Oxisols and Ultisols, bulk densities of 1.42 Mg m^{-3} and 1.38 Mg m^{-3}, respectively, restricted the root development of upland rice.

Outside of the upland ecosystems, rice is mostly grown in well-watered or saturated soils. Irrigated or wetland ecosystems account for more than half of the world's rice lands (Halwart and Gupta 2004). Soil flooding is harmful to most plant parts, but not for rice crops and other wetland species as long as plants are not submerged (Bailey-Serres and Colmer 2014). Suralta and Yamauchi (2008) have evaluated the adaptation of rice roots to aerobic and flood-prone environments. They found that the aerobic genotypes were as tolerant to flooding as the genotypes suitable for flood-prone areas as both sets of genotypes were able to adjust the level of aerenchyma development to effectively aid oxygen diffusion to the root tips.

For rice production under irrigated or flood-prone conditions, global climatic changes and the growing concern with water availability have stimulated the development of alternatives towards water-saving technologies. Such alternatives may favour the plant root system. One of the alternatives to save water when rice is grown in flooded areas is by the ground cover production system. In such a system, the soil is covered with a transparent plastic film after initial flooding, keeping the soil moisture close to saturation during the entire growing period, but with less irrigation during the development of the crop (Tao et al. 2015). When comparing the traditional flood-prone rice production system (TPRPS), the ground cover production system (GCRPS) increased root dry matter, length density and root surface area (Li et al. 2017; Zhang et al. 2017). Root features (dry matter, length density and surface area) were also positively correlated to crop yield and with the water use efficiency (i.e. grain yield divided by total water supply). Under the GCRPS before mid-tillering, adventitious roots growing were 31% longer and lateral roots 101% longer than those under TPRPS, and that was associated with the higher soil temperatures (Li et al. 2017). As a result, greater total root length and higher specific root length were observed. Root lengthening and branching may be affected by low temperatures as auxin transport, a hormone involved in both processes, is lowered under such circumstances (Li et al. 2017).

For maize crops, soil compaction damages the root system. According to Grzesiak et al. (2013), increase in the soil bulk density reduced root length and dry matter in maize plants, increasing the shoot/root ratio. Root penetration decreased as soil penetration resistance increases from 0.52 to 1.58 MPa (Grzesiak et al. 2013). Treatments to alleviate the effect of soil compaction on the development of maize crop include deep tillage to break the compacted layers, the use of genotypes more resistant to compaction and inoculation with mycorrhiza. Soil chiselling has been used as a strategy for alleviating compaction of a clayey soil which resulted in an increase of root density of maize plants (Nunes et al. 2015). In compacted soil, maize *cv.* Ankora that is sensitive to compaction had 14% reduction in seminal root length, while the most resistant cultivar (*cv.* Tina) had a reduction of only 8% (Grzesiak et al. 2013). As to the use of microorganisms in compacted soils, Miransari et al. (2007) demonstrated that the introduction of arbuscular mycorrhiza from different sources in sterilized and unsterilized soils significantly increased root dry weights for maize grown in compacted soils. The fungal hypha increased root surface area resulting in exploring higher volume of soil, increasing water and nutrient uptake.

Compaction of soil layers also reduces soil exploration by the root system of soybean crops (Cardoso et al. 2006; de Valadão et al. 2015). Roots growing in clayey Oxisols with penetration resistance of 1.48 and 1.84 MPa (soil water content = 0.28 $m^3\ m^{-3}$ and soil bulk density = 1.32 and 1.35 Mg m^{-3}) increased root diameter by 110 and 49%, respectively, in soil layers at a depth of 0.05–0.10 and 0.10–0.20 m (de Valadão et al. 2015). Greater root diameters of species growing in compacted soils does not seem to be a rule for all crops as for maize crops a reduction in root diameter was observed in plants growing in compacted soils (de Valadão et al. 2015). When comparing soybean root development in compacted and uncompacted clayey Oxisol, Cardoso et al. (2006) observed that the *cv.* Embrapa 4 had a higher root volume in compacted soil than the *cv.* BR16.

Sunflower roots are affected by soil compaction (Goodman and Ennos 1999), and the greater the soil penetration resistance (in a range of 1.60 and 1.85 MPa) the lower the crop yield (Bayhan et al. 2002). During seedling emergence, the increase in soil resistance to penetration delays emergence and decreases the percentage of seedling emergence (Bayhan et al. 2002). Soil compaction also reduced plant height, and while not reducing stem and flower head diameters, it resulted in the reduction of the crop yield of sunflower.

Goodman and Ennos (1999) reported that soil strength had a small effect on the shoot growth of sunflower. However, in sunflower subjected to increasing penetration resistance from 0.12 to 0.32 MPa, root system responded with a greater angle of spread than those grown in soil with low strength. Additionally, in compacted soils first-order lateral roots are thicker and tap roots tapered more rapidly. However, an important aspect of soil compaction for sunflower crops is the greater anchorage achieved in more compacted soils which is considered a positive feature due to the avoidance of lodging (Goodman and Ennos 1999). Puddling of clay soil through intensive tillage in the wet season damages soil structure and leads to soil compaction during the dry season. Rice straw mulch proved effective in increasing soil water content and reducing soil compaction in a salt-affected clay-textured soil, allowing a higher sunflower root biomass development at the 0–20 cm soil depth (Paul et al. 2021).

Mechanized field operations in areas cropped with oil palm increase soil bulk density and reduce the least limiting water range in the surface layer of the traffic zone (Sato et al. 2017). However, the effect of temporal variation of the soil water content can be detected in subsurface layers (20–40 and 40–60 cm), indicating effects on oil palm yield (Sato et al. 2017). To overcome the lack of available water, root system traits of some oil palm genotypes can help plants to uptake water and nutrients in a more efficient way (Nodichao et al. 2011). However, soil compaction can depress oil palm biomass and oil production (Chen and Weil 2010; Peltonen-Sainio et al. 2011). Oil palm genotypes that are tolerant to drought develop greater living fine root biomass (0.14 kg m^{-2}) and total fine root length (410 kg m^{-2}) in comparison with susceptible genotypes (0.09 kg m^{-2} and 238 kg m^{-2}, respectively) (Nodichao et al. 2011).

Rapeseed is the most important *Brassica* oilseed crop, and climate change and expansion into new production areas are exposing the crop to drought stress (Fletcher et al. 2015). However, little research has focused on the rapeseed root system with emphasis on root traits that help plants to survive under low soil water availability. Investigating genetic influence on response to drought in rapeseed revealed that root pulling force was most highly correlated with total root dry mass, taproot diameter, taproot length and branching zone length with no correlation between root pulling force and number of coarse secondary roots (Fletcher et al.

2015). On the other hand, it is important to consider that rapeseed crops have been used as a cover crop for improvement of the physical conditions of soil (Chen and Weil 2010, 2011), being representative of species with root system traits that allow growth under soil physical constraints. However, rapeseed shoot dry matter decreased by 63% as the soil penetration resistance increased, while root dry matter decreased by 50% (Chen and Weil 2010).

Under low-water and high-water availability in soil, the mustard root system is less affected than rapeseed. Mustard root length in subsoil (30–40 cm depth) decreased only 5 mm cm^{-3} under low water, while rapeseed root length decreased 15 mm cm^{-3}. The same tendency was observed to root surface area that was reduced by 2 mm^2 cm^{-3} for mustard and 15 mm^2 cm^{-3} for rapeseed at 30–40 cm depth under low-water availability. The number of tip roots also was less affected in mustard (reduced by 1,100 10^3 cm^{-3}) than rapeseed (3,600 10^3 cm^{-3}) in the subsoil under water stress (Liu et al. 2011b).

Further information on the effect of physical constraints on root depth and function are reported by (Oliveira and Fernandes 2022; Scanlan et al. 2022; So and Oliveira 2022; Vance and Milroy 2022).

10.4 Response of the Root System of Major Agricultural Crops to Soil Chemical Constraints

There is strong competition for edaphic resources below ground as the supply of nutrients is variable in both space and time (Brian and Helena 2001). Moreover, nutrient availability can frequently affect morphology, anatomy, topology, distribution and architecture of roots.

The chemical factors in the soil that may limit plant growth include deficiency or unavailability of essential elements and toxicity due to elements such as aluminium (Al), manganese (Mn) or other metallic cations. Salinity and the deficiency of silicon (Si), which is considered an essential element for some plant species and beneficial to others, may also affect plant growth (Foy 1992). Further information on chemical constraints that affect root growth and function are found in Butterly et al. (2022), Ma et al. (2022), Rengasamy et al. (2022), Scanlan et al. (2022) and Souza and Abrahão (2022).

10.4.1 Soil pH

Plant root responses to variations in soil pH are closely related to excess soluble Al and heavy metals and low phosphorus availability. The following information deals with responses in the rhizosphere for the interaction among pH, Al and phosphorus. The specific effects of Al, heavy metals/micronutrients and macronutrients are reviewed below.

Exudation of malate by soybean roots is influenced by the coordinated action of pH, phosphorus and Al availability (Liang et al. 2013). At solution pH 5.8, the genotypes of soybean that are more efficient in the phosphorus uptake displayed more malate exudation by roots. However, acidification (pH 4.3) of the nutrient solution together with supply of phosphorus resulted in a significant reduction in root malate exudation (Liang et al. 2013). On the other hand, adding Al to the nutrient solution with supply of phosphorus and low pH increased malate exudation (Liang et al. 2013). According to the authors, such results suggest that root malate exudation in soybean represents more than a simple response to the excess of soluble Al in acid soils. The exudation of malate in acid nutrient solutions deficient in phosphorus supply may also be involved in the mobilization of inorganic phosphorus complexes in the soil. Phosphorus supply in the nutrient solution also contributes to increase the tolerance of sunflower crops to high amounts of soluble Al in the soil (da Jesus and Azevedo Neto 2013).

In acid soils, attention has also been given to phosphorus deficiency. Phosphorus fertilizer placement close to the soil surface stimulates the development of roots in that layer. However, if soil water content declines in that layer, plants need to uptake phosphorus from the subsoil layer (Foy 1992). In this case, subsoil limitations such as acidity or compaction restrict root growth deeper in the soil which can lead to serious limitations in the absorption of water, phosphorus and other nutrients.

10.4.2 Aluminium Toxicity

Aluminium toxicity is discussed by Lynch and Wojciechowski (2015) as a subsoil constraint to root growth. Below a soil pH_{H2O} of ~ 4.8, Al solubility increases and Al^{3+} causes direct injuries to root apices, reducing root growth, which in turn decreases plant potential to absorb water and nutrients. Batista et al. (2013) reported that maize plants grown in solutions with Al had inhibited growth, few lateral roots and less development of the root system, which was associated with reduction in root dry mass, in number of lateral roots and in root length (Batista et al. 2013). In contrast with soils without soluble Al, maize roots grown in Al-rich soils present necrotic epidermis, larger aerenchyma volume and no "U" thickening of the endodermis cells and the pith parenchyma is poorly developed. Yet, the greatest changes were observed in the xylem cells as such cells were darker in colour, deformed (varying in diameter), without a developed secondary wall (characteristic of a still immature tissue) and smaller if compared to the plant without Al (Batista et al. 2013). Toxicity caused by Al reduces root growth in wheat (Khabaz-Saberi et al. 2014). Aluminium toxicity decreased non-tolerant wheat genotypes to ~10% relative root length, while the tolerant genotypes achieved ~60% relative root length. The negative effects of Al on root length and dry mass caused losses of wheat grain yield as the lowest relative yield was observed in non-tolerant genotypes (Khabaz-Saberi et al. 2014).

Clune and Copeland (1999) demonstrated that rapeseed plants (*Brassica napus* var. *napus* L.) are vulnerable to the presence of soluble Al in the soil. Exposure of rapeseed seedlings to 60 μM or higher of Al strongly inhibited root growth, with cellular damage observed mainly in the peripheral root cap cells. Such a condition led to reduced secondary root development and new lateral roots forming towards the apex of the tap root.

An important mechanism for plants to resist Al toxicity is the root exudation of organic acids in the rhizosphere (Chen et al. 2006). Organic acids such as citrate and malate are excreted by soybean, maize and wheat roots in the rhizosphere as a protective response to the stress caused by the excess of Al. Such acids chelate the Al^{3+}, reducing its toxicity to plant roots (Shen et al. 2005; Liao et al. 2006; Chen et al. 2006).

10.4.3 Heavy Metals, Toxic Elements and Micronutrients

In this review, we consider root responses associated with contamination and potential toxicity or ecotoxicity by arsenic (As), cadmium (Cd), chromium (Cr), cobalt (Co), copper (Cu), iron (Fe), lead (Pb), manganese (Mn), molybdenum (Mo), nickel (Ni), selenium (Se) and zinc (Zn). Although anthropic contamination by heavy metal is more expressive in topsoil, exceeding threshold values for heavy metals have also been reported in subsoil (Li and Shuman 1996; Ungaro et al. 2008). Unlike organic pollutants, heavy metals are not degraded naturally or by microbial communities and hence may persist in soils (Liu et al. 2015).

Heavy metal-induced oxidative stress results in the accumulation of reactive oxygen species (ROS) such as hydrogen peroxide (H_2O_2), superoxide ($^{\bullet}O_2^-$) and hydroxyl radical ($^{\bullet}OH$) (Mithöfer et al. 2004; Yadav 2010).

Heavy metals are absorbed by first few millimetres behind the root tip (Mithöfer et al. 2004). Once in the cortex, the metals are transported in the apoplastic space and also accumulate in the cell walls (Schützendübel and Polle 2002). Decrease in root elongation, number of roots, biomass and vessel diameter as well as damage on root tips and alterations in the structure of hypodermis and endodermis are the main response of roots to heavy metals (Ashraf et al. 2010). Reduced root growth and dry mass were observed in wheat crops subjected to soils rich in manganese or iron (Khabaz-Saberi et al. 2014).

Studies with rapeseed and black mustard demonstrated the use of such crops in phytoremediation of soils contaminated with heavy metals (Ehsan et al. 2014). The root system and enzyme exudation in the rhizosphere are mechanisms that allow the development of phytoremediator species in soils contaminated by heavy metal (Tangahu et al. 2011). As many of the heavy metal(loid)s are actually micronutrients to plants, toxicity or deficiency of such elements is discussed in the topics that follow.

Iron

Although a heavy metal, iron (Fe) is also an essential micronutrient for plant growth. High reactive Fe levels in soils may reduce uptake of nutrients (N, P, K and Mg) (Fageria et al. 2002). However, its solubility is extremely low, especially in aerated alkaline soils due to the formation of Fe hydroxides, oxyhydroxides and oxides (Broadley et al. 2012a). Iron deficiency in plants is not caused only by low iron solubility in soils but also due to presence of bicarbonate. For barley, sorghum and corn plants, bicarbonate presence promoted accumulation of organic acids in roots which decreased Fe uptake and translocation as well as root growth (Alhendawi et al. 1997). Flood or compaction decreases soil aeration and as a consequence, soluble reduced forms of Fe may be absorbed in large amounts, causing toxicity to plants (Fageria et al. 1990).

Due to reducing conditions, the chemical properties of flooded rice soils are entirely different from those of drained soils as anoxia may increase concentrations of Fe^{2+} which sometimes reaches toxic levels for rice plants (Fageria et al. 1990). Root length and dry weight of 40 rice cultivars decreased due to Fe excess (0.89 and 1.78 mM – optimum level at 0.09 mM) (Fageria and Rabelo 1987). Plants usually show bronzing of leaves if dissolved Fe in the rooting medium is in the range 300 to 500 mg kg^{-1} (Fageria et al. 1990; Suresh 2005). Under Fe toxicity, tolerant plants form a Fe plaque on the root, which acts as a barrier against Fe absorption by roots (Mahender et al. 2019).

Conversely, as iron is a micronutrient, the deficiency of Fe may damage plant development. Iron deficiency is associated with inhibition of root elongation, increase in the diameter of apical root zones and abundant root hair formation (Broadley et al. 2012b). Transfer cells (i.e. cells with wall ingrowths along the outer tangential walls as well as a higher cytoplasm-to-vacuole ratio) were reported in the hypodermis of Fe-deficient roots of soybean (*Glycine max* (L.) Merr., cv. Hawkeye) (Landsberg 1989), while in sugar beet (*Beta vulgaris* L. cv Kawemegamono), transfer cells were reported in nearly all rhizodermal cells from the subapical root swelling, including root hairs (Landsberg 1994).

The transfer cells most likely play a role in what is known as Strategy I for Fe acquisition. The transfer cells would enhance net excretion of protons and reducing capacity, increasing the ability of the roots to reduce ferric chelates, allowing mobilization of ferric iron (Brown and Von Jolley 1988; Landsberg 1989; Fageria et al. 2002). In Strategy II, found in grass species (Poaceae), plants cope with Fe deficiency by releasing phytosiderophores into the rhizosphere to chelate and solubilize metals that are soil bound (Brown and Von Jolley 1988; Ashraf et al. 2010; Broadley et al. 2012a). However, both rice and maize are thought to use a combination of both strategies for Fe uptake (Ricachenevsky and Sperotto 2014; Li et al. 2018). Notwithstanding the above findings about factors affecting Fe acquisition by roots in soils, their implications for uptake of subsoil Fe have not been reported. Subsoils are commonly more alkaline than topsoils, and restricted aeration of pores in subsoils may exacerbate the build-up of bicarbonate which impairs Fe uptake by root (Alhendawi et al. 1997).

Zinc

Zinc is an essential element for plant growth (Alloway 2013). It is a catalytic component of over 300 enzymes (Gupta et al. 2011) such as CuZn superoxide dismutase (Broadley et al. 2012a). Zinc deficiency is found in plants growing in calcareous or highly acid soils (Broadley et al. 2012a; Mertens and Smolders 2013). Severe Zn deficiency is characterized by root apex necrosis (Broadley et al. 2007). In rice, Zn deficiency affected the rate of crown root initiation by reducing the number of crown roots initiated (Nanda and Wissuwa 2016), and new adventitious root formation may decline by about 50–75% (Widodo et al. 2010). As Zn has limited mobility in the phloem, low Zn in the rhizosphere of root tips directly inhibits root elongation (Nable and Webb 1993). This is particularly relevant to subsoil root growth as the plant-available Zn levels are commonly lower in subsoil than topsoil (Bell and Dell 2008).

On the other hand, excess of Zn in wheat seedlings reduced root length (Li et al. 2012). The effect is even greater with increasing Zn concentration: for example, increasing Zn from 0.5 to 3 mM $ZnSO_4$ decreased root length from 63 to 11% of that in the control plants (Li et al. 2012). Moreover, the average level of lignification in xylem was greater in seedling roots under 3 mM Zn stress. Lignification causes cell wall rigidity which may have been responsible for inhibition of root elongation (Li et al. 2012). In rapeseed exposed to Zn at toxic levels, cross sections of the root tips showed decreased number of cortical and stellar cells (xylem and phloem were undersized) and a decreased diameter of root tip compared to the control with the epidermis being the most vulnerable tissue of the root tip as cells were deformed and shrunken (Mousavi Kouhi et al. 2016).

Cobalt

Cobalt (Co) is not an abundant element in soils with concentration ranging between 15 and 25 mg kg^{-1} (Pilon-Smits et al. 2009), and in plants its normal concentrations are as low as 0.1–10 µg g^{-1} dry weight (Khan and Khan 2010). Cobalt is essential for the bacterial fixation of nitrogen in root nodules of legumes (Alloway 2013). Cobalt has been reported to strongly bind to roots, entering plants through passive transport (Pilon-Smits et al. 2009). As Co has low mobility in plants, no strong gradient is found from stems to leaves (Palit et al. 1994). Cobalt toxicity to root growth of barley was affected by soil properties (Micó et al. 2008). The EC_{50} (effective concentration of Co added that reduced root length by 50%) values ranged from 45 to 863 mg kg^{-1}. Toxicity threshold values for near neutral and basic soils were greater than those obtained for acid soils which indicates that barley root was more sensitive to Co toxicity in acid soils (Micó et al. 2008).

For chickpea plants (*Cicer arietinum* L.) treated with 200 and 400 mg Co kg^{-1}, root growth was significantly affected (19.8 and 19.2 cm length, respectively) in comparison to control (21.3 cm), and root dry weights were similarly depressed (Khan and Khan 2010). Growth of lateral roots was also suppressed. Pod formation decreased by ~7% (Khan and Khan 2010).

Soybean also suffers from Co toxicity as all the three varieties of soybean tested (SL-688, PS-1347 and DS-9712) had shorter roots upon Co addition (Imtiyaz et al.

2014), but SL-688 was the most sensitive variety to increasing Co up to 150µM (Imtiyaz et al. 2014). The soybean roots were twisted at 10^{-4} to 10^{-3} M Co (Liu et al. 1995). Root growth was seriously inhibited in Co concentrations above 10^{-3} M (Liu et al. 1995).

However, for some plant species, Co input can result in positive effects to root system as its deficiency depresses N_2 fixation (Gad 2006). Cobalt is essential for the fixation of N_2 by rhizobia and other N_2-fixing microorganisms whether growing inside or outside root nodules (Fageria et al. 2002). When Co is not available, methionine synthesis is depressed which presumably leads to lower protein synthesis, thus contributing to the smaller nodules or rhizobia (Broadley et al. 2012b).

The addition of cobalt (8 ppm), applied as cobalt sulphate, to the soil with pea plants (*Pisum sativum* L.) increased root length from 7.5 cm (control without Co) to 10.2 cm (with Co) (Gad 2006). Both fresh and dry weight of roots were also increased (Gad 2006). Perhaps the most striking finding is that both macronutrient (N, P and K) and micronutrient contents (Fe, Mn and Zn) also increased within plants (Gad 2006). Therefore, addition of Co in the plant media could be magnified by the additional nutrient uptake by plants.

Addition of Co to low Co soils is important for nodulation in legumes (Ozanne et al. 1963; Palit et al. 1994) as it increases both nodule number and weight in pea (Gad 2006). The number of active nodules, red-pink coloured on the inside due to the presence of leghemoglobin, increased from 20 (out of 57 total nodules) to 40 (out of 86) (Gad 2006) which ultimately has a significant favourable effect on pod yield (Gad 2006).

Molybdenum

Molybdenum is a micronutrient (Alloway 2013) present in small amounts in soils (ranging from 0.2 to 36 mg kg^{-1}), and although variably mobile in xylem and phloem (Qin et al. 2017, the requirement of plants for Mo is lower than that for any of the other nutrients apart from nickel (Broadley et al. 2012a). Molybdenum is a metal component of enzymes, including the nitrogenase enzyme which is essential for N_2-fixing bacteria of legume nodules (Brodrick and Giller 1991; Broadley et al. 2012a). Legume crop species display poor nodulation under Mo deficiency (Gupta 1997). Hairy vetch (*Vicia villosa* Roth) plants treated with 0.5 mg Mo kg^{-1} had 62% more nodules in roots than the controls as well as a 54% increase in nodule weight (Alam et al. 2015). The total biomass, including root biomass, of Mo-treated hairy vetch plants was higher than that of control plants. However, nodule characteristics (number and weight) decreased in plants treated with 1.0 mg kg^{-1} Mo, and that may be attributed to the fact that at higher quantities plant growth is inhibited (Alam et al. 2015). Bacteria may promote root hair formation (Burdman et al. 1997) which may consequently increase nutrient uptake. Adding Mo to soybean plants also increased both root nodule numbers and weight (Jabbar and Saud 2012). Main root length, root system volume and dry weight of the roots were all enhanced by adding Mo to the soil (Liu et al. 2005).

Molybdenum toxicity in plants under most agricultural conditions is rare (Kaiser et al. 2005). The greatest concern with high Mo levels in crops is related to grazing or silage production as ruminant animals can suffer from molybdenosis, a disorder caused by excess Mo uptake that induces copper deficiencies (Kaiser et al. 2005).

Copper

Copper is tightly bound to compounds in soil organic matter, and its concentration in soils ranges from about 2 to 100 mg kg^{-1} with a mean of 30 mg kg^{-1} (Fageria et al. 2002). Excess of Cu altered the growth and morphology in crop species roots, inhibiting root growth as they were shorter and thicker than in the control seedlings (Przymusiński and Gwóźdź 1994). Tolerance indexes indicate that root growth was inhibited by 50% when exposed to 16.5 mg Cu L^{-1} (liquid solution) (Przymusiński and Gwóźdź 1994).

Copper (Cu) toxicity damages the plasma membrane and reduces the membrane potential, besides affecting cell elongation in roots by stimulating lignification which also explains the brown colour observed in roots treated with excess Cu (Arduini et al. 1995). However, when plants are not properly supplied with Cu, lignin concentration decreases and lignification of xylem vessels is impaired (Broadley et al. 2012a). Although the antioxidative system is apparently effective in common bean (*Phaseolus vulgaris* L.), Cu toxicity leads to the increased production of reactive oxygen species (ROS) (Weckx and Clijsters 1996).

In onion (*Allium cepa* L.), Cu toxicity directly limits cell division, thus affecting root elongation which may ultimately cease (Liu et al. 1994). Morphology of the maize roots subjected to difference concentrations of Cu varied after 48 h treatment, and mitotic index decreased with increased Cu concentration (Jiang et al. 2001). At 10^{-4} M, the roots appeared yellowish, while at 10^{-3} M they were yellow and twisted, and at 10^{-2} M roots were yellow-green and with a rotten appearance (Jiang et al. 2001).

Lead

In roots, the highest lead (Pb) concentrations can be found in root apices (Tung and Temple 1996). Other locations for Pb accumulation in corn roots were the hair zone, root primordia initials and the interface between the root and shoot where adventitious roots arise. In root apices of radish (*Raphanus sativus* L.), staining of cells for Pb is mostly in the cell walls, being less dense in the root cap if compared to other parts of the tissue (Lane and Martin 1977). At root apices, cells are young and therefore have thin cell walls (except in root cap cells) (Pourrut et al. 2011). Such characteristics along with the fact that lower pH in the apical area increases Pb solubility in the soil solution facilitates Pb uptake by roots (Pourrut et al. 2011).

Lead affects plant mineral uptake (Gopal and Rizvi 2008). The mineral imbalance within the tissues is related to the capacity of Pb to block the entry of cations such as Ca, Cu, Fe, Mg, Mn and Zn as well as anions such as NO_3^- (Sharma and Dubey 2005). Severe reductions in root growth, loss of apical dominance, an increase in root branching, the formation of swellings behind the root tips caused by the initiation of lateral roots and the bending of some root tips may also be caused by excess Pb (Kopittke et al. 2007). The inhibition and abnormalities of root growth in the presence of Pb might be related to disturbances in the water status of the cells which are known to directly affect both cell divisions (antimitotic effect) and elongation as well all stiffening of cell walls due to Pb incrustation (Przymusiński and Gwóźdź 1994; Shahid et al. 2011).

Several factors limit the transport of Pb from roots to aerial plant parts such as immobilization by negatively charged pectins within the cell wall, precipitation of insoluble Pb salts in intercellular spaces, accumulation in plasma membranes as well as sequestration in the vacuoles of rhizodermal and cortical cells (Pourrut et al. 2011). Accumulation of Pb in roots is also important in limiting the amounts of Pb that reaches the aerial parts of plants.

Arsenic

Arsenate is a chemical analogue of phosphate and competes for the same uptake carriers in the root plasmalemma of plants (Yadav 2010). In rice, increasing arsenate concentrations decreased root length, but cultivars varied in the extent of root inhibition (Dasgupta et al. 2004).

In tomato plant treated at lower concentrations of As (15 and 25 mg kg^{-1}), longer roots developed (~16 and ~14 cm, respectively), while at higher concentrations (50 and 100 mg kg^{-1}), shorter roots were observed (~12 and ~10 cm, respectively) (Miteva 2002). Although root weight was affected more in comparison to root length, the root growth of rice seedlings in terms of root length and root fresh weight were greatly inhibited at 50 mM As(III) as well as 500 mM As(V) (Shri et al. 2009). No root formation was observed at concentrations higher than 500 mM As(V). The roots turned black in the presence of 100 mM As(III) (Shri et al. 2009).

Inoculation of roots with arbuscular mycorrhizae fungus can exert protective effects on host plants by enhancing tolerance to As in crops such as lentil and sunflower (Garg and Singla 2011). The arbuscular mycorrhizae fungal symbiotic association enhances host plant phosphorus nutrition as external hyphae absorb phosphorus and transfer it to the host plant through fungal hyphae in root cortical cells (Smith et al. 2010).

Cadmium

The threshold value for Cd in agricultural soils in Finland is 1 mg kg^{-1} which is very similar to mean values of different national systems in Europe (Tóth et al. 2016). In rice, 100 μM Cd affected root architecture as adventitious root length was reduced, as was the formation of lateral root primordia in comparison with the control (Ronzan et al. 2018). Cadmium induced severe damage to the lateral root primordia, altering the planes of cell divisions, leading to altered root tips without a quiescent centre or with irregular definitions of the quiescent centre (Ronzan et al. 2018). Precocious differentiation of the aerenchyma of lateral roots was also observed. In barley, root biomass responded most sensitively to elevated Cd as 50% inhibition was observed at 30 μM Cd and contents of Mn, Mg and Mo in roots decreased (Brune and Dietz 1995). Root growth inhibition was also detected in barley seedlings exposed to a very low (2 μM) concentration of Cd and increased nearly linearly in a Cd concentration-dependent manner up to 25 μM Cd (Tamás et al. 2014).

A visible root swelling with well-developed root hairs was detected near root tips in treatments with Cd from 3 to 25 μM (Tamás et al. 2014). Such root swelling decreased with increasing Cd concentration, and at 25 μM Cd it was the smallest and root hairs were absent. Phytochelatins are small heavy metal complexing peptides in higher plants (Baycu 2002) considered to play a role in cellular Cd homeostasis and detoxification (Baycu 2002; Wójcik and Tukiendorf 2004). In plants, complexes of phytochelatin + Cd are transported and sequestered to the vacuole (Cobbett and Goldsbrough 2002). Although phytochelatin preference is for Cd, they are capable of binding to other elements such as Zn, As and Cu (Inouhe 2005).

10.4.4 Macronutrients

Root architecture may be directly affected by the availability of nitrogen (N), phosphorus (P) and potassium (K), even when root biomass does not decrease as much as the biomass in aerial parts due to deficiencies of macronutrients (Drew 1975; Osmont et al. 2007). In the paragraphs below, the influence of N, P and K on the plant root system will be further explained.

Nitrogen

Depletion of nitrogen (N) in soils is a problem that plant species commonly encounter. In environments where N is heterogeneously distributed, lateral root growth is induced wherever N levels are highest (Osmont et al. 2007). The response of the root system to nutrient heterogeneity is controlled by the presence of genes and specific transporters in plants (Zhang and Brian 1998). ANR1 is a MADS box gene induced by NO_3^- which leads to localized lateral root proliferation. Transgenic plants in which ANR1 was repressed were no longer able to respond to a nitrate-rich zone by lateral root proliferation (Zhang and Brian 1998).

Experiments with nitrate showed an increase in length and density of root hairs with the decrease of concentration below 1000 μM. At 2 μM, compared to 1000 μM NO_3^-, root hair length of tomato increased by a factor of 2, of rapeseed by a factor of 5 and of spinach by a factor of 9 (Foehse and Jungk 1983). The concentration of nitrate was found to exert a direct and highly localized effect upon the growth of lateral roots in barley (*Hordeum vulgare* L.) (Drew et al. 1973). Zones receiving 1000 μM nitrate showed an increase in the number and extension rate of both first- and second-order lateral roots, associated with a preferential accumulation of dry matter, compared with zones in 10 μM nitrate. The average number of laterals (both first and second order) per cm of parent root was two times greater in the presence of 1000 μM nitrate than of 10 μM. The average extension rates of first- and second-order laterals were more than five times longer in nitrate concentrations of 1000 than in 10 μM (Drew et al. 1973). The lateral roots of *Arabidopsis thaliana* (L.) Heynh. grown in uniformly high nitrate soil had reduced lateral root elongation throughout the root system (Zhang and Brian 1998). However, plants grown on a low nitrate soil will preferentially show higher root density and proliferate lateral roots within N-enriched soil patches (Zhang and Brian 1998; Kristensen and

Thorup-Kristensen 2007). Increased root depth by 0.2–0.4 m and density were found to be correlated with inorganic N placed in deep soil layers for deep-rooted crops such as white cabbage (*Brassica oleracea* L. convar. *capitata* (L.) Alef. var. *alba* DC 'Impala'), Chinese cabbage (*Brassica campestris* L. ssp. *pekinensis* 'Kasumi') and summer squash (*Cucurbita pepo* Duch. 'Ambassador') (Kristensen and Thorup-Kristensen 2007).

Higher root density in the subsoil is important for plant development in cases where the amounts of nitrogen on the topsoil are depleted or unavailable during plant growth. The deep-rooted *Lupinus pilosus* Murr. presents a root architecture likely to maximize nitrate capture. The higher root density and distribution of fine roots in the topsoil along with a deep taproot allowed such species to access nitrate available from the topsoil during the first 20–50 days as well as the subsoil nitrate from 60 to 100 days (Dunbabin et al. 2003). Varieties with greater root length density in the subsoil may reduce N leaching as N is absorbed by deep roots and additionally enable plants to access water in deeper soil horizons which could reduce the risk of drought stress (Yu et al. 2015).

A root system ideotype has been proposed for acquisition of nitrogen and water, named "steep, cheap, and deep" (SCD), which is an integration of architectural, anatomical and physiological phenotypes allowing a rapid exploitation of subsoil water and nitrate resources (Lynch 2013). According to Lynch (2013), for the maize root system, the SCD ideotype to improve nitrogen and water uptake is described as follows: (*i*) Primary roots: large diameter, few but long lateral roots and tolerance to cold soil temperatures; (*ii*) Seminal roots: many, with shallow growth angles, small diameter, many laterals and long root hairs. Alternatively, ideotype seminal roots should have intermediate number with steep growth angles, large diameter and few laterals coupled with abundant lateral branching of the initial crown roots; (*iii*) Crown roots: intermediate number with steep growth angles and few but long laterals; and (*iv*) Brace roots: one whorl of high occupancy, having a growth angle that is slightly shallower than the growth of crown roots, with few but long laterals.

Phenotypic ideotype aspects were described for maize but can be applied to other monocotyledonous crops and also dicotyledonous species. However, many aspects remain hypothetical and need to be validated (Lynch 2013).

Phosphorus

Low phosphorus limits crop production in many parts of the world but is a particular constraint in soils of large areas in Central Asia, Africa, North America and South America (Kvakić et al. 2018). In the Cerrado (Brazilian savannah), one of the world's major agricultural crop production areas, agricultural expansion was possible only due to correction of soil acidity with liming and supply of phosphorus and potassium as well as other techniques (Ribeiro and Walter 2008; Coutinho 2016). Phosphorus deficiency in soils may intensify with time due to land degradation (Lynch 2011). Besides, high rates of phosphorus fertilization are not common in the low-input farming systems adopted in poor nations. Therefore, efficiency in phosphorus absorption is an important feature of productive agriculture.

Plant strategies to uptake more phosphorus from the soil involve two basic themes: greater soil exploration by roots and mobilization of less available phosphorus pools within the rhizosphere (Lynch 2011). In maize grown in low phosphorus soils, plants had higher root/shoot ratio, increased root specific length, reduction in root diameter and severe reduction of root density which indicates a drastic reduction in above-ground growth in relation to roots (Mollier and Pellerin 1999; Fernandez and Rubio 2015). Maize plants subjected to phosphorus deficiency displayed roots with larger aerenchyma spaces in comparison to plants that were not phosphorus stressed (Fernandez and Rubio 2015). According to Mollier and Pellerin (1999), the changes observed in maize root are similar to those described for carbohydrate-deficient plants.

Phosphorus deficiency reduces root and above-ground biomass of soybean plants. Liao et al. (2006) demonstrated that phosphorus supply facilitates the development of crops more tolerant to soluble Al. On the other hand, with low phosphorus availability, root growth in soybean genotypes is inhibited when Al is added (Liang et al. 2013). Similar results regarding phosphorus deficiency were also found for sunflower crops (Eaton 1949) and black mustard (Eaton 1952).

Plants growing in P-deficient soils face a dilemma as they must explore the soil for nutrient uptake as well as maximize phosphorus use so the formation of lateral roots, lateral root growth and root hairs is induced while taproot is inhibited (Péret et al. 2011, 2014). However, for maize, lateral root formation is reduced under low P more in comparison to reductions in axial elongation (Mollier and Pellerin 1999). Root proliferation towards P-rich soil patches has also been reported (Hodge 2004).

The presence of microorganisms in the rhizosphere aids the development of sunflower plants in soil deficient in phosphorus. Plants that bear arbuscular mycorrhiza in their roots grow more rapidly in soil deficient in phosphorus than plants that are weakly colonized by such fungus (Thompson 1987). *Enterobacter* sp. Fs-11 is a bacteria species found colonizing roots of several crops (Shahid et al. 2012). *Enterobacter* sp. Fs-11 is an efficient phosphate-solubilizing bacteria, able to decrease pH of the medium and increase concentrations of soluble phosphorus, which contributes to improved plant growth (Shahid et al. 2012).

Phosphorus efficient rapeseed genotypes have the capacity to absorb more available phosphorus from the soil and translocate it into shoot than phosphorus-inefficient genotypes (Zhang et al. 2009). Exudation of malic and citric acids by rapeseed roots increases access to phosphorus bound to calcium (Zhang et al. 1997). Exudation of such acids is considered an adaptation of plants to phosphorus absorption in calcareous soils (Zhang et al. 1997).

As mentioned above (Sect. 10.4.1), malate exudation by roots is a mechanism that helps plants to deal with low phosphorus availability in acid soils. Carboxylate exudation in the rhizosphere is another strategy that allows plants to uptake less soluble P forms from soils (i.e. soluble P fertilizer is converted into less soluble forms over time) (Kabir et al. 2015). Chickpea plants promote depletion of less soluble P in the rhizosphere mainly in the subsoil layer. According to Kabir et al.

(2015), depletion of sparingly soluble P from the chickpea rhizosphere in the subsoil was linked with the greater levels of carboxylates in the rhizosphere. These findings indicate that modifications of rhizosphere and the deep rooting pattern of chickpea help plants to access P in the subsoil once topsoil dries out.

Potassium

In rice, root growth of all genotypes studied reduced under low K (Jia et al. 2008). Potassium deficiency can also affect root morphology and reduce root diameter (Carmeis Filho et al. 2017). However, moderate K deficiency increased the root length of the efficient genotypes, while at deficient and moderate K levels, all the efficient rice genotypes developed more fine roots (diameter <0.2 mm) than the inefficient ones (Jia et al. 2008). Rice plants under low level of K exhibited more favourable architecture for nutrient uptake as thin roots with small root diameter may enable roots to efficiently reach the nutrients in the soil (Carmeis Filho et al. 2017).

Reviewing the efficiency of K uptake and use, Rengel and Damon (2008) explain that genotypes efficient in K uptake may have a larger surface area of contact between roots and soil. However, increased root growth is not a rule for genotypic K efficiency as K-efficient potato genotype had half the root length of the K-inefficient cultivars (Trehan and Sharma 2002). Leaching of K in sandy soils may be a significant loss of K from the system; therefore, it may be beneficial to introduce genotypes that grow roots and uptake more K from deep soil layers (Rengel and Damon 2008).

10.4.5 Sodicity

Salinity and soil sodicity are associated with the presence of several salts and, more specifically, sodium (Na). Salinity and soil sodicity are limitations that may occur naturally in soils of arid, semi-arid and dry sub-humid regions or may be caused by improper management practices especially in arid regions (Rengasamy et al. 2022). Worldwide, up to 20% of the irrigated arable land is already salt affected (Mühling and Läuchli 2001). In semi-arid and arid regions, especially those under irrigation, typical concentrations of Na in the soil solution range from 50 to 100 mM (Broadley et al. 2012b). Although Na is an essential element for halophytes (plants that are able to survive and develop in saline soils) (Bueno González 2019) and plants that use C4 or CAM photosynthetic pathways, Na excess in the soil is detrimental for the growth of most crop plants (Pilon-Smits et al. 2009).

Potassium may be substituted by Na in some plant species so it raises the possibility of utilizing low-grade Na fertilizers which are less expensive than K fertilizers (Subbarao et al. 2003). Although K may be substituted by Na in some plant species, genotypic differences can be substantial for classifying whether a plant species may or not normally develop with Na replacing K (Broadley et al. 2012b).

Root growth of most crop plants is inhibited by soil salinity (Byrt et al. 2018). In sunflower, maize and cotton root growth and vigour are decreased under salt stress (Guo et al. 2015). In *Phaseolus vulgaris* roots, increases in Ca^{2+} concentration for salt-stressed plants from 0.5 mM to 5.0 mM resulted in a greater lignification of the protoxylem and metaxylem vessels as well as stronger lignification of the phloem fibre cell walls (Cachorro et al. 1993). Such alterations in xylem and phloem fibres may be responsible for inhibiting root growth (Cachorro et al. 1993).

Elongation rate of maize primary roots is inhibited after 2 days in saline conditions, and long-term inhibition of root growth by salinity induces hardening of cell walls in the growing root tips (Neumann et al. 1994).

For soybean, although lignin content was not affected by 50 and 100 mM of NaCl treatments, it increased by 72 and 90% after treatment with 150 and 200 mM of NaCl, respectively, in comparison with the control (Neves et al. 2010). Saline stress retarded primary xylem differentiation in soybean roots supplemented with 120 mM NaCl. Interestingly, the appearance of both proto- and metaxylem in the stressed roots at day 8 was similar to that in the 3-day-old seedlings rather than to that in the control roots at day 8 (Hilal et al. 1998).

For cassava (*Cassava esculenta* Crantz) cultivars (SC124, F01, SC205, K50 and C4), it was shown that such species grew better in the presence of 10 mM NaCl. Volume and surface area of fresh fibrous rootlets of three cassava cultivars (SC124, F01 and C4) increased at 10 mM NaCl. However, the growth of fibrous rootlets of cultivars was inhibited with further increase in NaCl concentration, starting at 20 mM depending on the cultivar, to a point that all five cassava cultivars could not grow at 100 mM NaCl (Cheng et al. 2018).

Changes in root architecture have been considered a response of plant species to overcome salinity problems (Osmont et al. 2007). These changes may occur through hydrotropism resulting from osmotic stress caused by excess salts on the surface layer, leading to root growth towards areas with greater water availability (Galvan-Ampudia et al. 2013; Li et al. 2015). However, the difference in root growth by altering root architecture may occur in response to sodium, which was termed as halotropism by Galvan-Ampudia et al. (2013).

As a plastic characteristic, root architecture may differ in response to environmental conditions (Osmont et al. 2007). There is evidence that abscisic acid (ABA) influences changes in root architecture (Deak and Malamy 2005). However, the halotropism observed in the roots points to auxin as a necessary phytohormone for triggering the response (Galvan-Ampudia et al. 2013).

Even in plants considered relatively tolerant to salinity and sodicity such as cotton (*Gossypium hirsutum* L.), the presence of Na reduces root diameter and affects root growth at the cellular level (reduction in stele cell area), leading to changes in root morphology (Dodd et al. 2010). The reduction in cotton crop root diameter by moderate salinity (25 to 100 mM NaCl) had already been noted by Kurth et al. (1986). However, the same authors pointed out that the addition of 10 mM Ca^{2+} to the medium improved root length and root mass even in the presence of moderate Na.

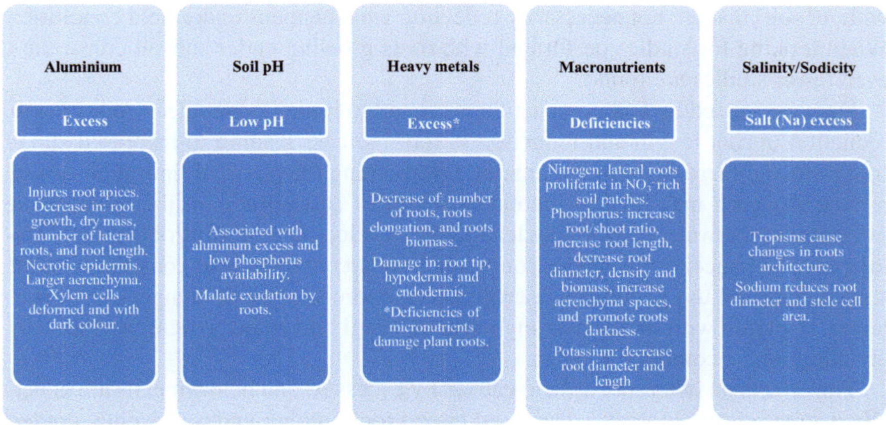

Fig. 10.3 The main effects of the soil chemical constraints on the root system

An interesting strategy developed to decrease the toxic effects caused by high salinity on plant growth is the use of plant growth-promoting bacteria as such beneficial microorganisms colonize the rhizosphere/endorhizosphere of plants and promote plant growth (Shrivastava and Kumar 2015). Wheat plants inoculated with bacterial isolate plants showed increase in root length and biomass under salinity stress (Tiwari et al. 2011).

Another interesting approach to the root system of plants growing in salt-affected soils concerns the phytoremediation potential of some plant species. The release of CO_2 and H^+ in the rhizosphere of nitrogen-fixing legumes, such as alfalfa, contributes to phytoremediation of calcareous sodic soils (Qadir et al. 2003). The release of CO_2 and H^+ in the rhizosphere favours the solubilization of Ca^{2+} in these soils. There is a replacement of Na^+ adsorbed in the colloid by Ca^{2+}, leaving Na^+ free in the soil solution for leaching (Qadir et al. 2003).

Figure 10.3 below summarizes the main general effects regarding the soil chemical constraints on the root system.

10.5 Summary

More research in the last two decades addressed the root system of crops, the so-called "hidden half" of plants. However, research has been uneven with more attention paid to some cultivated species than to others. For example, there are few studies on species such as sunflower, rapeseed and mustard. In addition, considering the limitations displayed by soils, root studies usually emphasize aspects of soil physics, as well as Al excess and NPK deficiency. Detailed studies on other soil restrictions such as heavy metal contamination and salinity/sodicity are scarce. Moreover, many studies on roots are performed under controlled conditions (with or without soil) that are not necessarily reflecting what happens under field conditions. When looking for studies performed with roots growing under subsoil constraints, even fewer studies are found.

Damage caused by soil compaction is related to limitations to deep root growth, reduction in root length and reduction in root biomass leading to increased shoot/root ratio and increase in root diameter. Overcoming the physical limitation caused by waterlogging may lead to the development of aerenchyma in the roots, while overcoming the water deficit may lead to the development of thin roots in drought-tolerant plant species. Strategies for plants to overcome physical constraints in the soils include the use genotypes or crop species more tolerant of compaction, inoculation of plants with arbuscular mycorrhizal fungi and management practices for breaking up the compacted layers.

The effects on plant root system caused by excess Al and heavy metals are generally decreased root biomass, decreased lateral root number and root length, epidermis necrosis, darkening and cell deformation. These effects are similar as those observed in soil with micronutrient deficiency to which increased root diameter and reduced lignin are added. As a result of macronutrient deficiency in the soil, lateral plant roots proliferate in nutrient-rich soil patches in response to N or P deficiency. Additionally, P deficiency leads to increases in root/shoot ratio, root length and root aerenchyma spaces, but decreases root diameter, density and biomass. Potassium deficiency reduces root length and diameter. Release of exudates into the rhizosphere and association with arbuscular mycorrhizal fungi and organisms such as enterobacteria are among the strategies for overcoming problems caused by excess Al and heavy metals and soils with poor amounts of macro- and/or micronutrients, but more research is needed on their effects on subsoil roots.

Ideal phenotypes of the root system are sought to ensure the maximum acquisition of water and nutrients by plants. In the case of phosphorus, the higher phosphorus availability on the soil surface means that the ideal phenotype has a higher number of roots in the surface soil layer. By contrast, the ideal phenotype of the root system for greater nitrogen acquisition is also related to roots that reach the subsoil more efficiently.

Salinity and sodicity also affect the root architecture with more roots being produced in areas with lower salt concentrations. Such changes occur due to hydrotropism whereby roots grow towards soil with greater water availability or due to halotropism where roots grow to escape from higher Na content in soil. However, these tropisms were investigated in only a few species and not always with plants in direct contact with the soil. Some plant species, such as N-fixing leguminous, make use of phytoremediation strategies by releasing CO_2 and H^+ in the rhizosphere which favours Ca solubilization in calcareous soils. Solubilized Ca promotes Na replacement from soil exchange complex, releasing the Na^+ to be leached. Such strategy, along with others mentioned in this chapter, represents a potential for the root system of plants to minimize constraints found not only in topsoil but also in the subsoil.

Although there is evidence that root systems respond to soil constraints as a way to overcome soil stresses, studies on root system response are still limited to a few species. There is still little information on the development of the root system of plant species in subsoils. Regarding the phenotypes considered ideal for enabling plant development under subsoil stresses, many of them still need to be confirmed under field conditions.

References

Abel GJ, Barakat B, Samir KC, Lutz W. Meeting the sustainable development goals leads to lower world population growth. Proc Natl Acad Sci. 2016;113:14294–9. https://doi.org/10.1073/pnas.1611386113.

Alam F, Kim TY, Kim SY, et al. Effect of molybdenum on nodulation, plant yield and nitrogen uptake in hairy vetch (*Vicia villosa* Roth). Soil Sci Plant Nutr. 2015;61:664–75. https://doi.org/10.1080/00380768.2015.1030690.

Alhendawi RA, Römheld V, Kirkby EA, et al. Influence of increasing bicarbonate concentrations on plant growth, organic acid accumulation in roots and iron uptake by barley, sorghum, and maize. J Plant Nutr. 1997;20:1731–53. https://doi.org/10.1080/01904169709365371.

Alloway BJ. Heavy metals and metalloids as micronutrients for plants and animals. In: Alloway BJ, editor. Heavy metals in soils. Environmental pollution, Vol 22. The Netherlands, Dordrecht: Springer; 2013. p. 195–209.

Arduini I, Godbold DL, Onnis A. Influence of copper on root growth and morphology of *Pinus pinea* L. and *Pinus pinaster* Ait. seedlings. Tree Physiol. 1995;15:411–5. https://doi.org/10.1093/treephys/15.6.411.

Ashraf M, Ahmad MSA, Ozturk M. Plant adaptation and phytoremediation. details missing. 2010.

Bailey-Serres J, Colmer TD. Plant tolerance of flooding stress - recent advances. Plant Cell Environ. 2014; https://doi.org/10.1111/pce.12420.

Batista MF, Moscheta IS, Bonato CM, et al. Aluminum in corn plants: influence on growth and morpho-anatomy of root and leaf. Rev Bras Ciência do Solo. 2013;37:177–87. https://doi.org/10.1590/S0100-06832013000100018.

Baycu G. Phytochelatin biosynthesis and cadmium detoxification. J Cell Mol Biol. 2002;1:45–55.

Bayhan Y, Kayisoglu B, Gonulol E. Effect of soil compaction on sunflower growth. Soil Tillage Res. 2002;68:31–8. https://doi.org/10.1016/S0167-1987(02)00078-8.

Bell RW, Dell B. Micronutrient for sustainable food, feed. Paris, France: Fibre and Bioenergy Production International Fertilizer Industry Association (IFA); 2008.

Bouillet J-P, Laclau J-P, Arnaud M, et al. Changes with age in the spatial distribution of roots of *Eucalyptus* clone in Congo. Ecol Manage. 2002;171:43–57. https://doi.org/10.1016/S0378-1127(02)00460-7.

Brian F, Helena L. The nutritional control of root development. Plant Soil. 2001;232:51–68. https://doi.org/10.1023/A:1010329902165.

Broadley M, Brown P, Cakmak I, et al. Function of nutrients. In: Marschner's mineral nutrition of higher plants. Elsevier; 2012a. p. 191–248.

Broadley M, Brown P, Cakmak I, et al. Beneficial elements. In: Marschner's mineral nutrition of higher plants. Elsevier; 2012b. p. 249–69.

Broadley MR, White PJ, Hammond JP, et al. Zinc in plants: Tansley review. New Phytol. 2007;173:677–702. https://doi.org/10.1111/j.1469-8137.2007.01996.x.

Brodrick SJ, Giller KE. Root nodules of Phaseolus: efficient scavengers of molybdenum for N2-fixation. J Exp Bot. 1991;42:679–86. https://doi.org/10.1093/jxb/42.5.679.

Brown JC, Von Jolley D. Strategy I and strategy II mechanisms affecting iron availability to plants may be established too narrow or limited. J Plant Nutr. 1988;11:1077–98. https://doi.org/10.1080/01904168809363870.

Brune A, Dietz KJ. A Comparative analysis of element composition of roots and leaves of barley seedlings grown in the presence of toxic cadmium, molybdenum, nickel, and zinc concentrations. J Plant Nutr. 1995;18:853–68. https://doi.org/10.1080/01904169509364943.
Bueno González M. Adaptation of halophytes to different habitats. In: Seed dormancy and germination [working title]. IntechOpen; 2019. p. 0–23.
Burdman S, Kigel J, Okon Y. Effects of *Azospirillum brasilense* on nodulation and growth of common bean (*Phaseolus vulgaris* L.). Soil Biol Biochem. 1997;29:923–9. https://doi.org/10.1016/S0038-0717(96)00222-2.
Butterly C, Amado T, Tang C. Chapter 3. Subsoil constraints - soil acidity and acidification. In: Oliveira T, Bell R, editors. Subsoil constraints for crop production. Cham: Springer; 2022. p. 53–91.
Byrt CS, Munns R, Burton RA, et al. Root cell wall solutions for crop plants in saline soils. Plant Sci. 2018;269:47–55. https://doi.org/10.1016/j.plantsci.2017.12.012.
Cachorro P, Ortiz A, Barceló AR, et al. Lignin deposition in vascular tissues of Phaseolus vulgaris roots in response to salt stress. Phyton (B Aires). 1993;33:33–40.
Cai H, Ma W, Zhang X, et al. Effect of subsoil tillage depth on nutrient accumulation, root distribution, and grain yield in spring maize. Crop J. 2014;2:297–307. https://doi.org/10.1016/j.cj.2014.04.006.
Cardoso EG, Zotarelli L, Piccinin JL, et al. Sistema radicular da soja em função da compactação do solo no sistema de plantio direto. Pesqui Agropecuária Bras. 2006;41:493–501. https://doi.org/10.1590/S0100-204X2006000300017.
Carmeis Filho A, Crusciol C, Nascente A, et al. Influence of potassium levels on root growth and nutrient uptake of upland rice cultivars. Rev Caatinga. 2017;30:32–44. https://doi.org/10.1590/1983-21252017v30n104rc.
Castro ASF, Moro MF, Menezes MOT. O complexo vegetacional da zona litorânea no Ceará: Pecém, São Gonçalo do Amarante. Acta Bot Brasilica. 2012;26:108–24. https://doi.org/10.1590/S0102-33062012000100013.
Chen R, Shen R, Gu P, et al. Response of rice (*Oryza sativa*) with root surface iron plaque under aluminium stress. Ann Bot. 2006;98:389–95. https://doi.org/10.1093/aob/mcl110.
Chen G, Weil RR. Penetration of cover crop roots through compacted soils. Plant Soil. 2010;331:31–43. https://doi.org/10.1007/s11104-009-0223-7.
Chen G, Weil RR. Root growth and yield of maize as affected by soil compaction and cover crops. Soil Tillage Res. 2011;117:17–27. https://doi.org/10.1016/j.still.2011.08.001.
Cheng Y-E, Dong M-Y, Fan X-W, et al. A study on cassava tolerance to and growth responses under salt stress. Environ Exp Bot. 2018;155:429–40. https://doi.org/10.1016/j.envexpbot.2018.07.022.
Cheng Y, Jiang Y, Wu Y, et al. Soil nitrogen status modifies rice root response to nematode-bacteria interactions in the rhizosphere. PLoS One. 2016;11:e0148021. https://doi.org/10.1371/journal.pone.0148021.
Clune TS, Copeland L. Effects of aluminium on canola roots. Plant Soil. 1999;216:27–33. https://doi.org/10.1023/A:1004789014255.
Cobbett C, Goldsbrough P. Phytochelatins and metallothioneins: Roles in heavy metal detoxification and homeostasis. Annu Rev Plant Biol. 2002;53:159–82. https://doi.org/10.1146/annurev.arplant.53.100301.135154.
Costa MCG. Distribuição e crescimento radicular em soqueiras de cana-de-açúcar: dois cultivares em solos com características distintas. Universidade de São Paulo; 2005.
Costa MCG, Mazza JA, Vitti GC, et al. Distribuição radicular, estado nutricional e produção de colmos e de açúcar em soqueiras de dois cultivares de cana-de-açúcar em solos distintos. Rev Bras Ciência do Solo. 2007;31:1503–14. https://doi.org/10.1590/S0100-06832007000600027.
Coutinho L. Biomas brasileiros. São Paulo, Brazil: Oficina de Textos; 2016.
Covre AM, Partelli FL, Gontijo I, et al. Distribuição do sistema radicular de cafeeiro conilon irrigado e não irrigado. Pesqui Agropecuária Bras. 2015;50:1006–16. https://doi.org/10.1590/S0100-204X2015001100003.

da Jesus DS, Azevedo Neto AD. Aluminum tolerance in sunflower plants is associated with phosphorus content in the toots. Commun Soil Sci Plant Anal. 2013;44:3423–30. https://doi.org/10.1080/00103624.2013.847453.
de Azevedo MCB, Chopart JL, de Medina C. Sugarcane root length density and distribution from root intersection counting on a trench-profile. Sci Agric. 2011;68:94–101. https://doi.org/10.1590/S0103-90162011000100014.
de Mello SLM, de Gonçalves JLM, Gava JL. Pre- and post-harvest fine root growth in *Eucalyptus grandis* stands installed in sandy and loamy soils. For Ecol Manage. 2007;246:186–95. https://doi.org/10.1016/j.foreco.2007.03.060.
de Valadão FCA, dos Weber OLS, Valadão Júnior DD, et al. Aubaçao fosfatada e compactação do solo: Sistema radicular da soja e do milho e atributos físicos do solo. Rev Bras Ciência do Solo. 2015;39:243–55. https://doi.org/10.1590/01000683rbcs20150144.
Dasgupta T, Hossain SA, Meharg AA, et al. An arsenate tolerance gene on chromosome 6 of rice. New Phytol. 2004;163:45–9. https://doi.org/10.1111/j.1469-8137.2004.01109.x.
Deak KI, Malamy J. Osmotic regulation of root system architecture. Plant J. 2005;43:17–28. https://doi.org/10.1111/j.1365-313X.2005.02425.x.
Dias LPR, Gatiboni LC, Brunetto G, et al. Distribuição e morfologia do sistema radicular de Eucalyptus dunnii em resposta à aplicação de fósforo. Rev Ciências Agroveterinárias. 2017;16:203–13. https://doi.org/10.5965/223811711632017203.
Dodd K, Guppy C, Lockwood P, et al. The effect of sodicity on cotton: Plant response to solutions containing high sodium concentrations. Plant Soil. 2010;330:239–49. https://doi.org/10.1007/s11104-009-0196-6.
Drew MC. Comparison of the effects of a localized supply of phosphate, nitrate, ammonium and potassium on the growth of the seminal root system and the shoot in barley. New Phytol. 1975;75:479–90. https://doi.org/10.1111/j.1469-8137.1975.tb01409.x.
Drew MC, Saker LR, Ashley TW. Nutrient supply and the growth of the seminal root system in barley: I. The effect of nitrate concentration on the growth of axes and laterals. J Exp Bot. 1973;24:1189–202. https://doi.org/10.1093/jxb/24.6.1189.
Dunbabin V, Diggle A, Rengel Z. Is there an optimal root architecture for nitrate capture in leaching environments? Plant Cell Environ. 2003;26:835–44. https://doi.org/10.1046/j.1365-3040.2003.01015.x.
Eamus D, Chen X, Kelley G, et al. Root biomass and root fractal analyses of an open *Eucalyptus* forest in a savanna of north Australia. Aust J Bot. 2002;50:31. https://doi.org/10.1071/BT01054.
Eaton S. Effects of phosphorus deficiency on growth and metabolism of sunflower. Bot Gaz. 1949;110:449–64.
Eaton S. Effect of phosphorus deficiency on growth and metabolism of black mustard. Bot Gaz. 1952;113:301–9.
Ehsan S, Ali S, Noureen S, et al. Citric acid assisted phytoremediation of cadmium by *Brassica napus* L. Ecotoxicol Environ Saf. 2014;106:164–72. https://doi.org/10.1016/j.ecoenv.2014.03.007.
Fageria NK, Baligar C, Clark RB. Micronutrients in crop production. Elsevier Inc.; 2002.
Fageria NK, Baligar VC, Wright RJ. Iron nutrition of plants: an overview on the chemistry and physiology of its deficiency and toxicity. Pesqui Agropecuária Bras. 1990;25:553–70.
Fageria NK, Rabelo NA. Tolerance of rice cultivars to iron toxicity. J Plant Nutr. 1987;10:653–61. https://doi.org/10.1080/01904168709363598.
Fan J, McConkey B, Wang H, et al. Root distribution by depth for temperate agricultural crops. F Crop Res. 2016;189:68–74. https://doi.org/10.1016/j.fcr.2016.02.013.
Fernandez MC, Rubio G. Root morphological traits related to phosphorus-uptake efficiency of soybean, sunflower, and maize. J Plant Nutr Soil Sci. 2015;178:807–15. https://doi.org/10.1002/jpln.201500155.
Fletcher RS, Mullen JL, Heiliger A, et al. QTL analysis of root morphology, flowering time, and yield reveals trade-offs in response to drought in *Brassica napus*. J Exp Bot. 2015;66:245–56. https://doi.org/10.1093/jxb/eru423.

Foehse D, Jungk A. Influence of phosphate and nitrate supply on root hair formation of rape, spinach and tomato plants. Plant Soil. 1983;74:359–68. https://doi.org/10.1007/BF02181353.

Foy CD. Soil chemical factors limiting plant root growth. In: Hatfield JL, Stewart BA, editors. Limitations to plant root growth. Advances in soil science. New York: Springer; 1992. p. 97–149. https://doi.org/10.1007/978-1-4612-2894-3_5.

Gad N. Increasing the efficiency of nitrogen fertilization through cobalt application to pea plant. Res J Agric Biol Sci. 2006;2:433–42.

Galvan-Ampudia CS, Julkowska MM, Darwish E, et al. Halotropism is a response of plant roots to avoid a saline environment. Curr Biol. 2013;23:2044–50. https://doi.org/10.1016/j.cub.2013.08.042.

Gao Y, Duan A, Qiu X, et al. Distribution of roots and root length density in a maize/soybean strip intercropping system. Agric Water Manag. 2010;98:199–212. https://doi.org/10.1016/j.agwat.2010.08.021.

Garg N, Singla P. Arsenic toxicity in crop plants: physiological effects and tolerance mechanisms. Environ Chem Lett. 2011;9:303–21. https://doi.org/10.1007/s10311-011-0313-7.

Gehrke V. Distribution of absorbing roots of coffee (*Coffea arabica* L.) and rubber (*Hevea brasiliensis* Muell Arg.) in mixed plantings in two ecological zones of Costa Rica. Tropical Center of Research and Graduate Training Turrialba; 1962.

Goodman A, Ennos A. The effects of soil bulk density on the morphology and anchorage mechanics of the root systems of sunflower and maize. Ann Bot. 1999;83:293–302. https://doi.org/10.1006/anbo.1998.0822.

Gopal R, Rizvi AH. Excess lead alters growth, metabolism and translocation of certain nutrients in radish. Chemosphere. 2008;70:1539–44. https://doi.org/10.1016/j.chemosphere.2007.08.043.

Grohmann F, Queiroz Neto J. Efeito da compactação artificial de dois solos limo-argilosos sobre a penetração das raízes de arroz. Bragantia. 1966;25:422–31.

Grzesiak S, Grzesiak MT, Hura T, et al. Changes in root system structure, leaf water potential and gas exchange of maize and triticale seedlings affected by soil compaction. Environ Exp Bot. 2013;88:2–10. https://doi.org/10.1016/j.envexpbot.2012.01.010.

Guo W-Q, Zhang P-T, Li C-H, et al. Recovery of root growth and physiological characters in cotton after salt stress relief. Chil J Agric Res. 2015;75:85–91. https://doi.org/10.4067/S0718-58392015000100012.

Gupta UC. Symptoms of molybdenum deficiency and toxicity in crops. In: Gupta UC, editor. Molybdenum in Agriculture. Cambridge: Cambridge University Press; 1997. p. 160–70.

Gupta B, Pathak GC, Pandey N. Induction of oxidative stress and antioxidant responses in *Vigna mungo* by zinc stress. Russ J Plant Physiol. 2011;58:85–91. https://doi.org/10.1134/S1021443711010079.

Halwart M, Gupta M. Culture of fish in rice fields. Malaysia, Penang: FAO; 2004.

Hamblin A, Hamblin J. Root characteristics of some temperate legume species and varieties on deep, free-draining entisols. Aust J Agric Res. 1985;36:63. https://doi.org/10.1071/AR9850063.

Henry A, Gowda VRP, Torres RO, et al. Variation in root system architecture and drought response in rice (*Oryza sativa*): Phenotyping of the *Oryza*SNP panel in rainfed lowland fields. Field Crop Res. 2011;120:205–14. https://doi.org/10.1016/j.fcr.2010.10.003.

Hilal M, Zenoff AM, Ponessa G, et al. Saline stress alters the temporal patterns of xylem differentiation and alternative oxidase expression in developing soybean roots. Plant Physiol. 1998;117:695–701. https://doi.org/10.1104/pp.117.2.695.

Hodge A. The plastic plant: root responses to heterogeneous supplies of nutrients. New Phytol. 2004;162:9–24. https://doi.org/10.1111/j.1469-8137.2004.01015.x.

Imtiyaz S, Agnihotri RK, Ahmad S, et al. Effect of cobalt and lead induced heavy metal stress on some physiological parameters in *Glycine max*. Int J Agric Crop Sci. 2014;7:26–34.

Inforzato R. Distribuição do sistema radicular da cana de açúcar var. 290, em solo tipo terra-roxa-legítima. Bragantia. 1957;16:1–13.

Inforzato R. Estudo do sistema radicular da soja (Glycine max (L.) Merril) em solo Latossolo Roxo adubado ou sem adubo. Bragantia. 1969;28:176–80.

Inforzato R, Reis A. Estudo comparativo do sistema radicular dos cafeeiros bourbon amarelo e mundo nôvo. Bragantia. 1963;22:741–50.

Inouhe M. Phytochelatins. Brazilian J Plant Physiol. 2005;17:65–78.

Itoh H, Hayashi S, Nakajima T, et al. Effects of soil type, vertical root distribution and precipitation on grain yield of winter wheat. Plant Prod Sci. 2009;12:503–13. https://doi.org/10.1626/pps.12.503.

Jabbar BKA, Saud HM. Effects of molybdenum on biological nitrogen fixation by combination of *Rhizobium* and *Azospirillum* in soybean under drip irrigation system. Int J Life Sci Biotechnol Pharma Res. 2012;1:63–77.

Jia Y, Yang X, Feng Y, Jilani G. Differential response of root morphology to potassium deficient stress among rice genotypes varying in potassium efficiency. J Zhejiang Univ Sci B. 2008;9:427–34. https://doi.org/10.1631/jzus.B0710636.

Jiang W, Liu D, Liu X. Effects of copper on root growth, cell division, and nucleolus of Zea mays. Biol Plant. 2001;44:105–9.

Jourdan C, Rey H. Architecture and development of the oil-palm *(Elaeis guineensis* Jacq.) root system. Plant Soil. 1997;189:33–48. https://doi.org/10.1023/A:1004290024473.

Kabir ME, Johansen C, Bell RW. Subsoil rhizosphere modification by chickpea under a dry topsoil: Implications for phosphorus acquisition. J Plant Nutr Soil Sci. 2015;178:904–13. https://doi.org/10.1002/jpln.201500021.

Kaiser BN, Gridley KL, Brady JN, et al. The role of molybdenum in agricultural plant production. Ann Bot. 2005;96:745–54. https://doi.org/10.1093/aob/mci226.

Khabaz-Saberi H, Barker SJ, Rengel Z. Tolerance to ion toxicities enhances wheat grain yield in acid soils prone to drought and transient waterlogging. Crop Pasture Sci. 2014;65:862. https://doi.org/10.1071/CP14011.

Khan MR, Khan MM. Effect of varying concentration of nickel and cobalt on the plant growth and yield of chickpea. Aust J Basic Appl Sci. 2010;4:1036–46.

Khoury CK, Bjorkman AD, Dempewolf H, et al. Increasing homogeneity in global food supplies and the implications for food security. Proc Natl Acad Sci. 2014;111:4001–6. https://doi.org/10.1073/pnas.1313490111.

Kirkegaard JA, Lilley JM. Root penetration rate - a benchmark to identify soil and plant limitations to rooting depth in wheat. Aust J Exp Agric. 2007;47:590. https://doi.org/10.1071/EA06071.

Kjellström C, Kirchmann H. The influence of soil volume on root and shoot parameters in container-grown oilseed rape. J Agron Crop Sci. 1992;168:69–71. https://doi.org/10.1111/j.1439-037X.1992.tb00979.x.

Kopittke PM, Asher CJ, Kopittke RA, et al. Toxic effects of Pb^{2+} on growth of cowpea (*Vigna unguiculata*). Environ Pollut. 2007;150:280–7. https://doi.org/10.1016/j.envpol.2007.01.011.

Kristensen H, Thorup-Kristensen K. Effects of vertical distribution of soil inorganic nitrogen on root growth and subsequent nitrogen uptake by field vegetable crops. Soil Use Manag. 2007;23:338–47. https://doi.org/10.1111/j.1473-2743.2007.00105.x.

Kurth E, Cramer GR, Läuchli A, et al. Effects of NaCl and $CaCl_2$ on cell enlargement and cell production in cotton roots. Plant Physiol. 1986;82:1102–6. https://doi.org/10.1104/pp.82.4.1102.

Kvakić M, Pellerin S, Ciais P, et al. Quantifying the limitation to world cereal production due to soil phosphorus status. Global Biogeochem Cycles. 2018;32:143–57. https://doi.org/10.1002/2017GB005754.

Laclau PB, Laclau JP. Growth of the whole root system for a plant crop of sugarcane under rainfed and irrigated environments in Brazil. Field Crop Res. 2009;114:351–60. https://doi.org/10.1016/j.fcr.2009.09.004.

Landsberg EC. Proton efflux and transfer cell formation as responses to Fe deficiency of soybean in nutrient solution culture. Plant Soil. 1989;114:53–61. https://doi.org/10.1007/BF02203081.

Landsberg EC. Transfer cell formation in sugar beet roots induced by latent Fe deficiency. Plant Soil. 1994;165:197–205. https://doi.org/10.1007/BF00008062.

Lane SD, Martin ES. A histochemical investigation of lead uptake in Raphanus sativus. New Phytol. 1977;79:281–6. https://doi.org/10.1111/j.1469-8137.1977.tb02206.x.

Larson JE, Funk JL. Seedling root responses to soil moisture and the identification of a below-ground trait spectrum across three growth forms. New Phytol. 2016;210:827–38. https://doi.org/10.1111/nph.13829.
Lecompte F, Pagès L, Ozier-Lafontaine H. Patterns of variability in the diameter of lateral roots in the banana root system. New Phytol. 2005;167:841–50. https://doi.org/10.1111/j.1469-8137.2005.01457.x.
Li C, Lei J, Zhao Y, et al. Effect of saline water irrigation on soil development and plant growth in the Taklimakan desert highway shelterbelt. Soil Tillage Res. 2015;146:99–107. https://doi.org/10.1016/j.still.2014.03.013.
Li Z, Shuman LM. Heavy metal movement in metal-contaminated soil profiles. Soil Sci. 1996;161:656–66. https://doi.org/10.1097/00010694-199610000-00003.
Li X, Yang Y, Zhang J, et al. Zinc induced phytotoxicity mechanism involved in root growth of *Triticum aestivum* L. Ecotoxicol Environ Saf. 2012;86:198–203. https://doi.org/10.1016/j.ecoenv.2012.09.021.
Li S, Zhou X, Chen J, Chen R. Is there a strategy I iron uptake mechanism in maize? Plant Signal Behav. 2018;13:e1161877. https://doi.org/10.1080/15592324.2016.1161877.
Li S, Zuo Q, Wang X, et al. Characterizing roots and water uptake in a ground cover rice production system. PLoS One. 2017;12:e0180713. https://doi.org/10.1371/journal.pone.0180713.
Liang C, Piñeros MA, Tian J, et al. Low pH, aluminum, and phosphorus coordinately regulate malate exudation through GmALMT1 to improve soybean adaptation to acid soils. Plant Physiol. 2013;161:1347–61. https://doi.org/10.1104/pp.112.208934.
Liao H, Wan H, Shaff J, et al. Phosphorus and aluminum interactions in soybean in relation to aluminum tolerance. Exudation of specific organic acids from different regions of the intact root system. Plant Physiol. 2006;141:674–84. https://doi.org/10.1104/pp.105.076497.
Lisanti S, Hall AJ, Chimenti CA. Influence of water deficit and canopy senescence pattern on *Helianthus annuus* (L.) root functionality during the grain-filling phase. Field Crop Res. 2013;154:1–11. https://doi.org/10.1016/j.fcr.2013.08.009.
Liu L, Gan Y, Bueckert R, et al. Rooting systems of oilseed and pulse crops I: temporal growth patterns across the plant developmental periods. Field Crop Res. 2011a;122:256–63. https://doi.org/10.1016/j.fcr.2011.04.002.
Liu L, Gan Y, Bueckert R, et al. Rooting systems of oilseed and pulse crops. II: vertical distribution patterns across the soil profile. Field Crop Res. 2011b;122:248–55. https://doi.org/10.1016/j.fcr.2011.04.003.
Liu D, Jiang W, Lu C, et al. Effects of copper sulfate on the nucleolus of *Allium cepa* root-tip cells. Hereditas. 1994;120:87–90. https://doi.org/10.1111/j.1601-5223.1994.00087.x.
Liu P, Yang YS, Xu GD, et al. The effect of molybdenum and boron in soil on the growth and photosynthesis of three soybean varieties. Plant Soil Environ. 2005;51:197–205.
Liu D, Zhai L, Jiang W, Wang W. Effects of Mg^{2+}, Co^{2+}, and Hg^{2+} on the nucleus and nucleolus in root tip cells of *Allium cepa*. Bull Environ Contam Toxicol. 1995;55:779–87. https://doi.org/10.1007/BF00203767.
Liu W, Zhang X, Liang L, et al. Phytochelatin and oxidative stress under heavy metal stress tolerance in plants. In: Reactive oxygen species and oxidative damage in plants under stress. Cham: Springer; 2015. p. 191–217.
Lynch J. Root architecture and plant productivity. Plant Physiol. 1995;109:7–13.
Lynch JP. Root phenes for enhanced soil exploration and phosphorus acquisition: tools for future crops. Plant Physiol. 2011;156:1041–9. https://doi.org/10.1104/pp.111.175414.
Lynch JP. Steep, cheap and deep: An ideotype to optimize water and N acquisition by maize root systems. Ann Bot. 2013;112:347–57. https://doi.org/10.1093/aob/mcs293.
Lynch JP, Wojciechowski T. Opportunities and challenges in the subsoil: pathways to deeper rooted crops. J Exp Bot. 2015;66:2199–210. https://doi.org/10.1093/jxb/eru508.
Ma Q, Bell R, Mattiello E. Chapter 12 Nutrient acquisition with particular reference to subsoil constraints. In: Oliveira T, Bell R, editors. Subsoil constraints for crop production. Cham: Springer; 2022. p. 289–321.

Mahender A, Swamy BPM, Anandan A, et al. Tolerance of iron-deficient and -toxic soil conditions in rice. Plants. 2019;8 https://doi.org/10.3390/plants8020031.

Manzur ME, Hall AJ, Chimenti CA. Root lodging tolerance in *Helianthus annuus* (L.): associations with morphological and mechanical attributes of roots. Plant Soil. 2014;381:71–83. https://doi.org/10.1007/s11104-014-2122-9.

Mayaki WC, Teare ID, Stone LR. Top and root growth of irrigated and nonirrigated soybeans. Crop Sci. 1976;16:92. https://doi.org/10.2135/cropsci1976.0011183X001600010023x.

Mengel DB, Barber SA. Development and distribution of the corn root system under field conditions. Agron J. 1974;66:341. https://doi.org/10.2134/agronj1974.00021962006600030002x.

Mertens J, Smolders E. Zinc. In: Alloway BJ, editor. Heavy metals in soils: trace metals and metalloids in soils and their bioavailability. Dordrecht: Springer; 2013. p. 465–93.

Micó C, Li HF, Zhao FJ, et al. Use of Co speciation and soil properties to explain variation in Co toxicity to root growth of barley (*Hordeum vulgare* L.) in different soils. Environ Pollut. 2008;156:883–90. https://doi.org/10.1016/j.envpol.2008.05.017.

Miotti AA, Costa MCG, Ferreira TO, et al. Profundidade e atributos físicos do solo e seus impactos nas raízes de bananeiras. Rev Bras Frutic. 2013;35:536–45. https://doi.org/10.1590/S0100-29452013000200024.

Miransari M, Bahrami HA, Rejali F, et al. Using arbuscular mycorrhiza to reduce the stressful effects of soil compaction on corn (*Zea mays* L.) growth. Soil Biol Biochem. 2007;39:2014–26. https://doi.org/10.1016/j.soilbio.2007.02.017.

Miteva E. Accumulation and effect of arsenic on tomatoes. Commun Soil Sci Plant Anal. 2002;33:1917–26. https://doi.org/10.1081/CSS-120004832.

Mithöfer A, Schulze B, Boland W. Biotic and heavy metal stress response in plants: evidence for common signals. FEBS Lett. 2004;566:1–5. https://doi.org/10.1016/j.febslet.2004.04.011.

Mollier A, Pellerin S. Maize root system growth and development as influenced by phosphorus deficiency. J Exp Bot. 1999;50:487–97. https://doi.org/10.1093/jxb/50.333.487.

Morita S, Abe J. Development of root systems in wheat and rice. In: Ito O, Johansen C, Adu-Gyamfi J, editors. Roots and nitrogen in cropping systems of the semi-arid tropics. Andhra Pradesh: Japan International Research Center for Agricultural Sciences - International Agricultural Series No.3; 1996. p. 185–98.

Moroni MT, Worledge D, Beadle CL. Root distribution of *Eucalyptus nitens* and *E. globulus* in irrigated and droughted soil. For Ecol Manage. 2003;177:399–407. https://doi.org/10.1016/S0378-1127(02)00410-3.

Motta ACV, Nick JA, Yorinori GT, et al. Distribuição horizontal e vertical da fertilidade do solo e das raízes de cafeeiro (*Coffea arabica* L.) cultivar Catuaí. Acta Sci Agron. 2006:28. https://doi.org/10.4025/actasciagron.v28i4.758.

Mousavi Kouhi SM, Lahouti M, Ganjeali A, et al. Anatomical and ultrastructural responses of *Brassica napus* after long-term exposure to excess zinc. Turkish J Biol. 2016;40:652–60. https://doi.org/10.3906/biy-1411-13.

Mühling KH, Läuchli E. Physiological traits of sodium toxicity and salt tolerance. In: Horst WJ, Schenk MK, Bükert A, et al., editors. Plant nutrition. Netherlands, Dordrecht: Springer; 2001. p. 378–9.

Muñoz-Romero V, Benítez-Vega J, López-Bellido L, et al. Monitoring wheat root development in a rainfed vertisol: tillage effect. Eur J Agron. 2010;33:182–7. https://doi.org/10.1016/j.eja.2010.05.004.

Nable RO, Webb MJ. Further evidence that zinc is required throughout the root zone for optimal plant growth and development. Plant Soil. 1993;150:247–53. https://doi.org/10.1007/BF00013021.

Nanda AK, Wissuwa M. Rapid crown root development confers tolerance to zinc deficiency in rice. Front Plant Sci. 2016;7:1–12. https://doi.org/10.3389/fpls.2016.00428.

Neumann PM, Azaizeh H, Leon D. Hardening of root cell walls: a growth inhibitory response to salinity stress. Plant Cell Environ. 1994;17:303–9. https://doi.org/10.1111/j.1365-3040.1994.tb00296.x.

Neves GYS, Marchiosi R, Ferrarese MLL, et al. Root growth inhibition and lignification induced by salt stress in soybean. J Agron Crop Sci. 2010;196:467–73. https://doi.org/10.1111/j.1439-037X.2010.00432.x.

Nodichao L, Chopart J-L, Roupsard O, et al. Genotypic variability of oil palm root system distribution in the field. Consequences for water uptake. Plant Soil. 2011;341:505–20. https://doi.org/10.1007/s11104-010-0663-0.

Nunes MR, Denardin JE, Pauletto EA, et al. Effect of soil chiseling on soil structure and root growth for a clayey soil under no-tillage. Geoderma. 2015;259–260:149–55. https://doi.org/10.1016/j.geoderma.2015.06.003.

Oliveira TS, Fernandes RBA. Chapter 6: Physical subsoil constraints of agricultural and forestry land. In: Oliveira T, Bell R, editors. Subsoil constraints for crop production. Cham: Springer; 2022. p. 125–160.

Osmont KS, Sibout R, Hardtke CS. Hidden branches: Developments in root system architecture. Annu Rev Plant Biol. 2007;58:93–113. https://doi.org/10.1146/annurev.arplant.58.032806.104006.

Otto R, Silva AP, Franco HCJ, et al. High soil penetration resistance reduces sugarcane root system development. Soil Tillage Res. 2011;117:201–10. https://doi.org/10.1016/j.still.2011.10.005.

Ozanne P, Greenwood E, Shaw T. The cobalt requirement of subterranean clover in the field. Aust J Agric Res. 1963;14:39. https://doi.org/10.1071/AR9630039.

Palit S, Sharma A, Talukder G. Effects of cobalt on plants. Bot Rev. 1994;60:149–81. https://doi.org/10.1007/BF02856575.

Partelli FL, Covre AM, Oliveira MG, et al. Root system distribution and yield of "Conilon" coffee propagated by seeds or cuttings. Pesqui Agropecuária Bras. 2014;49:349–55. https://doi.org/10.1590/S0100-204X2014000500004.

Paul PLC, Bell RW, Barret-Lennard EG, et al. Impact of rice straw mulch on soil physical properties, sunflower root distribution and yield in a salt-affected clay-textured soil. Agriculture. 2021;11(264):1–17. https://doi.org/10.3390/agriculture11030264.

Peltonen-Sainio P, Jauhiainen L, Laitinen P, et al. Identifying difficulties in rapeseed root penetration in farmers' fields in Northern European conditions. Soil Use Manag. 2011;27:229–37. https://doi.org/10.1111/j.1475-2743.2011.00331.x.

Péret B, Clément M, Nussaume L, Desnos T. Root developmental adaptation to phosphate starvation: Better safe than sorry. Trends Plant Sci. 2011;16:442–50. https://doi.org/10.1016/j.tplants.2011.05.006.

Péret B, Desnos T, Jost R, et al. Root architecture responses: in search of phosphate. Plant Physiol. 2014;166:1713–23. https://doi.org/10.1104/pp.114.244541.

Pierret A, Maeght JL, Clément C, et al. Understanding deep roots and their functions in ecosystems: an advocacy for more unconventional research. Ann Bot-London. 2016;118:621–35. https://doi.org/10.1093/aob/mcw130.

Pilon-Smits EA, Quinn CF, Tapken W, et al. Physiological functions of beneficial elements. Curr Opin Plant Biol. 2009;12:267–74. https://doi.org/10.1016/j.pbi.2009.04.009.

Pourrut B, Shahid M, Dumat C, et al. Lead uptake, toxicity, and detoxification in plants. In: Whitacre D, editor. Reviews of environmental contamination and toxicology, Vol. 213. Springer; 2011. p. 113–36.

Przymusiński R, Gwóźdź EA. Increased accumulation of the 16 × 103Mr polypeptide in lupin roots exposed to lead, copper and nitrite ions. Environ Exp Bot. 1994;34:63–8. https://doi.org/10.1016/0098-8472(94)90010-8.

Qadir M, Steffens D, Yan F, Schubert S. Proton release by N2-fixing plant roots: a possible contribution to phytoremediation of calcareous sodic soils. J Plant Nutr Soil Sci. 2003;166:14–22. https://doi.org/10.1002/jpln.200390007.

Qin SY, Sun XC, Hu CX, et al. Uptake, transport and distribution of molybdenum in two oilseed rape (*Brassica napus* L.) cultivars under different nitrate/ammonium ratios. J Zhejiang Univ Sci B. 2017;18:512–21. https://doi.org/10.1631/jzus.B1600249.

Raper CD, Barber SA. Rooting systems of soybeans. I differences in root morphology among varieties. Agron J. 1970;62:581. https://doi.org/10.2134/agronj1970.00021962006200050009x.
Rengasamy P, Lacerda C, Gheyi H. Chapter 4. Salinity, sodicity and alkalinity. In: Oliveira T, Bell R, editors. Subsoil constraints for crop production. Cham: Springer; 2022. p. 83–107.
Rengel Z, Damon PM. Crops and genotypes differ in efficiency of potassium uptake and use. Physiol Plant. 2008;133:624–36. https://doi.org/10.1111/j.1399-3054.2008.01079.x.
Ribeiro J. Manejo da cultura do girassol no Meio-Norte do Brasil. Embrapa Meio-Norte. Brazil, Teresina; 2008. p. 48.
Ribeiro J, Walter B. As principais fitofisionomias do bioma Cerrado. In: Sano S, Almeida S, Ribeiro J, editors. Ecologia e flora, Vol. 1. Brazil, Brasília: Enbrapa Cerrados; 2008. p. 151–212.
Ricachenevsky FK, Sperotto RA. There and back again, or always there? The evolution of rice combined strategy for Fe uptake. Front Plant Sci. 2014;5:1–5. https://doi.org/10.3389/fpls.2014.00189.
Ronchi CP, de Sousa Júnior JM, de Ameida WL, et al. Morfologia radicular de cultivares de café arábica submetidas a diferentes arranjos espaciais. Pesqui Agropecuária Bras. 2015;50:187–95. https://doi.org/10.1590/S0100-204X2015000300001.
Ronzan M, Piacentini D, Fattorini L, et al. Cadmium and arsenic affect root development in *Oryza sativa* L. negatively interacting with auxin. Environ Exp Bot. 2018;151:64–75. https://doi.org/10.1016/j.envexpbot.2018.04.008.
Sartori GMS, Marchesan E, De David R, et al. Growth and development of soybean roots according to planting management systems and irrigation in lowland areas. Ciência Rural. 2016;46:1572–8. https://doi.org/10.1590/0103-8478cr20151579.
Sato MK, de Lima HV, da Ferreira RLC, et al. Least limiting water range for oil palm production in Amazon region, Brazil. Sci Agric. 2017;74:148–56. https://doi.org/10.1590/1678-992x-2015-0408.
Scanlan C, Holmes K, RW B. Chapter 8. Sand and gravel subsoils. In: Oliveira T, Bell RS, editors. Subsoil Constraints for Crop Production. Cham: Springer; 2022. p. 179–98.
Schützendübel A, Polle A. Plant responses to abiotic stresses: Heavy metal-induced oxidative stress and protection by mycorrhization. J Exp Bot. 2002;53:1351–65. https://doi.org/10.1093/jxb/53.372.1351.
Seiler GJ. Primary and lateral root elongation of sunflower seedlings. Environ Exp Bot. 1994;34:409–18. https://doi.org/10.1016/0098-8472(94)90023-X.
Shahid M, Hameed S, Imran A, et al. Root colonization and growth promotion of sunflower (*Helianthus annuus* L.) by phosphate solubilizing Enterobacter sp. Fs-11. World J Microbiol Biotechnol. 2012;28:2749–58. https://doi.org/10.1007/s11274-012-1086-2.
Shahid M, Pinelli E, Pourrut B, et al. Lead-induced genotoxicity to *Vicia faba* L. roots in relation with metal cell uptake and initial speciation. Ecotoxicol Environ Saf. 2011;74:78–84. https://doi.org/10.1016/j.ecoenv.2010.08.037.
Sharma P, Dubey RS. Lead toxicity in plants. Brazilian J Plant Physiol. 2005;17:35–52. https://doi.org/10.1590/S1677-04202005000100004.
Sharma P, Pantuwan G, Ingram K, et al. Rainfed lowland rice roots: soil and hydrological effects. In: Kirk G, editor. Rice roots: nutrient and water use. Philippines, Manila: International Rice Research Institute; 1994.
Shen H, He LF, Sasaki T, et al. Citrate secretion coupled with the modulation of soybean root tip under aluminum stress. Up-regulation of transcription, translation, and threonine-oriented phosphorylation of plasma membrane H+-ATPase. Plant Physiol. 2005;138:287–96. https://doi.org/10.1104/pp.104.058065.
Shri M, Kumar S, Chakrabarty D, et al. Effect of arsenic on growth, oxidative stress, and antioxidant system in rice seedlings. Ecotoxicol Environ Saf. 2009;72:1102–10. https://doi.org/10.1016/j.ecoenv.2008.09.022.
Shrivastava P, Kumar R. Soil salinity: a serious environmental issue and plant growth promoting bacteria as one of the tools for its alleviation. Saudi J Biol Sci. 2015;22:123–31. https://doi.org/10.1016/j.sjbs.2014.12.001.

Smith SE, Christophersen HM, Pope S, et al. Arsenic uptake and toxicity in plants: integrating mycorrhizal influences. Plant Soil. 2010;327:1–21. https://doi.org/10.1007/s11104-009-0089-8.

Smith DM, Inman-Bamber NG, Thorburn PJ. Growth and function of the sugarcane root system. Field Crop Res. 2005;92:169–83. https://doi.org/10.1016/j.fcr.2005.01.017.

So H, Oliveira T. Chapter 7. Subsoil constraints of mined land and tailings. In: Oliveira T, Bell R, editors. Subsoil constraints for crop production. Cham: Springer; 2022. p. 161–77.

Souza J, Abrahão W. Chapter 5: Pyritic subsoils in acid sulfate soils and similar problems in mined areas with sulfidic rocks. In: Oliveira T, Bell R, editors. Subsoil constraints for crop production. Cham: Springer; 2022. p. 109–23.

Subbarao GV, Ito O, Berry WL, et al. Sodium - A functional plant nutrient. CRC Crit Rev Plant Sci. 2003;22:391–416. https://doi.org/10.1080/07352680390243495.

Suralta RR, Yamauchi A. Root growth, aerenchyma development, and oxygen transport in rice genotypes subjected to drought and waterlogging. Environ Exp Bot. 2008;64:75–82. https://doi.org/10.1016/j.envexpbot.2008.01.004.

Suresh S. Characteristics of soils prone to iron toxicity and management-A review. Agric Rev Res Commun Cent India. 2005;26:50.

Tamás L, Mistrík I, Alemayehu A. Low Cd concentration-activated morphogenic defence responses are inhibited by high Cd concentration-induced toxic superoxide generation in barley root tip. Planta. 2014;239:1003–13. https://doi.org/10.1007/s00425-014-2030-5.

Tangahu BV, Sheikh Abdullah SR, Basri H, et al. A review on heavy metals (As, Pb, and Hg) uptake by plants through phytoremediation. Int J Chem Eng. 2011;2011:1–31. https://doi.org/10.1155/2011/939161.

Tao Y, Zhang Y, Jin X, et al. More rice with less water – evaluation of yield and resource use efficiency in ground cover rice production system with transplanting. Eur J Agron. 2015;68:13–21. https://doi.org/10.1016/j.eja.2015.04.002.

Teixeira L, Furlan J Jr, Oliveira R, Bastos J. Pesquisa sobre o uso de engaço de dendê em dendezais adultos. Brazil, Belém: Embrapa Amazônia Oriental; 2001.

Thompson J. Decline of vesicular-arbuscular mycorrhizae in long fallow disorder of field crops and its expression in phosphorus deficiency of sunflower. Aust J Agric Res. 1987;38:847. https://doi.org/10.1071/AR9870847.

Thorup-Kristensen K, Halberg N, Nicolaisen M, et al. Digging deeper for Agricultural resources, the value of deep rooting. Trends Plant Sci. 2020;25:406–17. https://doi.org/10.1016/j.tplants.2019.12.007.

Tiwari S, Singh P, Tiwari R, et al. Salt-tolerant rhizobacteria-mediated induced tolerance in wheat (*Triticum aestivum*) and chemical diversity in rhizosphere enhance plant growth. Biol Fertil Soils. 2011;47:907–16. https://doi.org/10.1007/s00374-011-0598-5.

Tóth G, Hermann T, Da Silva MR, et al. Heavy metals in agricultural soils of the European union with implications for food safety. Environ Int. 2016;88:299–309. https://doi.org/10.1016/j.envint.2015.12.017.

Trehan SP, Sharma RC. Potassium uptake efficiency of young plants of three potato cultivars as related to root and shoot parameters. Commun Soil Sci Plant Anal. 2002;33:1813–23. https://doi.org/10.1081/CSS-120004824.

Tung G, Temple PJ. Uptake and localization of lead in corn (*Zea mays* L.) seedlings, a study by histochemical and electron microscopy. Sci Total Environ. 1996;188:71–85. https://doi.org/10.1016/0048-9697(96)05156-X.

Uga Y, Ebana K, Abe J, et al. Variation in root morphology and anatomy among accessions of cultivated rice (*Oryza sativa* L.) with different genetic backgrounds. Breed Sci. 2009;59:87–93. https://doi.org/10.1270/jsbbs.59.87.

Ungaro F, Ragazzi F, Cappellin R, et al. Arsenic concentration in the soils of the Brenta Plain (Northern Italy): Mapping the probability of exceeding contamination thresholds. J Geochemical Explor. 2008;96:117–31. https://doi.org/10.1016/j.gexplo.2007.03.006.

Vance WH, Milroy SP. Chapter 13. Water acquisition by roots from the subsoil: impact of physical constraints on the dynamics of water capture. In: Oliveira T, Bell R, editors. Subsoil constraints for crop production. Cham: Springer; 2022. p. 323–45.

Wang X, Zhou B, Sun X, et al. Soil tillage management affects maize grain yield by regulating spatial distribution coordination of roots, soil moisture and nitrogen status. PLoS One. 2015;10:e0129231. https://doi.org/10.1371/journal.pone.0129231.

Waser NM, Chittka L, Price MV, et al. Generalization in pollination systems, and why it matters. Ecology. 1996;77:1043–60. https://doi.org/10.2307/2265575.

Weckx JEJ, Clijsters HMM. Oxidative damage and defense mechanisms in primary leaves of *Phaseolus vulgaris* as a result of root assimilation of toxic amounts of copper. Physiol Plant. 1996;96:506–12. https://doi.org/10.1111/j.1399-3054.1996.tb00465.x.

Widodo JA, Broadley MR, Rose T, et al. Response to zinc deficiency of two rice lines with contrasting tolerance is determined by root growth maintenance and organic acid exudation rates, and not by zinc-transporter activity. New Phytol. 2010;186:400–14. https://doi.org/10.1111/j.1469-8137.2009.03177.x.

Wójcik M, Tukiendorf A. Phytochelatin synthesis and cadmium localization in wild type of *Arabidopsis thaliana*. Plant Growth Regul. 2004;44:71–80.

Yadav SK. Heavy metals toxicity in plants: An overview on the role of glutathione and phytochelatins in heavy metal stress tolerance of plants. South African J Bot. 2010;76:167–79. https://doi.org/10.1016/j.sajb.2009.10.007.

Yang C, Yang L, Yang Y, et al. Rice root growth and nutrient uptake as influenced by organic manure in continuously and alternately flooded paddy soils. Agric Water Manag. 2004;70:67–81. https://doi.org/10.1016/j.agwat.2004.05.003.

Yu P, Li X, White PJ, Li C. A large and deep root system underlies high nitrogen-use efficiency in maize production. PLoS One. 2015;10:1–17. https://doi.org/10.1371/journal.pone.0126293.

Yuan P, Ding G-D, Cai H-M, et al. A novel Brassica–rhizotron system to unravel the dynamic changes in root system architecture of oilseed rape under phosphorus deficiency. Ann Bot. 2016;118:173–84. https://doi.org/10.1093/aob/mcw083.

Zhang H, Brian GF. An Arabidopsis MADS box gene that controls nutrient-Induced changes in root architecture. Science. 1998;279(80):407–9. https://doi.org/10.1126/science.279.5349.407.

Zhang H, Huang Y, Ye X, et al. Genotypic differences in phosphorus acquisition and the rhizosphere properties of *Brassica napus* in response to low phosphorus stress. Plant Soil. 2009;320:91–102. https://doi.org/10.1007/s11104-008-9873-0.

Zhang Y, Liu M, Saiz G, et al. Enhancement of root systems improves productivity and sustainability in water saving ground cover rice production system. F Crop Res. 2017;213:186–93. https://doi.org/10.1016/j.fcr.2017.08.008.

Zhang F, Ma J, Cao Y. Phosphorus deficiency enhances root exudation of low-molecular weight organic acids and utilization of sparingly soluble inorganic phosphates by radish (*Raghanus sativus* L.) and rape (*Brassica napus* L.) plants. Plant Soil. 1997;196:261–4. https://doi.org/10.1023/A:1004214410785.

Chapter 11
Roots and Beneficial Interactions with Soil Microbes

Bruno Coutinho Moreira, Paulo Prates Júnior, Bernard Dell, and Maria Catarina Megumi Kasuya

Abstract Beneficial soil-borne bacteria and fungi are central to the performance of most plants. Knowledge of beneficial microorganisms and the processes in topsoils that favour the association of beneficial organisms with plants allows us to better manage soils for higher productivity and environment sustainability. This review describes the main groups of symbiotic and free-living organisms and explores how they contribute to plant and soil health in managed and natural ecosystems. Many field studies have investigated the biodiversity, ecology and function of beneficial organisms in relation to root distribution in topsoils and land management practices. There is scant information however on whether beneficial bacteria and fungi can persist and enhance root function in subsoils. Opportunities for enhancing beneficial plant-microbe interactions in the subsoil deserve scrutiny particularly as crop productivity is becoming more dependent on subsoil moisture with declines in rainfall in many parts of the world.

Keywords Biological nitrogen fixation · Climate change · Crop nutrition · Mycorrhiza · Rhizosphere · Root function · Soil carbon · Soil water · Subsoil · Topsoil

11.1 Introduction

Healthy plants need well-developed, functional shoots and roots. Roots provide anchorage in soil as well as the provision of essential supplies of water and mineral nutrients for plant growth. Roots contribute directly to socio-environmental

B. C. Moreira
Agricultural Engineering College, Federal University of San Francisco Valley, Petrolina, PE, Brazil
e-mail: bruno.moreira@univasf.edu.br

P. Prates Júnior · B. Dell
Microbiology Department, Federal University of Viçosa, Viçosa, MG, Brazil
e-mail: B.Dell@murdoch.edu.au

M. C. M. Kasuya
Agriculture and Forest Sciences, Murdoch University, Murdoch, WA, Australia
e-mail: mkasuya@ufv.br

T. S. d. Oliveira, R. W. Bell (eds.), *Subsoil Constraints for Crop Production*,
https://doi.org/10.1007/978-3-031-00317-2_11

services, such as soil structure, soil carbon storage and nutrient cycling. By the provision of carbon, roots also help to prime the diversity and biomass of soil microbes (bacteria and fungi), invertebrates and other soil-inhabiting organisms. Interactions between roots and microorganisms are fundamental to ecosystem function and plant physiology. These beneficial microorganisms can live in the rhizosphere, or in the phyllosphere or as endophytes within the plant body. They include mycorrhizal fungi, nitrogen-fixing bacteria, growth-promoting bacteria/fungi and disease-suppressive organisms. In the rhizosphere, microorganisms can solubilize and recycle nutrients in the soil, altering the availability of nutrients to plants (Marschner and Dell 1994; Kirkby et al. 2014; Pii et al. 2015). Microorganisms can also contribute to soil formation, structure and aggregate formation (Rashid et al. 2016).

There is an enormous range in the depth to which roots develop in subsoils, from about 0.5–1.5 m in some annual crop species such as wheat and soybean (Fan et al. 2016) to over 40 m in woody plants in water-limited environments (Dell et al. 1983; Schenk and Jackson 2002; Fan et al. 2017). In general, the length of fine roots declines from the topsoil to deeper soil horizons. Similarly, much of the biodiversity in soils occurs in the topsoil. Hence, microbes and soil fauna dominate soil organic matter (SOM) decomposition in upper soil horizons (Jackson et al. 2017). Microbial biomass declines with soil depth along with SOM and essential macronutrients such as N and P. Therefore, most research on roots and beneficial microorganisms has explored interactions within the top soil or the mixed cultivated horizon in agricultural soils (Roper et al. 2010). As a consequence, possible roles of microorganisms in root development and function in subsoils await discovery.

There is interest in breeding deeper-rooted annual crops particularly for regions experiencing reduced precipitation and more frequent drought events. The development of deep-rooted crops will require knowledge of root traits and environmental limitations of associated beneficial microorganisms. Subsoil constraints that can limit growth of root-associated and other soil fungi and bacteria include reduced porosity, hypoxia, subsoil salinity, acidity, alkalinity, metal toxicity (e.g. Al, Mn) and nutrient deficiency especially in highly weathered soils (mainly Cu, Ca, K, Mn, N, P, S, Mo, Zn and sometimes B and Mg) (Butterly et al. 2022; Costa and Coutinho 2022; Oliveira and Fernandes 2022; Ma et al. 2022; Rengasamy et al. 2022; Schaefer and Oliveira 2022).

Knowledge of beneficial microorganisms and the processes that occur in topsoils which favour the association of microorganisms with plants will allow us to better manage soils for higher productivity and environment sustainability. It also helps us to better explore opportunities for enhancing beneficial plant-microbe interactions in the subsoil in the future. This chapter describes the main beneficial interactions in the rhizosphere and considers opportunities for further research into subsoils in this field.

11.2 Mycorrhizal Associations

There are two main groups of mycorrhizal fungi in crop and wild plants in the world: arbuscular mycorrhizal fungi (AMF) and ectomycorrhizal fungi (EMF). AMF are among the most investigated group of plant growth-promotion microorganisms with particular focus in sustainable agriculture. The mutualistic association between AMF and terrestrial plants is over 400 million years old, and about 80% of terrestrial plants have this kind of symbiosis, including the majority of agricultural, horticultural and forestry species (Pozo and Azcón-Aguilar 2007; Smith and Read 2008). EMF are functionally important for the survival and growth of many trees and woody shrubs in such diverse ecosystems as tropical forests, Mediterranean-climate shrublands and cold temperate coniferous forests. Unlike the AMF, there is a huge diversity of EMF species, and many new taxa are yet to be described especially in the southern hemisphere. Not only is there high taxonomic diversity in EMF worldwide, but the ectomycorrhizal associations exhibit a diverse range of morphologies, size, colour, external appearance and the extent of hyphal penetration of the Hartig net, a network of fungal tissue, into the root. The phylogenetic analysis recently undertaken by Pena et al. (2017) suggests that the ectomycorrhizal symbiosis has arisen independently in around 80 fungal lineages.

Fungal hyphae external to all mycorrhizal roots facilitate exploration of litter and soil environments that would otherwise lie beyond the influence of the rhizosphere. The importance of these hyphae in nutrient acquisition for the host plants has been well documented (Marschner and Dell 1994; Plassard and Dell 2010; Garcia and Zimmermann 2014). The main physiological basis for this interaction is the bidirectional transfer of substances between the mycorrhizas and the colonized plants (Smith et al. 2010). Mycorrhizal roots can exploit extensive soil volumes (Smith and Read 2008), which are estimated to be over 100 times larger than those of non-mycorrhized plants (Sieverding 1991). In addition, the smaller diameters of the hyphae allow AMF, in particular, to have access to micropores in the soil (Grant et al. 2005), and their well-developed networks extend well beyond any P depletion zones in the rhizosphere (Smith et al. 2011), sites which non-mycorrhizal plants are not able to explore (Karagiannidis et al. 2011). The fungi, through their external hyphae, absorb large amounts of nutrients and water from the soil, especially when these compounds are scarce in the environment, and transport them to the host plants (Garcia and Zimmermann 2014). In contrast, plants can provide about 4–20% of produced photo-assimilates to meet the metabolic needs of the fungi (Smith and Read 2008; Smith et al. 2010). More recently, ectomycorrhizal fungal networks have been shown to provide substantial bidirectional carbon exchange between trees in tall forests of spruce, beech, larch and pine (Klein et al. 2016).

11.2.1 Benefits of Mycorrhizal Associations

Mycorrhizal associations provide many benefits to plants, the soil and the ecosystem. The symbiotic relationship between fungi and host plants can enhance nutrient storage and cycling, as well as improve soil structure, attributes that assist agricultural crops to be more sustainable as well as help to reduce environmental impacts of modern farming practices. In addition to facilitating the uptake of water and essential nutrients by roots, AMF and EMF have been shown to exhibit a range of attributes. These include increasing plant survival in soils contaminated with or naturally rich in heavy metals (Shahabivand et al. 2012; Rozpadek et al. 2014); promoting crop growth in saline soils (Chandrasekaran et al. 2014); increasing productivity as plant growth promotion agents (Azcón-Aguilar et al. 1997); biological control agents against phytopathogens (Pozo et al. 2002; Harrier and Watson 2004); and reducing the population of phytopathogenic nematodes in the mycorrhizosphere (Neeraj 2011).

The AMF in particular expedite the formation of soil aggregates through the extensive production of hyphae interacting with soil particles (Smith and Read 2008) and by the production of a glycoprotein, glomalin, which cements soil particles (Rillig et al. 2002; Rillig 2004; Yang et al. 2017). Enhanced soil aggregation increases soil macroporosity, which in turn enhances gas exchange as well as the percolation of water into the subsoil. In addition, the presence of glomalin may aid in the decontamination of toxic compounds, such as heavy metals (González-Chávez et al. 2004) and polycyclic aromatic hydrocarbons (PAH), which have high mutagenic properties (Gao et al. 2017).

11.2.2 Role in Carbon Cycling

Up to 20% of annual net primary productivity is allocated belowground (Hobbie 2006). Estimates of hyphal length range from 3 to 30–60 m per gram of soil (Jakobsen et al. 1992; Jones et al. 2009). Mycelia provide a conduit through which a majority of the carbon can enter the SOM pool (Godbold et al. 2006). Although the turnover of hyphae occurs at a faster rate than fine roots (Koide et al. 2011), the rate at which the EMF necromass decomposes is influenced by hyphal wall chemistry. The presence of recalcitrant melanin, for example, slows the decomposition of hyphae (Fernandez et al. 2016).

The role of mycorrhiza in global carbon accounting is gaining attention as ectomycorrhizal-associated plants store more carbon in soils than other plants. Averill (2016) argues that this is a result of competitive interaction between ectomycorrhizal plants and free-living microbial decomposers of soil N. Under N-limitation, the activity of the free-living decomposers is reduced, the cycling of soil carbon is slowed, and the storage of soil carbon increases. Using in-growth mesh bags under stands of eastern hemlock, Averill and Hawkes (2016) demonstrated that ectomycorrhizal roots and hyphae reduced soil carbon respiration rates by up to 67%.

There are key differences between ectomycorrhizal and arbuscular mycorrhizal trees that influence carbon and nitrogen cycling. This led Sulman et al. (2017) to hypothesize that *the ecosystem carbon balance is sensitive to the mycorrhizal strategies that plants use to acquire nutrients*. Leaf litter under ectomycorrhizal trees has a lower quality (Lin et al. 2017) and decomposes more slowly (Midgley et al. 2015) than under arbuscular mycorrhizal trees. Hence, there are greater forest floor carbon stocks in ectomycorrhizal forests (Lin et al. 2017). Climate alters the global pattern of carbon storage in mineral soils, with some arbuscular mycorrhizal forests having larger stores than ectomycorrhizal forests in tropical and subtropical zones but not in temperate zones (Lin et al. 2017). Overall, the type of mycorrhizal association, the populations of associated free-living decomposing organisms and the quality of the SOM (Taylor et al. 2016) all influence the extent of soil carbon and nitrogen cycling.

11.2.3 Mycorrhizal Associations and Nutrient Acquisition

The extent of emanating hyphae and rhizomorphs was used by Agerer (2001) to classify associations into soil exploration types: contact, short distance, medium distance and long distance. In Norway spruce stands, long distance exploration types were observed to be prevalent in mineral soil horizons (Scattolin et al. 2008) where access to some nutrients is more limited than in organic soil horizons. Hyphal morphology and extent appear to be functionally important as ectomycorrhizal without or with short emanating hyphae are reported to utilize soluble forms of N, such as ammonium and amino acids, whereas ectomycorrhiza with medium and long distance morphotypes that typically extend into organic layers have access to insoluble N compounds (Pena et al. 2017). Lilleskov et al. (2011) suggested that soil N availability is a factor influencing the distribution of fan and rhizomorph-forming EMF.

The ability of roots and mycorrhizal fungi to explore nutrient patches in soil is important for plant nutrition. In an urban garden site with 13 sympatric temperate tree species, Chen et al. (2016) showed that ectomycorrhizal trees responded to nutrient “hot spots” by producing more ectomycorrhizal hyphae, whereas arbuscular mycorrhizal trees produced more roots. Thus ectomycorrhizal trees have higher mycorrhizal foraging precision than root foraging precision arbuscular mycorrhizal trees within organic patches (Cheng et al. 2016).

11.2.4 Mycorrhizal Activity in Subsoils

Regarding the location of mycorrhiza and their extensive emanating soil hyphae, most studies in the field have explored aspects of distribution and function within the topsoil, and less attention has been given to mineral soil horizons or subsoils. It has been observed that ectomycorrhiza are often concentrated in topsoil that is rich in organic matter (Rosling et al. 2003) or that EMF diversity (Karlinski et al. 2013) and the amount of ectomycorrhizal extraradical hyphae (Wallander et al. 2004) decrease with soil depth. In Holm oak woodland sampled to a depth of 20 cm, about 75% of ectomycorrhizal root tips and 60% of EMF taxa were restricted to the top 10 cm of the soil profile, with the contact exploration type being dominant in the organic-rich horizon (Shahin et al. 2013). In contrast, in Norway spruce, the number of ectomycorrhizal root tips increased with soil depth from the organic-rich layer into mineral soil sampled to a depth of 75 cm (Scattolin et al. 2008). An explanation for the contrasting results has not yet been provided. In eucalypt plantations in Brazil, ectomycorrhizas have been observed to depths of 3–4 m (Bordron et al. 2019).

Studies on the distribution of AMF in deeper layers of soil are scarce, and few studies report on the vertical distribution of AMF in subsoil (Cardoso et al. 2003). The main limiting factor for the diversity and abundance of AMF in soil is root distribution. The distribution of roots in soil is the main condition for increasing the community of AMF, since these fungi are obligatory biotrophic. The AMF depend on photoassimilates from plants to survive (Douds Jr. et al. 2006; Douds Jr. et al. 2008), completing their life cycle only when associated with the roots of living plants as they are not able to survive without obtaining carbon from the plant (Azcón-Aguilar et al. 1997; Smith et al. 2010). Consequently, they cannot be multiplied separately in a defined culture medium (Douds Jr. et al. 2006), which hinders the large-scale propagation of these microorganisms.

The number of AMF spores decreases with increasing soil depth, but there are species that sporulate mainly in certain layers of soil (Oehl et al. 2005). This indicates that the specific physico-chemical characteristics of certain soil layers, such as O_2, pH, organic matter, calcium, phosphorus and nitrogen, are relevant factors in the vertical distribution of these microorganisms. Hence, agroforestry systems may favour the diversity and abundance of AMF by the distribution of roots in deeper layers of soil.

The distribution (vertical and horizontal), population size and diversity of ectomycorrhizas partially reflect soil and fungal traits. There is emerging evidence that niche partitioning can explain some of the observed differences in spatial distribution of ectomycorrhizal with soil depth (Taylor et al. 2016; Mujic et al. 2016). It has been proposed that some plants can adapt to base cation deficiencies by increasing the proliferation of EMF and subsoil mineral weathering by enhanced exudation of low molecular weight organic acids (Van Schöll et al. 2008). It is worth noting that the mycelium external to but connected to the roots may occupy a greater soil depth range and exhibit less vertical niche differentiation than the corresponding mycorrhizas (Genney et al. 2006).

The extent to which mycorrhizal fungi contribute to soil weathering and the provision of essential nutrients to plants remains to be quantified in spite of investigations into so-called rock-eating mycorrhizas (Van Schöll et al. 2008). Future studies into the subsoil and parent rock interface would need to take into account the bacterial communities as well as any symbiotic fungi that can persist in difficult environments. Wang et al. (2014) explored the first metre of soil profile under grassland and observed that the proportion of high efficiency mineral-dissolving bacteria increased with soil depth. From manipulating saprolite and limestone rock interfaces under beech trees, Nicolitch et al. (2017) suggested that tree roots may select specific subsoil bacterial communities to improve tree nutrition.

The degree to which mycorrhizal fungi can ameliorate effects of global climate change on vegetation or the potential impacts of climate change on populations of EMF are little understood. Indeed, there is a need for long-term studies in order to assess any mycorrhizal and ecosystem responses to global climate change (Mohan et al. 2014). He et al. (2017) manipulated the season of precipitation (decreased in the dry season, increased in the wet season) in a subtropical forest and found that the diversity of the fungal but not the bacterial community declined in the top 10 cm of the soil profile. Globally, ectomycorrhizal plants increased their biomass in response to elevated CO_2 regardless of the availability of nitrogen (Terrer et al. 2016). Furthermore, particular combinations of host genotype and ECM community improved the survival and growth of Colorado pinyon exposed to drought (Gehring et al. 2017).

11.2.5 Application of Mycorrhiza in Agriculture and Forestry

Given the advantages that mycorrhizas confer to their host plants, the production of mycorrhizal seedlings of various agronomic and forest species has been the subject of many studies. For some forest plantation species such as pines, seedlings are routinely inoculated in nurseries in order to promote growth after out-planting. However, in spite of the many benefits in different agricultural crops and farming systems, inoculation in large areas is still restricted because of the difficulty in producing bulk commercial inocula that are safe. As mentioned earlier, because they are obligate biotrophs, the AMF cannot be bulked up in the absence of a host (Berbara et al. 2006) unlike many ectomycorrhizal fungi. At present, spores or fragments of AMF in a carrier can be applied during nursery propagation for some horticultural and forestry crops. However, the challenge remains as to the most suitable inoculum type and delivery system for broadscale seeding of annual species such as soybean and maize.

Considering the increased efficiency of nutrient use, both in the surface layers and deeper layers of the soil, mycorrhizal associations are important biological tools to increase the absorption of mineral nutrients and water. These fungi are able to take advantage of trapping P in deeper compartments in tropical latosols, since there are roots at depth, able to provide photoassimilates for the mycorrhizal fungi to grow and function. This should facilitate the recycling of mineral nutrients to the surface soil layers.

Crop rotation and green manuring are agricultural practices that may favour arbuscular mycorrhizal fungi (Prates Júnior et al. 2019). The use of leguminous plants doubly colonized by AMF and rhizobia may result in higher availability of P and N to host plants, helping to alleviate nutrient shortages.

11.3 Biological Nitrogen-Fixing Bacteria

Biological nitrogen fixation (BNF) has great economic and ecological importance, since it is able to reduce the high costs and carbon footprints of inorganic nitrogen fertilizers. Furthermore, overuse of nitrogen fertilizers can cause environmental damage from leaching, contaminating soil and water resources mainly as nitrate (NO_3^-) (Graham and Vance 2003), and contribute to the greenhouse effect from nitrous oxide (N_2O) emissions (Spinelli et al. 2013).

Only a small portion of prokaryotes (rhizobia, cyanobacteria, azotobacteria, actinobacteria and archaea groups) has the ability to reduce atmospheric N_2, unavailable to the vast majority of living organisms, into NH_3 (ammonia), using the nitrogenase enzyme complex, followed by formation of organic nitrogenous compounds such as amides. The BNF capacity is present in free-living as well as associative microbes, among which we highlight the symbiosis between rhizobia and about 70% of legumes in the Fabaceae (Remigi et al. 2016). In addition to the Fabaceae, the genus *Parasponia*, belonging to the Cannabaceae, also establishes symbiotic relationship with rhizobia, forming nodules (Behm et al. 2014).

Among legumes and rhizobia, the symbiotic relationship varies from highly specific to promiscuous. In addition, in some cases, certain legume species may have greater dependence on rhizobia species, such as *Dimorphandra wilsonii* Rizzini (Fabaceae, Caesalpinioideae), which has BNF as an adaptive strategy for its growth in dystrophic soils (Fonseca et al. 2010). Thus, the survival and growth of some species of Fabaceae are related to the efficiency in BNF. These bacteria are mainly found in the superficial layers of the soil. Rupela et al. (1987) observed that "*Rhizobium* populations declined with soil depth and were highest (about 10^4 rhizobia g^{-1} soil) in the top 30 cm of the profile and lowest, but still present (10–10^3 rhizobia g^{-1} soil), at 90–120 cm, a depth where no nodules are found". They also observed that "populations fluctuated most in the top 5 cm, being reduced during periods of high soil temperature in summer and recovering after rains".

The BNF capacity provides competitive advantages for plants growing on nitrogen-depleted soil. These species have a potential for stress relief and revegetation of degraded areas (Franco and Faria 1997; Ferreira et al. 2012; Fatnassi et al. 2015), such as *Mimosa* rock species that associate with strains of *Burkholderia* (Araújo et al. 2017) and may contribute to higher biomass production and release of exudates, secretions and mucilages that stimulate the microbiota. Furthermore, the association of several legumes tolerant to high concentrations of salts in the soil with adapted rhizobia strains helps in the cultivation and supply of nitrogen in soils that would cause high salt stress for most plants (Bruning and Rozema 2013).

In terms of agronomics, *Bradyrhizobium japonicum* and *B. elkanii* were the most prominent taxa observed in studies begun in Brazil in the 1950s by Dr. Johanna Dobereiner, which led to the improvement of soybean (*Glycine max* Merr.) and the production of commercial rhizobia inoculants. This led to Brazil becoming an efficient producer of soybean (Dobereiner 1997). Application of nitrogen fertilizer to well-nodulated legume crops negatively affects the number and dry weight of nodules and impairs symbiotic performance, without improving crop yields (Kaschuk et al. 2016).

Considering the economic and nutritional importance of beans (*Phaseolus vulgaris* L.), especially in developing countries of South America and Africa, together with the high costs of mineral N fertilizers, the supply of N by nitrogen-fixing bacteria (NFB) is highly desirable. However, beans are promiscuous in the establishment of symbiosis with rhizobium species, but some of them inefficient in terms of N_2 fixation (Dall'Agnol and Ribeiro 2014), leading to the need for complementary nitrogen fertilization. It is necessary to collect and evaluate the symbiotic efficiency of strains adapted to the local conditions of cultivation, improving the rhizobia-bean symbiosis performance.

The gains in BNF may be indirect. For example, in Brazil some coffee plantations are managed with nodulated legumes, such as forage peanut (*Arachis pintoi* Krapov. & W.C. Greg.), calopogonium (*Calopogonium mucunoide* Desv.), *Stylosanthes guianensis* (Aubl.) Sw. and *Crotalaria juncea* L., as sources of nitrogen and organic matter (Matos et al. 2008; Araújo et al. 2013). In this case, the practice of green manuring is an alternative for the supply of nitrogen and contributes to the protection and improvements of the physical, chemical and biological conditions of the soil.

Furthermore, nitrogen-fixing actinobacteria in the genus *Frankia* form associations with more than 20 genera of dicotyledons (Benson and Silvester 1993) in the orders Fagales, Curcurbitales and Rosales, like *Casuarina* and *Alnus*. In these associations, root structures called actinorrhizas confer benefit to host plants that occupy soils that have a nitrogen shortage as a limiting factor for plant growth. In addition, an association between plants of the genus *Azolla* (Salvinaceae, Pteridophyta) and cyanobacteria of the genera *Anabaena* and *Nostoc* may occur, allowing *Azolla* to be used in animal feed, as a green manure and in the recovery of degraded areas.

Nitrogen-fixing bacteria also associate, without nodule formation, with other families of plants of economic importance, such as grasses (Poaceae), corn, sorghum, sugar cane and rice. In these plants, the bacteria colonize roots and aerial parts and perform the biological fixation of nitrogen. Bacteria of the genus *Azospirillum* commonly grow in the rhizosphere of corn and sorghum, increasing crop production. The bacterium *Gluconacetobacter diazotrophicus* (=*Acetobacter diazotrophicus*) can increase sugarcane growth (Sevilla et al. 2001), and the fixed N can provide up to 60% of plant demand (Urquiaga et al. 1992, 2012). The bacterium can also act as an antagonist of phytopathogenic bacteria and fungi.

Although huge advances have been made in the application of BNF to farming systems, challenges remain for research. This is particularly the case in sub-Saharan Africa where continuous cereal-based cropping has led to declines in soil fertility (Kermah et al. 2018), where crop rotation with grain legumes can enhance soil fertility and cereal yield (Franke et al. 2018) and where increasing production of grain legumes is an important strategy for food security in the region (Vanlauwe et al. 2014).

Free-living bacteria in soils or associated with the rhizosphere can enhance BNF in agriculture (Roper and Gupta 2016). The superficial layer of the soil contains cyanobacteria and lichens that can be an important source of nitrogen (Belnap 2002), but these microorganisms are photosynthetic and are limited to exposed surfaces. Furthermore, cyanobacteria can improve soil physico-chemical characteristics due to the production of biomass, enhanced water holding capacity and capability to fix atmospheric N_2 (Singh et al. 2016).

Nitrogenase, the enzyme responsible for BNF, requires a low oxygen potential for its activity similar to levels in deeper soil layers. Although the low oxygen potential favours nitrogen fixation, the process is costly energetically and depends on carbon sources to allow the growth of bacteria. The main sources of organic carbon in the deepest layers of the soil are from the root exudates; therefore, many of these microorganisms colonize the rhizospheric zone. Thus, it is interesting to stimulate root growth in deeper layers of the soil and to make use of crop rotation with species with deep root systems. Crop rotation also allows the accumulation of plant residues in the soil, favouring the stock of organic carbon particles that have a strong connection with the ecology and diversity of NFB (Wakelin et al. 2010).

In well-aerated soils, the formation of aggregates is important because it allows the formation of microsites with low oxygen. The maintenance of aggregates in soil depends on characteristics of the soil itself and the environment. However, management has a direct influence on the formation of aggregates and in the composition, diversity, abundance and activity of the free-living NFB community (Wakelin et al. 2010).

11.4 Plant Growth-Promoting Fungi

Plant growth-promoting fungi (PGPF) inhabit the rhizosphere or are plant endophytes, such as species of the genera *Trichoderma* (Shaw et al. 2016), *Penicillium* (Waqas et al. 2015), *Aspergillus* (Islam et al. 2014a), *Phoma* (Hossain et al. 2017) and some species of *Fusarium* (Islam et al. 2014b). PGPF can stimulate the systemic defence-induced response, compete and act as phytopathogen biocontrollers, solubilize phosphates, increase photosynthetic efficiency and the use of nitrogen and can improve germination, seedling vigour, biomass production and root hair development (Hossain et al. 2017). The dark septate (Ascomycota) fungi, like *Phialophora* sp. and *Phialocephala* sp., also deserve prominence and are recognized for forming septate and melanized hyphae, together with microsclerodes that grow inside plant root cells, promoting greater accumulation of nutrients and greater tolerance to biotic or abiotic stress.

Trichoderma spp. are capable of solubilizing inorganic phosphates and enhancing plant growth (Promwee et al. 2014) and are influenced by management practices such as fertilization, liming, ploughing and harrowing. The distribution of *Trichoderma* spp. in soil layers may be related to extension of the root system that provides exudates and energy for these fungi (Okoth et al. 2009). These fungi are highly interactive in roots and are antagonistic against soil-borne pathogens because they are excellent competitors, acting as mycoparasites, and can secrete antimicrobial substances.

Penicillium spp. are fungi that can colonize the rhizosphere of plants and play an important role in the cycling of organic matter and solubilization of inorganic phosphate. They are able to promote the growth of plants and improve survival under high soil salinity (Radhakrishnan et al. 2014). Compared to other genera of PGPF, *Penicillium* occurs in deeper layers of the soil and has lower density in the rhizosphere. Thus, *Penicillium* may be well adapted to subsoil conditions, where roots are less abundant, maintaining the ability to solubilize inorganic phosphate and degrade xenobiotic compounds.

Dark septate fungi can promote plant growth and facilitate stress tolerance under drought conditions (Li et al. 2018), indicating their potential for alleviating subsoil constraints because they can influence plant physiology and nutrient cycling (Vergara et al. 2018). These fungi can favour the occurrence of plant species in areas contaminated by heavy metals (Likar and Regvar 2013). They are abundant in soil samples and roots, however, as they are usually considered to be facultative biotrophic and are less dependent on the rhizosphere than AMF.

Among the PGPF, a representative that has gained prominence in many research groups around the world is the fungus *Piriformospora indica* (Basidiomycota, Sebacinaceae, Sebacinales). This fungus was isolated in the Thar Desert, India, and is considered a root endophyte that promotes plant growth (Verma et al. 1998). It has many characteristics similar to those found in AMF (Oelmüller et al. 2009; Selosse et al. 2009; Varma et al. 2012). Similar to AMF, *P. indica* can promote growth in a wide range of plants, but has the great advantage over AMF of being able to be cultivated in axenic cultures (Varma et al. 1999).

Inoculation of *P. indica* has demonstrated positive effects on plant development, including an increase in plant biomass production due to increased uptake of phosphate and nitrogen from soils with nutritional deficit (Yadav et al. 2010; Kumar and Rao 2012). The fungus assists plant survival in conditions of hydric and saline stress and high and low temperatures and increases the resistance of plants to toxins and heavy metals (Waller et al. 2005; Oelmüller et al. 2009; Varma et al. 2012).

Furthermore, the fungus has potential to be used as a biological control agent (Kumar et al. 2011). Examples are the control of *Fusarium subglutinans* f. sp. *ananas* on pineapple (Moreira et al. 2016), *F. culmorum* and *Blumeria graminis* on barley (Waller et al. 2005), *F. culmorum* and *F. graminearum* on wheat (Rabiey and Shaw 2016) and *Rhizoctonia solani* on rice plants (Nassimi and Taheri 2017).

Thus, the PGPF play a role in increasing crop production, alleviating nutrient depletion and suppressing plant pathogens. Many of these fungi have the ability to entangle particles within the hyphal network due to the production of polysaccharides that are released into the soil. This results in improvements in soil structure that promote water retention, build-up of organic matter and favourable environments for root penetration in the subsoil.

11.5 Plant Growth-Promoting Rhizobacteria

Plant growth-promoting rhizobacteria (PGPR) belong to various groups of bacteria that stimulate plant growth by producing phytohormones and providing nutrients such as P and N, as well as bestowing protection against agricultural diseases and pests, when they trigger induced systemic resistance or produce antibiotics. The rhizosphere, a layer of soil surrounding the root, is colonized by these bacteria, since it is rich in carbon sources and nutrients when compared to the soil volume distant from the root system, due to the release of exudates by the roots of plants.

In the rhizosphere, plants favour the growth of microorganisms, such as *Pseudomonas*, *Bacillus*, *Rhizobium* and *Azospirillum*, capable of producing phytohormones such as indoleacetic acid (IAA) and abscisic acid (ABA), as well as the synthesis of enzymes that control the production of phytohormones (Glick and Bashan 1997; Zahir et al. 2004; Angus et al. 2013; Hai-Bi et al. 2017).

Much of the inorganic phosphate in soil is strongly adsorbed to the reactive surfaces of soil minerals and SOM and is not readily available for uptake by roots (Malik et al. 2012). Several PGPR are able to solubilize inorganic phosphate adsorbed to soil particles and contribute to the degradation of organic matter, releasing phosphate by mineralization. Among the mechanisms that PGPR use for phosphate solubilization is the production of organic acids, common in bacteria of the genera *Burkholderia*, *Bacillus* and *Erwinia* (Bhattacharyya and Jha 2012). Several of these phosphate solubilizing bacteria have the ability to reduce pH, as well as increase the production of oxalic, lactic, citric, succinic, acetic and formic acids (Wei et al. 2018), which help in the release of *P* bound to oxides in the soil. The mineralization of *P* by the degradation of organic matter occurs by the release of alkaline and acidic phosphatases. Thus, PGPR help to mobilize phosphorus sources present in the soil and improve plant nutrition.

There are indirect mechanisms that allow PGPR to favour plant growth, especially when they act as antagonists or competitors of phytopathogenic microorganisms, producing antibiotics, bacteriocins, hydrocyanic acid (HCN) and lytic enzymes. These bacteria can induce systemic resistance and reduce the incidence of diseases caused by pathogens and parasites, and as well, they can inhibit the growth of phytopathogens by the production of siderophores capable of capturing Fe^{3+} molecules, rendering them unavailable to pathogens (Miethke and Marahiel 2007). By colonizing the plant root system, PGPR synthesize eliciting molecules, such as ethylene, salicylic acid and jasmonic acid, which induce plant genes to express phytopathogenic resistance components and, therefore, favour plant defence responses (Zahir et al. 2004). In addition, these PGPR may contribute to an increase in the activity of hydrolytic enzymes such as chitinases, peroxidases and glucanases (Maksimov et al. 2011) and reduce the population density of phytopathogens.

Some strains of PGPR can improve soil and plant health due to their contribution to disease-suppressive soils and also because their activity can minimize the use of non-renewable resources. Rhizobacteria, such as *Streptomyces* spp. (Actinobacteria), present in the rhizosphere of plants may act to inhibit or control pathogens, either by inducing systemic resistance or by metabolic changes such as the production of antimicrobials and siderophores (Dias et al. 2017). Actinobacteria are associated with disease-suppressive soil and are abundant in bulk soil, allowing these microorganisms to be less dependent on root growth and released exudates.

In terms of sustainability and use of marginal areas for agriculture, PGPR may be an important strategy in that they promote plant growth and increase salinity tolerance (Numan et al. 2018). Excess salt concentration in soil decreases agricultural productivity, but wheat plants inoculated with *Serratia* sp. had improved salt tolerance and increased shoot biomass (Singh and Jha 2016). This genus is considered a facultative anaerobic allowing distribution in anaerobic microsites within soil.

Inoculation with *Azospirillum* favours root growth and increases the surface of absorption due to the greater volume of explored soil. It occurs by inducing changes in the root system and increasing the amount of rootlets (Okon and Vanderleyden 1997). This pattern is mainly caused by phytohormones such as auxins, gibberellins and cytokinins (Patel and Saraf 2017).

The benefits for plants can also be indirect, through physical-chemical changes in the soil, resulting from their activity that favours aggregate formation. The release of different types of polysaccharide by these microorganisms favours the aggregation of particles and contributes to water and nutrient uptake by plant roots.

The success for greater understanding and use of bacteria that interact with roots of host plants is not only in the isolation, selection and testing under in vitro and field conditions but also in innovation and the ability to produce inoculants on a commercial scale at low cost, coupled with cultural changes to increase the use of biofertilizers and products for biological control. Greater presence of these microorganisms can increase beneficial effects in soil quality and plant growth, plant nutrition and disease suppression.

11.6 Plant Soil Feedback (PSF) and Plant Microbiome

Soil is a dynamic matrix that represents a large reservoir of biodiversity and enables interactions between plants and microorganisms. In natural conditions, plants have direct relationships with a great abundance and diversity of microorganisms, and there is evidence that together they form a metaorganism or holobiont, the sum of the host organism and its symbiotic microbiota (Vandenkoorhuyse et al. 2015).

In this perspective, when studying the characteristics of plants, it is necessary to consider the associated microbiota, considering the long co-evolutionary process that resulted in co-adaptation and increased fitness of both partners. Thus, the model plant-soil feedback helps to explain positive interferences, when the soil, under the influence of a certain plant, favours the growth of seedlings of the same species, or negative when the soil does not favour the growth of seedlings of the same species. Negative feedback occurs when plants have reduced growth capacity in soils that have previously been occupied by plants of the same species, either by nutrient depletion or increased incidence of pathogens (van der Putten et al. 2013). Positive feedback occurs when plants are able to grow in soils previously occupied by members of their species due to changes that may involve allelopathy, mutualism and litter deposition (van der Putten et al. 2013).

The plant soil feedback (PSF) has holistic aspects and drives plant function as ecosystem engineers (van Breemen and Finzi 1998), for the ability to modulate the availability of resources to other species and to change biotic and abiotic characteristics of the soil, creating new habitats. The soil acts as a "seed bank" of the rhizospheric microbiota and its physico-chemical properties determine the composition of the microbial community (Vandenkoorhuyse et al. 2015). The plants are able to alter soil properties and modulate the survival and growth of juvenile or nearby plants, and there may be differences between functional groups: grasses, shrubs, legumes, etc. Thus, soil can be considered as part of the extended phenotype of a plant, triggering evolutionary implications (van Breemen and Finzi 1998) at the level of interactions between organisms and on the landscape ecology scale.

It is recognized that some plants of the Fabaceae family, such as *Plathymenia reticulata* Benth. and *Melanoxylon brauna* Schott, grow better in soil obtained in the vicinity of an adult plant of the same species than in other types of substrate (authors' information) (Fig. 11.1). Studies indicate that arbuscular mycorrhizal fungi may be involved in increased survival and growth of these plants. However, it is necessary to explore the importance of other groups such as PGPR, rhizobia and endophytes as the mechanisms of co-adaptation to soil previously colonized by co-specific species. This is because not only may plants select specific communities of rhizobacteria; fungi may also benefit specific groups of bacteria in the mycorrhizosphere and expand their role in the PSF.

Fig. 11.1 *Plathymenia reticulata* Benth. after 180 days of greenhouse cultivation in soil from under an adult plant of the same species: (**a**) natural soil and (**b**) sterilized soil

In addition, fertilization can cause a cascade effect and modify the microbial diversity and function. For example, plant dry matter is higher when grown with microbial inoculum previously collected in unfertilized soil rather than fertilized (Revillini et al. 2016), indicating that the lower availability of N may select a microbial community adapted to the maximum increase of plant productivity.

The addition of mineral N in residue incorporated to soil mainly affects the fungal community and reduces soil aggregation, evidence that the fungi play a prominent role in the stabilization of aggregates which increases the sustainability of agroecosystems (Bossuyt et al. 2001). The greater or lesser availability of nutrients affects the composition and diversity of the microbial community as well as root exudation (Lagos et al. 2015). In addition, soil management practices, such as liming influences soil pH which has a large impact on the composition and diversity of the bacterial and fungi communities. Management also affects soil physical processes, such as soil aggregate formation which mediates many biological processes such as microbial respiration and microbial biomass.

Roots are an important component that integrate the fields of agronomy, management and soil microorganisms. The physico-chemical properties and the rhizosphere microbiota modulate the root system architecture, enabling the plants to increase the volume of soil explored for better use of water and nutrients. The release of exudates is controlled by the plant that is able to acidify the rhizosphere and decrease the growth of bacteria or release antifungal compounds, such as chitinases (Weisskopf et al. 2006).

It must be recognized that plants and microorganisms have gone through a long co-evolutionary process that allowed the specificity of interactions and differentiated responses to nutrient availability, such as N and P, to evolve. Some plants may be weakly responsive to phosphate fertilization, such as *Melanoxylon brauna* Schott, due to interaction with bacteria with high capacity of phosphate solubilization (Prates Júnior et al. 2021). Other plants, like *Plathymenia reticulata* Benth., are poorly responsive to nitrogen fertilization and may have considerable dependence on N-fixing bacteria (Prates Júnior et al. 2021). Plants can allocate more photoassimilates to specific regions of the root system, recruit specific microorganisms such as AMF and rhizobia and ensure more access to limited resources with P and N.

Inoculation of microorganisms is an important strategy to manipulate the microbial composition of the soil and interactions between plants and microorganisms, in order to increase the benefits to the plants in terms of productivity and suppression of diseases through plant-soil feedback mechanisms. However, it is fundamental to manage the diversity of microorganisms through agricultural practices, providing adequate soil conditions for the long-term persistence and function of beneficial microbiota. The PSF are important in terms of resource savings and greater sustainability of agroecosystems, since it opens up possibilities for a better understanding of abiotic and biotic soil factors and, consequently, suppression of diseases and use of symbiotic microorganisms in the agricultural and forestry sectors.

11.7 Final Considerations

In this chapter we have discussed beneficial microbes and root associations that are well-known to occur in topsoil. Surprisingly, the extent to which these microbes can persist and contribute to root development, plant growth and ecosystem functioning in subsoils is largely unknown. Changes in agricultural practices, such as the use of deep tillage to increase availability of subsoil nutrients and water (Kautz et al. 2013; Schneider et al. 2017), and the breeding of deep-rooted crops (Lynch and Wojciechowski 2015) should promote more research on the microbial communities in subsoils.

It is likely that all subsoil communities will be less diverse than in topsoils (Eilers et al. 2012), but they will contain specialized microbes adapted to deeper environments (Fierer et al. 2003). Eilers et al. (2012) sampled nine soil profiles to a depth of 180 cm in a forested montane watershed in Colorado and, from pyrosequencing of the 16 sRNA gene for bacteria and archaea, concluded that the microbial community composition was similar at depth regardless of landscape position even though it was highly variable in the surface horizons. In a study using phospholipid fatty acid (PLFA) analysis to assess the viable microbial biomass and community structure with depth in a low-carbon tropical soil, Stone et al. (2014) identified shifts in community structure with depth driven largely by a decline in fungal/bacterial ratios, an increase in gram-positive and actinobacteria markers and a decrease in gram-negative biomarkers.

A profitable area for further study is the importance of macropores or biopores in subsoils as habitats for beneficial soil microbes and roots. These structures, created by previous generations of sinker/tap roots or the activity of soil-dwelling invertebrates (Schaeffer and Oliveira 2022), provide preferred pathways for roots to enter into subsoil where conditions in the bulk soil such as high bulk density and soluble aluminium concentrations may be unfavourable for root growth. These linkages between the topsoil and subsoil may extend from a few metres to many tens of metres depending on the soil profile. Larger macropores, also called root channels, can be up to 20 cm in diameter and are prevalent in forests where roots must access water at depth during periods of drought (Dell et al. 1983). Not all soil pores contain roots, with only 5% of subsoil pores in a wheatfield containing roots compared to 20% in surface soil (White and Kirkegaard 2010). Furthermore, crop rotation can be used to enhance the formation of macropores, which opens up opportunities for altering the subsoil microbial communities through agronomic practices. For example, when chicory was used in rotation with barley and winter oilseed rape, the number of biopores was increased (Perkons et al. 2014).

Biopores become nutrient enriched through the accumulation of organic matter. Also, the organic matter that earthworms incorporate into biopores creates habitats for microorganisms which can facilitate SOM turnover in the subsoil (Hoang et al. 2017). A link between the accumulation of ammonium from activity of earthworms, acquisition by AMF and enhanced uptake by a fern has been suggested in south China (He et al. 2018). Also, subsoil biopores may have more labile P than in the bulk soil (Bauke et al. 2017). The extent to which microorganisms cohabit with roots in biopores at depths below 2 metres has not been explored.

The demand for technologies capable of increasing agricultural and forestry productivities that maintain the sustainability of the production system in an environmentally responsible way is growing. The use of microorganisms can contribute to the transition to more sustainable agricultural models with the possibility of reduced pesticide and synthetic fertilizer use, implying soil economic, social and environmental benefits.

Understanding the role of beneficial microorganisms in relation to nutrient cycling, BNF, phosphate solubilization in soils and biological control of pests and diseases, as well as their role in recovering degraded areas, has transformed the management practices traditionally used. In this way, it has developed the market for new agricultural products and is also a source of employment in various sectors of the economy. Research is far from reaching the full potential of using these microorganisms to enhance their use and convince traditional farmers to reduce the use of non-renewable inputs; however, we remain optimistic that it will happen in the near future.

References

Agerer R. Exploration types of ectomycorrhizae – a proposal to classify ectomycorrhizal mycelial systems according to their patterns of differentiation and putative ecological importance. Mycorrhiza. 2001;11:1070–114. https://doi.org/10.1007/s005720100108.

Angus AA, Lee A, Lum MRL, et al. Nodulation and effective nitrogen fixation of *Macroptilium atropurpureum* (siratro) by *Burkholderia tuberum*, a nodulating and plant growth promoting beta-proteobacterium, are influenced by environmental factors. Plant Soil. 2013;369(1–2):543–62. https://doi.org/10.1007/s11104-013-1590-7.

Araújo BS, Rodrigues MC, Rodrigues LB, et al. Nitrogen fertilization of coffee: organic compost and *Crotalaria juncea* L. Revi Ceres. 2013;60(6):842–51. https://doi.org/10.1590/S0034-737X2013000600013.

Araújo KS, de Carvalho F, Moreira FMS. Estirpes de *Burkholderia* promovem o crescimento de *Mimosa* spp. mas não o de *Macroptilium atropurpureum*. Rev Cienc Agron. 2017;48(1):41–8. https://doi.org/10.5935/1806-6690.20170005.

Averill C. Slowed decomposition in ectomycorrhizal ecosystems is independent of plant chemistry. Soil Biol Biochem. 2016;102:52–4. https://doi.org/10.1016/j.soilbio.2016.08.003.

Averill C, Hawkes CV. Ectomycorrhizal fungi slow soil carbon cycling. Ecol Lett. 2016;19:937–47. https://doi.org/10.1111/ele.12631.

Azcón-Aguilar C, Cantos M, Troncoso A, et al. Beneficial effect arbuscular mycorrhizas on acclimatization of micropropagated cassava plantlets. Sci Hortic-Amsterdam. 1997;72:63–71. https://doi.org/10.1016/S0304-4238(97)00120-9.

Bauke SL, von Sperber C, Siebers N, et al. Biopore effects on phosphorus biochemistry in subsoils. Soil Biol Biochem. 2017;111:157–65. https://doi.org/10.1016/j.soilbio.2017.04.012.

Behm JE, Geurts R, Kiers ET. *Parasponia*: a novel system for studying mutualism stability. Trends Plant Sci. 2014;19:757–63. https://doi.org/10.1016/j.tplants.2014.08.007.

Belnap J. Nitrogen fixation in biological soil crusts from Southeast Utah, USA. Biol Fert Soils. 2002;35:128–35. https://doi.org/10.1007/s00374-002-0452-x.

Benson DR, Silvester W. Biology of Frankia strains, actinomycete symbionts of actinorhizal plants. Microbiol Rev. 1993;57(2):293–319.

Berbara RLL, Souza FA, Fonseca HMAC. Fungos micorrízicos arbusculares: muito além da nutrição. In: Fernandes MS, editor. Nutrição Mineral de Plantas. 1st ed. Viçosa: Sociedade Brasileira de Ciência do Solo; 2006. p. 53–85.

Bhattacharyya PN, Jha DK. Plant growth-promoting rhizobacteria (PGPR): emergence in agriculture. World J Microb Biot. 2012;28:1327–50. https://doi.org/10.1007/s11274-011-0979-9.

Bordron B, Robin A, Oliveira IR, et al. Fertilization increases the functional specialization of fine roots in deep soil layers for young *Eucalyptus grandis* trees. Forest Ecol Manag. 2019;431:6–16. https://doi.org/10.1016/j.foreco.2018.03.018.

Bossuyt H, Denef K, Six J, et al. Influence of microbial populations and residue quality on aggregate stability. Appl Soil Ecol. 2001;16:195–208. https://doi.org/10.1016/S0929-1393(00)00116-5.

Bruning B, Rozema J. Symbiotic nitrogen fixation in legumes: perspectives for saline agriculture. Environ Exp Bot. 2013;92:134–43. https://doi.org/10.1016/j.envexpbot.2012.09.001.

Butterly CR, Amado TJC, Tang C. Chapter 3. Subsoil constraints – soil acidity and acidification. In: Oliveira TS, Bell RW, editors. Subsoil constraints for crop production. Switzerland: Springer; 2022. p. 53–91.

Cardoso IM, Boddington C, Janssen BH, et al. Distribution of mycorrhizal fungal spores in soils under agroforestry and monocultural coffee systems in Brazil. Agrofor Syst. 2003;58:33–43. https://doi.org/10.1023/A:1025479017393.

Chandrasekaran M, Boughattas S, Hu S, et al. A meta-analysis of arbuscular mycorrhizal effects on plants grown under salt stress. Mycorrhiza. 2014;24(8):611–25. https://doi.org/10.1007/s00572-014-0582-7.

Chen W, Koide RT, Adams TS, et al. Root morphology and mycorrhizal symbioses together shape nutrient foraging strategies of temperate trees. PNAS. 2016;113:8741–6. https://doi.org/10.1073/pnas.1601006113.

Cheng L, Chen W, Adams TS, et al. Mycorrhizal fungi and roots are complimentary in foraging within nutrient patches. Ecology. 2016;97:2815–23. https://doi.org/10.1002/ecy.1514.

Costa MCG, Coutinho ÍAC. Chapter 10. Root system aspects of agricultural crops under physical and chemical soil constraints. In: Oliveira TS, Bell RW, editors. Subsoil constraints for crop production. Switzerland: Springer; 2022. p. 225–61.

Dall'Agnol RF, Ribeiro RA, Delamuta JRM, et al. *Rhizobium paranaense* sp. nov., an effective N_2-fixing symbiont of common bean (*Phaseolus vulgaris* L.) with broad geographical distribution in Brazil. Int J Syst Evol Micr. 2014;64:3222–9. https://doi.org/10.1099/ijs.0.064543-0.

Dell B, Bartle JR, Tacey WH. Root occupation and root channels of jarrah forest subsoils. Aust J Bot. 1983;31:615–27. https://doi.org/10.1071/BT9830615.

Dias MP, Bastos MS, Xavier VB, et al. Plant growth and resistance promoted by *Streptomyces* spp. in tomato. Plant Physiol Bioch. 2017;118:475–93. https://doi.org/10.1016/j.plaphy.2017.07.017.

Dobereiner J. Biological nitrogen fixation in the tropics: social and economics contributions. Soil Biol Biochem. 1997;29:771–4. https://doi.org/10.1016/S0038-0717(96)00226-X.

Douds DD Jr, Nagahashi G, Pfeffer PE, et al. On-farm production of AM fungus inoculum in mixtures of compost and vermiculite. Bioresour Technol. 2006;97:809–18. https://doi.org/10.1016/j.biortech.2005.04.015.

Douds DD Jr, Nagahashi G, Reider C, et al. Choosing a mixture ratio for the on-farm production of AM fungus inoculum in mixtures of compost and vermiculite. Compost Sci Util. 2008;16(1):52–60. https://doi.org/10.1080/1065657X.2008.10702355.

Eilers KG, Debenport S, Anderson S, et al. Digging deeper to find unique microbial communities: the strong effect of depth on the structure of bacterial and archaeal communities I soil. Soil Biol Biochem. 2012;50:58–65. https://doi.org/10.1016/j.soilbio.2012.03.011.

Fan J, McConkey B, Wang H, et al. Root distribution by depth for temperate agricultural crops. Field Crop Res. 2016;189:68–74. https://doi.org/10.1016/j.fcr.2016.02.013.

Fan Y, Miguez-Macho G, Jobbágy EG, et al. Hydrologic regulation of plant rooting depth. PNAS. 2017;114:10572–677. https://doi.org/10.1073/pnas.1712381114.

Fatnassi IC, Chiboub M, Saadani O, et al. Impact of dual inoculation with *rhizobium* and PGPR on growth and antioxidant status of *Vicia faba* L. under copper stress. C R Biol. 2015;338(4):241–54. https://doi.org/10.1016/j.crvi.2015.02.001.

Fernandes R, Oliveira TS. Chapter 6: Physical subsoil constraints of agricultural and forestry land. In: Oliveira TS, Bell RW, editors. Subsoil constraints for crop production. Switzerland: Springer; 2022. p. 125–60.

Fernandez CW, Langley JA, Chapman S, et al. The decomposition of ectomycorrhizal fungal necromass. Soil Biol Biochem. 2016;93:38–49. https://doi.org/10.1016/j.soilbio.2015.10.017.

Ferreira PAA, Bomfeti CA, Soares BL, et al. Efficient nitrogen-fixing *Rhizobium* strains isolated from Amazonian soils are highly tolerant to acidity and aluminium. World J Microb Biot. 2012;28:1947–59. https://doi.org/10.1007/s11274-011-0997-7.

Fierer N, Schimel JP, Holden PA. Variations in microbial community composition through two soil depth profiles. Soil Biol Biochem. 2003;35:167–76. https://doi.org/10.1016/S0038-0717(02)00251-1.

Fonseca MB, França MGC, Zonta EGV. Crescimento inicial de *Dimorphandra wilsonii* (Fabaceae – Caesalpinioideae) em diferentes condições de fertilidade em solo de cerrado. Acta Bot Bras. 2010;24:322–7. https://doi.org/10.1590/S0102-33062010000200003.

Franco AA, Faria SM. The contribution of N_2-fixing tree legumes to land reclamation and sustainability in the tropics. Soil Biol Biochem. 1997;29:897–903. https://doi.org/10.1016/S0038-0717(96)00229-5.

Franke AC, van den Brand GJ, Vanlauwe B, et al. Sustainable intensification through rotations with grain legumes in Sub-Saharan Africa: a review. Agric Ecosyst Environ. 2018;261:172–85. https://doi.org/10.1016/j.agee.2017.09.029.

Gao Y, Zong J, Que H, et al. Inoculation with arbuscular mycorrhizal fungi increases glomalin related soil protein content and PAH removal in soils planted with *Medicago sativa* L. Soil Biol Biochem. 2017;115:148–51. https://doi.org/10.1016/j.soilbio.2017.08.023.

Garcia K, Zimmermann SD. The role of mycorrhizal associations in plant potassium nutrition. Front Plant Sci. 2014;5:337. https://doi.org/10.3389/fpls.2014.00337.

Gehring CA, Sthultz CM, Flores-Rentería L, et al. Tree genetics defines fungal partner communities that may confer drought tolerance. PNAS. 2017;114:11169–74. https://doi.org/10.1073/pnas.1704022114.

Genney DR, Anderson IA, Alexander IJ. Fine-scale distribution of pine ectomycorrhizas and their extrametrical mycelium. New Phytol. 2006;170:381–90. https://doi.org/10.1111/j.1469-8137.2006.01669.x.

Glick BR, Bashan Y. Genetic manipulation of plant growth-promoting bacteria to enhance biocontrol of phytopathogens. Biotechnol Adv. 1997;15:353–78. https://doi.org/10.1016/s0734-9750(97)00004-9.

Godbold DL, Hoosbeek MR, Lukac M, et al. Mycorrhizal hyphal turnover as a dominant process for carbon input into soil organic matter. Plant Soil. 2006;281:15–24. https://doi.org/10.1007/s11104-005-3701-6.

González-Chávez MC, Carrillo-Gonzáleza R, Wright SF, et al. The role of glomalin, a protein produced by arbuscular mycorrhizal fungi, in sequestering potentially toxic elements. Environ Pollut. 2004;130:317–23. https://doi.org/10.1016/j.envpol.2004.01.004.

Graham PH, Vance CP. Legumes: importance and constraints to greater use. Plant Physiol. 2003;131(3):872–7. https://doi.org/10.1104/pp.017004.

Grant C, Bittman S, Montreal M, et al. Soil and fertilizer phosphorus: effects on plant P supply and mycorrhizal development. Can J Plant Sci. 2005;85:3–14. https://doi.org/10.4141/P03-182.

Hai-Bi L, Rajesh KS, Pratiksha S, et al. Genetic diversity of nitrogen-fixing and plant growth promoting *Pseudomonas* species isolated from sugarcane rhizosphere. Front Microbiol. 2017;8:1268. https://doi.org/10.3389/fmicb.2017.01268.

Harrier LA, Watson CA. The potential role of arbuscular mycorrhizal (AM) fungi in the bioprotection of plants against soil-borne pathogens in organic and/or other sustainable farming systems. Pest Manag Sci. 2004;60:149–57. https://doi.org/10.1002/ps.820.

He D, Shen W, Eberwein J, et al. Diversity and co-occurrence network of soil fungi are more responsive than those of bacteria to shifts in precipitation seasonality in a subtropical forest. Soil Biol Biochem. 2017;115:499–510. https://doi.org/10.1016/j.soilbio.2017.09.023.

He X, Chen Y, Liu S, et al. Cooperation of earthworms and arbuscular mycorrhizae enhanced plant N uptake by balancing absorption and supply of ammonia. Soil Biol Biochem. 2018;116:351–9. https://doi.org/10.1016/j.soilbio.2017.10.038.

Hoang DTT, Bauke SL, Kuzyakov Y, et al. Rolling in the deep: priming effects in earthworm biopores in topsoil and subsoil. Soil Biol Biochem. 2017;114:59–71. https://doi.org/10.1016/j.soilbio.2017.06.021.

Hobbie EA. Carbon allocation to ectomycorrhizal fungi correlates with belowground allocation in culture studies. Ecology. 2006;87:563–9. https://doi.org/10.1890/05-0755.

Hossain M, Sultana F, Islam S. Plant growth-promoting fungi (PGPF): Phytostimulation and induced systemic resistance. In: Singh D, Singh H, Prabha R, editors. Plant-microbe interactions in agro-ecological perspectives. Singapore: Springer; 2017. p. 135–91. https://doi.org/10.1007/978-981-10-6593-4_6.

Islam S, Akanda AM, Sultana F, et al. Chilli rhizosphere fungus *Aspergillus* spp. PPA1 promotes vegetative growth of cucumber (*Cucumis sativus*) plants upon root colonisation. Arch Phytopathol PFL. 2014a;47:1231–8. https://doi.org/10.1080/03235408.2013.837633.

Islam S, Akanda AM, Prova A, et al. Growth promotion effect of *Fusarium* spp. PPF1 from Bermuda grass (*Cynodon dactylon*) rhizosphere on Indian spinach (*Basella alba*) seedlings are linked to root colonization. Arch Phytopathol PFL. 2014b;47:2319–31. https://doi.org/10.1080/03235408.2013.876745.

Jackson RB, Lajtha K, Crow SE, et al. The ecology of soil carbon: pools, vulnerabilities, and biotic and abiotic controls. Annu Rev Ecol Evol S. 2017;48:419–45. https://doi.org/10.1146/annurev-ecolsys-112414-054234.

Jakobsen I, Abbott LK, Robson AD. External hyphae of vesicular-arbuscular mycorrhizal fungi associated to *Trifolium subterraneum* L. 1. Spread of hyphae and phosphorus inflow into roots. New Phytol. 1992;120:371–80. https://doi.org/10.1111/j.1469-8137.1992.tb01077.x.

Jones DL, Nguyen C, Finlay RD. Carbon flow in the rhizosphere: carbon trading at the soil-root interface. Plant Soil. 2009;321:5–33. https://doi.org/10.1007/s11104-009-9925-0.

Karagiannidis N, Thomidis T, Lazari D, et al. Effect of three Greek arbuscular mycorrhizal fungi in improving the growth, nutrient concentration, and production of essential oils of oregano and mint plants. Sci Hortic-Amsterdam. 2011;129:329–34. https://doi.org/10.1016/j.scienta.2011.03.043.

Karlinski L, Rudawska M, Leski T. The influence of host genotype and soil conditions on ectomycorrhizal community of poplar clones. Eur J Soil Biol. 2013;58:51–8. https://doi.org/10.1016/j.ejsobi.2013.05.007.

Kaschuk G, Nogueira MA, Luca MJ, et al. Response of determinate and indeterminate soybean cultivars to basal and topdressing N fertilization compared to sole inoculation with *Bradyrhizobium*. Field Crop Res. 2016;195(15):21–7. https://doi.org/10.1016/j.fcr.2016.05.010.

Kautz T, Amelung W, Ewert F, et al. Nutrient acquisition from arable subsoils in temperate climates: a review. Soil Biol Biochem. 2013;57:1003–22. https://doi.org/10.1016/j.soilbio.2012.09.014.

Kermah M, Franke AC, Adjei-Nsiah S, et al. N_2-fixation and N contribution by grain legumes under different soil fertility status and cropping systems in the Guinea savannah of northern Ghana. Agric Ecosyst Environ. 2018;261:201–10. https://doi.org/10.1016/j.agee.2017.08.028.

Kirkby CA, Richardson AE, Wade LJ, et al. Nutrient availability limits carbon sequestration in arable soils. Soil Biol Biochem. 2014;68:402–9. https://doi.org/10.1016/j.soilbio.2013.09.032.

Klein T, Siegwolf RTW, Körner C. Belowground carbon trade among tall trees in temperate forests. Science. 2016;352(6283):342–4. https://doi.org/10.1126/science.aad6188.

Koide RT, Fernandez CW, Peoples MS. Can ectomycorrhizal colonization of *Pinus resinosa* roots affect their decomposition? New Phytol. 2011;191:508–14. https://doi.org/10.1111/j.1469-8137.2011.03694.x.

Kumar K, Rao IU. Morphophysiologicals problems in acclimatization of micropropagated plants in – ex vitro conditions – a review. J Ornam Hort Plant. 2012;2(4):271–83.

Kumar V, Sahai V, Bisaria VS. High-density spore production of *Piriformospora indica*, a plant growth-promoting endophyte, by optimization of nutritional and cultural parameters. Bioresour Technol. 2011;102:3169–75. https://doi.org/10.1016/j.biortech.2010.10.116.

Lagos ML, Maruyama F, Nannipieri P, et al. Current overview on the study of bacteria in the rhizosphere by modern molecular techniques: a mini–review. J Soil Sci Plant Nut. 2015;15:504–23. https://doi.org/10.4067/S0718-95162015005000042.

Li X, He X, Hou L, et al. Dark septate endophytes isolated from a xerophyte plant promote the growth of *Ammopiptanthus mongolicus* under drought condition. Sci Rep-UK. 2018;8(7896):1–11. https://doi.org/10.1038/s41598-018-26183-0.

Likar M, Regvar M. Isolates of dark septate endophytes reduce metal uptake and improve physiology of Salix caprea L. Plant Soil. 2013;370:593–604. https://doi.org/10.1007/s11104-013-1656-6.

Lilleskov EA, Hobbie EA, Horton TR. Conservation of ectomycorrhizal fungi: exploring the linkages between functional and taxonomic responses to anthropogenic N deposition. Fungal Ecol. 2011;4:174–83. https://doi.org/10.1016/j.funeco.2010.09.008.

Lin G, McCormack ML, Ma C, Guo D. Similar below-ground carbon cycling dynamics but contrasting modes of nitrogen cycling between arbuscular mycorrhizal and ectomycorrhizal forests. New Phytol. 2017;213:1440–51. https://doi.org/10.1111/nph.14206.

Lynch JP, Wojciechowski T. Opportunities and challenges in the subsoil: pathways to deeper rooted crops. J Exp Bot. 2015;66:2199–210. https://doi.org/10.1093/jxb/eru508.

Ma Q, Bell RW, Mattiello E. Chapter 12. Nutrient acquisition with particular reference to subsoil constraints. In: Oliveira TS, Bell RW, editors. Subsoil constraints for crop production. Switzerland: Springer; 2022. p. 289–321.

Maksimov IV, Abizgil'dina RR, Pusenkova LI. Plant growth promoting rhizobacteria as alternative to chemical crop protectors from pathogens (review). Appl Biochem Micro. 2011;47:333–45. https://doi.org/10.1134/S0003683811040090.

Malik MA, Marschner P, Khan KS. Addition of organic and inorganic P sources to soil – effects on P pools and microorganisms. Soil Biol Biochem. 2012;49:106–13. https://doi.org/10.1016/j.soilbio.2012.02.013.

Marschner H, Dell B. Nutrient uptake in mycorrhizal symbiosis. Plant Soil. 1994;159:89–102. https://doi.org/10.2307/42939410.

Matos ES, Mendonça ES, Lima PC, et al. Green manure in coffee systems in the region of Zona da Mata, Minas Gerais: characteristics and kinetics of carbon and nitrogen mineralization. Rev Bras Cienc Solo. 2008;32(5):2027–35. https://doi.org/10.1590/S0100-06832008000500024.

Midgley MG, Brzostek E, Phillips RP. Decay rates of leaf litters from arbuscular mycorrhizal trees are more sensitive to soil effects than litters from ectomycorrhizal trees. J Ecol. 2015;103:1454–63. https://doi.org/10.1111/1365-2745.12467.

Miethke M, Marahiel MA. Siderophore-based iron acquisition and pathogen control. Microbiol Mol Biol R. 2007;71:413–51. https://doi.org/10.1128/MMBR.00012-07.

Mohan JE, Cowden CC, Baas P, et al. Mycorrhizal fungi mediation of terrestrial ecosystem responses to global change: mini-review. Fungal Ecol. 2014;10:3–19. https://doi.org/10.1016/j.funeco.2014.01.005.

Moreira BC, Prates Júnior P, Jordão TC, et al. Effect of inoculation of symbiotic fungi on the growth and antioxidant enzymes' activities in the presence of *Fusarium subglutinans* f. sp. *ananas* in pineapple plantlets. Acta Physiol Plantarum. 2016;38:325–39. https://doi.org/10.1007/s11738-016-2247-y.

Mujic AB, Durall DM, Spatafora JW, et al. Competitive avoidance not edaphic specialization drives vertical niche partitioning among sister species of ectomycorrhizal fungi. New Phytol. 2016;209:1174–83. https://doi.org/10.1111/nph.13677.

Nassimi Z, Taheri P. Endophytic fungus *Piriformospora indica* induced systemic resistance against rice sheath blight via affecting hydrogen peroxide and antioxidants. Biocontrol Sci Tech. 2017;27(2):252–67. https://doi.org/10.1080/09583157.2016.1277690.

Neeraj KS. Organic amendments to soil inoculated arbuscular mycorrhizal fungi and *Pseudomonas fluorescens* treatments reduce the development of root-rot disease and enhance the yield of *Phaseolus vulgaris* L. Eur J Soil Biol. 2011;47:288–95. https://doi.org/10.1016/j.ejsobi.2011.07.002.

Nicolitch O, Colin Y, Turpault M-P, et al. Tree roots select specific bacterial communities in the subsurface critical zone. Soil Biol Biochem. 2017;115:109–23. https://doi.org/10.1016/j.soilbio.2017.07.003.

Numan M, Bashir S, Khan Y, et al. Plant growth promoting bacteria as an alternative strategy for salt tolerance in plants: a review. Microbiol Res. 2018;209:21–32. https://doi.org/10.1016/j.micres.2018.02.003.

Oehl F, Sieverding E, Ineichen K, et al. Community structure of arbuscular mycorrhizal fungi at different soil depths in extensively and intensively managed agroecosystems. New Phytol. 2005;165:273–83. https://doi.org/10.1111/j.1469-8137.2004.01235.x.

Oelmüller R, Sherameti I, Tripathi S, et al. *Piriformospora indica*, a cultivable root endophyte with multiple biotechnological applications. Symbiosis. 2009;49:1–17. https://doi.org/10.1007/s13199-009-0009-y.

Okon Y, Vanderleyden J. Root-associated *Azospirillum* species can stimulate plants. Appl Environ Microb. 1997;63:366–70.

Okoth AS, Okoth P, Muya E. Influence of soil chemical and physical properties on occurrence of *Trichoderma* spp. in Embu, Kenya. Trop Subtrop Agroecosyst. 2009;11:303–12.

Patel T, Saraf M. Biosynthesis of phytohormones from novel rhizobacterial isolates and their in vitro plant growth-promoting efficacy. Plant-Microorganism Interact. 2017;12:480–7. https://doi.org/10.1080/17429145.2017.1392625.

Pena R, Lang C, Lohaus G, et al. Phylogenetic and functional traits of ectomycorrhizal assemblages in top soil from different biogeographic regions and forest types. Mycorrhiza. 2017;27:233–45. https://doi.org/10.1007/s00572-016-0742-z.

Perkons U, Kautz T, Uteau D, et al. Root-length densities of various crops following crops with contrasting root systems. Soil Till Res. 2014;137:50–7. https://doi.org/10.1016/j.still.2013.11.005.

Pii Y, Mimmo T, Tomasi N, et al. Microbial interactions in the rhizosphere: beneficial influences of plant growth-promoting rhizobacteria on nutrient acquisition process. A review. Biol Fert Soils. 2015;51:403–15. https://doi.org/10.1007/s00374-015-0996-1.

Plassard C, Dell B. Phosphorus nutrition of mycorrhizal trees. Tree Physiol. 2010;30:1129–39. https://doi.org/10.1093/treephys/tpq063.

Pozo M, Azcón-Aguilar C. Unraveling mycorrhiza-induced resistance. Curr Opin Plant Biol. 2007;10:393–8. https://doi.org/10.1016/j.pbi.2007.05.004.

Pozo MJ, Cordier C, Dumas-Gaudot E, et al. Localized versus systemic effect of arbuscular mycorrhizal fungi on defense responses *Phytophthora infection* in tomato plants. J Exp Bot. 2002;53(368):525–34. https://doi.org/10.1093/jexbot/53.368.525.

Prates Júnior P, Moreira BC, da Silva MCS, et al. Agroecological coffee management increases arbuscular mycorrhizal fungi diversity. PLoS One. 2019;14(1):e0209093. https://doi.org/10.1371/journal.pone.0209093.

Prates Júnior P, Moreira BC, Silva MCS, et al. Mycorrhizal inoculation and phosphorus fertilization show contrasts on native species of the Brazilian Atlantic Forest and Cerrado. Rev Bras Cienc Solo. 2021;45:e0210013. https://doi.org/10.36783/18069657rbcs20210013.

Promwee A, Issarakraisila M, Intana W, et al. Phosphate solubilization and growth promotion of rubber tree (*Hevea brasiliensis* Muell. Arg.) by *Trichoderma* strains. J Agr Sci. 2014;6:8–20. https://doi.org/10.5539/jas.v6n9p8.

Rabiey M, Shaw MW. *Piriformospora indica* reduces *Fusarium* head blight disease severity and mycotoxin DON contamination in wheat under UK weather conditions. Plant Pathol. 2016;65:940–52. https://doi.org/10.1111/ppa.12483.

Radhakrishnan R, Kang S-M, Baek I-Y, et al. Characterization of plant growth promoting traits of *Penicillium* species against the effects of high soil salinity and root disease. Plant-Microorganism Interact. 2014;9:754–62. https://doi.org/10.1080/17429145.2014.930524.

Rashid MI, Mujawar LH, Shahzad T, et al. Bacteria and fungi can contribute to nutrients bioavailability and aggregate formation in degraded soils. Microbiol Res. 2016;183:26–41. https://doi.org/10.1016/j.micres.2015.11.007.

Remigi P, Zhu J, Young JPW, et al. Symbiosis within symbiosis: evolving nitrogen-fixing legume symbionts. Trends Microbiol. 2016;24:63–75. https://doi.org/10.1016/j.tim.2015.10.007.

Rengasamy P, de Lacerda CF, Gheyi HR. Chapter 4. Salinity, sodicity and alkalinity. In: Oliveira TS, Bell RW, editors. Subsoil constraints for crop production. Switzerland: Springer; 2022. p. 83–107.

Revillini D, Gehring CA, Johnson NC. The role of locally adapted mycorrhizas and rhizobacteria in plant-soil feedback systems. Funct Ecol. 2016;30:1086–98. https://doi.org/10.1111/1365-2435.12668.

Rillig MC. Arbuscular mycorrhizae, glomalin, and soil aggregation. Can J Soil Sci. 2004;84:355–63. https://doi.org/10.4141/S04-003.

Rillig MC, Wright SF, Eviner VT. The role of arbuscular mycorrhizal fungi and glomalin in soil aggregation: comparing effects of five plant species. Plant Soil. 2002;238:325–33. https://doi.org/10.1023/A:1014483303813.

Roper MM, Gupta VVSR. Enhancing non-symbiotic N_2 fixation in agriculture. Open Agric J. 2016;10:7–27. https://doi.org/10.2174/1874331501610010007.

Roper MM, Gupta VVSR, Murphy DV. Tillage practices altered labile soil organic carbon and microbial function without affecting crop yields. Aust J Soil Res. 2010;48:274–85. https://doi.org/10.1071/SR09143.

Rosling A, Landeweert R, Lindahl BD, et al. Vertical distribution of ectomycorrhizal fungal taxa in a podzol profile. New Phytol. 2003;159:775–83. https://doi.org/10.1046/j.1469-8137.2003.00829.x.

Rozpadek P, Wezowicz K, Stojakowska A, et al. Mycorrhizal fungi modulate phytochemical production and antioxidant activity of *Cichorium intybus* L. (Asteraceae) under metal toxicity. Chemosphere. 2014;112:217–24. https://doi.org/10.1016/j.chemosphere.2014.04.023.

Rupela OP, Toomsan B, Mittal S, et al. Chickpea rhizobium populations: survey of influence of season, soil depth and cropping pattern. Soil Biol Biochem. 1987;19(3):247–52. https://doi.org/10.1016/0038-0717(87)90005-8.

Scattolin L, Montecchio L, Mosca E, et al. Vertical distribution of the ectomycorrhizal community in the top soil of Norway spruce stands. Eur J For Res. 2008;127:347–57. https://doi.org/10.1007/s10342-008-0209-7.

Schaeffer C, Oliveira F. Chapter 15. Live subsoils: tropical regolith and biota interplays. In: Oliveira TS, Bell RW, editors. Subsoil constraints for crop production. Switzerland: Springer; 2022. p. 373–96.

Schenk HJ, Jackson RB. The global biogeography of roots. Ecol Monogr. 2002;72:311–28. https://doi.org/10.1890/0012-9615(2002)072[0311:TGBOR]2.0.CO;2.

Schneider F, Don A, Hennings I, et al. The effect of deep tillage on crop yield – what do we really know? Soil Till Res. 2017;174:193–204. https://doi.org/10.1016/j.still.2017.07.005.

Selosse MA, Dubois MP, Alvarez N. Do Sebacinales commonly associate with plant roots as endophytes? Mycol Res. 2009;113:1062–9. https://doi.org/10.1016/j.mycres.2009.07.004.

Sevilla M, Burris RH, Gunapala N, et al. Comparison of benefit to sugarcane plant growth and $^{15}N_2$ incorporation following inoculation of sterile plants with *Acetobacter diazotrophicus* wild-type and Nif- mutants strains. Mol Plant Microbe In. 2001;14:358–66. https://doi.org/10.1094/MPMI.2001.14.3.358.

Shahabivand S, Maivan HZ, Goltapeh EM, et al. The effects of root endophyte and arbuscular mycorrhizal fungi on growth and cadmium accumulation in wheat under cadmium toxicity. Plant Physiol Bioch. 2012;60:53–8. https://doi.org/10.1016/j.plaphy.2012.07.018.

Shahin O, Martin-St Paul N, Rambal S, et al. Ectomycorrhizal fungal diversity in *Quercus ilex* Mediterranean woodlands: variation among sites and over soil depth profiles in hyphal exploration types, species richness and community composition. Symbiosis. 2013;61:1–12. https://doi.org/10.1007/s13199-013-0252-0.

Shaw O, Cocq KL, Paszkiewicz K, et al. Transcriptional reprogramming underpins enhanced plant growth promotion by the biocontrol fungus *Trichoderma hamatum* GD12 during antagonistic interactions with *Sclerotinia sclerotiorum* in soil. Mol Plant Pathol. 2016;17:1425–41. https://doi.org/10.1111/mpp.12429.

Sieverding E. Vesicular-arbuscular mycorrhiza management in tropical agrosystems. Technical cooperation, Eschborn, Federal Republic of Germany; 1991. p. 371.

Singh RP, Jha PN. Alleviation of salinity-induced damage on wheat plant by an ACC deaminase-producing halophilic bacterium *Serratia* sp SL-12 isolated from a salt lake. Symbiosis. 2016;69:101–11. https://doi.org/10.1007/s13199-016-0387-x.

Singh JS, Kumar A, Rai AN, et al. Cyanobacteria: a precious bio-resource in agriculture, ecosystem, and environmental sustainability. Front Microbiol. 2016;7:529. https://doi.org/10.3389/fmicb.2016.00529.

Smith SE, Read DJ. Mycorrhizal Symbiosis. 3rd ed. New York: Academic Press; 2008.

Smith SE, Facelli E, Pope S, et al. Plant performance in stressful environments: interpreting new and established knowledge of the roles of arbuscular mycorrhizas. Plant Soil. 2010;326:3–20. https://doi.org/10.1007/s11104-009-9981-5.

Smith SE, Jakobsen I, Gronlund M, et al. Roles of arbuscular mycorrhizas in plant phosphorus nutrition: interactions between pathways of phosphorus uptake in arbuscular mycorrhizal roots have important implications for understanding and manipulating plant phosphorus acquisition. Plant Physiol. 2011;156:1050–7. https://doi.org/10.1104/pp.111.174581.

Spinelli D, Bardi L, Fierro A, et al. Environmental analysis of sunflower production with different forms of mineral nitrogen fertilizers. J Environ Manag. 2013;129:302–8. https://doi.org/10.1016/j.jenvman.2013.07.029.

Stone MM, DeForest JL, Plante AF. Changes in extracellular enzyme activity and microbial community structure with soil depth at the Luquillo Critical Zone Observatory. Soil Biol Biochem. 2014;75:237–47. https://doi.org/10.1016/j.soilbio.2014.04.017.

Sulman BN, Brzostek ER, Medici C, et al. Feedbacks between plant N demand and rhizosphere priming depend on type of mycorrhizal association. Ecol Lett. 2017;20:1043–53. https://doi.org/10.1111/ele.12802.

Taylor MK, Lankau RA, Wurzburger N. Mycorrhizal associations of trees have different indirect effects on organic matter decomposition. J Ecol. 2016;104:1576–84. https://doi.org/10.1111/1365-2745.12629.

Terrer C, Vicca S, Hungate BA, et al. Mycorrhizal association as a primary control of the CO_2 fertilization effect. Science. 2016;353(6294):72–3. https://doi.org/10.1126/science.aaf4610.

Urquiaga S, Cruz KHS, Boddey RM. Contribution of nitrogen fixation to sugar cane: nitrogen-15 and nitrogen-balance estimates. Soil Sci Soc Am J. 1992;56:105–14.

Urquiaga S, Xavier RP, Morais RF, et al. Evidence from field nitrogen balance and ^{15}N natural abundance data of the contribution of biological N_2 fixation to Brazilian sugarcane varieties. Plant Soil. 2012;356:5–21. https://doi.org/10.2136/sssaj1992.03615995005600010017x.

van Breemen N, Finzi IAC. Plant-soil interactions: ecological aspects and evolutionary implications. Biogeochemistry. 1998;42:1–19. https://doi.org/10.1023/A:1005996009413.

Van der Putten WH, Bardgett RD, Bever JD, et al. Plant-soil feedbacks: the past, the present and future challenges. J Ecol. 2013;101:265–76. https://doi.org/10.1111/1365-2745.12054.

Van Schöll L, Kuyper TW, Smits MM, et al. Rock-eating mycorrhizas: their role in plant nutrition and biogeochemical cycles. Plant Soil. 2008;303:35–47. https://doi.org/10.1007/s11104-007-9513-0.

Vandenkoorhuyse P, Quaiser A, Duhamel M, et al. The importance of the microbiome of the plant holobiont. New Phytol. 2015;206:1196–206. https://doi.org/10.1111/nph.13312.

Vanlauwe B, Coyne D, Gockwski J, et al. Sustainable intensification and the African smallholder farmer. Curr Opin Env Sust. 2014;8:15–22. https://doi.org/10.1016/j.cosust.2014.06.001.

Varma A, Verma S, Sudha SN, et al. *Piriformospora indica*, a cultivable plant-growth-promoting root endophyte. Appl Environ Microb. 1999;65:2741–4.

Varma A, Bakshi M, Lou B, et al. *Piriformospora indica*: a novel plant growth-promoting mycorrhizal fungus. Agr Res. 2012;1(2):117–31. https://doi.org/10.1007/s40003-012-0019-5.

Vergara C, Araujo KEC, Alves LS, et al. Contribution of dark septate fungi to the nutrient uptake and growth of rice plants. Braz J Microbiol. 2018;49:67–78. https://doi.org/10.1016/j.bjm.2017.04.010.

Verma S, Varma A, Rexer K-H, et al. *Piriformospora indica*, gen. et sp. nov., a new root-colonizing fungus. Mycologia. 1998;90:896–903. https://doi.org/10.1080/00275514.1998.12026983.

Wakelin AS, Gupta VVSR, Forrester ST. Regional and local factors affecting diversity, abundance and activity of free-living, N_2-fixing bacteria in Australian agricultural soils. Pedobiologia. 2010;53:391–9. https://doi.org/10.1016/j.pedobi.2010.08.001.

Wallander H, Göransson H, Rosengren U. Production, standing biomass and natural abundance of ^{15}N and ^{13}C in ectomycorrhizal mycelia collected at different soil depths in two forest types. Oecologia. 2004;139:89–97. https://doi.org/10.1007/s00442-003-1477-z.

Waller F, Achatz B, Baltruschat H, et al. The endophytic fungus *Piriformospora indica* reprograms barley to salt-stress tolerance, disease resistance, and higher yield. PNAS. 2005;102(38):13386–91. https://doi.org/10.1073/pnas.0504423102.

Wang Q, Cheng C, He L, et al. Characterization of depth-related changes in bacterial communities involved in mineral weathering along a mineral-rich soil profile. Geomicrobiol J. 2014;31:43–444. https://doi.org/10.1080/01490451.2013.848248.

Waqas M, Khan AL, Hamayun M, et al. Endophytic fungi promote plant growth and mitigate the adverse effects of stem rot: an example of *Penicillium citrinum* and *Aspergillus terreus*. J Plant Interact. 2015;10:280–7. https://doi.org/10.1080/17429145.2015.1079743.

Wei Y, Zhao Y, Shia M, et al. Effect of organic acids production and bacterial community on the possible mechanism of phosphorus solubilization during composting with enriched phosphate-solubilizing bacteria inoculation. Bioresour Technol. 2018;247:190–9. https://doi.org/10.1016/j.biortech.2017.09.092.

Weisskopf L, Abou-Mansour E, Fromin N, et al. White lupin has developed a complex strategy to limit microbial degradation of secreted citrate required for phosphate acquisition. Plant Cell Environ. 2006;29:919–27. https://doi.org/10.1111/j.1365-3040.2005.01473.x.

White RG, Kirkegaard JA. The distribution and abundance of wheat roots in a dense, structured subsoil – implications for water uptake. Plant Cell Environ. 2010;33:133–48. https://doi.org/10.1111/j.1365-3040.2009.02059.x.

Yadav V, Kumar M, Deep DK, et al. A phosphate transporter from the root endophytic fungus *Piriformospora indica* plays a role in the phosphate transport to the host plant. J Biol Chem. 2010;285:26532–44. https://doi.org/10.1074/jbc.M110.111021.

Yang Y, He C, Huang L, et al. The effects of arbuscular mycorrhizal fungi on glomalin-related soil protein distribution, aggregate stability and their relationships with soil properties at different soil depths in lead-zinc contaminated area. PLoS One. 2017;12(8):e0182264. https://doi.org/10.1371/journal.pone.0182264.

Zahir ZA, Arshad M, Frankenberger WT. Plant growth promoting rhizobacteria: applications and perspectives in agriculture. Adv Agron. 2004;81:97–168. https://doi.org/10.1016/S0065-2113(03)81003-9.

Chapter 12
Nutrient Acquisition with Particular Reference to Subsoil Constraints

Qifu Ma, Richard Willian Bell, and Edson Marcio Mattiello

Abstract With the wide adoption of conservation agriculture (minimal soil disturbance, stubble retention, crop rotation), soil nutrient stratification is becoming more prevalent especially for poorly mobile phosphorus (P), potassium (K), copper (Cu), zinc (Zn), and manganese (Mn) that concentrate in the fertilized topsoil (0–10 cm). In water-limited environments, surface soil drying limits root access to the topsoil nutrients, but the nutrients in moist subsoil may play a substantial role in crop nutrition and growth. Although the subsoil is generally lower in available nutrients and organic matter than the topsoil, there is strong evidence that subsoil can contribute significant amounts of nitrogen (N), P, and K taken up by crops. Placing fertilizers deeper in soil profiles increases plant nutrient efficiency in low rainfall regions, because deep fertilizing can induce deeper root growth and leave fertilizer-supplied nutrients in moist subsoil for longer periods during the growing season. However, the contribution of subsoil nutrition to crop growth is limited by subsoil constraints that restrict deeper rooting, including physical constraints, e.g. gravel layers and soil compaction, and chemical constraints, e.g. acidity, alkalinity, salinity, sodicity, nutrient deficiency, and element toxicity. On the other hand, crops and genotypes efficient in nutrient uptake under drought are likely to have an extensive, deep root system and thus a large surface area of contact between roots and soil. The uptake of soil water from moist subsoil and its release into dry topsoil by roots – hydraulic redistribution – may maintain the growth of fine roots and thus prolong nutrient uptake from drying surface soils. A good understanding of subsoil nutrient acquisition by crop species and their response to subsoil constraints is required for designing crop rotations and nutrient management programmes that allow for effective use of subsoil water and nutrients, especially in rainfed agriculture.

Q. Ma (✉) · R. W. Bell
Centre for Sustainable Farming Systems, Food Futures Institute, Murdoch University, Murdoch, WA, Australia
e-mail: q.ma@murdoch.edu.au; r.bell@murdoch.edu.au

E. M. Mattiello
Soils Department, Federal University of Viçosa, Viçosa, MG, Brazil
e-mail: mattiello@ufv.br

T. S. d. Oliveira, R. W. Bell (eds.), *Subsoil Constraints for Crop Production*,
https://doi.org/10.1007/978-3-031-00317-2_12

Keywords Deep fertilization · Nutrient acquisition · Root growth · Subsoil · Subsoil constraints · Topsoil

12.1 Introduction

To date, nutrient resources in the subsoil have been largely neglected in most agronomic and plant nutrition studies. With the increasing cost of nutrient supply as fertilizers, there is a need to establish the availability of nutrients stored in the subsoil for crop growth. Studies have reported that even when the topsoils have poor P or K status, yield increase may not be achieved by applying fertilizers due to the inability of plants to access nutrients in dry topsoil layers (Schachtschabel 1985; Jungk et al. 1993). Subsoil N, P, K, sulphur (S), calcium (Ca), magnesium (Mg), and micronutrients can make a significant contribution to crop nutrient uptake (Marschner 1995; Kautz et al. 2013; Scanlan et al. 2015b). The relevance of subsoil nutrients for crops is particularly high when the topsoil is dry or depleted in nutrients (Kuhlmann and Baumgärtel 1991; Kautz et al. 2013). Low availability of plant nutrients under drought can be improved by placing fertilizers deeper in soil profiles, where crops would have better access to soil moisture and deeper root growth for greater uptake and use of applied fertilizers (reviewed by Ma et al. 2009). Many studies have shown beneficial effects of deep fertilizer placement on crop growth and grain yield (Jarvis and Bolland 1991; Lotfollahi et al. 1997; Crabtree 1999; Hocking et al. 2003; Singh et al. 2005; Ma et al. 2007; Kabir et al. 2015). However, the effectiveness of subsoil nutrients or deep placement of fertilizers varies among crop species, due to different root distribution patterns, and is impaired by limiting edaphic conditions, such as subsoil compaction, acidity, salinity, nutrient deficiency, and toxicity, that restrict rooting depth (Lynch and Wojciechowski 2015). This chapter will review the impact of subsoil constraints on root growth and nutrient acquisition by crops and pastures with particular emphasis on minimal tillage and low rainfall environments. The association of morphological and physiological traits of plant roots (type, distribution, hydraulic redistribution, drought tolerance) with nutrient efficiency under soil nutrient heterogeneity will also be discussed.

12.2 Root Types and Distribution

Roots are important to the plant for a wide variety of processes, including nutrient and water uptake, anchorage and mechanical support, storage functions, and as the major interface between the plant and various biotic and abiotic factors in the soil environment (Smith and Smet 2012). In most plants, the first structure to emerge from the germinating seed is the radicle – primary root (Clowes 1961). In dicotyledons, e.g. canola and beans, the primary roots are often long-lived and form a prominent taproot that may continue to grow during the plant's lifecycle, with lateral roots emerging from the pericycle layer of the parent roots (Malamy and Benfey 1997a, b). The taproot can penetrate relatively deeply to acquire soil water, while shallow lateral roots would take up P and other nutrients that are less abundant in the lower soil profile (Zobel 2005). In monocotyledons, e.g. wheat and rice, the primary roots are often short-lived, and the emergence of multiple lateral and adventitious roots forms a fibrous root system (Aloni et al. 2006). Although cereals and other monocotyledons form primary and lateral roots in a manner roughly similar to dicotyledons, the overall root architecture is more complex in monocotyledons. In cereals, shoot-borne 'crown' roots that emerge from the below-ground stem nodes and 'brace' or 'adventitious' roots from the above-ground stem nodes constitute the majority of the monocotyledons root system (Smith and Smet 2012).

The size and distribution of root systems largely determine the ability of arable field crops to acquire soil water and nutrients. Overall, root volumes increase from the seedling stage to a maximum at anthesis in cereals, such as wheat (Rose et al. 2007; Gan et al. 2011; Ma et al. 2013), and at late flowering in broadleaf annual dicotyledons, such as chickpea, pea, lentil, and canola (Rose et al. 2007; Liu et al. 2010; Gan et al. 2011), and then decline to maturity. For a given crop, this pattern in root growth is similar under low- and high-water conditions and generally depends upon crop phenology (Gan et al. 2011). Due to the difference in temporal root growth among species, positional availability during the growing season would differentially affect nutrient acquisition and efficiency, e.g. post-anthesis nutrient supply is more important to canola than to wheat (Rose et al. 2007). Differences in the spatial distribution of roots can also affect nutrient acquisition. Root volumes decrease with the increase in soil depths. On average, about 44% of roots are located in the top 20 cm, 70% in the top 40 cm, and 90% in the top 60 cm. Below 60 cm, there can be significant differences between the abilities of species to extend their root systems. For example, 5% of the roots of pulses may be found below this depth, compared with 12% for wheat, 13% for mustard, and 18% for canola (Gan et al. 2011). Deeper rooting is particularly beneficial for the plant to obtain water and nutrients from greater soil depths during periods of drought (McKenzie et al. 2009; Gaiser et al. 2012; Lynch and Wojciechowski 2015). However, deep rooting and nutrient acquisition are often impaired by a range of physical and chemical constraints in the subsoil (see below). The knowledge of root growth and its interactions with subsoil constraints is useful for developing more productive, sustainable, and resilient cropping systems.

12.3 Root Growth in Heterogeneous Soil

Most soils are inherently heterogeneous, and consequently nutrient availability is variable over relatively small distances in space and over time (Jackson and Caldwell 1993; Farley and Fitter 1999; Lark et al. 2004). Except for soil N derived from the fixation of atmospheric N_2, plant nutrients in the soil arise mainly from mineral weathering of parental materials, with smaller amounts from atmospheric deposition. The interaction of nutrients derived from weathering and atmospheric deposition with soil processes, such as nutrient leaching and biological cycling, results in the formation of vertical and horizontal nutrient gradients within the soil profile (Giehl and von Wirén 2014). For example, P availability increases in the topsoil because organic matter cycling is more concentrated in this soil stratum (Lynch and Brown 2001). In contrast, the concentrations of S may increase in deeper soil profiles due to SO_4 leaching (Jobbágy and Jackson 2001).

Roots entering the subsoil meet an environment remarkably different from the topsoil, e.g. the subsoil has higher bulk density and lower carbon and plant nutrient concentrations than the topsoil. On farmland, mineralization happens relatively quickly in the topsoil, particularly following cultivation, and nutrients are typically placed on or near the surface during fertilizer applications. In the subsoil, however, nutrient release from chemical weathering or accumulation due to leaching have a greater effect on nutrient availability (Kautz et al. 2013). Soil type can also have a large impact on how nutrients are distributed throughout the soil profile. For example, in some texture-contrast soils, K and S contents are much lower in the top sandy horizons than in the underlying clay and gravel layers (Robson et al. 1992), or the A2 horizon can be low in nutrients relative to both the topsoil above and the clay subsoil below (e.g. P, Hall et al. 2010). Under many climatic conditions, nutrient availability in the topsoil also declines during the growing season because low soil water content becomes a limiting factor for nutrient delivery to the root surface and/or leaching during the early part of the season depletes the topsoil of plant available nutrients. Under these conditions, nutrient uptake from the subsoil can become particularly important for plant growth. For example, in alfalfa the root mass in the subsoil, which represents ~3% of the total root mass, has been observed to take up >60% of the total plant nutrients. Similarly, in spring wheat 40–50% of P and K uptake in the later stages of growth can be derived from the subsoil (Marschner 1995).

In addition to the intrinsic soil nutrient heterogeneity, management practices such as zero tillage and the shallow banding of fertilizers at seeding have increased nutrient stratification in the surface soil layer (Ma et al. 2009) and may impair soil nutrient availability and root uptake in rainfed agriculture (Robson and Taylor 1987). For example, surface application of P and K fertilizers and limited incorporation of crop residues into the soil are largely responsible for vertical stratification of P and K nutrients in the topsoil (Mackay et al. 1987; Morrison and Chichester 1994; Howard et al. 1999; Duiker and Beegle 2006). Placing fertilizers in bands with crop rows in the same position year after year or having them offset from the previous-year rows in wide-row spacing or for disease/weed management may also cause horizontal heterogeneity in topsoil nutrient distributions.

Plant roots are able to adapt to localized nutrient availability in two ways: (*i*) by increasing the rate of nutrient uptake in the enriched zone, a physiological response that is rapid and reversible, (*ii*) through root proliferation (more lateral roots) in the zone of nutrient availability, a morphological response that is slow and irreversible. These adaptive responses by root systems are the major mechanisms by which plants cope with soil nutrient heterogeneity (Robinson 1994, 1996; Dunbabin et al. 2001a, b; Giehl and von Wirén 2014). The ability of plants to alter their root uptake and morphology in response to nutrient heterogeneity is both species and nutrient specific. For example, there tends to be greater plasticity in graminaceous (e.g. wheat, maize) than leguminous species (e.g. faba bean, chickpea) (Li et al. 2014). Root proliferation in concentrated bands of nutrients occurs for N and P, but not for K (Drew 1975). Some studies have observed that higher plant uptake of nutrients in nutrient-rich patches can lead to greater overall nutrient uptake and plant biomass than under homogeneous conditions with the same overall nutrient supply (Jackson and Caldwell 1996; Wijesinghe et al. 2001). Roots grown with banded nutrient supplies have also been observed to penetrate into deeper soil layers and exhibit greater drought resistance (Singh et al. 2005). Although this is not a uniform response (Hutchings and John 2004), fertilizer placement can alter root distribution more than the total root length in wheat (Alston 1976) and chickpea (Li et al. 2014), and soil nutrient stratification directly suppresses root growth in the deficient patches (Ma et al. 2007).

12.4 Root Growth Under Drought

Drought can be defined as a period of low precipitation (rain and snow) over an extended period. In agriculture, drought also occurs when there is insufficient soil moisture to meet crop demand in the growing season, e.g. during hot periods of low precipitation. Root elongation in drying soils is limited by a combination of water stress and mechanical impedance (Hinsinger et al. 2009; Bengough et al. 2011). Severe water stress, i.e. when water potential is significantly lower in the soil than in the plant, will result in tissue dehydration, loss of cell turgor, and restriction of root growth (Rich and Watt 2013). Soil strength also increases during drying, making it harder for plant roots to penetrate the soil (see below). However, the often heterogeneous nature of water distribution within the soil profile and the plastic nature of root systems mean that while drought is generally detrimental to overall plant health and crop yields, many species can withstand extended periods with limited water input (Rich and Watt 2013).

In dry soils, despite increased mechanical impedance, root growth is often less depressed than shoot growth, leading to an increase in root/shoot dry weight ratio. For example, the root/shoot ratio in maize seedlings increased from 1.45 without drought to 5.79 under drought conditions (Sharp et al. 1988). In rice, the drought-induced increase in root/shoot ratio is closely associated with high contents of soluble sugars in the roots via an increase in leaf sucrose-phosphate synthase and root invertase activity and more sucrose transport from leaves to roots (Xu et al. 2015). A faster osmotic adjustment of the roots compared to the shoot may also be involved (Schildwacht 1988). Drought-tolerant plants also tend to develop a deeper root system, allowing roots to extend into still moist subsoils and maintain adequate water supply to the shoots (Gregory 2006; Ingram and Malamy 2010). Some species also develop strategies to increase the ability of roots to extract soil nutrients, for example, chickpea increases the exudation of carboxylates to increase access to the subsoil P when the topsoil dries out during plant growth (Kabir et al. 2015). Soil drying will also stimulate the development of root hairs, which increases total root surface area and water and nutrient uptake (Marschner 1995; Wang and Yamauchi 2006). Increasing root system access to water deep in the soil profile may be a promising way to enhance water and nutrient capture and drought tolerance in water-limited environments (Dodd et al. 2011; Jin et al. 2013; White et al. 2013).

12.5 Root Growth with Subsoil Constraints

While plants may have strategies to increase root growth into the subsoils under water-limited conditions, this growth can be limited by a range of both physical and chemical constraints that are commonly present in subsoil environments.

12.5.1 Physical Constraints

A range of physical constraints may occur in subsoils, but the most common ones are compacted and hardsetting layers. Soil compaction typically occurs as a result of the long-term application of compressive forces, especially from heavy machinery (Batey 2009; Oliveira and Fernandes 2022), and is one of the major physical factors hindering root penetration into the deep soil horizons (Materechera et al. 1992; Valentine et al. 2012). Hardsetting layers (Fig. 12.1), where soil would turn to a hard, structureless mass during drying, can also significantly restrict root growth and development (Daniells 2012). In a wet and compacted soil, the elongation of roots may be limited by mechanical soil impedance and hypoxia, whereas in a dry and compacted soil, root growth is impaired by soil strength and water deficit (Lynch and Wojciechowski 2015). In many species, root elongation is suppressed when penetrometer resistance of the soil is >2 MPa, air-filled volume is <10%, and the matric potential is lower than −1.5 MPa (Bengough et al. 2011). In compacted soils, roots are normally shorter, thicker, and more irregularly shaped than the thinner, fibrous roots that develop under noncompacted conditions (Masle 2002). This is because radial thickening can make roots more resistant to buckling (Materechera et al. 1992; Clark et al. 2008; Whalley et al. 2008; Bengough et al. 2011) and increase root penetration in soil with high strength. There can be considerable variation between the ability of different species/genotypes to penetrate into compacted soils, and this genetic diversity may help identify root traits able to overcome the constraints of soil compaction. For example, wheat (e.g. Botwright Acuña and Wade 2012) and lupins (e.g. Chen et al. 2014) differ in their response to soil mechanical impedance. Although screening for faster root growth alone is unlikely to identify wheat genotypes suited to soils with physical constraints, genotypic variation in root depth is related to differences in root traits that confer an ability to penetrate high-strength soils (Botwright Acuña et al. 2012).

Fig. 12.1 Eucalyptus root growth on a hardsetting soil in Espírito Santo, Brazil. No or few roots grow deeper than 30 cm deep. (Courtesy of Nairam F. Barros)

In addition to altering their physiology to facilitate growth through compacted soils, roots may also alter their growth to take advantage of lines of least resistance, such as soil biopores and cracks (White and Kirkegaard 2010). Biopores are mainly created by root penetration (McCallum et al. 2004; Bodner et al. 2014) and earthworm movement (Lamandé et al. 2003; van Schaik et al. 2014) and can play a crucial role in allowing root growth into deep soil horizons (Kautz et al. 2013; Han et al. 2017) and increasing nutrient adsorption (Perkons et al. 2014). In arable fields, biopore formation is strongly affected by the root penetration capacity of the species grown (Materechera et al. 1992). For example, species characterized by large root diameters (e.g. lucerne) are capable of the greatest penetration into compacted subsoils (McCallum et al. 2004). The number of earthworm burrows in the subsoil can be ninefold higher than in the topsoil (Ehlers 1975), and the drilosphere, defined as a 2 mm wide zone around the earthworm burrows, represents a microsite often enriched in soil organic matter and nutrients, with a generally higher accessibility for roots (Brown et al. 2000; Kautz et al. 2013). However, the effects of biopores on nutrient extraction depend on biopore size and plant root characteristics, and further research is needed to develop direct methods to quantify biopore-root-shoot processes and root phenotyping for detecting genetic variation in response to biopore systems (Han et al. 2017).

Many arable soils contain significant amounts of gravel (>2 mm), ranging from <10% to >50% by volume, which can act as a constraint to root growth. The distribution of gravelly soils varies highly from region to region (Scanlan et al. 2022). For example, gravelly soils are common in the Mediterranean areas of Western Europe (Poesen and Lavee 1994), and three out the 18 Mha of agricultural lands in the southwest of Western Australia are lateritic gravelly soils (Bowden 2014). The presence of gravels increases soil bulk density and reduces soil water availability and may restrict root growth if gravels are in a closed packing arrangement. Total root length and root penetration decrease with increasing gravel content, and the growth of roots may be reduced by 40–75% in gravelly soils (Lal and Shukla 2004). High gravel contents also affect the interpretation of soil quality indicators as most of the indicators are measured after the gravels are removed by sieving (Bowden 2014). Therefore, for gravelly subsoils the gravel content should be quantified to give true indications of subsoil condition and to improve soil management.

12.5.2 *Chemical Constraints*

Nutrient Deficiencies

Subsoils are generally lower in available nutrients and organic matter than topsoils and in some instances may contribute little to crop nutrition (Table 12.1). Nutrients in subsoils are also less uniformly distributed than in the topsoil, and thus the distances between nutrient sources and roots are greater (Kautz et al. 2013). Nutrient deficiencies in the subsoil may be a major limitation to the yields of dryland crops, where continued root growth and function are essential to enable the crops to extract water and nutrients from the subsoil once the topsoil is dried or available nutrients are used up (Rengasamy 2000). Low nutrient levels in soils can induce changes in root morphology and reduce or prevent root growth (Giehl et al. 2014). For example, root length, branching, and root hairs in cereals and legumes are more affected by N than is the root dry weight (Baligar et al. 1998; Fageria et al. 2011; Gruber et al. 2013). The formation of cluster roots (i.e. very dense and short rootlets

Table 12.1 Soil properties of a typical Oxisol under coffee plantation in Espírito Santo, Brazil

Soil depth	pH	P	K	S	B	Zn	Ca	Mg	Al	SB	CEC	V	OM
(cm)	(H_2O)	(mgdm^{-3})					(cmolcdm^{-3})					(%)	(g kg^{-1})
0–10	5.58	13.5	105	3.1	0.85	2.07	2.63	0.92	0.00	3.82	7.42	51.5	27.1
10–20	4.89	10.4	31	0.1	0.77	1.16	1.18	0.30	0.10	1.56	4.76	32.8	14.9
20–40	4.48	3.5	19	2.8	0.50	0.27	1.13	0.28	0.39	1.46	4.36	33.5	10.9
40–60	4.27	1.0	4	6.6	0.45	0.35	0.94	0.19	0.69	1.14	4.54	25.1	10.9
60–100	4.27	0.8	4	13.8	0.45	0.27	0.86	0.17	0.69	1.04	3.64	28.6	8.1
100–150	4.13	2.2	2	34.2	0.45	0.39	0.53	0.19	0.69	0.73	3.33	21.9	5.4

Note: *SB:* sum of bases, *CEC:* cation exchange capacity at pH 7.0, *V:* base saturation, *OM:* organic matter

typically arranged in a bottle brush-like manner with abundant root hairs) in a diverse range of plant species is the dramatic example of a close coupling of morphological and physiological adaptations to P deficiency (Lambers et al. 2006). Cluster roots not only significantly increase root surface area but also release large amounts of carboxylates that help mobilize P from sparingly soluble pools (Shane et al. 2004; Veneklaas et al. 2003). Soil K deficiency stops root growth completely within 10–12 days of planting in wheat (Tennant 1976), and low K supply causes greater reduction in root growth relative to shoot growth in wheat, i.e. a decreased root-to-shoot ratio (Ma et al. 2013). Low Mg depresses photosynthesis and hence Mg deficiency may restrict subsoil root growth (Cakmak et al. 1994). Low concentrations of plant available B, Zn, Cu, and Mn also limit root growth in subsoils (Bell and Dell 2008). Since these micronutrients are largely immobile in the phloem, the uptake from topsoil may be unable to meet the demand for root growth when subsoil levels are low (Loneragan et al. 1987; Loneragan 1988; Nable and Webb 1993). To date, few studies have examined the subsoil levels of micronutrients or the significance of subsoil micronutrient uptake (Bell and Dell 2008), and this aspect of crop nutrition is not only under reported but also largely ignored.

Soil Acidity

Acid soils occupy 30% of the world's ice-free land area (Baligar et al. 1998) in both the tropical and temperate belts, and as much as 50% of the world's potential arable crops are negatively affected by soil acidity (Matsumoto and Sinaguru 2008). In acid soils ($pH_w < 5.5$), the solubility of aluminium (Al^{3+}) and manganese (Mn^{2+}) increases and can reach toxic concentrations once pH falls below 5.3 for Mn^{2+} and 4.8 for Al^{3+} (Marschner 1995). Soil acidity and Al^{3+} toxicity often increase with soil depth (Table 12.1). High Al^{3+} concentrations can damage root apices and restrict the elongation of the main axis and lateral roots (Taylor 1988; Klotz and Horst 1988). This inhibits root penetration, particularly into the subsoil, leading to a shallow root system and reduced utilization of subsoil water and nutrients (Marschner 1991). As a result, crop Al^{3+} toxicity is often most noticeable in the seasons with a terminal drought, since plant roots have limited access to stored water in the subsoil for grain filling (Tang et al. 2003). For Al^{3+} tolerance, inducible carboxylate exudation is an important mechanism of Al^{3+} exclusion in several important crops, and its physiological and genetic perspectives are already deployed in crop breeding programmes in the tropics (Lynch and Wojciechowski 2015). In contrast to Al^{3+}, Mn^{2+} absorbed by the roots is readily transported to the shoots, and therefore symptoms of Mn^{2+} toxicity are first seen on the shoots (Marschner 1995). Genotypic variation in Mn^{2+} tolerance may be related to antioxidant mechanisms and subcellular compartmentation of excess Mn (Gonzalez et al. 1998; Gonzalez and Lynch 1999).

Low P availability is also an inherent challenge to root growth in acid subsoils. In zero tillage systems, available P is concentrated in the topsoil because of low P mobility and the lack of soil mixing, which leads to root biomass accumulation and greater microbiological activity in the topsoil (Lynch and Brown 2001). Plant adaptations to low P soils consist of mechanisms to increase P foraging, especially from the P-enriched topsoil; mechanisms to increase P availability in rhizosphere; and strategies to reduce internal P requirements (Vance et al. 2003; Lynch 2011; Richardson et al. 2011). In acid subsoils, low Ca availability may also restrict root growth because Ca is poorly mobile in the phloem, and demand for Ca in root apices must be satisfied directly from the soil adjacent to the growing roots (Marschner 1995). Tissue Ca requirement can vary among genotypes (Spehar and Galwey 1997), and certain genotypes with reduced Ca requirement would be better able to tolerate acidic subsoil conditions (Lynch and Wojciechowski 2015).

Alkalinity/Sodicity

Soils are alkaline if the pH_w is >7.5 in a 1:5 soil:water suspension, with a further distinction between calcareous alkaline soils (pH_w 7.5–8.5) and alkaline sodic soils ($pH_w > 8.5$) (Rengasamy and Olsson 1991). Soil alkalinity is common in semi-arid and arid climates where evaporation exceeds rainfall, causing the accumulation of carbonates and bicarbonates (Marschner 1995). Alkaline sodic soils occur when a high proportion of Na is present relative to other cations (Ca, Mg, and K). Sodicity can lead to dispersion or disintegration of clay aggregates into individual particles when soils become wet, creating significant soil structural problems, and is associated with Na and B toxicity (Marschner 1995). Sodicity often occurs in the subsoil (Cochrane et al. 1994), e.g. McArthur (2004) found that 57% of the soils from 152 reference sites were sodic but only 21% had sodic topsoils throughout the agricultural area in Western Australia.

Soil alkalinity adversely affects plant growth mainly due to its effect on soil nutrient availability, Fe in particular, but also P, Zn, and Mn availabilities (Naidu and Rengasamy 1993; Marschner 1995). The high bicarbonate concentrations in alkaline soils can also directly inhibit root growth by decreasing cell elongation and thereby shoot growth (Tang et al. 1993). Some species are better adapted to alkaline soils than others and are capable of obtaining Fe, P, Zn, and Mn despite low availability under alkalinity, and their root extension growth is less affected by high bicarbonate concentrations (Tang et al. 1993). Although agriculture is an acidifying process, the timescale over which alkaline soils are neutralized by agricultural processes could be hundreds of years. The sodicity of the surface soils can be reduced by applying gypsum, but the economic amelioration of sodic layers in the subsoil is currently difficult (Dang et al. 2010).

Salinity

Saline soils are abundant in semi-arid and arid regions and occur as a result of evaporation and transpiration of saline underground water or due to salt input from rainfall or irrigation in particular (Marschner 1995). Soil salinity directly affects plant growth by causing osmotic stress and nutrient disorders when Na and Cl uptake depresses uptake of macronutrients and micronutrients. Root elongation can be immediately depressed at high NaCl and low Ca concentrations (Cramer et al. 1988). This reduced root growth under salinity restricts water and nutrient extraction from the subsoils and causes poor crop yield, particularly for dryland cropping regions (Rengasamy et al. 2003). Substantial differences in salt tolerance exist among and within crop species and is generally related to a plant's ability to exclude toxic ions from tissue cytoplasm, produce compatible solutes, and balance the competing osmotic needs for water acquisition and ion exclusion (Munns and Tester 2008). Growing tolerant genotypes on saline soils is a practical and relatively simple approach to salinity management and reflects the shift to a strategy of 'tailoring the plants to fit the soil' from the older strategy of 'tailoring the soil to fit the plant'. However, this strategy is more problematic on saline-sodic or saline-alkaline soils due to the occurrence of multiple additional constraints (B toxicity, K, P, Zn, Mn, Fe deficiencies) (Naidu and Rengasamy 1993).

12.6 Subsoil Nutrient Acquisition

Although more than two-thirds of soil nutrients can be found in subsoils (Kautz et al. 2013), the information regarding nutrient acquisition from the subsoil is scant due to a lack of adequate and simple methods for subsoil investigation. Estimation of subsoil nutrient acquisition requires destructive soil samplings, which is laborious and time-consuming. Moreover, to fully assess subsoil nutrient cycling and how nutrients move from the subsoil to the plant and then back again, long-term studies are required. However, research has mostly used short-term experiments, and consequently our current knowledge regarding subsoil processes is limited. Based on the existing literature, nutrient acquisition from arable subsoils may be conceptualized into three major process components: (*i*) mobilization from the subsoil, (*ii*) translocation to the shoot and long-term accumulation in the Ap horizon, and (*iii*) re-allocation to the subsoil (Fig. 12.2., Kautz et al. 2013). Due to the complexity of these processes at various temporal and spatial scales, the quantitative estimation of nutrient acquisition from the subsoil may require the linking of field experiments with mathematical modelling approaches (see the Sect. 12.7). The discussion below outlines what is currently known regarding subsoil nutrient acquisition and highlights areas where future research is required.

12.6.1 Nutrient Mobilization in Subsoils

Nutrient mobilization is the first major process that occurs during nutrient acquisition from the subsoil and can be affected by a number of factors. Biopores are of particular interest as they contribute to air movement into the soil, increase water infiltration, reduce water runoff and soil erosion, serve as preferential pathways for

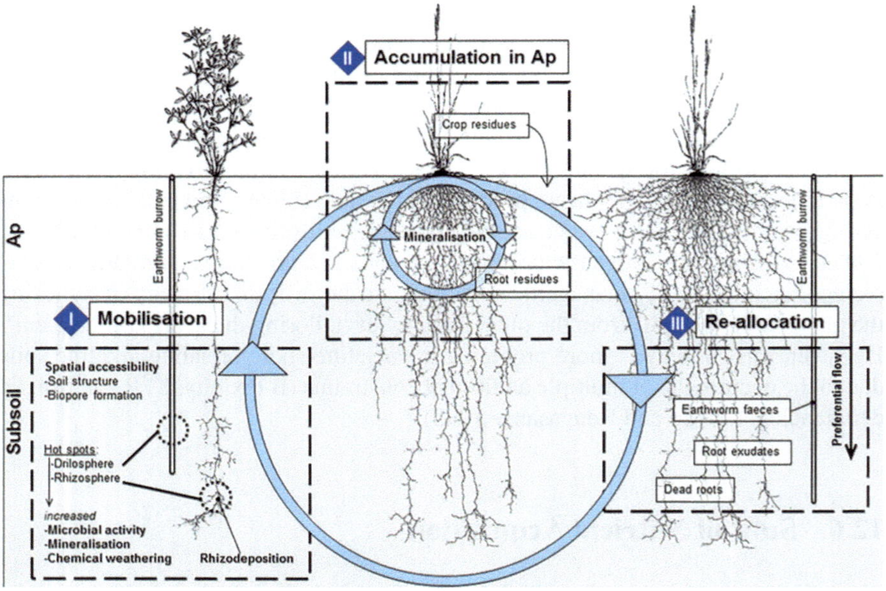

Fig. 12.2 A conceptual model of subsoil nutrient acquisition. (Courtesy of Kautz et al. 2013).

root elongation, and facilitate the acquisition of water and nutrients from the subsoil (McGrath et al. 2010; Kautz 2015). Under certain soil conditions, e.g. hardsetting clay B-horizons, a large proportion (80%) of all subsoil roots are preferentially located in the soil within 1 mm of biopores (Pierret et al. 1999). The relevance of biopores in enabling crops to access subsoil resources was demonstrated by using a nylon mesh sheet buried horizontally to allow root growth into deeper soil layers only through previously inserted holes, which mimicked biopores in a compacted soil. The resultant increase in leaf area index and plant height of five barley genotypes was related to the number of holes in the restricting mesh during a dry summer season (McKenzie et al. 2009).

Rhizodeposition is also likely to be an important factor in nutrient mobilization in subsoils, particularly because of lower substrate availability and consequently lower microbial activity in subsoil environments (Kautz et al. 2013). Mycorrhizal abundance is also known to decrease with increasing soil depth (Oehl et al. 2005; Yang et al. 2010), although it is not clear whether mycorrhizal fungi have a significant impact in nutrient mobilization from subsoils (Moreira et al. 2022). A steep gradient of microbial biomass also exists from the topsoil to subsoil with less availability of carbon and other nutrient sources in deeper soil layers (e.g. Ekschmitt et al. 2008), and there is a need to improve our understanding about the role of microbes in subsoil nutrition.

In the subsoil, N can be mobilized from organic compounds, but little is known about the kinetics of N release from the organic pools (Kautz et al. 2013). Subsoil nitrate is potentially an effective N reserve for wheat late in the season when there is no water available in the topsoil. The ability of plant roots to recover N from the subsoil is high, and N uptake from subsoil after anthesis can increase grain protein concentration of wheat (Lotfollahi et al. 1997). By comparison, the amount of specifically bound NH_4 deposited in the interlayers of 2:1 clay minerals could be as much as 150–850 mg kg^{-1} in agricultural soils, and its share of the total N content increases with increasing soil depth (Scherer 1993). Mobilization of specifically bound NH_4 is closely related to plant N uptake and affected by soil water content (Mengel and Scherer 1981). Under wet conditions, clay minerals may expand, which facilitates the release of NH_4 ions from the interlayers. If the topsoil dries out during the summer, plant N demand may be at least partially (35–40%) supplied from the subsoil NH_4 (Mengel and Scherer 1981).

The overall proportion of P in the subsoil ranges from 25% to 70% of the total P in the profile (e.g. Godlinski et al. 2004; Barej et al. 2014), but little is known about P acquisition from the subsoil. Weathering from primary minerals is probably a principal chemical process responsible for the release of inorganic P in the subsoil of immature soils, whereas P adsorbed by the clay fraction will be the major form of P in many mature soils. Increases in carboxylate concentrations in the subsoil rhizosphere of chickpea roots appeared to increase the plant availability of subsoil P when drying of the topsoil occurred (Kabir et al. 2015). Depletion of rhizosphere P fractions in the subsoil was associated with increased rhizosphere carboxylate concentrations and increased plant P uptake. Kabir (2012) estimated that 36% of total chickpea P uptake and 30% of the P acquired by grain were from the subsoil when the topsoil dried 3 weeks before flowering. Access to soil organic P by the plant and arbuscular mycorrhiza was reviewed by Vance et al. (2003) and Bucher (2007), but many questions remain open, especially for subsoils. Although biopore formation promotes root penetration into deep soil layers and enhances the use of subsoil P by plants, biopores may contribute to less than 2.5% of the total P stock in the subsoil, too little to sustain nutrition of the current crop (Barej et al. 2014). However, it is likely that new roots use these pores for extending into the subsoil while using some of the P in the biopore linings and then growing through the biopores into the surrounding subsoil to access further P (Kautz et al. 2013).

Most of the soil K is incorporated in K-bearing primary minerals, e.g. micas and feldspars, and not directly available for plant uptake. In most agricultural soils, phyllosilicates are reported to be more relevant for K release than feldspars (Öborn et al. 2005; Andrist-Rangel et al. 2006). In contrast, organic matter is hardly involved in K release by mineralization as K is present as a free ion in plant tissue and is generally released from plant residues early in the decomposition process, unlike P and N which are chemically bound (Römheld and Kirkby 2010). In the rhizosphere, a significant release of interlayer-K can be caused by root activities, e.g. excretion of proton and organic acids (Hinsinger et al. 1993; Wang et al. 2000) and cation exchange (Moritsuka et al. 2004). In soils with a fine texture, i.e. a high percentage of clay minerals, up to 35–70 kg K ha^{-1} $year^{-1}$ can be released from the solid phase, whereas in soils with coarse texture, lower release rates are expected (Simonsson et al. 2007). By comparison, exchangeable K is electrostatically bound as an outer-sphere complex to the surfaces of clay minerals and humic substances (Barre et al.

2008) and makes up only 1–2% of the total soil K in rapid equilibrium with soil solution K (Römheld and Kirkby 2010). According to Kuhlmann and Barraclough (1987), winter wheat could acquire 50% of its K from the subsoil. Although soil exchangeable K is used widely as a measure to determine soil K availability and to predict K fertilization needs of crops, its suitability and reliability is unsatisfactory in soils that contain 2:1 layer silicates and have the ability to retain K as is the case of some flooded soils used for rice production (Dobermann et al. 1996).

12.6.2 Nutrient Translocation From Subsoils

Considerable amounts of N, P, and K that have been absorbed by the roots from the subsoil are translocated to shoots and topsoil roots. If these nutrients are not removed during harvest, they are deposited in the topsoil as residues, where following mineralization they are available for nutrient uptake by next crops or returned to the subsoil by leaching or the activity of soil biota (Kautz et al. 2013). This plant 'uplift' can be a dominant process affecting the vertical distribution of most of the plant nutrients and occurs in all terrestrial systems (Jobbágy and Jackson 2001, 2004). Several studies have reported the potential for subsoil nutrients to be redistributed by agricultural crops to the topsoil (Kuhlmann et al. 1989; Kuhlmann 1990; Richards et al. 1995; Witter and Johansson 2001; Haberle et al. 2006; Barej et al. 2014). For example, the results from 22 sites (deep loess grey podsolic soils) showed that the subsoil contributed about 1/3 of total N uptake in winter wheat and the plants took up 152 kg N ha^{-1} from a high-N subsoil, when unfertilized, but only 31–39 kg N ha^{-1} from a low-N subsoil, when fertilized with 170 kg N ha^{-1} (Kuhlmann et al. 1989). In P nutrition, the subsoil delivered >30% of total P uptake of wheat (e.g. Kuhlmann and Baumgärtel 1991) and chickpea (Kabir 2012). On a calcareous loess soil, significant P was taken up by crops from 50 to 100 cm soil depth (Garz et al. 2000). With 34 experiments on loess-parabrown soils in Northern Germany, Kuhlmann (1990) found the contribution of subsoil K increased from 8% at first node stage to 35% at ear emergence of spring wheat. The significance of subsoil K was also related to root architecture, e.g. deep rooting forage crops *Cichorium intybus* and *Medicago sativa* acquired 56 and 67%, respectively, of K from the subsoil on clayey loam soil, whereas *Lolium perenne* with a low rooting density in the subsoil acquired just 42% of K from the subsoil under the same field conditions (Witter and Johansson 2001).

12.6.3 Nutrient Re-allocation to Subsoils

There are several processes by which nutrients can be re-allocated into subsoils. In cropping fields, anecic earthworms will feed on organic substances, e.g. plant residues at the soil surface, and then deposit the undecomposed residues as well as faeces and mucus deeper in the soil profile (Kautz et al. 2013). Nutrients dissolved in water can also be washed into the subsoil, particularly under saturated conditions when water is able to move preferentially via biopores or soil cracks. The walls of earthworm burrows are laterally less permeable to aqueous solutions than the bulk soil, which substantially enhances vertical water fluxes (Bastardie et al. 2005; Watt et al. 2006). Preferential flow via the macropores can lead to a particularly beneficial supply of water and nutrients to the roots growing through the bulk subsoil (Kautz et al. 2013).

12.7 Modelling Subsoil Nutrient Acquisition

The challenges in measuring subsoil nutrients and the complexities of the processes involved in nutrient acquisition from the subsoil have led to the use of simulation models for quantification of subsoil nutrient acquisition and nutrient balance by different crops and cropping systems. Process-based models (PBM) are widely used because they apply an integrated approach to assess soil and plant processes and their responses to climatic conditions and farming management (reviewed by Kautz et al. 2013). PBMs simulate physiological processes and describe metabolism and crop growth in terms of mass variables per unit area of land (Vos et al. 2010). Substantial progress has been made in modelling growth and development processes of crops with regard to soil and climate factors by using CropSyst (Stöckle et al. 2003), DSSAT (Jones et al. 2003), APSIM (Keating et al. 2003; Scanlan et al. 2015a, b), and STICS (Brisson et al. 2003). However, PBMs exhibit major shortcomings with respect to nutrient uptake from the subsoil: (*i*) most of these models only consider water stress and N, although a few PBMs have tested the dynamics and uptake of subsoil P (Chen et al. 2008) and K (Scanlan et al. 2015a, b); (*ii*) the physical and chemical differences among soil horizons are neglected; (*iii*) processes such as biopore formation or nutrient re-allocation and accumulation in the drilosphere are often ignored (Kautz et al. 2013).

Functional-structural plant models (FSPM) (Godin and Sinoquet 2005) were developed to account explicitly for plant structural features in the prediction of crop growth and provide a link between structural features of root and soil architectures with nutrient and water uptake from heterogeneous subsoils (Kautz et al. 2013). The FSPMs need detailed physical parameters and a spatially explicit description of soil and root structures and typically consider the impact of 3D variable soil properties for root growth, solute, and water uptake. Using this type of model, Dunbabin et al. (2004) demonstrated that the plasticity of root development to varying soil conditions, e.g. spatially variable N, is crucial to predict the functionality of a root system. Thus, FSPMs offer the opportunity to link root plasticity, dynamic soil conditions, root uptake processes, and flow and transport processes in a heterogeneous soil. The application of such models to a range of environmental conditions, subsoil structures, and nutrient distribution patterns in the soil profile would greatly improve the understanding of conditions under which nutrient uptake from the subsoil is significant for crop growth and perhaps how to enhance the utility of models for crops (Kautz et al. 2013).

12.8 Crop Response to Deep Fertilization

In rainfed agriculture, fertilizer placement can play a large role in nutrient uptake efficiency, especially when drought restricts nutrient availability at the late stages of plant growth. Compared with conventional fertilizer placement (broadcasting or banding fertilizers with the seed), deep placement or deep banding generally refers to placing fertilizers from 30 mm below the seed with tines at seeding (Jarvis and Bolland 1990) to 150–200 mm below the soil surface in a coulter-knife operation prior to seeding (Borges and Mallarino 2001). Other practices include using a para plough or straight-shanked deep ripper tines to apply fluid fertilizers through the profile to 400 mm or the same amount of nutrients from 200 to 400 mm in the profile (Doudle 2002; Doudle and Wilhelm 2002a, b).

Shallow and localized nutrient supply may reduce nutrient uptake, crop growth, and yield in water-limited environments, where the topsoil dries out quickly at high temperature and high evaporation rates (Fig. 12.3.). In the dried topsoil, plant nutrients become less available at the root surface, because nutrient transport by mass flow [nitrogen (N) and sulphur (S)] or diffusion (P and K) is restricted by water deficit (Marschner 1995; Seiffert et al. 1995). Placing fertilizers in the subsoil would not only induce deeper root growth but also leave the nutrients in moist soil for longer periods of the growing season and thereby enhance nutrient uptake and utilization, especially for plant species that still acquire soil nutrients at late growth

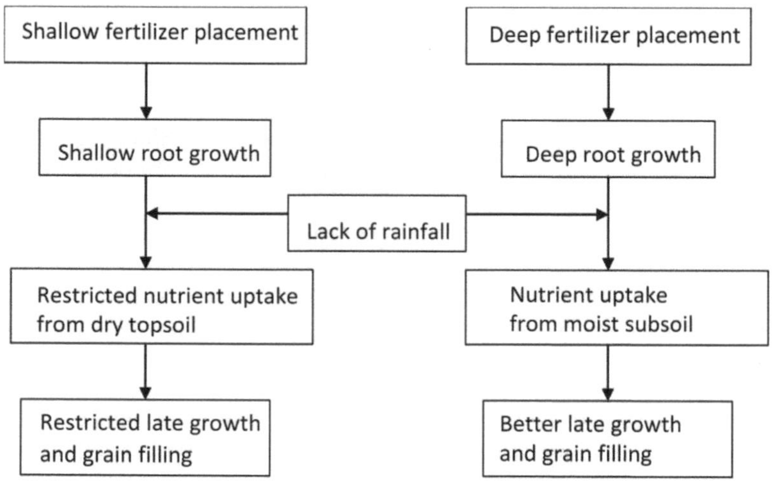

Fig. 12.3 Schematic comparison of crop responses between shallow and deep placements of fertilizers in water-limited environments. (Courtesy of Ma et al. 2009)

stage (reviewed by Ma et al. 2009). Deep fertilizer placement can also avoid the toxic effects of fertilizer applied with, or close to, the seed (Zhang and Rengel 1999, 2002).

Numerous studies have reported yield responses to deep fertilizer placement in wheat (Alston 1976; Nable and Webb 1993; Sander and Eghball 1999; Singh et al. 2005; Wilhelm 2005), lupins (Jarvis and Bolland 1990, 1991; Crabtree et al. 1998; Brennan 1999; Crabtree 1999; Scott et al. 2003), canola (Grewal et al. 1997; Hocking et al. 2003; Wilhelm 2005), maize (Mallarino and Murrell 1998; Mallarino et al. 1999), cotton (Tupper and Ebelhar 1994; Singh et al. 2005), lucerne (Simpson and Lipsett 1973; Teutsch et al. 2000; Singh et al. 2005), and rice (Wu et al. 2017). However, other studies have found little or no benefit of deep-placed nutrients (Hudak et al. 1989; Reeves and Mullins 1995; Bolland and Jarvis 1996; McCutcheon and Rzewnicki 2001; Vyn and Janovicek 2001) or the results that varied between seasons or locations (Eck and Fanning 1961; McConnell et al. 1986; Mullins et al. 1997). This inconsistency could be due to the interaction of fertilizer placement with soil and climatic conditions and crop species (see details below).

12.8.1 Soil Types

Soil profiles with sandy topsoil, e.g. sand-over-clay profile (Frischke and Doudle 2003) or deep sand profile (Jarvis and Bolland 1991; Brennan 1999; Crabtree 1999), have consistently shown positive growth and yield responses in rainfed crops to deep nutrient placement in southern Australia, because sandy topsoils are particularly prone to drying in the Mediterranean climate zone. Yield increases in response to deep-placed nutrients are also reported on other soils, including loamy sands and sandy loams (Jarvis and Bolland 1991), grey/brown Vertosol, and red Kandosol (Singh et al. 2005). However, where deep placement of fertilizers occurs into subsoils with constrained conditions, e.g. a sodic horizon, deep placement may not be effective as poor root growth becomes the major yield-limiting factor, e.g. in the Edillilie/Wanilla district of South Australia (Davenport et al. 2003). Similarly, crop responses to deep banding P and K fertilizers on a range of soil types in North America did not appear to be related to soil type per se, rather the topsoil fertility and nutrient distribution down the profile (Mallarino et al. 1999; Borges and Mallarino 2001).

Where soil type has a significant impact on plant response to deep-placed nutrients, this generally occurs due to the impact of soil chemical and physical properties (e.g. pH, cation exchange capacity, clay content, P sorption). Sandy soils have low cation exchange capacity and poor water holding capacity. Drainage and evaporation from the highly porous substrates mean that soil moisture levels decrease with increasing proximity to the surface layer, particularly in dry seasons. As a result, shallow fertilizer placement in sandy soils (drilling fertilizers with the seed or top-dressing) would be ineffective in low rainfall areas or seasons with below-average rainfall. By comparison, drilling fertilizers with the seed in heavier soils is likely to be more efficient than banding deeper in the profiles, due to high cation exchange and water holding capacities. Indeed, Lynch and Brown (2001) report that for well-watered soils, increasing root density close to the soil surface boosts P uptake by concentrating roots in the layer with highest available P. In heavy soils, top-dressed P is poorly effective, while deep banding may cause P deficiency during early growth before seedling roots reach the deep-placed P. In the case of subsoils with chemical toxicities such as boron or salt, increasing root proliferation into the subsoils by placing fertilizers deep may even cause crop damage.

12.8.2 Plant Species

The patterns of nutrient uptake among crops differ over the growing season, which may affect their responses to deep fertilizer in a low rainfall environment. A field study with canola in southern New South Wales, for example, showed that placing nutrients 30 mm below the seed produced higher seed yield than the with-seed placement on yellow podsolic and red gradational soils (Hocking et al. 2003). Improved canola yield in the Mediterranean-type climate of South Australia was achieved by deep placement (400 mm) of fluid fertilizers containing P, N, Zn, Cu, and Mn (Wilhelm 2005). Maximum P and K accumulation occurred during late flowering in canola (Rose et al. 2007) but peaked just prior to anthesis in wheat (Rose et al. 2007; Ma et al. 2013), suggesting a greater importance of deep-placed fertilizer to meet the post-flowering demand for nutrients in canola than wheat under water-limited conditions.

Lupins, which are sensitive to low winter temperatures, grow slowly and take up only 10–20% of the total P uptake by anthesis. Consequently, they rely less on nutrient redistribution from vegetative organs to the grain than do cereal crops (e.g. wheat, sorghum, maize) in the Mediterranean climates of southern Australia. Therefore, lupins are more dependent than cereals on taking up soil P late in the growing season for grain filling. Lupins are also deep-rooted, with a smaller proportion of the total root length in topsoil than is the case for most other crops grown in south-western Australia. All of these factors might account for a better response by lupin than by wheat to deep-placed P fertilizer (Jarvis and Bolland 1990). Large variability also exists in root morphology across the lupin germplasm (Clements et al. 1993), and responses of root growth and nitrate uptake to heterogeneous nitrate supply differ between *Lupinus angustifolius* (dominant tap root and primary lateral system) and *L. pilosus* (minor tap root and well-developed lateral system) (Dunbabin et al. 2001a, b).

The most consistent yield increases due to deeper placement of P fertilizer have been reported in fodder crops as they grow year-round (in warmer climates) and are frequently defoliated and hence rely on continual P supply for biomass production. In situations where soil profiles are stratified and the topsoil dries, pasture legumes, such as lucerne and barrel medic, have responded to deep-placed P in both temperate and tropical climates with higher dry matter yields (Scott 1973; Simpson and

Lipsett 1973; Singh et al. 2005). Pasture legumes contain over 70% of total root length in the top 200 mm of the soil compared with wheat and lupins that contain less than 50% of total root length in this zone (Hamblin and Hamblin 1985). As a consequence, under drought conditions, pasture legumes would be more susceptible to shortage of nutrient supply when fertilizers are placed in the topsoil.

12.8.3 Tillage Practice

Previous fertilizer applications would be another factor determining the likelihood of crop response to newly applied, deep-placed fertilizers. Repeated and regular addition of P-K fertilizers progressively increases the residual levels in the soil profile. Therefore, an economic response to deep-placed P-K fertilizers is only likely if locally calibrated soil tests suggest a response (i.e. low soil test value) to conventionally placed P (Bolland and Jarvis 1996; Mallarino et al. 1999; Borges and Mallarino 2000; Buah et al. 2000a, b; Scott et al. 2003) or K (Bordoli and Mallarino 1998; Buah et al. 2000a, b; Vyn and Janovicek 2001). Bolland and Jarvis (1996) reported that on sandy soils susceptible to P-leaching, lupin grain yield from deep-placed P would exceed that of top-dressed P when soil-tested P for the 100–200 mm horizon was deficient (<10–15 mg P kg^{-1} soil). Soil profiles that contain high levels of P in the topsoil are less likely to respond to deep-placed P regardless of late season topsoil drying, because plants can accumulate sufficient P for maximum growth and yield prior to flowering.

Singh et al. (2005) also found that a wheat crop grown on stored moisture in northern Australia increased yields in response to deep P application but was unresponsive to conventional P application: yields continued to increase at a higher rate of P application to the subsoil. Generally, critical soil test values account for nutrients only in the top 100–150 mm of soil profile, because of its relevance to conventional fertilizer practice. The subsoil commonly contains larger amounts of clay than the topsoils and requires higher inputs of P fertilizer to overcome the associated higher P sorption capacity of subsoil. Therefore, new soil testing calibrations for subsoils are required before deep fertilization becomes a common practice.

Cultivation of the nutrient-stratified soils can redistribute nutrients in the profiles and aid nutrient uptake by crops (Takker and Walker 1993; Bolland and Brennan 2006). As a result, response to deep-placed P fertilizer is unlikely if the soil has been recently cultivated (Schultz 1975).

Some studies, however, have reported responses to deep-placed K on the soils that had optimum or high soil K tests (Mallarino et al. 1999; Borges and Mallarino 2000). Responses to deep-placed K also depend upon the tillage systems (e.g. zero tillage, zone tillage, or mulch tillage) and the subsequent magnitude of soil mixing (Vyn et al. 1999, 2002). On soils where a compacted traffic pan hinders crop root growth into the subsoil, amelioration by deep mechanical loosening can lead to enhanced root growth and nutrient uptake (Himmelbauer et al. 2010). Deep soil preparation may be particularly useful for perennial crops, e.g. coffee plantation in

Fig. 12.4 Deep soil preparation at 1.5 m depth, 0.8 m width plus mineral and organic fertilization, including biochar on a hardsetting soil for a coffee plantation in Espírito Santo, Brazil

the presence of severe subsoil chemical and physical constraints (Fig. 12.4.). However, deep mechanical loosening can cause a potential destruction of continuous biopores in the subsoil and thus may be recommended for severely compacted subsoils only (Kautz et al. 2013).

Application of gypsum on the soil surface followed by its leaching into acidic subsoils can lead to an increase in root growth and higher absorption of water and nutrients by plants (Carvalho and van Raij 1997; Caires et al. 2003, 2011). In Brazil, the surface application of gypsum in coffee plantations has been a common practice to improve root growth (Fig. 12.5) and nutrient availability in the subsoil. Gypsum increased available Ca, K, and Mg concentrations and decreased the extractable Al^{3+} concentration in the deeper layers (Table 12.2). On soybean farms, it is root growth in the subsoil that largely determines yield due to increased water and nutrient uptake, while surface soil analyses have often poorly explained the yield responses.

Fig. 12.5 Gypsum application on the soil surface (left) and root growth (right) of coffee plants in Minas Gerais-Brazil. (Source: AP Agrícola)

Table 12.2 Soil chemical characteristics after 11 years of gypsum application at the rates of nil or 9.4 t ha^{-1} on a typical Oxisol in a coffee plantation in Minas Gerais state, Brazil (Tanure 2016)

Soil depth (cm)	pH (H_2O)	S[a]	P[b]	K[b]	Ca^{2+c}	Mg^{2+c}	Al^{3+c}	CEC	OM[d]	Clay
		-------($mgdm^{-3}$)-------			--------- (cmolc dm^{-3}) --------				––(gkg^{-1})––	
	Nil gypsum									
0–5	4.68	14.6	1.37	86.7	0.11	0.12	3.01	11.9	66.6	800
5–10	4.67	13.2	0.70	59.7	0.30	0.08	2.74	10.1	50.6	810
10–20	4.85	13.8	0.47	44.3	0.07	0.07	2.81	8.99	44.5	790
20–40	5.13	13.4	0.20	26.7	0.05	0.04	1.42	6.33	27.7	750
40–60	5.12	13.5	0.07	15.7	0.05	0.03	1.19	5.19	21.6	790
60–100	5.25	15.2	0.10	17.7	0.06	0.04	1.06	4.06	19.4	780
100–200	5.52	16.9	0.00	8.0	0.05	0.01	1.09	3.35	3.9	650
	9.4 t gypsum ha^{-1}									
0–5	5.05	41.8	9.13	214	2.32	0.55	0.29	10.38	42.2	820
5–10	5.12	35.4	3.57	355	5.79	0.43	0.59	8.81	38.3	830
10–20	5.22	32.3	1.33	276	1.83	0.35	0.55	8.52	35.8	820
20–40	5.27	35.5	0.40	218	1.47	0.25	0.10	6.41	25.8	810
40–60	5.46	36.2	0.10	141	1.78	0.23	0.00	5.44	20.7	800
60–100	5.70	25.8	0.03	48	2.18	0.20	0.00	5.86	16.8	800
100–200	5.08	16.6	0.03	11	1.54	0.37	0.00	3.90	12.9	830

Note: [a]$Ca(H_2PO_4)_2.H_2O$ in HOAc 2 mol L^{-1}
[b]Mehlich-1
[c]KCl extraction 1 mol L^{-1}
[d]*OM:* organic matter by Walkley-Black method; *CEC:* cation exchange capacity

12.8.4 Hydraulic Redistribution

Hydraulic redistribution is a process whereby roots absorb soil water from a zone with high-water potential (generally the subsoil) and release some at night into soil with low water potential (generally the topsoil) (Caldwell et al. 1998). One of the benefits of hydraulic redistribution is that it may maintain functioning fine roots in the dry topsoil and prolong the uptake of nutrients until soil conditions become either more favourable for the resumption of topsoil root activity or too unfavourable for survival (Huang 1999). Hydraulic redistribution has been reported in wheat, maize, lucerne, oats, soybean, rye, cotton, and canola (Blum and Johnson 1992; Wan et al. 2000; Valizadeh et al. 2003; Zegada-Lizarazu and Iijima 2004; Rose et al. 2008) and tree species such as eucalyptus (Brooksbank et al. 2011). Wheat plants reportedly lifted more water in response to increased localized supply of P fertilizer (compared with nil fertilizer), which improved P uptake (Valizadeh et al. 2003). Using a split-pot design, Prieto et al. (2012) demonstrated that hydraulic redistribution favoured the selective placement of roots in nutrient-rich patches, as well as nutrient capture under drought, a process that may secure nutrient supply and maintain plant performance during drought periods. Significant genetic variations of hydraulic redistribution exist among drought-susceptible and drought-tolerant maize hybrids, with a strong relationship observed between hydraulic redistribution and drought tolerance characteristics (Caldwell et al. 1998). Although canola roots also hydraulically lift water during the period from vegetative growth to late flowering, the continuation of nutrient uptake by roots from dry soil appears to be cultivar- and nutrient-specific, and not necessarily associated with hydraulic redistribution (Rose et al. 2008). More studies are needed to verify the linkage of active hydraulic redistribution with root growth and more efficient nutrient foraging processes in heterogeneous soils.

12.9 Conclusion

Considerable nutrient stocks are present in the subsoil and can contribute to crop growth and yield production. However, plant access to subsoil nutrients is limited by compaction, low organic C, and low microbial activity, along with restricted root growth by chemical constraints, e.g. acidity, alkalinity, salinity, sodicity, and both nutrient deficiency and toxicity. In addition, our knowledge regarding nutrient acquisition from arable subsoils is still scant, especially in terms of relevant long-term processes. The evaluation of subsoil impacts on crop nutrition mostly comes from short-term or microcosm experiments with disturbed soil. The contribution of subsoil nutrients may vary from <10 to >70% of total plant uptake for certain soil nutrients, particularly when a dryer topsoil induces deeper root growth and therefore more utilization of water and nutrients from the subsoil. Placing fertilizer in the subsoil has been shown to increase nutrient availability and plant uptake in low rainfall environments, but the effectiveness of deep fertilizer placement is dependent on soil types, crop species, and agronomic practices. With ongoing global climate warming and more frequent summer drought in agricultural regions, the exploitation of subsoil water and nutrients could be of increased future relevance. Long-term field studies, together with mathematical modelling, are required to quantify the significance of acquisition of subsoil nutrients for crop nutrition, particularly in low-input and rainfed farming systems.

References

Aloni R, Aloni E, Langhans M. Role of cytokinin and auxin in shaping root architecture: Regulating vascular differentiation, lateral root initiation, root apical dominance and root gravitropism. Ann Bot. 2006;97:883–93. https://doi.org/10.1093/aob/mcl027.

Alston AM. Effects of depth of fertilizer placement on wheat grown under three water regimes. Aust J Agric Res. 1976;27:1–10. https://doi.org/10.1071/AR9760001.

Andrist-Rangel Y, Simonsson M, Andersson S, et al. Mineralogical budgeting of potassium in soil: a basis for understanding standard measures of reserve potassium. J Plant Nutr Soil Sci. 2006;169:605–15. https://doi.org/10.1002/jpln.200621972.

Baligar VC, Beaver WV, Ahlrichs JL. Nature and distribution of acid soils in the world. In: Schaffert RE, editor. Proceedings of a workshop to develop a strategy for collaborative research and dissemination of technology in sustainable crop production in acid savannas and other problem soils of the world, May 4–6, 1998. Purdue University Press; 1998. p. 1–12.

Barej JAM, Patzold S, Perkons U, et al. Phosphorus fractions in bulk subsoil and its biopore systems. Eur J Soil Sci. 2014;65:553–61. https://doi.org/10.1111/ejss.12124.

Barre P, Montagnier C, Chenu C, et al. Clay minerals as a soil potassium reservoir: observation and quantification through X-ray diffraction. Plant Soil. 2008;302:213–20. https://doi.org/10.1007/s11104-007-9471-6.

Bastardie F, Ruy S, Cluzeau D. Assessment of earthworm contribution to soil hydrology: a laboratory method to measure water diffusion through burrow walls. Biol Fertil Soils. 2005;41:124–8. https://doi.org/10.1007/s00374-004-0817-4.

Batey T. Soil compaction and soil management – a review. Soil Use Manag. 2009;25:335–45. https://doi.org/10.1111/j.1475-2743.2009.00236.x.

Bell RW, Dell B. Micronutrients in sustainable food, feed, fibre and bioenergy production. Paris: IFA; 2008.

Bengough AG, Mckenzie BM, Hallett PD, et al. Root elongation, water stress, and mechanical impedance: a review of limiting stresses and beneficial root tip traits. J Exp Bot. 2011;62:59–68. https://doi.org/10.1093/jxb/erq350.

Blum A, Johnson JW. Transfer of water from roots into dry soil and the effect on wheat water relations and growth. Plant Soil. 1992;145:141–9. https://doi.org/10.1007/BF00009550.

Bodner G, Leitner D, Kaul HP. Coarse and fine root plants affect pore size distributions differently. Plant Soil. 2014;380:133–51. https://doi.org/10.1007/s11104-014-2079-8.

Bolland MDA, Brennan RF. Phosphorus, copper and zinc requirements of no-till wheat crops and methods of collecting soil samples for soil testing. Aust J Exp Agric. 2006;46:1051–9. https://doi.org/10.1071/EA05024.

Bolland MDA, Jarvis RJ. Effectiveness of different methods of applying superphosphate for lupins grown on sandplain soils. Aust J Exp Agric. 1996;36:707–15. https://doi.org/10.1071/EA9960707.

Bordoli JM, Mallarino AP. Deep and shallow banding of phosphorus and potassium as alternatives to broadcast fertilization for no-till corn. Agron J. 1998;90:27–33. https://doi.org/10.2134/agronj1998.00021962009000010006x.

Borges R, Mallarino AP. Grain yield, early growth, and nutrient uptake of no-till soybean as affected by phosphorus and potassium placement. Agron J. 2000;92:380–8. https://doi.org/10.2134/agronj2000.922380x.

Borges R, Mallarino AP. Deep banding phosphorus and potassium fertilizers for corn managed with ridge tillage. Soil Sci Soc Am J. 2001;65:376–84. https://doi.org/10.2136/sssaj2001.652376x.

Botwright Acuña TL, Wade LJ. Genotype × environment interactions for root depth of wheat. Field Crops Res. 2012;137:117–25. https://doi.org/10.1016/j.fcr.2012.08.004.

Botwright Acuña TL, He X, Wade LJ. Temporal variation in root penetration ability of wheat genotypes through thin wax layers in contrasting water regimes and in the field. Field Crops Res. 2012;138:1–10. https://doi.org/10.1016/j.fcr.2012.09.018.

Bowden B. The impact of gravel on crop management. Media Release. COGGO – Council of Grain Grower Organisation; 2014.

Brennan RF. Lupin grain yields and fertiliser effectiveness are increased by banding manganese below the seed. Aust J Exp Agric. 1999;39:595–603. https://doi.org/10.1071/EA98179.

Brisson N, Gary C, Justes E, et al. An overview of the crop model STICS. Eur J Agron. 2003;18:309–32. https://doi.org/10.1016/S1161-0301(02)00110-7.

Brooksbank K, Veneklaas EJ, White DA, et al. The fate of hydraulically redistributed water in a semi-arid zone eucalyptus species. Tree Physiol. 2011;31:649–58. https://doi.org/10.1093/treephys/tpr052.

Brown GG, Barois I, Lavelle P. Regulation of soil organic matter dynamics and microbial activity in the drilosphere and the role of interactions with other edaphic functional domains. Eur J Soil Biol. 2000;36:177–98. https://doi.org/10.1016/S1164-5563(00)01062-1.

Buah S, Polito TA, Killorn R. No-tillage corn response to placement of fertilizer nitrogen, phosphorus, and potassium. Commun Soil Sci Plant Anal. 2000a;31:3121–33. https://doi.org/10.1080/00103620009370655.

Buah SSJ, Polito TA, Killorn R. No-tillage soybean response to banded and broadcast and direct and residual fertilizer phosphorus and potassium applications. Agron J. 2000b;92:657–62. https://doi.org/10.2134/agronj2000.924657x.

Bucher M. Functional biology of plant phosphate uptake at root and mycorrhiza interfaces. New Phytol. 2007;173:11–26. https://doi.org/10.1111/j.1469-8137.2006.01935.x.

Caires EF, Blum J, Barth G, et al. Changes in chemical soil characteristics and soybean response to lime and gypsum applications in a no-tillage system. Rev Bras Ciênc Solo. 2003;27:275–86. (in Portuguese, with abstract in English).

Caires EF, Garbuio FJ, Churka S, et al. Use of gypsum for crop grain production under a subtropical no-till cropping system. Agron J. 2011;103:1804–14. https://doi.org/10.2134/agronj2011.0192.

Cakmak I, Hengeler C, Marschner H. Partitioning of shoot and root dry matter and carbohydrates in bean plants suffering from phosphorus, potassium and magnesium deficiency. J Exp Bot. 1994;45:1245–50. https://doi.org/10.1093/jxb/45.9.1245.

Caldwell MM, Manwaring JH, Richards JH. Hydraulic lift: consequences of water efflux from the roots of plants. Oecologia. 1998;113:151–61. https://doi.org/10.1007/s004420050363.

Carvalho MCS, van Raij B. Calcium sulphate, phosphogypsum and calcium carbonate in the amelioration of acid subsoils for root growth. Plant Soil. 1997;192:37–48. https://doi.org/10.1023/A:1004285113189.

Chen W, Dunbabin V, Bell R, et al. Simulating and understanding root growth using rootmap to guide phosphorus fertiliser placement in wide row lupin cropping systems. In: Palta JA, Berger JB, editors. 'Lupins for health and wealth' proceedings of the 12th international Lupin conference, 14–18 Sept 2008, Fremantle, Western Australia; 2008. p. 368–72.

Chen YL, Palta J, Clements J, et al. Root architecture alteration of narrow-leafed lupin and wheat in response to soil compaction. Field Crops Res. 2014;165:61–70. https://doi.org/10.1016/j.fcr.2014.04.007.

Clark LJ, Ferraris S, Price AH, et al. A gradual rather than abrupt increase in soil strength gives better root penetration of strong layers. Plant Soil. 2008;307:235–42. https://doi.org/10.1007/s11104-008-9602-8.

Clements JC, White PF, Buirchell BJ. The root morphology of *Lupinus angustifolius* in relation to other *Lupinus* species. Aust J Agric Res. 1993;44:1367–75. https://doi.org/10.1071/AR9931367.

Clowes FAL. Apical meristems. Oxford: Blackwell; 1961.

Cochrane HR, Scholz G, van Vreeswyk AME. Sodic soils in Western Australia. Aust J Soil Res. 1994;32:359–88. https://doi.org/10.1071/SR9940359.
Crabtree WL. Deep placement of Mn fertiliser on a sandy soil increased grain yield and reduced split seed in *Lupinus angustifolius*. Plant Soil. 1999;214:9–14. https://doi.org/10.1023/A:1004373430488.
Crabtree WL, Robson AD, Ritchie GSP. Drying of surface soil decreased *Lupinus angustifolius* root length and Mn uptake in a split pot experiment. Aust J Agric Res. 1998;49:1119–24. https://doi.org/10.1071/A98015.
Cramer GR, Epstein E, Läuchli A. Kinetics of root elongation of maize in response to short-term exposure to NaCl and elevated calcium concentration. J Exp Bot. 1988;39:1513–22. https://doi.org/10.1093/jxb/39.11.1513.
Dang YP, Dalal RC, Buck SR, et al. Diagnosis, extent, impacts and management of subsoil constraints in the northern grains cropping region of Australia. Aust J Soil Res. 2010;48:105–19. https://doi.org/10.1071/sr09074.
Daniells IG. Hardsetting soils: a review. Soil Res. 2012;50:349–59. https://doi.org/10.1071/sr11102.
Davenport D, Braun J, Growden B. Deep placed nutrients at Edillilie. In: Eyre Peninsula farming systems 2003 summary. South Australian Research and Development Institute; 2003. p. 164.
Dobermann A, Cassman KG, Cruz PCS, et al. Fertilizer inputs, nutrient balance, and soil-nutrient supplying power in intensive, irrigated rice systems. II. Effective soil K-supplying capacity. Nutr Cycl Agroecosyst. 1996;46:11–21. https://doi.org/10.1007/bf00210220.
Dodd IC, Whalley WR, Ober ES, et al. Genetic and management approaches to boost UK wheat yields by ameliorating water deficits. J Exp Bot. 2011;62:5241–8. https://doi.org/10.1093/jxb/err242.
Doudle S. Subsoil nutrition demonstration at Buckleboo. In: Eyre Peninsula Farming Systems 2002 Summary. South Australian Research and Development Institute; 2002. p. 120–2.
Doudle S, Wilhelm N. Subsoil nutrition – what kind works, where and how deep does it need to be? In: Eyre Peninsula Farming Systems 2002 Summary. South Australian Research and Development Institute; 2002a. p. 115–9.
Doudle S, Wilhelm N. Trace element application methods on clay. In: Eyre Peninsula farming systems 2002 summary. South Australian Research and Development Institute; 2002b. p. 127–8.
Drew MC. Comparison of the effects of a localized supply of phosphate, nitrate, ammonium and potassium on the growth of the seminal root system, and the shoot, in barley. New Phytol. 1975;75:479–90. https://doi.org/10.1111/j.1469-8137.1975.tb01409.x.
Duiker SW, Beegle DB. Soil fertility distributions in long-term no-till, chisel/disk and moldboard plow/disk systems. Soil Tillage Res. 2006;88:30–41. https://doi.org/10.1016/j.still.2005.04.004.
Dunbabin V, Rengel Z, Diggle A. The root growth response to heterogeneous nitrate supply differs for *Lupinus angustifolius* and *Lupinus pilosus*. Aust J Agric Res. 2001a;52:495–503. https://doi.org/10.1071/AR00098.
Dunbabin V, Rengel Z, Diggle A. *Lupinus angustifolius* has a plastic uptake response to heterogeneously supplied nitrate while *Lupinus pilosus* does not. Aust J Agric Res. 2001b;52:505–12. https://doi.org/10.1071/AR00099.
Dunbabin VM, Rengel Z, Diggle AJ. Simulating form and function of root systems: efficiency of nitrate uptake is dependent on root system architecture and the spatial and temporal variability of nitrate supply. Funct Ecol. 2004;18:204–11. https://doi.org/10.1111/j.0269-8463.2004.00827.x.
Eck HV, Fanning C. Placement of fertilizer in relation to soil moisture supply. Agron J. 1961;53:335–8. https://doi.org/10.2134/agronj1961.00021962005300050021x.
Ehlers W. Observations on earthworm channels and infiltration on tilled and untilled loess soil. Soil Sci. 1975;119:242–9. https://doi.org/10.1097/00010694-197503000-00010.
Ekschmitt K, Kandeler E, Poll C, et al. Soil-carbon preservation through habitat constraints and biological limitations on decomposer activity. J Plant Nutri Soil Sci. 2008;171:27–35. https://doi.org/10.1002/jpln.200700051.

Fageria NK, Moreira A, Coelho AM. Yield and yield components of upland rice as influenced by nitrogen sources. J Plant Nutri. 2011;34:361–9. https://doi.org/10.1080/01904167.2011.536878.

Farley RA, Fitter AH. Temporal and spatial variation in soil resources in a deciduous woodland. J Ecol. 1999;87:688–96. https://doi.org/10.1046/j.1365-2745.1999.00390.x.

Frischke B, Doudle S. Innovative deep nutrient application with fluid fertilisers. In: Eyre Peninsula farming systems 2003 summary; 2003. p. 116–7. South Australian Research and Development Institute.

Gaiser T, Perkons U, Kupper PM, et al. Evidence of improved water uptake from subsoil by spring wheat following lucerne in a temperate humid climate. Field Crops Res. 2012;126:56–62. https://doi.org/10.1016/j.fcr.2011.09.019.

Gan Y, Liu L, Cutforth H, et al. Vertical distribution profiles and temporal growth patterns of roots in selected oilseeds, pulses and spring wheat. Crop Pasture Sci. 2011;62:457–66. https://doi.org/10.1071/CP10406.

Garz J, Schliephake W, Merbach W. Changes in the subsoil of long-term trials in Halle (Saale), Germany, caused by mineral fertilization. J Plant Nutr Soil Sci. 2000;163:663–8. https://doi.org/10.1002/1522-2624(200012)163:6%3C663::AID-JPLN663%3E3.0.CO;2-X.

Giehl RFH, von Wirén N. Root nutrient foraging. Plant Physiol. 2014;166:509–17. https://doi.org/10.1104/pp.114.245225.

Giehl RFH, Gruber BD, von Wirén N. It's time to make changes: modulation of root system architecture by nutrient signals. J Exp Bot. 2014;6:769–78. https://doi.org/10.1093/jxb/ert421.

Godin C, Sinoquet H. Functional-structural plant modelling. New Phytol. 2005;166:705–8. https://doi.org/10.1111/j.1469-8137.2005.01445.x.

Godlinski F, Leinweber P, Meissner R, et al. Phosphorus status of soil and leaching losses: results from operating and dismantled lysimeters after 15 experimental years. Nutr Cycl Agroecosyst. 2004;68:47–57. https://doi.org/10.1023/B:FRES.0000012235.80656.cd.

Gonzalez A, Lynch JP. Subcellular and tissue Mn compartmentation in bean leaves under Mn toxicity stress. Aust J Plant Physiol. 1999;26:811–22. https://doi.org/10.1071/PP99030.

Gonzalez A, Steffen KL, Lynch JP. Light and excess manganese – implications for oxidative stress in common bean. Plant Physiol. 1998;118:493–504. https://doi.org/10.1104/pp.118.2.493.

Gregory P. Plant roots. London: Blackwell; 2006. https://doi.org/10.1002/9780470995563.

Grewal HS, Lu Z, Graham RD. Influence of subsoil zinc on dry matter production, seed yield and distribution of zinc in oilseed rape genotypes differing in zinc efficiency. Plant Soil. 1997;192:181–9. https://doi.org/10.1023/A:1004228610138.

Gruber BD, Giehl RFH, Friedel S, et al. Plasticity of the Arabidopsis root system under nutrient deficiencies. Plant Physiol. 2013;163:161–79. https://doi.org/10.1104/pp.113.218453.

Haberle J, Svoboda P, Krejcová J. Uptake of mineral nitrogen from subsoil by winter wheat. Plant Soil Environ. 2006;52:377–84. https://doi.org/10.17221/3455-PSE.

Hall DJM, Jones HR, Crabtree WL, et al. Claying and deep ripping can increase crop yields and profits on water repellent sands with marginal fertility in southern Western Australia. Aust J Soil Res. 2010;48:178–87. https://doi.org/10.1071/SR09078.

Hamblin AP, Hamblin J. Root characteristics of some temperate legume species and varieties on deep, free-draining entisols. Aust J Agric Res. 1985;36:63–72. https://doi.org/10.1071/AR9850063.

Han E, Kautz T, Huang N. Dynamics of plant nutrient uptake as affected by biopore-associated root growth in arable subsoil. Plant Soil. 2017;415:145–60. https://doi.org/10.1007/s11104-016-3150-4.

Himmelbauer ML, Loiskandl W, Rousseva S. Spatial root distribution and water uptake of maize grown on field with subsoil compaction. J Hydrol Hydromech. 2010;58:163–74. https://doi.org/10.2478/v10098-010-0015-z.

Hinsinger P, Elsass F, Jaillard B, et al. Root-induced irreversible transformation of a trioctahedral mica in the rhizosphere of rape. J Soil Sci. 1993;44:535–45. https://doi.org/10.1111/j.1365-2389.1993.tb00475.x.

Hinsinger P, Bengough AG, Vetterlein D, et al. Rhizosphere: biophysics, biogeochemistry and ecological relevance. Plant Soil. 2009;321:117–52. https://doi.org/10.1007/s11104-008-9885-9.
Hocking PJ, Mead JA, Good JA, et al. The response of canola (*Brassica napus* L.) to tillage and fertiliser placement in contrasting environments in southern New South Wales. Aust J Exp Agric. 2003;43:1323–35.
Howard DD, Essington ME, Tyler DD. Vertical phosphorus and potassium stratification in no-till cotton soils. Agron J. 1999;91:266–9. https://doi.org/10.2134/agronj1999.00021962009100020014x.
Huang B. Water relations and root activities of *Buchloe dactyloides* and *Zoysia japonica* in response to localized soil drying. Plant Soil. 1999;208:179–86. https://doi.org/10.1023/A:1004472314631.
Hudak C, Stehouwer R, Johnson J. An evaluation of K rate, placement and tillage systems for soybeans. J Fertil Iss. 1989;6:25–31.
Hutchings MJ, John EA. The effects of environmental heterogeneity on root growth and root/shoot partitioning. Ann Bot. 2004;94:1–8. https://doi.org/10.1093/aob/mch111.
Ingram PA, Malamy JE. Root system architecture. Adv Bot Res. 2010;55:75–117. https://doi.org/10.1016/B978-0-12-380868-4.00002-8.
Jackson RB, Caldwell MM. The scale of nutrient heterogeneity around individual plants and its quantification with geostatistics. Ecology. 1993;74:612–4. https://doi.org/10.2307/1939320.
Jackson RB, Caldwell MM. Integrating resource heterogeneity and plant plasticity: modelling nitrate and phosphate uptake in a patchy soil environment. J Ecol. 1996;84:891–903. https://doi.org/10.2307/2960560.
Jarvis RJ, Bolland MDA. Placing superphosphate at different depths in the soil changes its effectiveness for wheat and lupin production. Fertil Res. 1990;22:97–107. https://doi.org/10.1007/BF01116183.
Jarvis RJ, Bolland MDA. Lupin grain yields and fertiliser effectiveness are increased by banding superphosphate below the seed. Aust J Exp Agric. 1991;31:357–66. https://doi.org/10.1071/EA9910357.
Jin KM, Shen JB, Ashton RW, et al. How do roots elongate in a structured soil? J Exp Bot. 2013;64:4761–77. https://doi.org/10.1093/jxb/ert286.
Jobbágy EG, Jackson RB. The distribution of soil nutrients with depth: global patterns and the imprint of plants. Biogeochemistry. 2001;53:51–77. https://doi.org/10.1023/A:1010760720215.
Jobbágy EG, Jackson RB. The uplift of soil nutrients by plants: biogeochemical consequences and across scales. Ecology. 2004;85:2380–9. https://doi.org/10.1890/03-0245.
Jones JW, Hoogenboom G, Porter CH, Boote KJ, Batchelor WD, Hunt LA, Wilkens PW, Singh U, Gijsman AJ, Ritchie JT. The DSSAT cropping system model. Eur J Agron. 2003;18:235–65. https://doi.org/10.1016/S1161-0301(02)00107-7.
Jungk A, Claassen N, Schulz V, et al. Pflanzenverfügbarkeit der phosphatvorräte ackerbaulich genutzter böden. Z Pflanzenernähr Bodenkd. 1993;156:397–406. https://doi.org/10.1002/jpln.19931560504.
Kabir ME. Phosphorus nutrition of chickpea under dry topsoil conditions as in the High Barind Tract of Bangladesh. PhD thesis. Perth: Murdoch University; 2012.
Kabir ME, Johansen C, Bell RW. Subsoil rhizosphere modification by chickpea under a dry topsoil: implications for phosphorus acquisition. J Plant Nutr Soil Sci. 2015;178:904–13. https://doi.org/10.1002/jpln.201500021.
Kautz T. Research on subsoil biopores and their functions in organically managed soils: A review. Renew Agric Food Syst. 2015;30:318–27. https://doi.org/10.1017/S1742170513000549.
Kautz T, Amelung W, Ewert F, et al. Nutrient acquisition from arable subsoils in temperate climates: a review. Soil Biol Biochem. 2013;57:1003–22. https://doi.org/10.1016/j.soilbio.2012.09.014.
Keating BA, Carberry PS, Hammer GL, et al. An overview of APSIM, a model designed for farming systems simulation. Eur J Agron. 2003;18:267–88. https://doi.org/10.1016/S1161-0301(02)00108-9.

Klotz F, Horst WJ. Genotypic differences in aluminium tolerance of soybean (*Glycine max*. L.) as affected by ammonium and nitrate-nitrogen nutrition. J Plant Physiol. 1988;132:702–7. https://doi.org/10.1016/S0176-1617(88)80232-3.

Kuhlmann H. Importance of the subsoil for the K nutrition of crops. Plant Soil. 1990;127:129–36. https://doi.org/10.1007/BF00010845.

Kuhlmann H, Barraclough PB. Comparison between the seminal and nodal root systems of winter wheat in their activity for N and K uptake. Z Pflanzenernähr Bodenkd. 1987;150:24–30. https://doi.org/10.1002/jpln.19871500106.

Kuhlmann H, Baumgärtel G. Potential importance of the subsoil for the P and Mg nutrition of wheat. Plant Soil. 1991;137:259–66. https://doi.org/10.1007/BF00011204.

Kuhlmann H, Barraclough PB, Weir AH. Utilization of mineral nitrogen in the subsoil by winter wheat. Z Pflanzenernähr Bodenkd. 1989;152:291–5. https://doi.org/10.1002/jpln.19891520305.

Lal R, Shukla MK. Physical property of gravelly soils. In: Principles of soil physics. CRC Press/Angus & Robertson; 2004. p. 571.

Lamandé M, Hallaire V, Curmi P, et al. Changes of pore morphology, infiltration and earthworm community in a loamy soil under different agricultural managements. Catena. 2003;54:637–49. https://doi.org/10.1016/S0341-8162(03)00114-0.

Lambers H, Shane MW, Cramer MD, et al. Root structure and functioning for efficient acquisition of phosphorus: matching morphological and physiological traits. Ann Bot. 2006;98:693–713. https://doi.org/10.1093/aob/mcl114.

Lark RM, Milne AE, Addiscott TM, et al. Scale- and location-dependent correlation of nitrous oxide emissions with soil properties: an analysis using wavelets. Eur J Soil Sci. 2004;55:611–27. https://doi.org/10.1111/j.1365-2389.2004.00620.x.

Li H, Ma Q, Li H, et al. Root morphological responses to localized nutrient supply differ among crop species with contrasting root traits. Plant Soil. 2014;376:151–63. https://doi.org/10.1007/s11104-013-1965-9.

Liu LP, Gan YT, Bueckert RA, et al. Fine root distribution of oilseeds and pulses in semiarid environment. Crop Sci. 2010;50:222–6. https://doi.org/10.2135/cropsci2009.03.0156.

Loneragan JF. Distribution and movement of manganese in plants. In: Graham RD, Hannam RJ, Uren NC, editors. Manganese in soils and plants. Dordrecht: Kluwer Academic Publishers; 1988. p. 113–24.

Loneragan JF, Kirk GJ, Webb MJ. Translocation and function of zinc in roots. J Plant Nutr. 1987;10:1247–54. https://doi.org/10.1080/01904168709363653.

Lotfollahi M, Alston AM, McDonald GK. Effect of nitrogen fertiliser placement on grain protein concentration of wheat under different water regimes. Aust J Agric Res. 1997;48:241–50. https://doi.org/10.1071/A96066.

Lynch JP. Root phenes for enhanced soil exploration and phosphorus acquisition: tools for future crops. Plant Physiol. 2011;156:1041–9. https://doi.org/10.1104/pp.111.175414.

Lynch JP, Brown KM. Topsoil foraging - an architectural adaptation of plants to low phosphorus availability. Plant Soil. 2001;237:225–37. https://doi.org/10.1023/A:1013324727040.

Lynch JP, Wojciechowski T. Opportunities and challenges in the subsoil: pathways to deeper rooted crops. J Exp Bot. 2015;66:2199–210. https://doi.org/10.1093/jxb/eru508.

Ma Q, Rengel Z, Bowden B. Heterogeneous distribution of phosphorus and potassium in soil influences wheat growth and nutrient uptake. Plant Soil. 2007;291:301–9. https://doi.org/10.1007/s11104-007-9197-5.

Ma Q, Rengel Z, Rose T. The effectiveness of deep placement of fertilisers is determined by crop species and edaphic conditions in Mediterranean-type environments: a review. Aust J Soil Res. 2009;47:19–32. https://doi.org/10.1071/SR08105.

Ma Q, Scanlan C, Bell R, et al. The dynamics of potassium uptake and use, leaf gas exchange and root growth throughout plant phenological development and its effects on seed yield in wheat (*Triticum aestivum*) on a low-K sandy soil. Plant Soil. 2013;373:373–84. https://doi.org/10.1007/s11104-013-1812-z.

Mackay AD, Kladivko EJ, Barber SA, et al. Phosphorus and potassium uptake in conservation tillage systems. Soil Sci Soc Am J. 1987;51:970–4. https://doi.org/10.2136/sssaj1987.03615995005100040027x.

Malamy JE, Benfey PN. Down and out in *Arabidopsis*: the formation of lateral roots. Trends Plant Sci. 1997a;2:390–6. https://doi.org/10.1016/S1360-1385(97)90054-6.

Malamy JE, Benfey PN. Organization and cell differentiation in lateral roots of *Arabidopsis thaliana*. Development. 1997b;124:33–44.

Mallarino AP, Murrell TS. No-till corn grain yield responses to band applications of potassium. Better Crops. 1998;82:4–6.

Mallarino AP, Bordoli JM, Borges R. Phosphorus and potassium placement effects on early growth and nutrient uptake of no-till corn and relationships with grain yield. Agron J. 1999;91:37–45. https://doi.org/10.2134/agronj1999.00021962009100010007x.

Marschner H. Mechanism of adaptation of plants to acid soils. Plant Soil. 1991;134:1–20. https://doi.org/10.1007/BF00010712.

Marschner H. Mineral nutrition of higher plants. 2nd ed. London: Academic Press; 1995.

Masle J. High soil strength: mechanical forces at play on root morphogenesis and in root: shoot signalling. In: Waisel Y, Eshel A, Beeckman T, et al., editors. Plant roots: the hidden half. 3rd ed. CRC Press/Taylor & Francis Group; 2002. p. 807–19.

Materechera SA, Alston AM, Kirby JM, et al. Influence of root diameter on the penetration of seminal roots into a compacted subsoil. Plant Soil. 1992;144:297–303. https://doi.org/10.1007/BF00012888.

Matsumoto H, Sinaguru M. Advances in the aluminum toxicity and tolerance of plants for increased productivity in acid soils. In: Dubois AN, editor. Soil contamination: new research. New York: Nova Science; 2008. p. 1–42.

McArthur WM. Reference soils of south-western Australia. Perth: Australian Soil Science Society of Australia Inc. (WA Branch); 2004.

McCallum MH, Kirkegaard JA, Green TW, et al. Improved subsoil macroporosity following perennial pastures. Aust J Exp Agric. 2004;44:299–307. https://doi.org/10.1071/EA03076.

McConnell SG, Sander DH, Peterson GA. Effect of fertilizer phosphorus placement depth on winter wheat yield. Soil Sci Soc Am J. 1986;50:148–53. https://doi.org/10.2136/sssaj1986.03615995005000010028x.

McCutcheon J, Rzewnicki P. Placement of P and K on corn. In: Special circular. Wooster: Ohio Agricultural Research and Development Center/Ohio State University; 2001. p. 80–1.

McGrath GS, Hinz C, Sivapalan M, et al. Identifying a rainfall event threshold triggering herbicide leaching by preferential flow. Water Resour Res. 2010;46:W02513. https://doi.org/10.1029/2008WR007506.

McKenzie BM, Bengough AG, Hallett PD, et al. Deep rooting and drought screening of cereal crops: a novel field-based method and its application. Field Crops Res. 2009;112:165–71. https://doi.org/10.1016/j.fcr.2009.02.012.

Mengel K, Scherer HW. Release of nonexchangeable (fixed) soil ammonium under field conditions during the growing season. Soil Sci. 1981;131:226–32. https://doi.org/10.1097/00010694-198104000-00006.

Moreira BC, Prates P Jr, Dell B, et al. Roots and beneficial interactions with soil microbes. In: Oliveira TS, Bell RW, editors. Subsoil constraints for crop production. 1st ed. Springer; 2022. p. 263–87.

Moritsuka N, Yanai J, Kosaki T. Possible processes releasing nonexchangeable potassium from the rhizosphere of maize. Plant Soil. 2004;258:261–8. https://doi.org/10.1023/B:PLSO.0000016556.79278.7f.

Morrison JE, Chichester FW. Tillage system effects on soil and plant nutrient distributions on vertisols. J Prod Agric. 1994;7:364–73. https://doi.org/10.2134/jpa1994.0364.

Mullins GL, Burmester CH, Reeves DW. Cotton response to in-row subsoiling and potassium fertilizer placement in Alabama. Soil Tillage Res. 1997;40:145–54. https://doi.org/10.1016/S0167-1987(96)01070-7.

Munns R, Tester M. Mechanisms of salinity tolerance. Annu Rev Plant Biol. 2008;59:651–81. https://doi.org/10.1146/annurev.arplant.59.032607.092911.

Nable RO, Webb MJ. Further evidence that zinc is required throughout the root zone for optimal plant growth and development. Plant Soil. 1993;150:247–53. https://doi.org/10.1007/BF00013021.

Naidu R, Rengasamy P. Ion interactions and constraints to plant nutrition in Australian sodic soils. Aust J Soil Res. 1993;31:257–73. https://doi.org/10.1071/SR9930801.

Öborn I, Andrist-Rangel Y, Askegaard M, et al. Critical aspects of potassium management in agricultural systems. Soil Use Manag. 2005;21:102–12. https://doi.org/10.1111/j.1475-2743.2005.tb00114.x.

Oehl F, Sieverding E, Ineichen K, et al. Community structure of arbuscular mycorrhizal fungi at different soil depths in extensively and intensively managed agroecosystems. New Phytol. 2005;165:273–83. https://doi.org/10.1111/j.1469-8137.2004.01235.x.

Oliveira TS, Fernandes RBA. Physical subsoil constraints of agricultural and forestry land. In: Oliveira TS, Bell RW, editors. Subsoil constraints for crop production. 1st ed. Springer; 2022. p. 125–60.

Perkons U, Kautz T, Uteau D, et al. Root-length densities of various annual crops following crops with contrasting root systems. Soil Tillage Res. 2014;137:50–7. https://doi.org/10.1016/j.still.2013.11.005.

Pierret A, Moran CJ, Pankhurst CE. Differentiation of soil properties related to the spatial association of wheat roots and soil macropores. Plant Soil. 1999;211:51–8. https://doi.org/10.1023/A:1004490800536.

Poesen J, Lavee H. Rock fragments in top soils: significance and processes. Catena. 1994;23:1–28. https://doi.org/10.1016/0341-8162(94)90050-7.

Prieto I, Amas C, Pugnaire FI. Water release through plant roots: new insights on its consequences at the plant and ecosystem level. New Phytol. 2012;193:830–41. https://doi.org/10.1111/j.1469-8137.2011.04039.x.

Reeves DW, Mullins GL. Subsoiling and potassium placement effects on water relations and yield of cotton. Agron J. 1995;87:847–52. https://doi.org/10.2134/agronj1995.00021962008700050011x.

Rengasamy P. Subsoil constraints and agricultural productivity. J India Soc Soil Sci. 2000;48:674–82.

Rengasamy P, Olsson KA. Sodicity and soil structure. Aust J Soil Res. 1991;29:935–52. https://doi.org/10.1071/SR9910935.

Rengasamy P, Chittleborough D, Helyar K. Root-zone constraints and plant-based solutions for dryland salinity. Plant Soil. 2003;257:249–60. https://doi.org/10.1023/A:1027326424022.

Rich SM, Watt M. Soil conditions and cereal root system architecture: review and considerations for linking Darwin and Weaver. J Exp Bot. 2013;64:1193–208. https://doi.org/10.1093/jxb/ert043.

Richards JE, Bates TE, Sheppard SC. Changes in forms and distribution of soil phosphorus due to long-term crop production. Can J Soil Sci. 1995;75:311–8. https://doi.org/10.4141/cjss95-045.

Richardson AE, Lynch JP, Ryan PR, et al. Plant and microbial strategies to improve the phosphorus efficiency of agriculture. Plant Soil. 2011;349:121–56. https://doi.org/10.1007/s11104-011-0950-4.

Robinson D. Tansley Review No. 73. The responses of plants to non-uniform supplies of nutrients. New Phytol. 1994;127:635–74. https://doi.org/10.1111/j.1469-8137.1994.tb02969.x.

Robinson D. Variation, co-ordination and compensation in root systems in relation to soil variability. Plant Soil. 1996;187:57–66. https://doi.org/10.1007/BF00011657.

Robson AD, Taylor AC. The effect of tillage on the chemical fertility of soil. In: Cornish PS, Pratley JE, editors. Tillage – new directions in Australian agriculture. Melbourne: Inkata Press; 1987. p. 308–17.

Robson AD, Longnecker NE, Osborne LD. Effects of heterogeneous nutrient supply on root growth and nutrient uptake in relation to nutrient supply on duplex soils. Aust J Exp Agric. 1992;32:879–86. https://doi.org/10.1071/EA9920879.

Römheld V, Kirkby EA. Research on potassium in agriculture: needs and prospects. Plant Soil. 2010;335:155–80. https://doi.org/10.1007/s11104-010-0520-1.

Rose TJ, Rengel Z, Ma Q, et al. Differential accumulation patterns of phosphorus and potassium by canola cultivars compared to wheat. J Plant Nutr Soil Sci. 2007;170:404–11. https://doi.org/10.1002/jpln.200625163.

Rose TJ, Rengel Z, Ma Q, et al. Hydraulic lift by canola plants aids P and K uptake from dry topsoil. Aust J Agric Res. 2008;59:38–45. https://doi.org/10.1071/AR07146.

Sander DH, Eghball B. Planting date and phosphorus fertiliser effects on winter wheat. Agron J. 1999;91:707–12. https://doi.org/10.2134/agronj1999.914707x.

Scanlan CA, Huth NI, Bell RW. Simulating wheat growth response to potassium availability under field conditions with sandy soils. I Model development. Field Crops Res. 2015a;178:109–24. https://doi.org/10.1016/j.fcr.2015.03.022.

Scanlan CA, Bell RW, Brennan RF. Simulating wheat growth response to potassium availability under field conditions with sandy soils. II. Effect of subsurface potassium on grain yield response to potassium fertilizer. Field Crops Res. 2015b;178:125–34. https://doi.org/10.1016/j.fcr.2015.03.019.

Scanlan CA, Holmes KW, Bell RW. Sand and gravel subsoils. In: de Oliveira TS, Bell RW, editors. Subsoil Springer; 2022. pp x-xx.

Schachtschabel P. Beziehung zwischen dem durch K-Düngung erzielbaren Mehrertrag und dem K-Gehalt der Böden nach Feldversuchen in der Bundesrepublik Deutschland. Z Pflanzenernähr Bodenkd. 1985;148:439–58. https://doi.org/10.1002/jpln.19851480408.

Scherer HW. Dynamics and availability of the non-exchangeable NH_4-N – a review. Eur J Agron. 1993;2:149–60. https://doi.org/10.1016/S1161-0301(14)80124-X.

Schildwacht PM. Changes in the osmotic potential of the root as factor in the decrease in the root-shoot ratio of *Zea mays* plants under water stress. Plant Soil. 1988;111:271–5. https://doi.org/10.1007/BF02139952.

Schultz JE. Effect on wheat yield and grain phosphorus content of deep placement and foliar application of phosphate. In: Agricultural record. Department of Agriculture South Australia; 1975. p. 51–3.

Scott BJ. The response of barrel medic pasture to topdressed and placed superphosphate in central western New South Wales. Aust J Exp Agric Anim Husb. 1973;13:705–10. https://doi.org/10.1071/EA9730705.

Scott BJ, Carpenter DJ, Braysher BD, et al. Phosphorus fertiliser placement for lupins in southern New South Wales. Aust J Exp Agric. 2003;43:79–86. https://doi.org/10.1071/EA01201.

Seiffert S, Kaselowsky J, Jungk A, et al. Observed and calculated potassium uptake by maize as affected by soil water content and bulk density. Agron J. 1995;87:1070–7. https://doi.org/10.2134/agronj1995.00021962008700060007x.

Shane MW, Cramer MD, Funayama-Noguchi S, et al. Developmental physiology of cluster-root carboxylate synthesis and exudation in harsh hakea: expression of phosphoenolpyruvate carboxylase and the alternative oxidase. Plant Physiol. 2004;135:549–60. https://doi.org/10.1104/pp.103.035659.

Sharp R, Silk W, Hsiao T. Growth of the maize primary root at low water potentials. I. Spatial distribution of expansive growth. Plant Physiol. 1988;87:50–7. https://doi.org/10.1104/pp.87.1.50.

Simonsson M, Andersson S, Andrist-Rangel Y, et al. Potassium release and fixation as a function of fertilizer application rate and soil parent material. Geoderma. 2007;140:188–98. https://doi.org/10.1016/j.geoderma.2007.04.002.

Simpson JR, Lipsett J. Effects of surface moisture supply on the subsoil nutritional requirements of lucerne (*Medicago sativa* L.). Aust J Agric Res. 1973;24:199–209. https://doi.org/10.1071/AR9730199.

Singh DK, Sale PWG, Routley RR. Increasing phosphorus supply in subsurface soil in northern Australia: rationale for deep placement and the effects with various crops. Plant Soil. 2005;269:35–44. https://doi.org/10.1007/s11104-004-2475-6.

Smith S, Smet ID. Root system architecture: insights from *Arabidopsis* and cereal crops. Philos T R Soc B. 2012;367:1441–52. https://doi.org/10.1098/rstb.2011.0234.

Spehar CR, Galwey NW. Screening soya beans [*Glycine max* (L.) Merrill] for calcium efficiency by root growth in low-Ca nutrient solution. Euphytica. 1997;94:113–7. https://doi.org/10.1023/A:1002985720197.

Stöckle CO, Donatelli M, Nelson R. CropSyst, a cropping systems simulation model. Eur J Agron. 2003;18:289–307. https://doi.org/10.1016/S1161-0301(02)00109-0.

Takker PN, Walker CD. The distribution and correction of zinc deficiency. In: Robson AD, editor. Zinc in soils and plants. Dordrecht: Kluwer Academic Publishers; 1993. p. 151–65.

Tang C, Buirchell B, Longnecker NE, et al. Variation in the growth of lupin species and genotypes in alkaline soil. Plant Soil. 1993;156:513–6. https://doi.org/10.1007/BF00025096.

Tang C, Rengel Z, Diatloff E, et al. Responses of wheat and barley to liming on a sandy soil with subsoil acidity. Field Crops Res. 2003;80:235–44. https://doi.org/10.1016/S0378-4290(02)00192-2.

Tanure LPP. Dynamics of rare-earth elements, uranium and thorium by the use of phosphogypsum in tropical soils. PhD thesis. Vicosa: Universidade Federal de Viçosa; 2016.

Taylor GJ. The physiology of aluminum tolerance in higher plants. Commun Soil Sci Plant Anal. 1988;19:1179–94. https://doi.org/10.1080/00103628809368004.

Tennant D. Root growth of wheat. I. Early patterns of multiplication and extension of wheat roots including effects of levels of nitrogen, phosphorus and potassium. Aust J Agric Res. 1976;27:183–96. https://doi.org/10.1071/AR9760183.

Teutsch CD, Sulc RM, Barta AL. Banded phosphorus effects on alfalfa seedling growth and productivity after temporary waterlogging. Agron J. 2000;92:48–54. https://doi.org/10.2134/agronj2000.92148x.

Tupper GR, Ebelhar MW. Results of correcting subsoil problems. In: Proceedings of beltwide cotton conferences, 5–8 January 1994, San Diego, California, USA; 1994. p. 1524–6.

Valentine TA, Hallett PD, Binnie K. Soil strength and macropore volume limit root elongation rates in many UK agricultural soils. Ann Bot. 2012;110:259–70. https://doi.org/10.1093/aob/mcs118.

Valizadeh GR, Rengel Z, Rate AW. Response of wheat genotypes efficient in P utilisation and genotypes responsive to P fertilisation to different P banding depths and watering regimes. Aust J Agric Res. 2003;54:59–65. https://doi.org/10.1071/AR02040.

van Schaik L, Palm J, Klaus J, et al. Linking spatial earthworm distribution to macropore numbers and hydrological effectiveness. Ecohydrology. 2014;7:401–8. https://doi.org/10.1002/eco.1358.

Vance CP, Uhde-Stone C, Allan DL. Phosphorus acquisition and use: critical adaptations by plants for securing a nonrenewable resource. New Phytol. 2003;157:423–47. https://doi.org/10.1046/j.1469-8137.2003.00695.x.

Veneklaas EJ, Stevens J, Cawthray GR, et al. Chickpea and white lupin rhizosphere carboxylates vary with soil properties and enhance phosphorus uptake. Plant Soil. 2003;248:187–97. https://doi.org/10.1023/A:1022367312851.

Vos J, Evers JB, Buck-Sorlin GH, et al. Functional-structural plant modelling: a new versatile tool in crop science. J Exp Bot. 2010;61:2101–15. https://doi.org/10.1093/jxb/erp345.

Vyn TJ, Janovicek KJ. Potassium placement and tillage system effects on corn response following long-term no till. Agron J. 2001;93:487–95. https://doi.org/10.2134/agronj2001.933487x.

Vyn T, Janovicek K, Bruulsema T. Banded potash boosts no-till corn yield. Better Crops Plant Food. 1999;83:8–9.

Vyn TJ, Galic DM, Janovicek KJ. Corn response to potassium placement in conservation tillage. Soil Tillage Res. 2002;67:159–69. https://doi.org/10.1016/S0167-1987(02)00061-2.

Wan C, Xu W, Sosebee RE, et al. Hydraulic lift in drought-tolerant and –susceptible maize hybrids. Plant Soil. 2000;219:117–26. https://doi.org/10.1023/A:1004740511326.

Wang H, Yamauchi A. Growth and function of roots under abiotic stress soils. In: Huang B, editor. Plant-environment interactions. 3rd ed. Taylor and Francis Group, LLC, New York: CRC Press; 2006. p. 271–319.

Wang J, Zhang F, Zhang X, et al. Release of potassium from K-bearing minerals: effect of plant roots under P deficiency. Nutr Cycl Agroecosyst. 2000;56:45–52. https://doi.org/10.1023/A:1009894427550.

Watt M, Silk WK, Passioura JB. Rates of root and organism growth, soil conditions, and temporal and spatial development of the rhizosphere. Ann Bot. 2006;97:839–55. https://doi.org/10.1093/aob/mcl028.

Whalley WR, Watts CW, Gregory AS, et al. The effect of soil strength on the yield of wheat. Plant Soil. 2008;306:237–47. https://doi.org/10.1007/s11104-008-9577-5.

White RG, Kirkegaard JA. The distribution and abundance of wheat roots in a dense, structured subsoil: implications for water uptake. Plant Cell Environ. 2010;33:133–48. https://doi.org/10.1111/j.1365-3040.2009.02059.x.

White PJ, George TS, Gregory PJ, et al. Matching roots to their environment. Ann Bot. 2013;112:207–22. https://doi.org/10.1093/aob/mct123.

Wijesinghe DK, John EA, Beurskens S, et al. Root system size and precision in nutrient foraging: responses to spatial pattern of nutrition supply in six herbaceous species. J Ecol. 2001;89:972–83. https://doi.org/10.1111/j.1365-2745.2001.00618.x.

Wilhelm NS. Deep placement of nutrients - few excuses left not to recommend it. In: Eyre Peninsula farming systems 2005 summary. South Australian Research and Development Institute; 2005. p. 126–7.

Witter E, Johansson G. Potassium uptake from the subsoil by green manure crops. Biol Agric Hortic. 2001;19:127–41. https://doi.org/10.1080/01448765.2001.9754917.

Wu M, Li G, Li W, et al. Nitrogen fertilizer deep placement for increased grain yield and nitrogen recovery efficiency in rice grown in subtropical China. Front Plant Sci. 2017;8:1227. https://doi.org/10.3389/fpls.2017.01227.

Xu W, Cui K, Xu A, et al. Drought stress condition increases root to shoot ratio via alteration of carbohydrate partitioning and enzymatic activity in rice seedlings. Acta Physiol Plant. 2015;37:9. https://doi.org/10.1007/s11738-014-1760-0.

Yang FY, Li GZ, Zhang DE, et al. Geographical and plant genotype effects on the formation of arbuscular mycorrhiza in *Avena sativa* and *Avena nuda* at different soil depths. Biol Fertil Soils. 2010;46:435–43. https://doi.org/10.1007/s00374-010-0450-3.

Zegada-Lizarazu W, Iijima M. Hydrogen stable isotope analysis of water acquisition ability of deep roots and hydraulic lift in sixteen food crop species. Plant Prod Sci. 2004;7:427–34. https://doi.org/10.1626/pps.7.427.

Zhang XK, Rengel Z. Gradients of pH, ammonium, and phosphorus between the fertiliser band and wheat roots. Aust J Agric Res. 1999;50:365–73. https://doi.org/10.1071/A98134.

Zhang XK, Rengel Z. Temporal dynamics of gradients of phosphorus, ammonium, pH, and electrical conductivity between a di-ammonium phosphate band and wheat roots. Aust J Agric Res. 2002;53:985–92. https://doi.org/10.1071/AR01149.

Zobel RW. Tertiary root systems. In: Zobel RW, Wright SF, editors. Roots and soil management: interactions between roots and the soil. Madison: ASA, CSSA, SSSA; 2005. p. 35–6. https://doi.org/10.2134/agronmonogr48.c3.

Chapter 13
Water Acquisition by Roots From the Subsoil: Impact of Physical Constraints on the Dynamics of Water Capture

Wendy H. Vance and Stephen P. Milroy

Abstract Physical subsoil constraints, such as high soil strength, low porosity or unfavourable pore characteristics, impair crop water use, either through effects on water availability or the ability of the crop to access the water. By reducing the capacity of the soil to store water or by impeding infiltration or drainage, physical subsoil constraints can alter the availability of water to the crop. By delaying root exploration, reducing ultimate rooting depth or reducing the efficiency with which water is extracted from a soil zone, they can reduce the crop's ability to access water present. The resultant impact on crop water use is modulated by factors including the amount and distribution of rainfall, the soil's water holding capacity and the depth and severity of the constraint. While the processes by which subsoil constraints influence crop water uptake are generally well-understood, important aspects still need clarification or quantification. There are still many questions regarding processes of water transfer from the bulk soil to the roots' vascular elements. New knowledge will need to be effectively linked with our understanding of water uptake at the scale of the crop or soil profile. There is also a need to improve knowledge of the influence of agronomic management on pore size distribution, continuity and stability in terms of their influence on root system development. Finally, simulation studies that evaluate the interaction of access to water with differing soil types and climatic zones will provide important extrapolation to allow the agronomic importance of subsoil constraints to be quantified in the context of inter-annual variation in rainfall distribution.

W. H. Vance (✉)
Centre for Sustainable Farming Systems, Food Futures Institute, Murdoch University, Murdoch, WA, Australia
e-mail: W.Vance@murdoch.edu.au

S. P. Milroy
Centre for Crop and Food Innovation, Food Futures Institute, Murdoch University, Murdoch, WA, Australia
e-mail: S.Milroy@murdoch.edu.au

T. S. d. Oliveira, R. W. Bell (eds.), *Subsoil Constraints for Crop Production*,
https://doi.org/10.1007/978-3-031-00317-2_13

Keywords Root depth · Root distribution · Soil pores · Soil water · Subsoil constraint · Water uptake

13.1 Introduction

The availability of an appropriate supply of water is fundamental to crop productivity. On the one hand, subsoil constraints impact the amount and dynamics of water available to plants and, on the other hand, constrain the capacity of the plant to access the water that is there. In the context of optimum plant growth, physical subsoil constraints are those which impede water movement or root growth into or within subsoil layers, thus constraining access to both water and nutrients. The definition of the subsoil varies; usage differs between regions, soil types and disciplines. Here we adopt the functional definition of Oliveira and Bell (2022) given in Chap. 1: the subsoil is the zone below the topsoil (up to 0.2 m depth) where cultivation, fertiliser placement and soil sampling do not normally occur.

Water moves along the soil-plant-atmosphere continuum due to differences in water potential. Firstly, transpiration is driven by the difference in water vapour pressure in the atmosphere compared to that inside the leaves; the water loss generates a lower water potential in the leaf. This then drives water movement through the plant from the roots (Lambers et al. 2008). Similarly, depletion of soil water due to uptake by the roots generates a gradient in soil water potential that drives the movement of water from the bulk soil to the plant roots (Bengough 2003). Soil structure, porosity, pore size distribution and pore continuity can all affect both the storage of water in the soil and the movement of water from the bulk soil to the roots. Further, roots have to be able to adequately explore the soil volume and maintain effective contact with the soil matrix to enable water uptake by the root. Any changes in soil condition that alter these traits have a fundamental impact on the crop water use and hence crop performance (Tinker 1976; Hamblin 1986; Ritchie 1981; Bengough 2003; Jin et al. 2013; Ahmed et al. 2018).

The impact of subsoil impediments and their alleviation on water uptake and hence yield of crops have been extensively studied. For example, Schneider et al. (2017) conducted a meta-analysis of 1530 yield comparisons across 67 experimental sites, where some form of deep cultivation had been applied. The strongest drivers of yield benefit were as follows: the presence of soil layers which restricted root growth, the surface soil texture and limited water availability. These factors interacted so that the benefit of disrupting a root-restricting layer was generally greater under conditions with limited water availability. This analysis emphasizes the importance of restricted access to water as a process through which subsoil constraints limit yield.

The main components influencing water supply and demand by a crop are outlined in Fig. 13.1. These interact over time to determine the water balance of the crop and the significance of the water balance for final crop performance. In this chapter, we will use a field-based definition of the plant-available water-holding

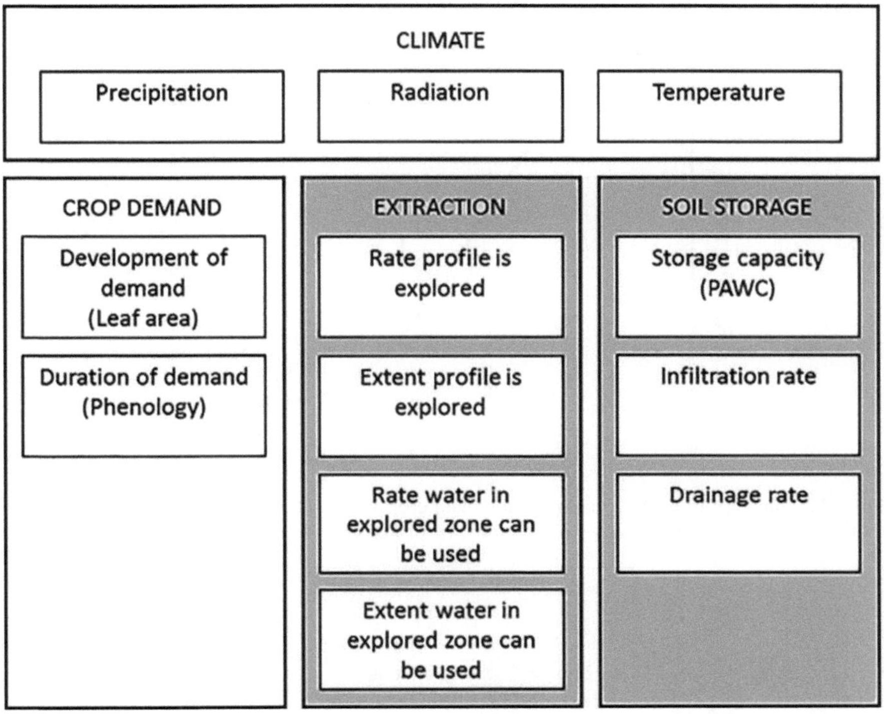

Fig. 13.1 Components contributing to the water supply and demand of a crop. Components grouped in shaded boxes may be influenced by subsoil constraints

capacity (PAWC) of a soil (Dalgliesh and Foale 2005). The PAWC is that water held between the drained upper limit (DUL) and the crop lower limit (CLL). The DUL is the water content when wet soil has drained under gravity to a relatively stable state. The CLL is the water content beyond which the crop is unable to extract further water (Dalgliesh and Foale 2005). The DUL is a function of soil characteristics only, whereas the CLL is determined by the interaction of crop and soil traits (Hochman et al. 2001; Dalgiesh and Foale 2005). Subsoil constraints may alter a number of characteristics that impact on either DUL or CLL and hence PAWC. Extraction of water by the crop root system will be considered in terms of the conceptual framework presented by Meinke et al. (1993) based on the work of Passioura (1983) and Monteith (1986). The exploration of the soil profile is described by the rate of vertical root descent and the maximum rooting depth achieved (Fig. 13.2). Once roots enter a given soil layer, water extraction can be described by an exponential decline with time from the DUL to the CLL for that layer (Fig. 13.3). Each of these processes can be modified by subsoil constraints or their amendment. There are a diverse range of potential interventions which aim to improve water acquisition by plants and which can be particularly valuable in the case of subsoil constraints. These interventions aim to improve soil structure,

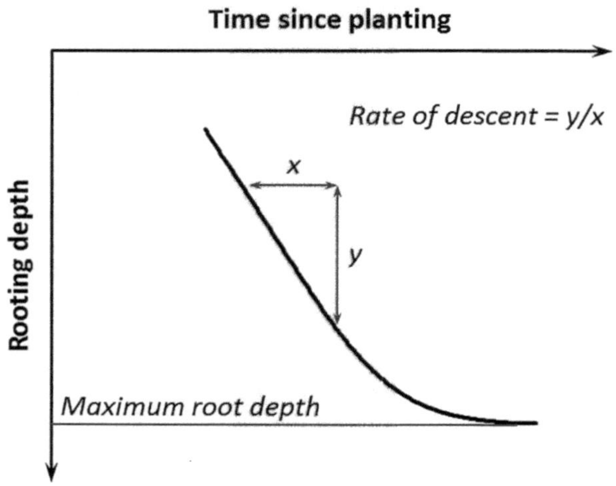

Fig. 13.2 Increase in rooting depth of an annual crop with time since planting. The increase in rooting depth often slows or stops around the time of flowering or the start of seed filling

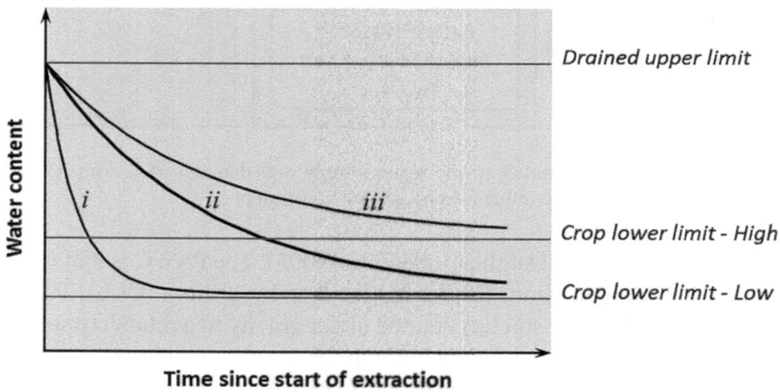

Fig. 13.3 Extraction of water from a layer in a profile after that layer is accessed: (*i*) high rate of extraction and low lower limit, (*ii*) low rate of extraction and low lower limit and (*iii*) low rate of extraction and high (less dry) lower limit

increase root elongation into the subsoil, increase plant-available water or improve access to the water present. Management tools may either involve direct interventions or agronomic management approaches to change the physical structure of the soil, or they may utilize better adapted plant genotypes to improve crop performance under constrained conditions (Table 13.1). Important aspects of the impact of physical subsoil constraints on root system development and function have been discussed in Chap. 6 (Oliveira and Fernandes 2022) and Chap. 8 (Scanlan et al. 2022). For completeness, some of these ideas will be briefly mentioned first.

Table 13.1 Interventions to improve water acquisition by plants in soil with physical subsoils constraints

Intervention	Mechanism	Impact	References
Physical interventions: to change the physical structure or the soil			
Subsoil cultivation	Removal of plough pan or hard layer impeding vertical root exploration. Improved infiltration	Earlier root exploration of deeper layers. Increased plant-available water-holding capacity. Increased available water	Hamza and Anderson (2003, 2008), Sadras et al. (2005), Mohanty et al. (2007), Barraclough and Weir (1988), Tardieu (1994)
Subsoil cultivation	Removal of impeding layer increases root length density in lower layers	Improved root proliferation at depth	Munkholm et al. (2008), Chen et al. (2014)
Compaction	Imposed compaction, reduce excessive drainage	Increased availability of water	Singh et al. (2014)
Agronomic interventions: to improve root elongation into subsoil			
Crop rotation: biological subsoiling	Large rooted species penetrate hard layers	Increased root length density at depth	Guaman et al. (2016)
Crop rotation: primer plants	Species with vigorous root systems penetrate deeper into subsoil	Increased stability and continuity of biopores. Improved root proliferation in subsoil	Yunusa and Newton (2003)
Amelioration	Subsoil manuring: organic matter amendment to subsoil clay layer	Improved soil physical properties and increased root growth in layer	Gill et al. (2009), Sale et al. (2018)
Amelioration	Deep placement of lime or gypsum	Improved soil aggregate structure or stability	Baldock et al. (1994), Vance et al. (1998)
Plant genotype differences: strategies for root growth and water uptake			
Root growth	Preferential root growth into pores and cracks	Root elongation deeper into profile	Volkmar (1996), Hatano et al. (1988), White and Kirkegaard (2010)
Root types	Root types that penetrate pores and make contact with pore walls	Improved exploration of profile. Reduce resistance at soil-root interface	Athmann et al. (2013)
Anchorage	Root hairs provide anchorage to allow penetration of soil matrix	Improved root exploration through pores and within soil matrix	Bengough et al. (2011), Jin et al. (2013)
Hydraulic redistribution	Redistribution of water in soil matrix *via* roots due to water potential gradients	Water within soil matrix is potentially more available	Prieto et al. (2012)

(continued)

Table 13.1 (continued)

Physical interventions: to change the physical structure or the soil			
Intervention	**Mechanism**	**Impact**	**References**
Root-soil contact	Root hairs behind root tips increase surface area Mucilage production increases water content of the rhizosheath	Increased rate of water absorption May increase hydraulic conductivity at root-soil interface during drying	Carminati et al. (2010), Carminati et al. (2017a, b); Ahmed et al. (2018)
Penetration ability	Differences in root penetration through hard soils, re-entry to soil matrix from pores	Root elongation deeper into profile	Materechera et al. (1991), Clark et al. (2003), Hirth et al. (2005), Botright Acuña et al. (2007), Bengough et al. (2011)
Root system architecture	Optimises distribution between surface and subsurface	Increased root length and density in subsoil	Palta et al. (2011), Wasson et al. (2012), Lynch and Wojciechowski (2015)

13.2 Conditions for Optimum Root Growth and Function

For optimum crop performance, the key soil physical properties, temperature, water status, aeration and soil mechanical resistance (soil strength), must not limit root growth or function (Letey 1958; Boone 1988; McKenzie et al. 2011). These primary factors affect root growth directly (Letey 1958), but they are interrelated and interact to a significant degree (Miller 1986; Zou et al. 2000; Iijima and Kato 2007; Bengough et al. 2011). Soil texture, bulk density, structure and structural stability and porosity characteristics have an indirect effect on root growth *via* their influence on the primary factors (Letey 1958; MacEwan et al. 2010). In this section, we will briefly outline the empirically derived parameters that define suitable conditions for root growth.

13.2.1 Temperature

Like most plant functions, root growth responds strongly to temperature. Thus, soil temperature, which varies with time of year, with time of day, with depth in the soil and with management, is a fundamental governor of root system development (Boone 1988; McMichael and Burke 1996). Root growth can occur over a relatively broad range of soil temperatures. However, each species has an optimum range, with the rate of root extension and development being reduced progressively by temperatures either above or below this range (Kaspar and Bland 1992; McMichael and Burke 1996; Misra 1999; Gregory 2007). This has a direct impact on the ability of the plant to explore the soil profile and hence access soil resources.

13.2.2 Aeration

In general terms, soil aeration needs are met when the soil porosity is adequate to admit a supply of oxygen that meets the needs of the root system for maintenance, growth and function. However, for a given soil in a given condition, soil aeration is the direct complement to water status. The more pore space that is occupied by water, the less can be occupied by air. The aeration requirement varies between species and with the ability of the plant to supply oxygen to the active roots *via* internal means (e.g. by the development of aerenchyma). However, to ensure plant development is not penalized, the soil should have an air-filled porosity of 10–15% with at least 10% of the gas in the pore space being oxygen (Dexter 1988; da Silva et al. 1994; Bengough 2003; McKenzie et al. 2011). The critical air-filled porosity also depends on the gas diffusion rates in the particular soil with well-structured soils being better able to supply oxygen to the root compared to apedal soils (MacEwan et al. 2010 modified from Pierce et al. 1983). The air-filled porosity that limits root growth also varies with texture, ranging from 14–20% for sandy soils to 10–13% with increasing clay percentage (MacEwan et al. 2010). Finally, for aeration requirements to be met, there is also a need for continuity of transmission pores (i.e. pores >30 μm equivalent diameter) to allow gas exchange with the atmosphere.

13.2.3 Water Status

Soil water status impairs root growth and function in two principal ways: poor aeration in wet soils and increased soil strength in dry soils. As indicated in the previous paragraph, high soil moisture status can result in poor soil aeration when too much of the pore space becomes filled with water, particularly in soils with low porosity or impaired pore continuity. On the other hand, as the soil dries, the greatest impact on root elongation is through the increased soil strength. This is discussed in the following paragraph. Further, as the root and soil dry, shrinkage may result in reduced contact between the root and soil contributing to increased hydraulic resistance at the interface and inhibition of the function of the root in water uptake. However, the significance of this for hydraulic resistance is still unclear (Ahmed et al. 2018). Root hairs and mucilage may play an important role in maintaining connectivity (Carminati et al. 2017a, b). Finally, for a water supply that does not restrict root growth, the pore size distribution should be made up of greater than 15% macropores (i.e. transmission pores >30 μm equivalent diameter) and greater than 20% mesopores (0.2–30 μm equivalent diameter) (Cockroft and Olsson 1997). The hydraulic conductivity of the bulk soil *per se* rarely restricts water supply to a level that would impair growth (Dexter 1988).

13.2.4 Mechanical Resistance

Soil mechanical resistance is a critical characteristic governing the rate of root growth. A penetrometer resistance of 2 MPa is a useful representation of the soil strength above which root elongation will be impeded (da Silva et al. 1994; Clark et al. 2003). In reality, root growth elongation slows as resistance increases beyond a species-specific threshold until the resistance is great enough to stop further elongation (McKenzie et al. 2011). The resistances at which elongation begins to slow and at which growth stops vary among species. While species differ in their ability to exert axial root growth pressure, differences are more clearly related to root diameter (Boone 1988; Materechera et al. 1991; Misra 1997; Clark et al. 2003).

Soil mechanical resistance can vary with a number of characteristics, including texture and compaction (Kirkegaard et al. 1992; Iijima and Kato 2007). Importantly it also varies substantially with soil water content because soil strength increases exponentially as the soil dries (Stirzaker et al. 1996; Iijima and Kato 2007; Bengough et al. 2011). However, Bengough and co-workers (2011) highlight that even in wet soils, with matric potentials as high as −100 to −200 kPa, mechanical resistance can still be high enough to decrease root elongation rates by 50%.

When any of the above requirements are not met, there will be poor root system growth and development and therefore a reduced capacity of the crop to access soil resources. While soil physical properties may limit root growth either in the topsoil or the subsoil, these constraints have been identified as particularly problematic for root growth in subsoils (Adcock et al. 2007; MacEwan et al. 2010; McDonald et al. 2013; Lynch and Wojciechowski 2015), especially when the topsoil contains insufficient water and nutrients (Wong and Asseng 2007).

13.3 Soil Water Availability

Inhospitable subsoils commonly combine a number of constraints that can reduce water infiltration and drainage rates as well as the soil's water storage capacity due to lower total porosity, an altered pore size distribution and reduced pore continuity (Stirzaker et al. 1996; Bengough 2003; Gregory 2007; MacEwan et al. 2010; Lipiec et al. 2012; Gao et al. 2016; Pires et al. 2017). The net result is a change in the amount of water available for use by the crop and potentially a change in the timing of when it is available.

Compaction is frequently found to reduce water movement into the subsoil. Mossadeghi-Bjorklund et al. (2016) showed a significant reduction in hydraulic conductivity in the 0.3–0.5 m depth layer as a result of an imposed compaction treatment. This was associated with a reduction in the density of macropores, and there was also evidence that the compaction may have disrupted macropore continuity. Conversely, Hamza and Anderson (2003, 2008) achieved dramatic increases in infiltration rates due to subsoil cultivation across a number of soils with compacted subsoil, although the conductivity of the subsoil was not measured. The outcome was an increased soil water content in the top 500 mm of the soil. Mohanty et al. (2007) also found a significant benefit in terms of water storage in response to subsoil cultivation on a Vertisol, although the increase was small.

While the usual impact of compaction on water dynamics is negative due to the retarded water movement into the soil resulting in lower water availability, this is not always the case. Thus, Singh et al. (2014) reported that on a highly permeable soil, imposing a compaction treatment reduced excessive drainage. This improved water retention as well as reducing nitrate losses via leaching.

Water movement in texture contrast soils can present particular problems. The saturated hydraulic conductivity of the subsoils may be less than 1–2 mm day^{-1} (Belford et al. 1992; Dracup et al. 1992; Eastham and Gregory 2000). The low rate of water movement through the subsoil results in water accumulating in the more permeable layers above, leading to temporary perched water tables with the potential to cause waterlogging (Dracup et al. 1992; Zhang et al. 2004). In the Mediterranean-type climate of Western Australia, this typically occurs during winter when rainfall often exceeds the potential evapotranspiration rate. Because crop evapotranspiration (ET) is close to the potential at this time, the differences in soil water status have little direct influence on crop water use (Eastham and Gregory 2000). However, the outcome is likely to vary, depending on the dynamics of water availability versus the pattern of crop demand with development, as well as any secondary effects due to the impact of waterlogging on the crop, such as root system damage or impaired nutrient uptake.

Not only can subsoil constraints alter water supply by influencing infiltration and drainage, they have also been shown to alter the plant-available water-holding capacity (PAWC) of certain layers and thus the PAWC of the whole soil profile. Lipiec and co-workers (2012) showed that compaction led to less water being held in the plant-available range within the subsoil. Water in micropores of less than 0.5 μm radius is held at low potential (very negative) and so is unavailable to the plant. Compaction reduced the total porosity in subsurface aggregates and decreased the volume of pores with 1–3 μm radius but increased the volume of pores with a radius less than 0.3 μm. The overall result was a reduction in the volume of pores holding water in the available range (ca. −10 to −1500 kPa). Similarly, Babalola and Lal (1977a) found that the PAWC of a gravel layer was reduced with increased gravel percentage. They found lower total porosity and differences in pore size distribution, which may have contributed to the differences in PAWC.

The benefit of removing a compaction layer has also been demonstrated at the crop scale. Sadras et al. (2005) found that deep cultivation reduced the lower limit of extraction measured for some soil layers under a wheat crop but had little influence on the drained upper limit, thus increasing the PAWC. That is, the amount of water available to the crop was increased. As these experiments focused at the crop level, the mechanisms underlying the change were not explored. It is therefore unclear to what extent the effects were due to alterations in water holding characteristics of the soil or to the effectiveness of root exploration within the layer.

Thus, overall, there is evidence that subsoil constraints can influence the storage of water in the profile through changes in infiltration rate, drainage rate and PAWC. However, the significance of such changes for crop performance will depend on the rainfall pattern and the dynamics of crop water demand.

13.4 Rate at Which Roots Explore the Soil Profile

Regardless of whether plant roots are in the subsoils or surface, they need to overcome the resistance forces of the soil to penetrate the soil matrix (Bengough et al. 2011; Jin et al. 2013). Soil mechanical resistance is often greater in the subsoil than in the soil surface due to overburden pressure, the presence of fewer roots and less fauna to create biopores, the potential presence of gravel layers and the lack of disturbance by tillage (Unger 1979; Jin et al. 2013; Gao et al. 2016).

The pressure required for a plant root to penetrate the soil is the sum of the radial pressure required to expand a cavity and the axial pressure to overcome the frictional resistance at the soil-root surface along the root (Bengough et al. 2011; Jin et al. 2013). The root-soil friction is reduced by the sloughing off of border cells and the production of mucilage at the root tips (Iijima et al. 2004; Gregory 2007). To allow the tip to advance against mechanical impedance, root hairs behind the root tip play an important role in providing anchorage. This is also important for allowing the root to bend and change directions or to grow across an existing crack in the soil and re-enter the soil matrix (Bengough et al. 2011; Jin et al. 2013).

There are a number of soil factors whose influence on root growth is via their contribution to soil mechanical resistance. Taylor and Ratliff (1969) generated variation in resistance through differences in soil water potential and bulk density. They were able to derive a single relationship between resistance and root elongation rate of peanut across both sources of variation, emphasizing that the importance of the component variables was their contribution to mechanical resistance. The response curve suggested that a penetrometer resistance of 2.0 MPa would reduce the root extension rate by 50%.

The maximum rate at which roots penetrate downward into the soil (Fig. 13.2) is to some degree characteristic of a species (Dardanelli et al. 1997). For wheat, field measurements indicate a rate of descent of around 1.2–1.3 mm $°C^{-1}$ day^{-1} (mm per degree-day) across sands and structured clay, and for winter and spring, genotypes (Kirkegaard and Lilley 2007; Thorup-Kristensen et al. 2009). For soybean and maize, a rate of 2.0 mm $°C^{-1}$ day^{-1} has been derived (Ordonez et al. 2018). Physical subsoil constraints, such as compaction or high gravel percentage, impede the descent of the roots and hence delay the time at which crops can access water held at different depths in the soil profile.

At the crop level, while high mechanical resistance due to compacted or gravel layers may retard the descent of the roots into the soil, the passage of roots through soil biopores and cracks means that the impact is not as great as might be calculated based on uniformly strong soil (Stirzaker et al. 1996). White and Kirkegaard (2010) found that in a well-structured, high-strength soil, 85–100% of wheat roots in subsurface layers were in pores and cracks, with multiple roots occupying single voids. However, only 5% of pores were occupied by roots, which may reflect a lack of pore continuity, precluding roots from using a proportion of pores for vertical exploration. Under controlled conditions, Hatano and co-workers (1988) demonstrated the importance of macropores for root growth by showing a correlation between the spatial distribution of pores and the distribution of maize roots across a number of soil types. Importantly, the data showed that the proportion of roots penetrating macropores, rather than the soil matrix, was higher in the soils with higher bulk density and lower water content, that is, in stronger soils. Volkmar (1996) also demonstrated the increased dependence of root penetration on macropores as soils became drier. Interestingly, Stirzaker et al. (1996) demonstrated that the frequency of roots in pores of strong soil was three to four times higher than might have been expected based on probability.

While there is good evidence for the greater importance of soil pores for root extension in strong soils, the impact that this has on access to soil resources is likely to differ among species. For example, Athmann et al. (2013) found that the mode of contact between a root and the pore wall appears to differ between species. Barley with its fibrous root system and oilseed rape with its taproot system exhibited different strategies. In barley, seminal roots made contact with the pore wall, growing few laterals but having many long root hairs. In oilseed rape, on the other hand, roots grew vertically down the centre of the pores, the root hairs were shorter and contact with the pore wall was made by the lateral roots. There is little information available on how other root system characteristics influence the ability of a plant to utilize pores in strong soil. It could be suggested that variation among species (or among genotypes within a species) in terms of the number of primary axes, the degree of branching and the root width could influence the ability of roots to grow into available pores and hence access resources in zones of high soil strength.

Both soil compaction and gravel content impede the descent of the rooting front through the impact of high mechanical resistance on the elongation of individual roots (Babalola and Lal 1977a, b; Taylor and Brar 1991; Popova et al. 2016). Field experiments assessing the benefits of deep cultivation can provide useful comparisons, demonstrating the impact of mechanical resistance in retarding the exploration of the profile. In an early study, Barraclough and Weir (1988) examined the response of winter wheat to subsoil cultivation to remove a layer of high mechanical resistance. The 'plough pan' lay beneath the zone of cultivation with a peak penetrometer resistance at approximately 0.35 m. Four months after sowing, roots had reached 1.2 m in the treatment that received subsoil cultivation but only 0.40 m where the plough pan had remained. After this, however, vertical root penetration in the untreated soil was rapid. Tardieu (1994) cited earlier work in which it was found that the vertical penetration of maize roots through a clay loam soil was seriously retarded by an imposed compaction treatment until the drying profile began to develop cracks which allowed extension of roots through the layer of high mechanical resistance. In both examples, root exploration of deeper layers was delayed until the constraint of the layer with high mechanical resistance was overcome. However, such responses can show marked inter-annual variation. Rengasamy and Reid (1993a, b) showed that subsoil cultivation at 0.3 m could dramatically improve the vertical penetration rate of faba bean roots on a compacted silt loam. However, in the first year, roots in the compacted treatment had reached ca. 0.35 m by flowering, whereas where subsoil cultivation was used, roots had reached ca. 0.55 m, but in the second year, there was no difference in rooting depth until the beginning of pod filling.

The delay in vertical root exploration is strongly associated with access to soil resources. Radford et al. (2001) presented a good example of the impact of compaction on retardation of the depth of soil water extraction. Comparing three intentionally compacted treatments to the control, they demonstrated that the most severe compaction treatment delayed the time at which the crop accessed water at a depth of 1.0 m by as much as 50 days. In terms of the depth of profile that could be utilized by the crop at a given time, at 50 days after sowing (DAS), the crop on the compacted treatments was extracting to between 0.4 and 0.6 m, while on the non-compacted treatments, extraction had reached around 0.9 m. The most severely compacted treatment did not access water from 0.9 m until 100 DAS, at which time the non-compacted control was extracting from a depth of around 1.3 m.

The delayed exploration of the soil due to subsoil constraints means that a smaller proportion of the water in the profile is available to the plant at any given time. Thus, there is a reduced capacity to continue optimal crop growth if the soil water is not replenished by irrigation or rainfall. The significance of this for yield will therefore depend to a very great degree on the pattern of water inputs relative to the temporal development of crop demand and the PAWC of the soil.

13.5 Maximum Depth of Soil Exploration

Under unconstrained conditions, the maximum depth of soil exploration by the crop is determined by the rate of vertical root growth and the duration of root growth (Fig. 13.2). In annual crops, root growth, and hence vertical exploration, typically ceases sometime around the time of flowering or the start of seed filling (Dardanelli et al. 1997). A summary of typical roots depths for annual and perennial crops is reported by Costa and Coutinho (2022). As well as impeding the rate of exploration of the soil profile, subsoil constraints can limit the maximum depth of soil to which roots explore and hence from which water can be extracted. There are two ways in which this occurs: the constraint in a layer may be of sufficient magnitude to render it impenetrable to roots, thus dictating the maximum rooting depth, or the descent of roots may be delayed by constraints to such a degree that the soil is not fully explored before ontogenetic factors effectively stop root growth.

Soil strata of particularly high mechanical resistance, whether due to compaction, high gravel content, cemented gravels or in some cases high clay subsoils, can present an absolute limit to root exploration (e.g. Dracup et al. 1992; Wong et al. 2009; Khan et al. 2016). In such cases, the maximum amount of water available to a crop is limited to that stored above the impenetrable layer. The impact that this has on water use and crop performance depends on the interaction of the depth of the constraint, soil water holding capacity above the constraint, the pattern of rainfall and crop management (Wong and Asseng 2006, 2007). The depth at which an impenetrable layer lies can vary markedly even within a field (Wong et al. 2008), which has consequences for the amount of water available to the crop, and hence crop yield, as well as the risk of drainage and leaching of nutrients (Wong et al. 2006). In rainfed systems, the spatial variability in yield induced by impenetrable barriers is usually most marked in wet years. In dry seasons, crop growth across the site is more likely to be limited uniformly by water deficit, but in wet seasons, the profile is more likely be filled and the water accessible to the crop at any given position in the field becomes a reflection of the depth of the impenetrable layer (Wong and Asseng 2006). Simulation analysis of the interaction of seasonal rainfall and agronomic inputs allows management to be varied spatially to reflect crop yield potential, as well as drainage and leaching risks (Wong and Asseng 2006). It can also provide important information to support the decision of whether correction of the subsoil condition is warranted.

In other situations, the depth of soil water extraction may be limited not by an absolute barrier but due to the rate of root descent being reduced. If root descent has been retarded sufficiently, ontogenetic limitations on the duration of significant root growth may mean exploration ceases before the roots have exploited the depth which a non-constrained root system might achieve. In the study of Radford et al. (2001), described earlier, although the rate of descent of the extraction front was delayed by the treatment with an intermediate degree of compaction, it ultimately reached the same depth as that in the non-compacted treatment. In the most severely compacted treatment, the rate of descent early in development was retarded to such an extent that although the subsequent rate of exploration was not different from that in the moderate compaction treatments, the ultimate depth of extraction was some 0.3 m shallower than the non-constrained treatment.

13.6 Efficiency of Extraction From a Soil Layer

Once a soil layer has been accessed by the root system, the rate at which water is removed and the amount of water that can be removed (Fig. 13.3) are influenced by both plant traits and soil properties (Meinke et al. 1993; Dardanelli et al. 1997). Subsoil constraints can reduce the efficiency with which water is extracted by the crop from any given soil volume in terms of either the rate of extraction or the proportion of the soil water ultimately accessed by the crop. In addition to soil-based mechanisms, such as changes to water potential, content and movement, there are three groups of plant-based mechanisms by which this can occur: changes in the amount and disposition of roots within a layer, effects at the root/soil interface and alterations to morphology, anatomy and chemistry of individual roots.

Since soil mechanical resistance reduces root elongation rate, the amount of roots in a layer of high resistance is typically less than if that layer had had lower resistance. Root length density (RLD) of potato has been shown to differ dramatically between plots of contrasting soil resistance (Parker et al. 1989). Thus, crops such as potato, broccoli and lettuce have been shown to be highly responsive to treatments to remove layers with high mechanical resistance such as plough pans (Montagu et al. 1998; Guaman et al. 2016). The results of Guaman et al. (2016) are of particular interest. Inter-row subsoil cultivation reduced the soil penetration resistance between 0 and 0.6 m in a soil with a strong plough pan, resulting in a 70% increase in the RLD in these layers. However, 'biological subsoiling' (the use of rotation crops to correct the plough pan) did not reduce the observed penetration resistance, but the RLD was still improved by the same amount. Further, a combination of the two treatments resulted in a 120% increase in RLD relative to the control.

Importantly, the removal of a layer with high mechanical resistance often results in greater root density in layers deeper in the soil. Differences in resistances in soil layers, around 0.3 m, have been shown to alter the RLD of wheat throughout the profile (Munkholm et al. 2008; Chen et al. 2014). However, in the work of Vocanson et al. (2006), the extent of such benefits appeared to differ with weather and genotype. Interpreting the mechanisms underlying field experiments such as these is somewhat difficult due to possible feedforward effects: That is, removing compaction improves root proliferation which provides access to a greater amount of soil resources, thus improving crop growth, which in turn leads to more root proliferation. There would be a particular value in combining field-based studies with more mechanistic exploration of the factors contributing to the responses.

Within a given layer, the impact of mechanical resistance on water extraction is greater than might be expected on the basis of the difference in average RLD. This relates again to the importance of macropores and soil structure for root growth. Root distribution within a soil layer is not uniform. Examining a range of structured soil types under controlled traffic, Logsdon and Allmaras (1991) found significant clumping of maize and soybean roots in all layers and under all tillage methods. In this example, the roots were not constrained to biopores or significant cracks. The non-uniform and non-random distribution of roots in soils means that more of the soil volume is a greater distance from the nearest root than would otherwise be the case (Tardieu 1988; Logsdon and Allmaras 1991). This means that on the basis of geometry alone, water is less readily extracted from the soil (De Willigen 1987). As a result, residual water remains in unexplored parts of the soil matrix (Pardo et al. 2000; Amato and Ritchie 2002). Thus, the extraction of water from the soil is slower and not all water is extracted from the soil matrix (Passioura 1991). In high strength soil, the dependence of roots on soil pores is greater (Hatano et al. 1988; Volkmar 1996) because of the greater difficulty of penetrating the bulk soil. At the same time, the number of pores is lower and distribution less uniform (Hatano et al. 1988; Kim et al. 2010; Berisso et al. 2012). As a result, the non-uniformity of root distribution can be expected to be higher and the rate and extent of water extraction to be lower.

White and Kirkegaard (2010), working with a well-structured, high-strength soil, found that virtually all roots below a depth of 0.6 m occupied pores or cracks.

However, in this case, distribution did not differ significantly from random. An estimation of the potential rate of water extraction suggested that the density and distribution of roots was not likely to limit water uptake. This was not consistent with the significant proportion of residual water left by an adjacent crop even though it was growing under water limitation (Kirkegaard et al. 2007). The authors suggested that the discrepancy may have been due to limitations to hydraulic conductivity at the soil-root interface. Mechanisms that govern the rate of water transfer from the soil to the root require significant further research. In reviewing hydraulic processes in plant water uptake and their significance for the yield of water-limited grain crops, Ahmed and co-workers (2018) emphasized the need to better understand the transfer of water between the bulk soil and the root vascular tissue as well as the significance of the underlying processes at the plant and crop level. In particular, they stressed our limited knowledge of the role of mucilage, root hairs, soil-root contact and aquaporins. Both root hairs and root exudates are significant in overcoming high hydraulic resistance at the soil-root interface. Mucilage appears to play an important role in maintaining contact between the soil and the root as soil water declines and roots shrink (Ahmed et al. 2014; Carminati et al. 2017a). Similarly, it has been demonstrated that the presence of root hairs results in a smaller drop in water potential across the soil-root interface as the soil dries (Carminati et al. 2017b). In high-strength soils, the importance of these mechanisms is likely to be accentuated, because contact between the root and the soil matrix may be either particularly high or low. In the study of White and Kirkegaard (2010), mentioned earlier, most roots in the subsoil of a well-structured, high-strength soil were in pores and contacted the soil primarily via root hairs. By contrast, roots in cracks were adpressed to the soil surface and had few root hairs. Indeed, the density of root hairs could be related to the proportion of the root surface that was in contact with the soil. If high soil strength limits root hair elongation (Haling et al. 2014), their function in overcoming limited contact between the root and the soil may be reduced. The results of Haling et al. (2014) are consistent with the limited penetration of the pore wall observed by White and Kirkegaard (2010). Thus, the need for research into mechanisms influencing the rate of water transfer across the soil-root interface is of particular significance to the question of the influence subsoil constraints on crop water supply.

13.7 Consequences for Seasonal Crop Water Use

The magnitude of the impact of a subsoil constraint on crop water use, and hence yield, is modulated by many factors, including the amount and distribution of rainfall, the PAWC of the soil and the depth and severity of the constraint. The world's major food crops, rice, wheat, maize and potato, are each grown across a wide range of climatic zones. For example, in Australia alone, wheat is grown in subtropical climates on predominantly stored soil water, in temperate climates with equi-seasonal rainfall and in Mediterranean-type climates with winter dominant rainfall

(Fig. 13.4.). In addition to climatic type, inter-annual variation also needs to be considered.

The influence of the pattern of rainfall relative to the time course of crop demand in modulating the significance of a subsoil constraint is demonstrated in the results of Rengasamy and Reid (1993b). Subsoil cultivation to remove compaction increased total crop ET of faba bean in one case but not in the other two. In one case, adequate rainfall late in the season met crop demand for water, and so no stress appears to have been encountered. In another, it appears that soil water was fully depleted in both the compacted and deep cultivated soils. These three possible outcomes were conceptualized as generalized cases by Ahmed et al. (2018).

The impact of a subsoil constraint on water supply and crop performance is also dependent on soil type. Sadras et al. (2005) found contrasting responses to the removal of subsoil compaction in different positions in the landscape (lower versus higher positions) that differed in soil type. They used deep cultivation to remove a compaction layer in a landscape with sandy-loam soil on low-lying land and sandy ridges. The treatments had little effect on water use (ET) of a wheat crop on the sandy-loam soils but increased ET by 30–40% on the sandy soil with transpiration increasing by up to 90%. The responses in crop growth were consistent with measured transpiration.

The complexity of the interaction between factors that alter the impact of subsoil constraints has been explored using dynamic simulation models of crop growth and yield. Wong and Asseng (2007) used a model to estimate the yield benefit to wheat crops from correcting subsoil constraints in sandy soils in a Mediterranean-type environment. They found that in lower rainfall regions or dry seasons, the benefit of correcting a constraint was small, as root and crop growth were limited by the depth of soil wetting before the constraint was reached. In wetter years and regions, the

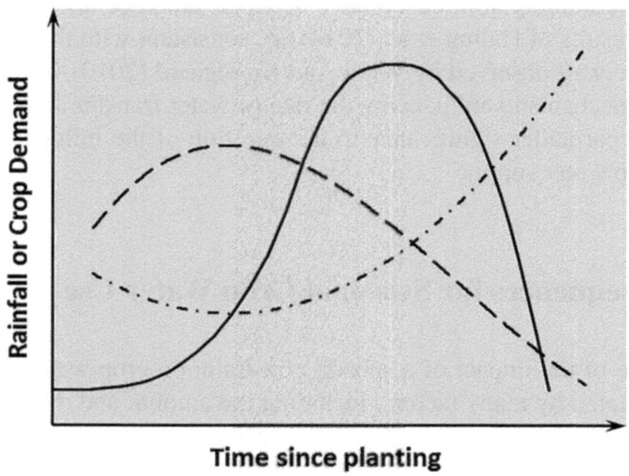

Fig. 13.4 Increase and then decline in water demand by an annual crop (solid line) compared to generalized rainfall patterns for a Mediterranean-type climate (dashed line) and a subtropical climate (dot-dash line)

benefit was greater, because the constraint stopped the roots from accessing the water that would otherwise have been available for growth. They also found a greater average response on coarse texture soil, because the wetting front reached greater depths for a given amount of rainfall and the roots needed to penetrate deeper to access the same amount of water. Such analyses are important to support the decision to make investment in correcting a subsoil constraint. Lilley and Kirkegaard (2007) used the same model to examine the benefit to wheat of access to water located deep in the rooting zone. Working in a somewhat wetter, less strongly Mediterranean environment with heavier soils, they drew similar conclusions to Wong and Asseng (2007). They also found that the yield benefit of access to water deep in the profile was significantly altered by the amount of water available at sowing. This points to the possible impact of agronomic decisions such as crop sequence (via the extent to which the previous crop depleted the soil water reserve) and summer fallow management on the benefit arising from correcting a subsoil constraint and raises the general question of the interaction between subsoil constraints and any agronomic practices that alter water availability or demand.

For cereal crops grown on stored soil moisture and with limited in-season rainfall, it is important to ensure adequate water supply remains to allow good grain size as a component of yield (Cornish and Lymbery 1987; Richards and Passioura 1989; Passioura 2006). Pre-anthesis water use supports the development of potential yield in terms of canopy development, the number of ears per unit area and the number of kernels per ear. Post-anthesis water use supports the filling of the reproductive sink. Thus, while the impact of subsoil constraints on total crop ET will be of primary importance, it is conceivable that delaying the timing of water use by the crop may also influence yield, yield components and quality. However, there is limited field-based information on such effects by subsoil constraints.

In summary, the influence of subsoil constraints on crop water use is known to be influenced by the amount and distribution of rainfall, soil type and the depth of the constraint. Given the complexity of the interactions and the importance of the timing of water supply to yield development in many crops, there is a need to develop a broader picture of the likely impact of subsoil constraints under different conditions. Simulation modelling will be a useful tool in this regard to allow likely benefits to be evaluated for different soil types, in different rainfall environments and in the context of inter-annual variation in rainfall and temperature.

13.8 Future Research Needs

While there is a significant amount of understanding about the mechanisms by which subsoil constraints influence the uptake of water by crops, there are still important aspects that need clarification or quantification. Three key areas are outlined here.

13.8.1 Soil-Root Interface

Ahmed et al. (2018) outlined a suite of questions that still need to be resolved regarding the mechanisms of water transfer from the bulk soil to the roots' vascular elements. In addition, our understanding of the movement of water across the soil-root interface is not well integrated with our understanding of water uptake at the scale of the crop and soil profile. Linking these different process levels, and, in particular, defining how the processes at the interface are influenced by subsoil constraints, may contribute important information to our understanding of how crops respond to declining water availability under adverse soil conditions. The extent to which processes at the soil-root interface differ among species and the consequences of this for crop performance have also received little attention.

13.8.2 Crop Management and Soil Pores

Given the apparent importance of soil pores for the development and function of crop root systems, there is a need to strengthen our knowledge of the influence of crop management practices on pore size distribution, pore continuity and pore stability. This should not only focus on the influence of different cultivation methods and traffic management but should also encompass the interaction with crop rotation, cover management and soil chemical amendment, which may influence the rate of development and persistence of pores of different sizes.

13.8.3 Quantifying Importance

At the whole crop level, subsoil constraints modify the availability of water both spatially and temporally. At the same time, they also influence the capacity of the crop to access that water, again, both in space and time. There are few studies which attempt to explore both sets of processes under the same conditions and so identify their relative importance and the extent to which they interact. The impact of these processes on the total amount of water used by a crop and the pattern of usage over the crop cycle will determine the magnitude of the impact on crop performance. Simulation studies, such as those of Wong and Asseng (2007) or Lilley and Kirkegaard (2007) that have evaluated the influence of access to water in the context of differing cropping systems, management, soil types and climatic zones, provide an important extrapolation to allow the agronomic importance of these processes to be seen in the context of inter-annual variation in rainfall. There would be particular value in applying this approach specifically to subsoil constraints, if the appropriate processes could be robustly captured in simulation routines. This is not facile. A substantial amount of research would need to be conducted to adequately quantify responses in the component process and allow for calibration and validation. Integration and analysis at this level is fundamental to allow the extensive knowledge of the influence of soil constraints on soil and plant processes to be used to inform agronomic decision-making.

References

Adcock D, McNeill AM, McDonald GK, Armstrong RD. Subsoil constraints to crop production on neutral and alkaline soils in south-eastern Australia: a review of current knowledge and management strategies. Aust J Exp Agric. 2007;47:1245–61.

Ahmed MA, Kroener E, Holz M, Zarebanadkouki M, Carminati A. Mucilage exudation facilitates root water uptake in dry soils. Funct Plant Biol. 2014;41:1129–37.

Ahmed MA, Passioura J, Carminati A. Hydraulic processes in roots and the rhizosphere pertinent to increasing yield of water-limited grain crops: a critical review. J Exp Bot. 2018;69:3255–65.

Amato M, Ritchie JT. Spatial distribution of roots and water uptake of maize (*Zea mays* L.) as effected by soil structure. Crop Sci. 2002;42:773–80.

Athmann M, Kautz T, Pude R, Kopke U. Root growth in biopores – evaluation with in situ endoscopy. Plant Soil. 2013;371:179–90.

Babalola O, Lal R. Subsoil gravel horizon and maize root growth. I. Gravel concentration and bulk density effects. Plant Soil. 1977a;46:337–46.

Babalola O, Lal R. Subsoil gravel horizon and maize root growth. II. Effects of gravel size, inter-gravel texture and natural gravel horizon. Plant Soil. 1977b;46:347–57.

Baldock JA, Aoyama M, Oades JM, Susanto GCD. Structural amelioration of a South Australian red-brown earth using calcium and organic amendments. Soil Res. 1994;32:571–94.

Barraclough PB, Weir AH. Effects of a compacted subsoil layer on root and shoot growth, water use and nutrient uptake of winter wheat. J Agric Sci. 1988;110:207–16.

Belford RK, Dracup M, Tennant D. Limitations to growth and yield of cereal and lupin crops on duplex soils. Aust J Exp Agric. 1992;32:929–45.

Bengough AG. Root growth and function in relation to soil structure, composition, and strength. In: de Kroon H, Visser EJW, editors. Root ecology. Berlin/Heidelberg: Springer; 2003. p. 151–72.

Bengough AG, McKenzie BM, Hallett PD, Valentine TA. Root elongation, water stress, and mechanical impedance: a review of limiting stresses and beneficial root tip traits. J Exp Bot. 2011;62:59–68.

Berisso FE, Schjonning P, Keller T, Lamande M, Etana A, de Jong LW, Iversen BV, Arvidsson J, Forkman J. Persistent effects of subsoil compaction on pore size distribution and gas transport in a loamy soil. Soil Tillage Res. 2012;122:42–51.

Boone FR. Weather and other environmental factors influencing crop responses to tillage and traffic. Soil Tillage Res. 1988;11:283–324.

Botwright Acuña TL, Pasuquin E, Wade LJ. Genotypic differences in root penetration ability of wheat through thin wax layers in contrasting water regimes and in the field. Plant Soil. 2007;301:135–49.

Carminati A, Moradi AB, Vetterlein D, Vontobel P, Lehmann E, Weller U, Vogel HJ, Oswald SE. Dynamics of soil water content in the rhizosphere. Plant Soil. 2010;332:163–76.

Carminati A, Benard P, Ahmed MA, Zarebanadkouki M. Liquid bridges at the root-soil interface. Plant Soil. 2017a;417:1–15.

Carminati A, Passioura JB, Zarebanadkouki M, Ahmed MA, Ryan P, Watt M, Delhaize E. Root hairs enable high transpiration rates in drying soils. New Phytol. 2017b;216:771–181.

Chen YL, Palta J, Clements J, Buirchell B, Siddique KHM, Rengel Z. Root architecture alteration of narrow-leafed lupin and wheat in response to soil compaction. Field Crops Res. 2014;165:61–70.

Clark LJ, Whalley WR, Barraclough PB. How do roots penetrate strong soil? Plant Soil. 2003;255:93–104.

Cockroft B, Olsson KA. Case study of soil quality in south-eastern Australia: management of structure for roots in duplex soils. In: Gregorich EG, Carter MR, editors. Soil quality for crop production and ecosystem health. Amsterdam: Elsevier; 1997. p. 339–50.

Cornish PS, Lymbery JR. Reduced early growth of direct drilled wheat in southern New South Wales: causes and consequences. Aust J Exp Agric. 1987;27:869–80.

Costa MCG, Coutinho IAC. Root systems of agricultural crops and their response to physical and chemical constraints. In: Oliveira TS, Bell RW, editors. Subsoil constraints for crop production. Springer; 2022. p. 225–61.

da Silva AP, Kay BD, Perfect E. Characterisation of the least limiting water range of soils. Soil Sci Soc Am J. 1994;58:1775–81.

Dalgliesh NP, Foale M. Soil matters – monitoring soil water and nutrients in dryland farming. CSIRO Agricultural Production Systems Research Unit, Australia. 2005. https://www.apsim.info/Portals/0/APSoil/SoilMatters/. Accessed 20 June 2019.

Dardanelli JJ, Bachmeier OA, Sereno R, Gil R. Rooting depth and soil water extraction patterns of different crops in a silty loam Haplustoll. Field Crops Res. 1997;54:29–38.

de Willigen P. Uptake potential of non-regularly distributed roots. J Plant Nutr. 1987;10:1273–80.

Dexter AR. Advances in characterisation of soil structure. Soil Tillage Res. 1988;11:199–238.

Dracup M, Belford RK, Gregory PJ. Constraints to root growth of wheat and lupin crops in duplex soils. Aust J Exp Agric. 1992;32:947–61.

Eastham J, Gregory PJ. The influence of crop management on the water balance of lupin and wheat crops on a layered soil in a Mediterranean climate. Plant Soil. 2000;221:239–51.

Gao W, Whalley WR, Tian Z, Liu J, Ren T. A simple model to predict soil penetrometer resistance as a function of density, drying and depth in the field. Soil Tillage Res. 2016;155:190–8.

Gill JS, Sale PWG, Peries R, Tang C. Changes in soil physical properties and crop root growth in dense sodic subsoil following incorporation of organic amendments. Field Crops Res. 2009;114:137–46.

Gregory PJ. Plant roots: growth, activity and interaction with soils. Oxford: Blackwell Publishing; 2007.

Guaman V, Bath B, Hagman J, Gunnarsson A, Persson P. Short time effects of biological and inter-row subsoiling on yield of potatoes grown a loamy sand, and on penetration resistance, root growth and nitrogen uptake. Eur J Agron. 2016;80:55–65.

Haling RE, Brown LK, Bengough AG, Valentine TA, White PJ, Young IM, George TS. Root hair length and rhizosheath mass depend on soil porosity, strength and water content in barley genotypes. Planta. 2014;239:643–51.

Hamblin AP. The influence of soil structure on water movement, crop root growth, and water uptake. Adv Agron. 1986;38:95–158.

Hamza MA, Anderson WK. Responses of soil properties and grain yields to deep ripping and gypsum application in a compacted loamy sand soil contrasted with a sandy clay loam soil in Western Australia. Aust J Agric Res. 2003;54:273–82.

Hamza MA, Anderson WK. Combinations of ripping depth and tine spacing for compacted sandy and clayey soils. Soil Tillage Res. 2008;99:213–20.

Hatano R, Iwanaga K, Okajima H, Sakuma T. Relationship between the distribution of soil macropores and root elongation. Soil Sci Plant Nutr. 1988;34:535–46.

Hirth JR, McKenzie BM, Tisdall JM. Ability of seedling roots of *Lolium perenne* L. to penetrate soil from artificial biopores is modified by soil bulk density, biopore angle and biopore relief. Plant Soil. 2005;272:327–36.

Hochman Z, Dalgliesh NP, Bell K. Contributions of soil and crop factors to plant available soil water capacity of annual crops on black and grey Vertosols. Aust J Agric Res. 2001;52:955–61.

Iijima M, Kato J. Combined soil physical stress of soil drying, anaerobiosis and mechanical impedance to seedling root growth of four crop species. Plant Prod Sci. 2007;10:451–9.

Iijima M, Higuchi T, Barlow PW. Contribution of root cap mucilage and presence of an intact root cap in maize (*Zea mays*) to the reduction of soil mechanical impedance. Ann Bot. 2004;94:473–7.
Jin K, Shen J, Ashton RW, Dodd IC, Parry MAJ, Whalley WR. How do roots elongate in a structured soil? J Exp Bot. 2013;64:4761–77.
Kaspar TC, Bland WL. Soil temperature and root growth. Soil Sci. 1992;154:290–9.
Khan FS, Zaman QU, Chang YK, Farooque AA, Schumann AW, Madani A. Estimation of the root zone depth above a gravel layer (in wild blue berry fields) using electromagnetic induction method. Precis Agric. 2016;17:155–67.
Kim H, Anderson SH, Motavalli PP, Gantzer CJ. Compaction effects on soil macropore geometry and related parameters for an arable field. Geoderma. 2010;160:244–51.
Kirkegaard JA, Lilley JM. Root penetration rate – a benchmark to identify soil and plant limitations to rooting depth in wheat. Aust J Exp Agric. 2007;47:590–602.
Kirkegaard JA, So HB, Troedson RJ. The effect of soil strength on the growth of pigeonpea radicles and seedlings. Plant Soil. 1992;140:65–74.
Kirkegaard JA, Lilley JM, Howe GN, Graham JN. Impact of subsoil water use on wheat yield. Aust J Agric Res. 2007;58:303–15.
Lambers H, Chapin FS III, Pons TL. Plant physiological ecology. 2nd ed. New York: Springer; 2008.
Letey J. Relationship between soil physical properties and crop production. Adv Soil Sci. 1958;1:277–94.
Lilley JM, Kirkegaard JA. Seasonal variation in the value of subsoil water to wheat: simulation studies in southern New South Wales. Aust J Agric Res. 2007;58:1115–28.
Lipiec J, Hajnos M, Świeboda R. Estimating effects of compaction on pore size distribution of soil aggregates by mercury porosimeter. Geoderma. 2012;179-180:20–7.
Logsdon SD, Allmaras RR. Maize and soybean root clustering as indicated by root mapping. Plant Soil. 1991;131:169–76.
Lynch JP, Wojciechowski T. Opportunities and challenges in the subsoil: pathways to deeper rooted crops. J Exp Bot. 2015;66:2199–210.
MacEwan RJ, Crawford DM, Newton PJ, Clune TS. High clay contents, dense soils, and spatial variability are the principal subsoil constraints to cropping the higher rainfall land in south-eastern Australia. Soil Res. 2010;48:150–66.
Materechera SA, Dexter AR, Alston AM. Penetration of very strong soils by seedling roots of different plant species. Plant Soil. 1991;135:31–41.
McDonald GK, Taylor JD, Verbyla A, Kuchel H. Assessing the importance of subsoil constraints to yield of wheat and its implications for yield improvement. Crop Pasture Sci. 2013;63:1043–65.
McKenzie BM, Tisdall JM, Vance WH. Soil physical quality. In: Glinski J, Horabik J, Lipiec J, editors. Encyclopedia of agrophysics. Dordrecht: Springer; 2011. p. 770–7.
McMichael BL, Burke JJ. Temperature effects on root growth. In: Waisel Y, Eshel A, Kafkafi U, editors. Plant roots: the hidden half. 2nd ed. New York: Marcel Dekker; 1996. p. 383–96.
Meinke H, Hammer GL, Want P. Potential soil water extraction by sunflower on a range of soils. Field Crops Res. 1993;32:59–81.
Miller DE. Root systems in relation to stress tolerance. HortScience. 1986;21:963–70.
Misra RK. Maximum axial growth pressure of the lateral roots of pea and eucalypt. Plant Soil. 1997;188:161–70.
Misra RK. Root and shoot elongation of rhizotron-grown seedlings of *Eucalyptus nitens* and *Eucalyptus globulus* in relation to temperature. Plant Soil. 1999;206:37–46.
Mohanty M, Bandyopadhyay KK, Painuli DK, Ghosh PK, Misrah AK, Hati KM. Water transmission characteristics of a Vertisol and water use efficiency of rainfed soybean (*Glycine max* (L.) Merr.) under subsoiling and manuring. Soil Tillage Res. 2007;93:420–8.
Montagu KD, Conroy JP, Francis GS. Root and shoot response of field-grown lettuce and broccoli to compact subsoil. Aust J Agric Res. 1998;49:89–97.
Monteith JL. How do crops manipulate water supply and demand? Philos Trans R Soc Lond Ser A. 1986;316:245–59.

Mossadeghi-Bjorkland M, Arvidsson J, Keller T, Koestel J, Lamande M, Larsbo M, Jarvis N. Effects of subsoil compaction on hydraulic properties and preferential flow in a Swedish clay soil. Soil Tillage Res. 2016;156:91–8.

Munkholm LJ, Hansen EM, Olesen JE. The effect of tillage intensity on soil structure and winter wheat root/shoot growth. Soil Use Manag. 2008;24:392–400.

Oliveira TS, Bell RW. Introduction to subsoil constraints for crop production. In: Oliveira TS, Bell RW, editors. Subsoil constraints for crop production. Springer; 2022. p. 1–10.

Oliveira TS, Fernandes RBSA. Physical subsoil constraints of agricultural and forestry land. In: Oliveira TS, Bell RW, editors. Subsoil constraints for crop production. Springer; 2022. p. 125–60.

Ordonez RA, Castellano MJ, Hatfield JL, Helmers MJ, Licht MA, Liebman M, Dietzel R, Martinez-Feria R, Iqbal J, Puntel LA, Cordova SC, Togliatti K, Wright EE, Archontoulis SV. Maize and soybean root front velocity and maximum depth in Iowa, USA. Field Crops Res. 2018;215:122–31.

Palta JA, Chen X, Milroy SP, Rebetzke GJ, Dreccer MF, Watt M. Large root systems: are they useful in adapting wheat to dry environments? Funct Plant Biol. 2011;38:347–54.

Pardo A, Amato M, Quaglietta Chiaranda F. Relationships between soil structure, root distribution and water uptake of chickpea (*Cicer arietinum* L.). Plant growth and water distribution. Eur J Agron. 2000;13:39–45.

Parker CJ, Carr MKV, Jarvis NJ, Evans MTB, Lee VH. Effects of subsoil loosening and irrigation on soil physical properties, root distribution and water uptake of potatoes (*Solanum tuberosum*). Soil Tillage Res. 1989;13:267–85.

Passioura JB. Roots and drought resistance. Agric Water Manag. 1983;7:265–80.

Passioura JB. Soil structure and plant growth. Aust J Soil Res. 1991;29:717–28.

Passioura J. Increasing crop productivity when water is scarce - from breeding to field management. Agric Water Manag. 2006;80:176–96.

Pierce FJ, Larson WE, Dowdy RH, Graham WAP. Productivity of soils: assessing long-term changes due to erosion. J Soil Water Conserv. 1983;38:39–44.

Pires LF, Borges JAR, Rosa JA, Cooper M, Heck RJ, Passoni S, Roque WL. Soil structure changes induced by tillage systems. Soil Tillage Res. 2017;165:66–79.

Popova L, van Dusschoten D, Nagel KA, Fiorani F, Mazzolia B. Plant root tortuosity: an indicator of root path formation in soil with different composition and density. Ann Bot. 2016;118:685–98.

Prieto I, Armas C, Pugnaire FI. Water release through plant roots: new insights into its consequences at the plant and ecosystem level. New Phytol. 2012;193:830–41.

Radford BJ, Yule DF, McGarry D, Playford C. Crop responses to applied soil compaction and to compaction repair treatments. Soil Tillage Res. 2001;61:157–66.

Rengasamy JI, Reid JB. Root system modification of faba bean (*Vicia faba* L.), and its effects on crop performance. 1. Responses of root and shoot growth to subsoiling, irrigation and sowing date. Field Crops Res. 1993a;33:175–96.

Rengasamy JI, Reid JB. Root system modification of faba bean (*Vicia faba* L.), and its effects on crop performance. 2. Role of water stress. Field Crops Res. 1993b;33:197–215.

Richards RA, Passioura JB. A breeding program to reduce the diameter of the major xylem vessel of the seminal roots of wheat and its effect on grain yield in rain-fed environments. Aust J Agric Res. 1989;40:943–50.

Ritchie JT. Water dynamics in the soil-plant-atmosphere system. Plant Soil. 1981;58:81–96.

Sadras VO, O'Leary GJ, Roget DK. Crop responses to compacted soil: capture and efficiency in the use of water and radiation. Field Crops Res. 2005;91:131–48.

Sale PW, Gill JS, Peries RR, Tang C. Crop responses to subsoil manuring. I. Results in south-western Victoria from 2009 to 2012. Crop Pasture Sci. 2018;70:44–54.

Scanlan CA, Holmes KW, Bell RW. Sand and gravel subsoils. In: Oliveira TS, Bell RW, editors. Subsoil constraints for crop production. Springer; 2022. p. 179 –98.

Schneider F, Don A, Hennings I, Schmittmann O, Seidel SJ. The effect of deep tillage on crop yield – what do we really know? Soil Tillage Res. 2017;174:193–204.

Singh N, Dhillon BS, Sandhu PS. Alleviation of soil physical constraints for improved crop production: a review. Ecol Environ Conserv. 2014;20:29–36.

Stirzaker RJ, Passioura JB, Wilms Y. Soil structure and plant growth: impact of bulk density and biopores. Plant Soil. 1996;185:151–62.

Tardieu F. Analysis of the spatial variability of maize root density. Distance between roots. Plant Soil. 1988;107:267–72.

Tardieu F. Growth and functioning of roots and root systems subjected to soil compaction. Towards a system with multiple signalling? Soil Tillage Res. 1994;30:217–53.

Taylor HM, Brar GS. Effect of soil compaction on root development. Soil Tillage Res. 1991;19:111–9.

Taylor HM, Ratliff LF. Root elongation rates of cotton and peanuts as a function of soil strength and soil water content. Soil Sci. 1969;108:113–9.

Thorup-Kristensen K, Cortasa MS, Loges R. Winter wheat roots grow twice as deep as spring wheat roots, is this important for N uptake and N leaching losses? Plant Soil. 2009;322:101–14.

Tinker PB. Transport of water to plant roots in soil. Philos Trans R Soc Lond Ser B. 1976;273(927):445–61.

Unger PW. Effects of deep tillage and profile modification on soil properties, root growth, and crop yields in the United States and Canada. Geoderma. 1979;22:275–95.

Vance WH, Tisdall JM, McKenzie BM. Residual effects of surface applications of organic matter and calcium salts on the subsoil of a red-brown earth. Aust J Exp Agric. 1998;38:595–600.

Vocanson A, Roger-Estrade J, Boizard H, Jeuffroy M-H. Effects of soil structure on pea (*Pisum sativum* L.) root development according to sowing date and cultivar. Plant Soil. 2006;281:121–35.

Volkmar KM. Effects of biopores on the growth and N-uptake of wheat at three levels of soil moisture. Can J Soil Sci. 1996;76:453–8.

Wasson AP, Richards RA, Chatrath R, Misra SC, Sai Prasad SV, Rebetzke GJ, Kirkegaard JA, Christopher J, Watt M. Traits and selection strategies to improve root systems and water uptake in water-limited wheat crops. J Exp Bot. 2012;63:3485–98.

White RG, Kirkegaard JA. The distribution and abundance of wheat roots in a dense, structured subsoil – implications for water uptake. Plant Cell Environ. 2010;33:133–48.

Wong MTF, Asseng S. Determining the causes of spatial and temporal variability of wheat yields at sub-field scale using a new method of upscaling a crop model. Plant Soil. 2006;283:203–15.

Wong MTF, Asseng S. Yield and environmental benefits of ameliorating subsoil constraints under variable rainfall in a Mediterranean environment. Plant Soil. 2007;297:29–42.

Wong MTF, Asseng S, Zhang H. A flexible approach to managing variability in grain yield at within-field to farm scales. Precis Agric. 2006;7:405–17.

Wong MTF, Asseng S, Robertson MJ, Oliver YM. Mapping subsoil acidity and shallow soil across a field with information from yield maps, geophysical sensing and the grower. Precis Agric. 2008;9:3–15.

Wong MTF, Oliver YM, Robertson MJ. Gamma-radiometric assessment of soil depth across a landscape not measurable using electromagnetic surveys. Soil Sci Soc Am J. 2009;73:1261–7.

Yunusa IAM, Newton PJ. Plants for amelioration of subsoil constraints and hydrological control: the primer-plant concept. Plant Soil. 2003;257:261–81.

Zhang H, Turner NC, Poole ML. Yield of wheat and canola in the high rainfall zone of south-western Australia in years with and without a transient perched water table. Aust J Agric Res. 2004;55:461–70.

Zou C, Sands R, Buchan G, Hudson I. Least limiting water range: a potential indicator of physical quality of forest soils. Soil Res. 2000;38:947–58.

Chapter 14
Deep Soil Carbon: Characteristics and Measurement with Particular Bearing on Kaolinitic Profiles

Podjanee Sangmanee, Bernard Dell, David Henry, and Richard Harper

Abstract Deep soils are located in most continents of the world. Soil carbon measurements are invariably made from the surface horizons, whereas much larger carbon stores occur to depths of many meters, with plant roots providing the main source of carbon. This root biomass persists long after land is deforested for agricultural and other pursuits or forests are killed by pests and fires and may represent a considerable carbon store at the global scale. The impacts on these carbon stores of reforestation or climate change are mostly unknown as the estimation of root biomass and carbon dynamics is challenging in deep soils. This chapter explores deep soil carbon from the perspectives of its definition, source, and persistence; methodologies available to study deep soil carbon; and the effect of land-use change on this carbon store.

Keywords Carbon measurement · Deep carbon · Kaolinitic profiles · Organic matter composition · Regolith · Rooting depth

14.1 Definitions

Deep soil is the entire upper weathering layer of the earth's crust, which can be tens of meters deep (Ramaan 1928; Richter and Markewitz 1995). This chapter defines deep soil according to the original work on deep soil carbon in south-western Australia, where soils commonly have profiles greater than 5 m in depth where there is little influence of above-ground input of carbon (Harper and Tibbett 2013).

P. Sangmanee
Faculty of Agriculture, Uttaradit Rajabhat University, Uttaradit, Thailand

B. Dell · R. Harper (✉)
Agriculture and Forest Sciences, Murdoch University, Murdoch, WA, Australia
e-mail: B.Dell@murdoch.edu.au; R.Harper@murdoch.edu.au

D. Henry
Murdoch University, Murdoch, WA, Australia
e-mail: D.Henry@murdoch.edu.au

T. S. d. Oliveira, R. W. Bell (eds.), *Subsoil Constraints for Crop Production*,
https://doi.org/10.1007/978-3-031-00317-2_14

However, soil layers that lie above this arbitrary 5 m can be considered as a transition zone that contributes to the understanding of deep soils. Therefore, this zone is included in the review for some perspectives in relation to deep soil.

Historically, soil organic carbon has been measured as a proxy for soil organic matter, which is complex, may change over time, and is difficult to measure directly. As a result, soil organic matter has been defined differently according to methodologies used in order to understand dynamics of organic matter, and these mostly focus on surface soil (Rossell et al. 1998). For example, organic matter can be defined by physical (size and density), biological (soil microbial activity, microbial respiration), and chemical (humin, humic, and fulvic acids) fractionations. Nevertheless, soil organic matter is well understood by the general definition given by the Soil Science Society of America (1979) as "the organic fraction of soil, including plant, animal and microbial residues, fresh and at all stages of decomposition, and the relatively resistant soil humus".

The definition of soil organic carbon used in this chapter embraces the carbon components at the molecular scale, exclusive of decaying tissues, in size greater than 500 μm (more generally defined in the size range between 100 and 2000 μm). Organic carbon differing in sizes is partitioned using chemical techniques. For example, low-molecular-weight compounds (LMWC) are apolar to moderately polar volatile compounds that can be readily extracted using an organic solvent. Compared to LMWC, macro-organic carbon (MOC) comprises higher-molecular-weight compound that are non-volatile and needs extensive fragmentation before analysis.

14.2 Possible Sources of Organic Carbon and Its Occurrence in Deep Soils

14.2.1 Plant Roots

Roots play a key role in organic carbon translocation from the top of the vegetation canopy to the deepest root tips, and they also transport water and mineral nutrients upwards in the xylem (Brantley et al. 2011). Carbohydrates synthesized by photosynthesis facilitate the growth of vegetation as well as organisms in the rhizosphere communities (Bundt et al. 2001; Leake et al. 2004). Storage of carbon in soil is mainly contributed by root biomass, by root exudates, and by microbial and invertebrate communities in the root zone. The main pathways of carbon capture and release in soil have been reviewed by Kell (2011, 2012). Leaching may redistribute some carbon independently from root distribution.

The study of deep roots has gained more attention after the meta-analyses of Canadell et al. (1996) and Jackson et al. (1996). Examples include studies up to 9 m in central Cambodia (Ohnuki et al. 2008) and 10 m in south-eastern Brazil (Laclau et al. 2013) and 20 m in central Texas, USA (Bleby et al. 2010). Root architectures have evolved to exploit deep soils as adaptations to survive environmental stress. For example, in deep granitic regolith in Western Australia, sinker roots follow macropores to depths of ≥40 m (Dell et al. 1984). These authors observed much greater concentrations of soil organic matter in the macropores than in the bulk soil. The ecosystem within the macropore sheath was investigated in a duplex (i.e. texture contrast such as sand over clay) soil between the surface and 80 cm depth in south-eastern Australia by Pankhurst et al. (2002). The soil in the macropore sheath zone had higher organic carbon and microbial population compared to the bulk soil. Moreover, throughout the studied depth, roots located within the macropore sheath were observed to support microbial communities that were quantitatively and functionally diverse.

The attributes of deep roots vary with plant species, soil characteristics, and land use and have been given particular attention in the literature. Rhizoliths, the calcified roots in loess deposition underlying limestones and sandstones, were investigated down to 2.5 m in Hungary (Huguet et al. 2013), 3 m in Serbia (Gocke et al. 2014), and 9 m in Germany (Gocke et al. 2011). The inorganic and organic forms were radiocarbon dated to >3000 years BP (Gocke et al. 2011). Furthermore, lipids and alkanes were identified in rhizoliths (Huguet et al. 2012, 2013; Gocke et al. 2014). However, the amounts of rhizolith carbon stored in subsoils and deep soils have not been evaluated.

Besides rhizoliths, fresh live roots were observed in 2.5–18 m deep profiles in the Chinese loess plateau by Wang et al. (2015). The storage of organic carbon in deep soil (5–21 m) was estimated under forest (47 ± 0.43 kgm^{-2}) and permanent cropland (38 ± 0.47 kgm^{-2}), but forms of carbon have not been characterized.

14.2.2 Other Living Sources of Carbon

Microbial biomass, the living microbial component in the soil ecosystem, is another source of soil carbon. Soil microorganisms responsible for litter decomposition and organic matter formation in surface soil have been studied intensively over many decades (Prescott 2010). By contrast, the role and abundance of microorganisms in deep soil are less well investigated. The implication of deep soil microorganisms was revealed when fungal remnants in the form of glucan were discovered in paddy soil sampled at a depth of 40–43 m in Japan (Kotake et al. 2013).

Microorganisms in groundwater indicate another potential source of deep soil carbon. These microorganisms share 6–40% of the prokaryotic biomass on earth, but this biomass hidden within the terrestrial subsurface has only been marginally explored to date (Griebler and Lueders 2009). Investigation of groundwater at depths of 15–90 m showed that overall bacterial abundance remained high (10^5 cell ml^{-1}), even though the level of organic nutrients varied and dissolved oxygen was very low (<0.40 μg L^{-1}) (Roudnew et al. 2012). However, microbial functional diversity is influenced by land use and season due to water quality, nutrient availability, and other factors (Korbel et al. 2013). For instance, microbial activity of groundwater at 10–30 m differed among irrigated cropping, non-irrigated cropping, and grazing land uses (Korbel et al. 2013). In addition to microorganisms, stygofauna or aquatic animals contribute carbon at depth, and their abundance and richness also vary with different agricultural landscapes (Korbel et al. 2013). Recently, amino acids of bacterial debris and lignin-derived phenols were identified in DOC of groundwater collected from 76 m depth in a fractured rock zone (Shen et al. 2015).

Other living organisms such as earthworms and termites are often overlooked when quantifying deep soil carbon. However, a study on mineral exploration observed that vertical bioturbation of termites (*Tumulitermes tumuli*) exceeded 4 m, transporting gold from the subsoil for nest construction in the Western Australia Goldfield (Stewart et al. 2012). In addition, earthworms play a role in shallower subsoils at 2 m depth (Major et al. 2010). The giant Gippsland earthworm (*Megascolides australis*), endemic to an area of approximately 440 km^2 of South Gippsland (Victoria, Australia), is an example of soil megafauna with an average weight of 200 g and length of 3 m that occupies depths of 1.5 m in clay soils. Land-use activities have led to vertical migration of the earthworm, but maximum depths have not been recorded (Van Praagh and Yen 2010). The burrows created by these living organisms facilitate preferential flow of organic material from surface layers to subsoils or perhaps into deep soils in some situations.

14.3 Stability of Deep Soil Carbon

The study of deep root decomposition is limited due to the challenge of sampling without contamination. Even though the carbon in deep soil is hard to access, the stabilities of deep soil were postulated and conclusions drawn from short-time ($\leq$1 year) incubation of subsoils.

For example, ancient buried carbon was lost when incubated with fresh plant-derived carbon (Fontaine et al. 2007). Similarly, Zhang et al. (2015) demonstrated that the addition of sucrose triggered the degradation of labile and recalcitrant native loess carbon. A rhizosphere-positive priming effect was observed in trees grown under greenhouse conditions by Dijkstra and Cheng (2007), resulting in a net loss of soil carbon due to chemical interactions of soil and tree roots. The priming effect is generally promoted by simple compounds, which are favourable substrates for increasing microbial activity. However, the mechanism of exudate-induced native carbon loss is more complex. Recently, Keiluweit et al. (2015) found that the production of root exudates, such as oxalic acid, can cause the loss of mineral protected-carbon, because the exudate, acting as a ligand, liberated carbon through complexation and dissolution reactions of the protective mineral phase. As a result, the process promotes microbial access and subsequent loss of mineral-protected carbon. By contrast, the addition of pyrogenic carbon materials in the form of black carbon, charcoal, or biochar influences the stability of native carbon and may lead to positive or negative priming effects (Hernandez-Soriano et al. 2015; Keith et al. 2009; McClean et al. 2016; Rittl et al. 2015; Weng et al. 2015).

The insertion of pyrogenic carbon into deep soil could occur from the in situ burning of woody roots or leaching of burnt carbon by facilitated transport through macropores. Burnt roots have been observed down to 3 m in a bauxitic profile at Weipa, in northern Australia (Eggleton and Taylor 2008). As there is no research on pyrogenic carbon in deep soil, the interaction between deep soil carbon and black carbon in relation to their stability should be investigated in future research.

14.4 Age of Deep Soil Carbon

Soil organic matter is the product of an ongoing process, and contamination by recent carbon leads to an underestimation of carbon age when determined by radiocarbon dating (Trumbore 2000). With less disturbance, the age of deep soil organic carbon tends to be a more realistic estimate than for carbon in shallow soil. Understanding the residence time of carbon in deep soil could help predict the longevity of carbon storage and shed light on the global carbon balance.

Research has tended to focus on the rapid cycling of fine roots while ignoring the longevity of large woody root systems in deep soil, even though large amounts of carbon are allocated belowground. For example, Amazon trees aged from 200 to 1400 years old were radiocarbon dated (Jefferey et al. 1998), but the residence time of carbon in roots was not assessed. The recycling of old carbon by fresh live roots was observed in boreal forests by Helmisaari et al. (2015), where 3-month-old fine roots, which developed in ingrowth cores, showed an apparent ^{14}C age between 1 and 20 years.

The residence time of woody roots has rarely been studied. However, modelling approaches may be useful to predict the cycling of coarse biomass (Galbraith et al. 2013). Interestingly, the woody biomass residence time of Australian tropical forests was highest (104 years) among 177 tropical forests across the world, and the heavily weathered soil was considered to be the main factor of the long residence time (Galbraith et al. 2013). The range of estimated mean residence times is very large, ranging from tens to thousands of years, as predicted from three forest ecosystems of eastern China (Zhang et al. 2010).

On the other hand, similar degradation rates of coarse woody roots and stumps were observed in a field experiment in an Irish midlands forest by Olajuyigbe et al. (2010). However, the decay rate determined from carbon respiration varies with microbial activity, and an over simplification of carbon flux from coarse woody debris was demonstrated by Forrester et al. (2015), due to an inferred flux of microbial CO_2 emission.

14.5 Methodology to Study Deep Soil Carbon

Many methods have been proposed for carbon quantification and characterization. Quantification has often focused on simply determining total organic carbon (TOC) content with samples by wet or dry combustion methods used in many laboratories. However, methods based on remote sensing (Angelopoulou et al. 2019) and a range of spectroscopic measurements (Izaurralde et al. 2013) are also being applied for quantifying soil carbon. Method principles, advantages, and disadvantages are summarized by Chatterjee et al. (2009) and Rossell et al. (1998).

Characterization of soil carbon is usually directed to determining the nature and identity of the organic compounds in the soil that form the pool of TOC. Although it is possible to apply some spectroscopic techniques directly to soil samples (e.g. mid-infrared spectroscopy and solid-state nuclear magnetic resonance), these will only provide information on the functional groups present, not the identity of specific compounds. Furthermore, these approaches can be hampered by small concentration of organic material, the complex nature of the soil matrix, and interference from inorganic components. Consequently, characterization often relies initially on extraction of organic material from soil. This is usually followed by separation of individual compounds using a chromatographic technique and finally identification using a spectroscopic technique, such as mass spectrometry. This section focuses on the methods that the authors consider to have the most potential for determining deep soil carbon. The range of methods is briefly described below.

14.5.1 Carbon Quantification Methods

Dry Combustion Method

The principal steps of the dry combustion method are as follows: (*i*) the soil carbon is converted to CO_2 by oxidizing the sample at a high temperature, (*ii*) CO_2 gas is separated from other gases by either a chromatographic system or selective traps, and (*iii*) the concentration of CO_2 is detected by thermal conductivity, mass spectroscopy, or infrared spectroscopy (Chatterjee et al. 2009). The main advantages are the direct measurement of C in the sample and the very low quantification limit. Therefore, this method is acknowledged as a benchmark method of carbon determination (Chatterjee et al. 2009; De Vos et al. 2007; Mikhailova et al. 2003). This method was used to detect concentrations as low as 0.01% TOC (Harper and Tibbett 2013).

Wet Digestion Method

Wet digestion methods, such as the Walkley-Black method and Heanes method, are conventional methods that require limited apparatus and are inexpensive compared to the dry combustion method. Principally, the oxidizable matter is oxidized by an excess $K_2Cr_2O_7$ solution and the reaction accelerated by adding H_2SO_4. The amount of oxidizable carbon in the sample is directly related to the amount of dichromate consumed according to the following equation.

$$2H_2Cr_2O_7 + 6H_2SO_4 + 3C \rightarrow 2Cr_2\left(SO_4\right)_3 + 3CO_2 + 8H_2O$$

This in turn can be determined by ferrous sulphate titration of the excess dichromate. However, the method generally suffers from variation in recovery due to insufficient heat being generated during the reactions. Consequently, the tube digestion method has been developed in a number of laboratories (Bartlett and Ross 1988; Edson and Mills 1955; Schollenberger 1927; Tyurin 1931) in which external heat is applied and the digestion time is extended. Application of external heat (135 °C for 30 min) was found by Heanes (1984) to be optimum, and this modification has been widely employed for analysis of Australian soils. Note that recalcitrant forms of C, such as charcoal, may not be oxidized, and also some inorganic compounds may be oxidized. Thus, the method may over- or underestimate the true organic C content of soil.

Near-Infrared Spectroscopy

Infrared spectroscopy is a non-invasive tool for identifying functional groups that interact with energy in the infrared region of the electromagnetic spectrum. The infrared spectrum is divided into near- (14000–4000 cm^{-1}), mid- (4000–400 cm^{-1}), and far- (400–10 cm^{-1}) infrared regions, but those regions that can identify functional groups are restricted to the mid- and near-infrared. Near-infrared region (14000–4000 cm^{-1}) spectroscopy measures the absorption of the C-H, N-H, O-H, C=O, S-H, CH_2, and C-C groups of organic compounds (Berardo et al. 2005; He and Hu 2013). The potential of the method depends on a developed model for the calibration process, which normally needs chemometric techniques to extract the useful information from NIR spectra (Viscarra Rossel et al. 2006). The estimation of soil carbon by NIR has been reviewed by Bellon-Maurel and McBratney (2011) and Reeves (2010). The method has limitations for C-C groups as vibrations with low dipole variations are poorly detected.

14.5.2 *Soil Organic Matter Characterization Methods*

Functional and molecular scale characterization techniques are essential to understand deep soil carbon. Several methods are currently being employed for characterization of organic matter associated with bulk soils and organic carbon fractions.

Mid-Infrared Spectroscopy

Mid-infrared spectroscopy (MIR) is being employed for characterization and identification of samples, because peaks obtained from interactions between functional groups and energy are more distinct compared to the NIR region. Functional groups of soil constituents identified by MIR are listed in Table 14.1.

Table 14.1 Group frequencies of soil constituents

Components organics	Band regions (cm^{-1})
O-H stretching of carboxylic acids, phenols, alcohols	3500–3200
N-H stretching of amines, amides	3400–3200
Aromatic C-H stretching	3150–3000
Aliphatic C-H stretching	2970–2820
C = O stretching of carboxylic acids, amides, ketones	1750–1630
Salts of carboxylic acids	
Asymmetric COO^- stretching	1650–1540
Symmetric COO^- stretching	1450–1360
C-H bending of $–CH_2$- and $–CH_3$-	1465–1440
C-O stretching, O-H bending of -COOH	1250–1200
C-O stretching of polysaccharide	1170–950
Components inorganics	
Clay minerals and oxides O-H stretching of structural OH O-H bending of structural OH Si-O-Si stretching	3750–3300 950–820 1200–970
Sorbed water O-H stretching O-H bending	3600–3300 1650–1620
Carbonates – Asym CO_3 – Sym CO_3	1600–1300 900–670
Phosphates	1200–1100
Sulphates	680–600

Source: Johnston and Aochi (1996)

Chromatographic Technique Coupled with Mass Spectroscopy

Conventional gas chromatography/mass spectrometry (GC/MS) is being used to identify organic carbon by many researchers. Basically, volatile organic carbon compounds are separated over a gas chromatograph and a compound subsequently identified by a mass spectrometer. Therefore, fragmentation of non-volatile carbon into a volatile form is required before GC/MS analysis, and this can be undertaken by thermochemolysis or pyrolysis and is reviewed by Derenne and Quénéa (2015) and Shadkami and Helleur (2010).

14.6 Carbon Components as a Tool for Identifying Sources of Soil Organic Carbon

Gas chromatography/mass spectrometry (GC/MS) facilitates the molecular characterization and identification of carbon compounds in soil. Compounds detected by GC/MS may vary depending on organic carbon precursors, mainly plant molecules and microbial products. Secondary variations of compounds are dependent on sample preparation techniques as well as procedures.

Here, we briefly review the organic matter sources identified from carbon components of LMWC and MOC. Both forms of carbon are considered, because carbon observed in kaolinitic regoliths is generally heterogeneous. Table 14.2. provides a summary of compounds derived from soil organic carbon, their origin, and environmental significance. It is apparent that the contribution of deep soil carbon can be interpreted from multi-compounds obtained from chromatographic identification.

14.7 Deep Soils and Deep Carbon

14.7.1 Deep Soils in the World

Deep soils formed from in situ intense weathering and aeolian processes are widespread in the world (Alavi Nezhad Khalil Abad et al. 2014; Bowler 1976; Ayling and McGowan 2006: Chevrier et al. 2006; Gaudin et al. 2011; Jones 1985; Lancaster 2009; Stuut et al. 2009; Modenesi-Gauttieri et al. 2011). Terrestrial sediments such as loess deposits cover vast areas in north-west China (Wang et al. 2013), the Great Plain of North America (Bettis et al. 2003), the European Loess Belt (Vasiljević et al. 2014), and parts of Oceania such New Zealand (Raeside 1964). Deep kaolinitic profiles are abundant in many parts of the world, and this regolith material is explored further below.

Kaolinite genesis can be categorized into two groups according to the type of rocks. Primary kaolins are derived from in situ primary rocks and hydrothermal

Table 14.2 Summary of compounds of soil organic matter, their possible origin, and environmental significance

Assignment/compound	Origin/environmental significance	References
Alkanes (C_7-C_{14})	Lipid, usually microbial, completely degraded material, algal derived from organic matter in natural water	Almendros et al. 1996 Buurman et al. 2005, 2007; Frazier et al. 2003
6- and 7-monomethylalkanes, C_{17} n-alkane	Cyanobacteria	Hoshino and George 2015
Alkanes and alkenes (C_{14}-C_{26})	Lipid, usually plant cutin, suberin, or waxes	Buurman et al. 2005, 2007
Alkanes and alkenes (C_{25}-C_{33} odd)	Higher plant waxes, fungi	Otto and Simpson 2007
Odd or even dominance on the mid-chain and long alkanes and alkene	Relative of non-degraded material over microbial material	Buurman et al. 2005, 2007
Alkanols (C22-C32 even)	Higher plant waxes, suberin	van Bergen et al. 1998; Otto and Simpson 2007
Alkanol (C26 dominant)	Constituent in many grasses	van Bergen et al. 1998
Alkanoic acids	Derive directly from plant or products from oxidation of other compounds such as alkanes and alkanols or lipids	van Bergen et al. 1998
Branched-chain alkanoic acids	Molecular evidence for microbial activity	van Bergen et al. 1998
Alkyl esters	Wax esters	van Bergen et al. 1998
Phytols, phytanols, sterols	Polar waxes	Franco et al. 2000
Alkanones	In situ microbial oxidation of other lipid components	van Bergen et al. 1998
β-Sitosterol, stigmasterol, canpesterol	Steroids of plants	Otto and Simpson 2007
Ergosterol	Steroids of fungi	Otto and Simpson 2007
Cholesterol	Steroids of animals, fungi, and plants	Otto and Simpson 2007
Cyclopentenone, cyclohexenes	Polysaccharide	Hatcher et al. 2001; Page et al. 2002
Furan, ethanoic acid	Polysaccharide	Heckman et al. 2014
Anhydrosugar	Polysaccharide	Page et al. 2002
Acetic acid	Polysaccharide	Hatcher et al. 2001

(continued)

Table 14.2 (continued)

Assignment/compound	Origin/environmental significance	References
Levoglucosan, levomanosan	Polysaccharide/relative undecomposed cellulose or microbial polysaccharide	Sollins et al. 1996 Helfrich et al. 2006
Alkylbenzene	Polysaccharide/ lignin	Nierop et al. 2001
4-Ethenylphenol, 4-ethenyl-2-methoxyphenol	Angiosperm lignin-cellulose, the precursors are *p*-coumaric acid and ferulic acid, respectively	van Bergen et al. 1998
Di-, trimethoxy benzene, trimethoxy toluene	Carbohydrates, tannins	Frazier et al. 2003
Methyl phenol, methoxy phenol	Lignin	Heckman et al. 2014
Vanillic acid	Lignin of grasses	Sáiz-Jiménez and De Leeuw 1986
Styrene	Lignin	Ralph and Hatfield 1991
Methoxy benzene, methoxy benzoic acid, methyl ester	Proteins	Frazier et al. 2003
Pyridine, methyl pyridine, benzonitrile, acetylbenzonitrile, indole, methyl indole, and diketodipyrole	Nitrogenous compound Different origins including vegetal and microbial protein	Heckman et al. 2014; van Bergen et al. 1998 Schulten and Schnitzer 1997; van Bergen et al. 1998
9-Octadecenamide	Compound extracted from polar wax of non-wetting sand under eucalyptus trees	Franco et al. 2000
Glucose, mannose, sucrose	Carbohydrates of all organisms	Otto and Simpson 2007
Trehalose	Carbohydrate of fungi	Otto and Simpson 2007
Galactosamine, glucosamine, mannosamine	Amino sugars of bacteria and fungi	Otto and Simpson 2007
Phenols	Lignin, tannin, protein, polysaccharide	Nierop et al. 1999; Hatcher et al. 2001
Benzophenone	Found in humic acids in soils under pine forest	González-Vila et al. 1987

alteration of volcanic rocks, while secondary kaolins are associated with sedimentary rocks (Ekosse 2010). Locations of deep kaolinitic profiles are summarized in Table 14.3. Several deep kaolinitic profiles in Africa, such as in Mozambique (Pekkala et al. 2008) and Sierra Leonne (Warnsloh 2011), are being mined.

Table 14.3 Locations of deep kaolinitic profiles in tropical and subtropical regions

Location	Country	Parent rock/Genesis	Depth (m)	Reference
Asia/Oceania				
Darling range, Western Australia	Australia	In situ weathering of granite	20–150	Anand and Paine (2002)
Fusui, Guangxi Province, Youjiang Basin	China	Intense chemical weathering of sedimentary rock	20	Yu et al. (2014)
Northern Guizhou Province	China	Deposition of limestone	5–17	Gu et al. (2013)
Kerala, southern India	India	Intense weathering of khondalites and subsequently transported and deposited input into lakes	30	Nakagawa et al. (2006)
Johor, southern peninsular	Malaysia	Weathering of granite	24	Alavi Nezhad Khalil Abad et al. (2014)
North/South America				
Georgia Piedmont Province	USA	Weathering of granite	10	White et al. (2001)
Amazon basin	Brazil	Origin/genesis of kaolin is under debate between weathering originating from sediments or sediment	10–60	Bonotto et al. (2007); da Costa et al. (2009); Montes et al. (2002)
Acoculco	Mexico	Alteration of the dacitic lavas and pyroclastic deposits	200	López-Hernández et al. (2009)
Africa				
Cunene complex, southern Angola	Angola	Transformation of basic anorthosites and gabbros	40	Saviano et al. (2005)
Djebel Debbagh, North-Eastern Algeria	Algeria	Burial genesis and deformation of limestone	60–200	Renac and Assassi (2009)
Makoro kaolin deposit, South-Eastern Botswana	Botswana	Alteration of feldspathic arenites	50	Ekosse (2000)
Balkouin, Central Burkina Faso	Burkina Faso	In situ lateritization process of granitic rocks	12	Giorgis et al. (2014)
Burundi	Burundi	Weathering of basalt	1–3	Schirrmeister and Störr (1994)
Mayouom, Western Cameroon	Cameroon	Hydrothermal alteration of feldspar- and Mica-rich rocks	10–13	Njoya et al. (2006)
Bana, north-West Cameroon	Cameroon	Weathering of granite	20	Wouatong et al. (1996)
Nsimi, South Cameroon	Cameroon	Deep weathering from physical erosion of granite	38	Braun et al. (2012)

(continued)

Table 14.3 (continued)

Location	Country	Parent rock/Genesis	Depth (m)	Reference
Mada region, south-East Cameroon	Cameroon	The basement rock is constituted by serpentinites which are intrusive in mica schists and quartzites	21	Ndjigui et al. (2009)
Kombelcha	Ethiopia	In situ weathering of granite	10	Fentaw and Mengistu (1998)
Bombowha	Ethiopia	Hydrothermal and in situ weathering of pegmatites and granites	11	Fentaw and Mengistu (1998)
Mahlangatsha Mountains	Swaziland	In situ weathering of granites, gneisses, and acid volcanic	18–73	Hunter and Urie (1966)
Sidi El Bader	Tunisia	Alteration of sandstone	100	Felhi et al. (2008)
Buwambo deposit, Kampala	Uganda	Weathering of granite	12	Nyakairu et al. (2001)

14.7.2 Deep Soils in South-Western Australia

Soils in south-western Australian landscapes have been intensively surveyed and studied by Anand and Paine (2002). Briefly, this region is characterized by a deep weathering profile formed on the Archean granites and gneisses of the Yilgarn Craton (Gilkes et al. 1973). Landscapes having soils that extend up to 150 m deep were described by Anand and Paine (2002), but typically the soils are shallower than this (McArthur 1991).

The two deep weathered soil profiles generally found in south-western Australia are a lateritic profile and a deep sand profile. The lateritic profile is mostly distributed across the Darling Range, the south-western part of the Yilgarn Craton where the Jarrah (*Eucalyptus marginata*) forest ecosystem has evolved (Dell and Havel 1989). A typical lateritic profile averages about 20 m in thickness (Anand and Paine 2002) and consists of six horizons: ferruginous (top soil and duricrust), mottled zone, pallid zone, saprolite, and bedrock. The bedrock is mostly granite, but veins of mafic rocks such as dolerite often occur.

Kaolinite is the predominant clay mineral in south-western Australia, resulting from intense weathering of granite (Viscarra Rossel 2011). Singh and Gilkes (1992) characterized clays from this region and found that approximately 80% of all clay types was kaolinite. Kaolinite occurs in the mottled, pallid, and saprolite zones at depths ranging from 2 to 50 m below the soil surface (Gilkes et al. 1973; McArthur 1991) and tends to increase with depth (Sadleir and Gilkes 1976). Kaolinite interacts with other soil materials, leading to the coating of sand-size quartz and filling of channels with iron oxide, and stained kaolin occurs throughout the regolith (Kew and Gilkes 2007; Kew et al. 2010).

Chemical and physical attributes of kaolinite are affected by the weathering process, resulting in isomorphic substitution of Fe^{3+} and Al^{3+}. Consequently, kaolinite contains 2.5% of Fe_2O_3 in its structure and has a poor X-ray crystallinity index of 5.4 (Singh and Gilkes 1992). The kaolinite in south-western Australia is higher in surface area (35 m^2 g^{-1}) and cation exchange capacity (56.7 $mmol_c$ kg^{-1}) compared to standard kaolinites that normally have a surface area of 10 m^2 g^{-1} and cation exchange capacity of 4.8 $mmol_c$ kg^{-1} (Singh and Gilkes 1992).

14.7.3 Deep Roots and Land-Use Change in South-Western Australia

Most woody vegetation in these dry sclerophyll forests exhibit dimorphic root systems: a shallow lateral root system supplying water and nutrients in the wet season, with groundwater taken up by the deep tap roots or sinker roots during the dry season (Dell and Havel 1989; Dawson and Pate 1996). The different bedrocks influence the characteristics of sinker root penetration. In doleritic profiles, numerous large and fine roots penetrate up to 40 m into deep clay horizons without root channels, whereas in the denser granitic profiles, roots access vertical root channels (recharge channels or macropores) to access water in deep soil profiles (Dell et al. 1984). These root channels may allow tree roots to grow deeper into the pallid zone of granitic profiles than in doleritic profiles (Johnston et al. 1983), but further study is needed before a definitive conclusion can be drawn.

Like in many parts of the world, extensive land-use changes have taken place over the last 100 years, with deforestation of deep-rooted native vegetation to grow annual pastures and crops, and in some areas, reforestation with *Eucalyptus* spp. in recent decades. Unlike the deep-rooted perennial native vegetation, the rooting depth of Gatton panic (*Megathyrsus maximus*) pasture was 5.3 m after 5 years of planting, and a root depth of 1.5 m was observed in annual crops such as barley (*Hordeum vulgare*) and lupin (*Lupinus angustifolius*) (Ward et al. 2015). By contrast, root systems of 7-year-replanted eucalypts exploited soil water to at least 8–10 m depth on several agricultural soils (Harper et al. 2009). Thus, the change in land use has resulted in a substantial decrease in the proportion of the soil profile being accessed by roots over the past century. The impact that this has had on the dynamics of deep soil carbon in the region is unknown.

In recent decades, extensive areas of forest with lateritic soils rich in aluminium hydroxide minerals have been mined for bauxite (McArthur 1991). Where the vertical root channels become occluded during mining, the taproots and sinker roots of the revegetation species may be unable to penetrate below the depth of machine ripping (Szota et al. 2007), due to the hostile regolith. It would be interesting to explore whether root channels also become occluded under agricultural practices where the soil is annually tilled.

In south-western Australia, the study of deep roots after land-use change has been overlooked. Deep roots are affected by anthropogenic activities, such as mining, road cutting, and agricultural activity (Figs. 14.1, 14.2 and 14.3). Opportunistic

Fig. 14.1 Karragullen gravel mine (32° 5′ 51″S, 16° 7′ 14″E): (**a**) a face in mine floor approximately 4 m height, (**b**) a bunch of roots penetrating through the duricrust zone, (**c**) tree roots present at the bottom of the pit. (Photographs by P. Sangmanee)

observations of deep roots reveal that dead deep roots differ in their extent of decomposition across the landscape. Furthermore, char materials produced by fire may contribute to the subsoil carbon pool (Fig. 14.3b). The persistence of woody roots at depth and their contribution to global soil organic carbon stock accounting have yet to be investigated for this region.

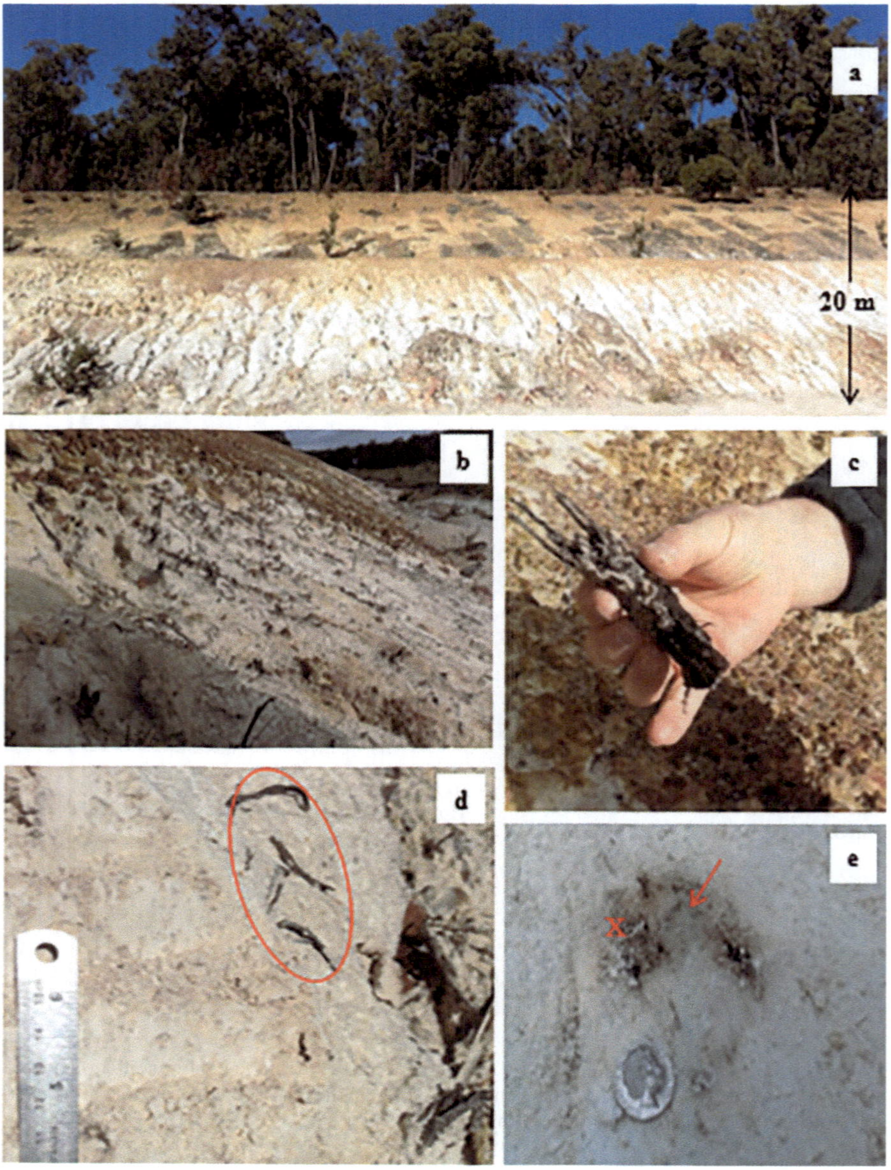

Fig. 14.2 Roadcut at Mundaring (32° 53′ 40″S, 116° 13′ 36″E): (**a**) profile approximately 20 m height showing pallid zone; (**b**) dead root fragments could be observed ubiquitously at the surface of cutting; (**c**) fragile woody root; (**d**) presence of woody roots at surface of cutting, scale in centimetres; (**e**) root surrounded with stain when observed closely, coin for scale. (Photographs by P. Sangmanee)

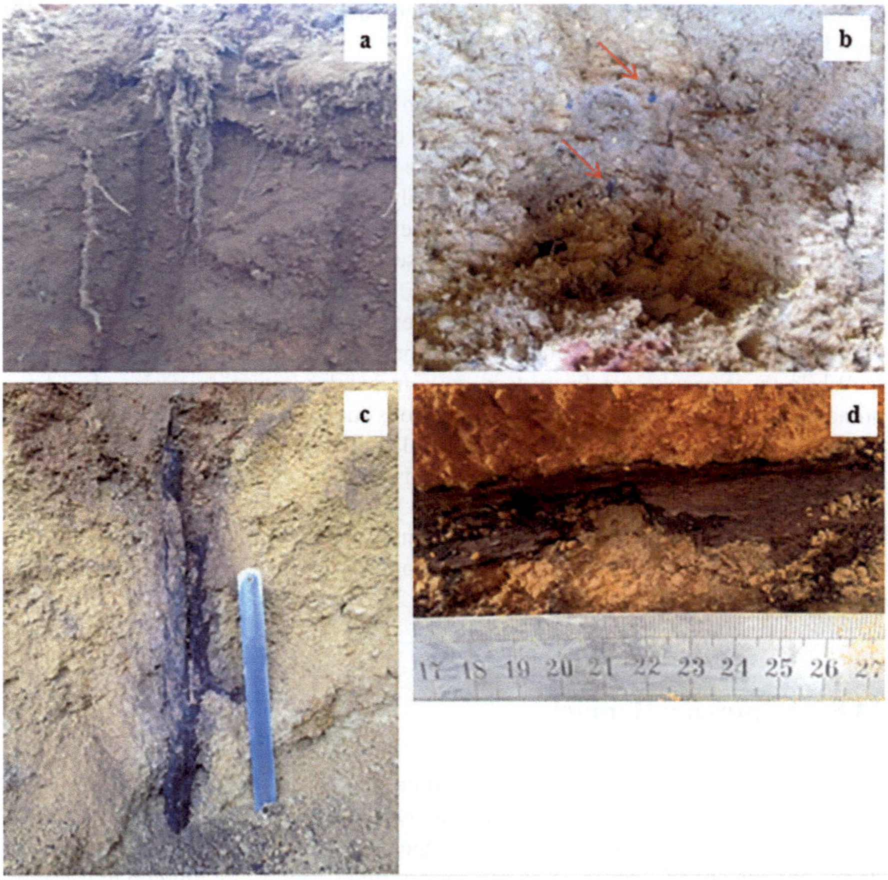

Fig. 14.3 Agricultural grassland, Kalamunda (31° 58′ 13′S, 116° 7′ 38′E): (**a**) grass root penetrate down to 0.3 m; (**b**) charred materials 1 cm diameter size were observed at 0.5 m depth; (**c**) woody root located at 1 m, 0.3 m-ruler for scale; (**d**) decaying root of native Jarrah (*E. marginata*) was found at 1.5 m depth. Scale in centimetres. (Photographs by P. Sangmanee)

14.7.4 Deep Carbon Storage and Composition in South-Western Australia

Globally, the monitoring of soil carbon storage is limited to 0.3 m despite many soils being much deeper than this. Thus, measurements of soil carbon content were made in the deep soils at five sites across south-western Australia (Harper and Tibbett 2013). Soil carbon occurred in amounts of two to five times more than would normally be reported in the surface horizons, to depths of up to 38 m depth. This was despite the land having been deforested for up to 80 years. This raises questions about the dynamics of these deeper carbon stores under the influence of both land-use and climate change.

Techniques were consequently developed to quantitatively and qualitatively determine small concentrations of carbon in deep soils (Sangmanee 2016; Sangmanee et al. 2017). There were marked differences in the composition of carbon compounds with depth. Near the surface (0–1 m), these were macromolecular organic compounds derived from lignin, polysaccharides, proteins, and terpenes, whereas at depth (11–12 m, 18–19 m), there were low-molecular-weight compounds, such as 13-docosenamide, 13-docosenoate, xanthone, and benzophenone. The deeper compounds were likely derived from the roots of the previous forest, whereas the surface soils are affected by current land use. The in situ decomposition of deep roots was revealed by the pyridine compound.

While these results are from deep soils in south-western Australia, they have broader, global implications as shown in regions such as the Amazon (Nepstad et al. 1994), where deep soils also occur. In particular, the results demonstrate that carbon contained in deep soils should be considered in global carbon accounting and that determining the fate of this carbon should be prioritized particularly given ongoing land-use change and climate change. Additionally, developing land-use systems that enhance the storage of carbon in deep soils appears to be a promising approach to enhancing carbon mitigation (Kell 2011, 2012).

14.8 Summary Remarks

Deep soil is potentially a large reservoir of carbon in terrestrial systems. Although deep soils are abundant in many regions, their carbon content to depth has been little studied. The implications of land-use change and climate change on these carbon stores, and the consequences on carbon accounting systems, are unknown. It is possible, however, that land-use systems could be developed that store considerably more carbon at depth (e.g. Kell 2011, 2012), given how the stores are caused predominantly by plant root activity.

In general, soil carbon content declines with depth as the influence of aboveground vegetation weakens, and there is likely to be a wide range of carbon compounds differing in size and concentration varying within the soil profile. In order to estimate carbon storage in deep soil, including understanding its origin and stability, some consideration of methods used for the study of carbon will be necessary. So far, approaches for studying deep soil quantitatively and qualitatively are poorly developed, and this is an area that needs development.

References

Alavi Nezhad Khalil Abad SV, Mohamad ET, Komoo I. Dominant weathering profiles of granite in Southern peninsular Malaysia. Eng Geol. 2014;183:208–15. https://doi.org/10.1016/j.enggeo.2014.10.019.

Almendros G, Sanz J, Velasco F. Signatures of lipid assemblages in soils under continental Mediterranean forests. Eur J Soil Sci. 1996;47(2):183–96. https://doi.org/10.1111/j.1365-2389.1996.tb01389.x.

Anand RR, Paine M. Regolith geology of the Yilgarn craton, Western Australia: implications for exploration. Aust J Earth Sci. 2002;49(1):3–162. https://doi.org/10.1046/j.1440-0952.2002.00912.x.

Angelopoulou T, Tziolas N, Balafoutis A, et al. Remote sensing techniques for soil organic carbon estimation: a review. Remote Sens. 2019;11:676. https://doi.org/10.3390/rs11060676.

Ayling BF, McGowan HA. Niveo-eolian sediment deposits in coastal South Victoria land, Antarctica: indicators of regional variability in weather and climate. Arct Antarct Alp Res. 2006;38(3):313–24. https://doi.org/10.1657/1523-0430(2006)38[313:NSDICS]2.0.CO;2.

Bartlett RJ, Ross DS. Colorimetric determination of oxidizable carbon in acid soil solutions. Soil Sci Soc Am J. 1988;52(4):1191–2. https://doi.org/10.2136/sssaj1988.03615995005200040055x.

Bellon-Maurel V, McBratney A. Near-infrared (NIR) and mid-infrared (MIR) spectroscopic techniques for assessing the amount of carbon stock in soils–critical review and research perspectives. Soil Biol Biochem. 2011;43(7):1398–410. https://doi.org/10.1016/j.soilbio.2011.02.019.

Berardo N, Pisacane V, Battilani P, et al. Rapid detection of kernel rots and mycotoxins in maize by near-infrared reflectance spectroscopy. J Agric Food Chem. 2005;53(21):8128–34. https://doi.org/10.1021/jf0512297.

Bettis EA, Muhs DR, Roberts HM, et al. Last glacial loess in the conterminous USA. Quat Sci Rev. 2003;22:1907–46. https://doi.org/10.1016/S0277-3791(03)00169-0.

Bleby TM, McElrone AJ, Jackson RB. Water uptake and hydraulic redistribution across large woody root systems to 20 m depth. Plant Cell Environ. 2010;33(12):2132–48. https://doi.org/10.1111/j.1365-3040.2010.02212.x.

Bonotto DM, Fujimori K, Moreira-Nordemann L. Determination of weathering rate of the Morro do Ferro Th-REEs deposit, Brazil using U-isotope method. Appl Radiat Isot. 2007;65(5):474–81. https://doi.org/10.1016/j.apradiso.2006.11.003.

Bowler JM. Aridity in Australia: age, origins and expression in aeolian landforms and sediments. Earth Sci Rev. 1976;12(2–3):279–310. https://doi.org/10.1016/0012-8252(76)90008-8.

Brantley SL, Megonigal JP, et al. Twelve testable hypotheses on the geobiology of weathering. Geobiology. 2011;9(2):140–65. https://doi.org/10.1111/j.1472-4669.2010.00264.x.

Braun J-J, Marechal J-C, et al. Elemental weathering fluxes and saprolite production rate in a central African lateritic terrain (Nsimi, South Cameroon). Geochim Cosmochim Acta. 2012;99:243–70. https://doi.org/10.1016/j.gca.2012.09.024.

Bundt M, Widmer F, Pesaro M, et al. Preferential flow paths: biological 'hot spots' in soils. Soil Biol Biochem. 2001;33(6):729–38. https://doi.org/10.1016/S0038-0717(00)00218-2.

Buurman P, Peterse F, Almendros Martin G. Soil organic matter chemistry in allophanic soils: a pyrolysis-GC/MS study of a Costa Rican andosol catena. Eur J Soil Sci. 2007;58(6):1330–47. https://doi.org/10.1111/j.1365-2389.2007.00925.x.

Buurman P, Van Bergen P, Jongmans A, Meijer E, Duran B, Van Lagen B. Spatial and temporal variation in podzol organic matter studied by pyrolysis-gas chromatography/mass spectrometry and micromorphology. Eur J Soil Sci. 2005;56(2):253–70. https://doi.org/10.1111/j.1365-2389.2004.00662.x.

Canadell J, Jackson R, Ehleringer J, et al. Maximum rooting depth of vegetation types at the global scale. Oecologia. 1996;108(4):583–95. https://doi.org/10.1007/BF00329030.

Chatterjee A, Lal R, Wielopolski L, Martin MZ, et al. Evaluation of different soil carbon determination methods. Crit Rev Plant Sci. 2009;28(3):164–78. https://doi.org/10.1080/07352680902776556.

Chevrier V, Mathé PE, Rochette P, et al. Iron weathering products in a CO_2 + (H_2O or H_2O_2) atmosphere: implications for weathering processes on the surface of Mars. Earth Planet Sci Lett. 2006;244:501–14. https://doi.org/10.1016/j.gca.2006.06.1368.
da Costa ML, Sousa DJL, Angélica RS. The contribution of lateritization processes to the formation of the kaolin deposits from Eastern Amazon. J S Am Earth Sci. 2009;27(2–3):219–34. https://doi.org/10.1016/j.jsames.2008.11.008.
Dawson TE, Pate JS. Seasonal water uptake and movement in root systems of Australian phreatophytic plants of dimorphic root morphology: a stable isotope investigation. Oecologia. 1996;107(1):13–20. https://doi.org/10.1007/BF00582230.
De Vos B, Lettens S, Muys B, et al. Walkley–black analysis of forest soil organic carbon: recovery, limitations and uncertainty. Soil Use Manag. 2007;23(3):221–9. https://doi.org/10.1111/j.1475-2743.2007.00084.x.
Dell B, Bartle J, Tacey W. Root occupation and root channels of jarrah forest subsoils. Aust J Bot. 1984;31(6):615–27. https://doi.org/10.1071/BT9830615.
Dell B, Havel JJ. The jarrah forest, an introduction. In: Dell B, Havel JJ, Malajczuk N, editors. The jarrah forest: a complex mediterranean ecosystem. Dordrecht: Kluwer Academic Publ; 1989. p. 1–10. https://doi.org/10.1007/978-94-009-3111-4.
Derenne S, Quénéa K. Analytical pyrolysis as a tool to probe soil organic matter. J Anal Appl Pyrolysis. 2015;111:108–20. https://doi.org/10.1016/j.jaap.2014.12.001.
Dijkstra FA, Cheng W. Interactions between soil and tree roots accelerate long-term soil carbon decomposition. Ecol Lett. 2007;10(11):1046–53. https://doi.org/10.1111/j.1461-0248.2007.01095.x.
Edson SN, Mills RH. Soil assay, colorimetric field test for organic matter in mineral soils. J Agr Food Chem. 1955;3:852–3.
Eggleton RA, Taylor G. Impact of fire on the Weipa bauxite, Northern Australia. Aust J Earth Sci. 2008;55(sup1):S83–6. https://doi.org/10.1080/08120090802438266.
Ekosse G. The Makoro kaolin deposit, Southeastern Botswana: its genesis and possible industrial applications. Appl Clay Sci. 2000;16(5–6):301–20. https://doi.org/10.1016/S0169-1317(99)00059-9.
Ekosse G-IE. Kaolin deposits and occurrences in Africa: geology, mineralogy and utilization. Appl Clay Sci. 2010;50(2):212–36. https://doi.org/10.1016/j.clay.2010.08.003.
Felhi M, Tlili A, Gaied ME, et al. Mineralogical study of kaolinitic clays from Sidi El Bader in the far North of Tunisia. Appl Clay Sci. 2008;39(3–4):208–17. https://doi.org/10.1016/j.clay.2007.06.004.
Fentaw HM, Mengistu T. Comparison of Kombelcha and Bombowha kaolins of Ethiopia. Appl Clay Sci. 1998;13(2):149–64. https://doi.org/10.1016/S0169-1317(98)00017-9.
Fontaine S, Barot S, Barre P, et al. Stability of organic carbon in deep soil layers controlled by fresh carbon supply. Nature. 2007;450(7167):277–80. https://doi.org/10.1038/nature06275.
Forrester JA, Mladenoff DJ, D'Amato AW, et al. Temporal trends and sources of variation in carbon flux from coarse woody debris in experimental forest canopy openings. Oecologia. 2015;179(3):889–900. https://doi.org/10.1007/s00442-015-3393-4.
Franco CMM, Clarke PJ, Tate ME, et al. Hydrophobic properties and chemical characterisation of natural water repellent materials in Australian sands. J Hydrol. 2000;231–232:47–58. https://doi.org/10.1016/S0022-1694(00)00182-7.
Frazier SW, Nowack KO, Goins KM, et al. Characterization of organic matter from natural waters using tetramethylammonium hydroxide thermochemolysis GC-MS. J Anal Appl Pyrolysis. 2003;70(1):99–128. https://doi.org/10.1016/S0165-2370(02)00098-0.
Galbraith D, Malhi Y, Affum-Baffoe K, et al. Residence times of woody biomass in tropical forests. Plant Ecol Divers. 2013;6(1):139–57. https://doi.org/10.1080/17550874.2013.770578.
Gaudin A, Dehouck E, Mangold N. Evidence for weathering on early Mars from a comparison with terrestrial weathering profiles. Icarus. 2011;216(1):257–68. https://doi.org/10.1016/j.icarus.2011.09.004.

Gilkes R, Scholz G, Dimmock G. Lateritic deep weathering of granite. J Soil Sci. 1973;24(4):523–36. https://doi.org/10.1111/j.1365-2389.1973.tb02319.x.
Giorgis I, Bonetto S, Giustetto R, et al. The lateritic profile of Balkouin, Burkina Faso: geochemistry, mineralogy and genesis. J Afr Earth Sci. 2014;90:31–48. https://doi.org/10.1016/j.jafrearsci.2013.11.006.
Gocke M, Gulyás S, Hambach U, et al. Biopores and root features as new tools for improving paleoecological understanding of terrestrial sediment-paleosol sequences. Palaeogeogr Palaeoclimatol Palaeoecol. 2014;394:42–58. https://doi.org/10.1016/j.palaeo.2013.11.010.
Gocke M, Pustovoytov K, Kühn P, et al. Carbonate rhizoliths in loess and their implications for paleoenvironmental reconstruction revealed by isotopic composition: δ13C, 14C. Chem Geol. 2011;283(3–4):251–60. https://doi.org/10.1016/j.chemgeo.2011.01.022.
González-Vila F, Almendros G, Martin F. An evaluation of the differences in the composition of humic acids in soils under oak and pine forest by GC-MS after mild degradation. Plant Soil. 1987;103(1):83–8. https://doi.org/10.1007/BF02370671.
Griebler C, Lueders T. Microbial biodiversity in groundwater ecosystems. Freshw Biol. 2009;54(4):649–77. https://doi.org/10.1111/j.1365-2427.2008.02013.x.
Gu J, Huang Z, Fan H, et al. Mineralogy, geochemistry, and genesis of lateritic bauxite deposits in the Wuchuan–Zheng'an–Daozhen area, Northern Guizhou Province, China. J Geochem Explor. 2013;130:44–59. https://doi.org/10.1016/j.gexplo.2013.03.003.
Harper R, Smettem K, Carter J, et al. Drought deaths in *Eucalyptus globulus* (Labill.) plantations in relation to soils, geomorphology and climate. Plant Soil. 2009;324(1–2):199–207. https://doi.org/10.1007/s11104-009-9944-x.
Harper R, Tibbett M. The hidden organic carbon in deep mineral soils. Plant Soil. 2013;368(1–2):641–8. https://doi.org/10.1007/s11104-013-1600-9.
Hatcher PG, Dria KJ, Kim S, et al. Modern analytical studies of humic substances. Soil Sci. 2001;166(11):770–94. https://doi.org/10.1097/00010694-200111000-00005.
He W, Hu H. Prediction of hot-water-soluble extractive, pentosan and cellulose content of various wood species using FT-NIR spectroscopy. Bioresour Technol. 2013;140:299–305. https://doi.org/10.1016/j.biortech.2013.04.115.
Heanes DL. Determination of total organic-C in soils by an improved chromic acid digestion and spectrophotometric procedure. Commun Soil Sci Plant Anal. 1984;15(10):1191–213. https://doi.org/10.1080/00103628409367551.
Heckman K, Throckmorton H, Clingensmith C, Vila FJG, Horwath WR, Knicker H, Rasmussen C. Factors affecting the molecular structure and mean residence time of occluded organics in a lithosequence of soils under ponderosa pine. Soil Biol Biochem. 2014;77:1–11. https://doi.org/10.1016/j.soilbio.2014.05.028.
Helfrich M, Ludwig B, Buurman P, et al. Effect of land use on the composition of soil organic matter in density and aggregate fractions as revealed by solid-state 13C NMR spectroscopy. Geoderma. 2006;136(1):331–41. https://doi.org/10.1016/j.geoderma.2006.03.048.
Helmisaari H-S, Leppälammi-Kujansuu J, Sah S, et al. Old carbon in young fine roots in boreal forests. Biogeochem. 2015;125(1):37–46. https://doi.org/10.1007/s10533-015-0110-7.
Hernandez-Soriano MC, Kerré B, Goos P, et al. Long-term effect of biochar on the stabilization of recent carbon: soils with historical inputs of charcoal. GCB Bioenergy. 2015;8(2):371–81. https://doi.org/10.1111/gcbb.12250.
Hoshino Y, George SC. Cyanobacterial inhabitation on Archean rock surfaces in the Pilbara craton, Western Australia. Astrobiology. 2015;15(7):559–74. https://doi.org/10.1089/ast.2014.1275.
Huguet A, Gocke M, Derenne S, et al. Root-associated branched tetraether source microorganisms may reduce estimated paleotemperatures in subsoil. Chem Geol. 2013;356:1–10. https://doi.org/10.1016/j.chemgeo.2013.07.017.
Huguet A, Wiesenberg GLB, Gocke M, et al. Branched tetraether membrane lipids associated with rhizoliths in loess: Rhizomicrobial overprinting of initial biomarker record. Org Geochem. 2012;43:12–9. https://doi.org/10.1016/j.orggeochem.2011.11.006.

Hunter DR, Urie JG. The origin of kaolin deposits, Mahlangatsha mountains, Swaziland. Econ Geol. 1966;61(6):1104–14. https://doi.org/10.2113/gsecongeo.61.6.1104.
Izaurralde RC, Rice CW, Wielopolski L, et al. Evaluation of three field-based methods for quantifying soil carbon. PLoS One. 2013;8(1):e55560. https://doi.org/10.1371/journal.pone.0055560.
Jackson RB, Canadell J, Ehleringer JR, et al. A global analysis of root distributions for terrestrial biomes. Oecologia. 1996;108(3):389–411. https://doi.org/10.1007/BF00333714.
Jefferey QC, Higuchi N, Schimel J. Ancient trees in Amazonia. Nature. 1998;391(6663):135–6. https://doi.org/10.1038/34325.
Johnston CT, Aochi YO. Fourier transform infrared and Raman spectroscopy. In: Sparks DL, Page AL, Helmke PA, Loeppett RH, Tabatabai MA, Johnston CT, Summer ME, editors. Methods of soil analysis: chemical methods, part 3. Wisconsin: Soil Science Society of America Inc; 1996. p. 274–5. https://doi.org/10.2136/sssabookser5.3.c10.
Johnston CD, Hurle DH, Hudson DR et al. Water movement through preferred paths in lateritic profiles of the darling plateau, Western Australia, CSIRO Australia. Div. Groundwater Research, Tech. Pap. No.1; 1983.
Jones MJ. The weathered zone aquifers of the basement complex areas of Africa. Q J Eng Geol Hydrogeol. 1985;18(1):35–46. https://doi.org/10.1144/GSL.QJEG.1985.018.01.06.
Keiluweit M, Bougoure JJ, Nico PS, Pett-Ridge J, Weber PK, Kleber M. Mineral protection of soil carbon counteracted by root exudates. Nat Clim Chang. 2015;5(6):588–95. https://doi.org/10.1038/nclimate2580.
Keith H, Mackey BG, Lindenmayer DB. Re-evaluation of forest biomass carbon stocks and lessons from the world's most carbon-dense forests. Proc Natl Acad Sci. 2009;106(28):11635–40. https://doi.org/10.1073/pnas.0901970106.
Kell DB. Breeding crop plants with deep roots: their role in sustainable carbon, nutrient and water sequestration. Ann Bot. 2011;108:407–18. https://doi.org/10.1093/aob/mcr175.
Kell DB. Large-scale sequestration of atmospheric carbon via plant roots in natural and agricultural ecosystems: why and how. Philos Trans R Soc Lond Ser B Biol Sci. 2012;367(1595):1589–97. https://doi.org/10.1098/rstb.2011.0244.
Kew G, Gilkes R. Properties of regolith beneath lateritic bauxite in the darling range of South Western Australia. Soil Res. 2007;45(3):164–81. https://doi.org/10.1071/SR06128.
Kew G, Gilkes R, Evans D. Relationships between fabric, water retention, and strength of hard subsoils in the South of Western Australia. Soil Res. 2010;48(2):167–77. https://doi.org/10.1071/SR09080.
Korbel K, Hancock P, Serov P, et al. Groundwater ecosystems vary with land use across a mixed agricultural landscape. J Environ Qual. 2013;42(2):380–90. https://doi.org/10.2134/jeq2012.0018.
Kotake T, Hamamoto S, Saito T, et al. Characterization of alkali-soluble polysaccharides in deep subsoil layers. Soil Sci Plant Nutr. 2013;59(6):871–6. https://doi.org/10.1080/00380768.2013.853280.
Laclau J-P, da Silva EA, Lambais GR, et al. Dynamics of soil exploration by fine roots down to a depth of 10 m throughout the entire rotation in Eucalyptus grandis plantations. Front Plant Sci. 2013;4:1–12. https://doi.org/10.3389/fpls.2013.00243.
Lancaster N. Aeolian features and processes. In: Young R, Norby L, editors. Geological monitoring. Boulder, Colorado: Geological Society of America; 2009. p. 1–25. https://doi.org/10.1130/2009.monitoring(01).
Leake J, Johnson D, Donnelly D, et al. Networks of power and influence: the role of mycorrhizal mycelium in controlling plant communities and agroecosystem functioning. Can J Bot. 2004;82(8):1016–45. https://doi.org/10.1139/b04-060.
López-Hernández A, García-Estrada G, Aguirre-Díaz G, et al. Hydrothermal activity in the Tulancingo-Acoculco caldera complex, Central Mexico: exploratory studies. Geothermics. 2009;38(3):279–93. https://doi.org/10.1016/j.geothermics.2009.05.001.

Major J, Lehmann J, Rondon M, et al. Fate of soil-applied black carbon: downward migration, leaching and soil respiration. Glob Chang Biol. 2010;16(4):1366–79. https://doi.org/10.1111/j.1365-2486.2009.02044.x.

McArthur WM. Reference soils of South-Western Australia. 1st ed. Perth, Western Australia, Australia: Department of Agriculture; 1991.

McClean GJ, Meredith W, Cross A, et al. The priming potential of environmentally weathered pyrogenic carbon during land-use transition to biomass crop production. GCB Bioenergy. 2016;8(4):805–17. https://doi.org/10.1111/gcbb.12293.

Mikhailova EA, Noble RRP, Post CJ. Comparison of soil organic carbon recovery by Walkley-black and dry combustion methods in the Russian Chernozem. Commun Soil Sci Plant Anal. 2003;34(13–14):1853–60. https://doi.org/10.1081/CSS-120023220.

Modenesi-Gauttieri MC, de Toledo MCM, Hiruma ST, et al. Deep weathering and landscape evolution in a tropical plateau. Catena. 2011;85(3):221–30. https://doi.org/10.1016/j.catena.2011.01.006.

Montes CR, Melfi AJ, Carvalho A, et al. Genesis, mineralogy and geochemistry of kaolin deposits of the Jari River, Amapá state, Brazil. Clay Clay Miner. 2002;50(4):494–503. https://doi.org/10.1346/000986002320514217.

Nakagawa M, Santosh M, Yoshikura S, et al. Kaolin deposits at Melthonnakkal and Pallipuram within Trivandrum block, Southern India. GR. 2006;9(4):530–8. https://doi.org/10.1016/j.gr.2005.11.010.

Ndjigui P-D, Bilong P, Bitom D. Negative cerium anomalies in the saprolite zone of serpentinite lateritic profiles in the Lomié ultramafic complex, South-East Cameroon. J Afr Earth Sci. 2009;53(1):59–69. https://doi.org/10.1016/j.jafrearsci.2008.09.002.

Nepstad DC, Carvalho CR, Davidson EA, et al. The role of deep roots in the hydrological and carbon cycles of Amazonian forests and pastures. Nature. 1994;372:666–9. https://doi.org/10.1038/372666a0.

Nierop KG, Buurman P, de Leeuw JW. Effect of vegetation on chemical composition of H horizons in incipient podzols as characterized by 13C NMR and pyrolysis-GC/MS. Geoderma. 1999;90(1):111–29. https://doi.org/10.1016/S0016-7061(98)00095-0.

Nierop KGJ, Pulleman MM, Marinissen JCY. Management induced organic matter differentiation in grassland and arable soil: a study using pyrolysis techniques. Soil Biol Biochem. 2001;33(6):755–64. https://doi.org/10.1016/S0038-0717(00)00223-6.

Njoya A, Nkoumbou C, Grosbois C, et al. Genesis of Mayouom kaolin deposit (Western Cameroon). Appl Clay Sci. 2006;32(1–2):125–40. https://doi.org/10.1016/j.clay.2005.11.005.

Nyakairu GW, Koeberl C, Kurzweil H. The Buwambo kaolin deposit in Central Uganda: mineralogical and chemical composition. Geochem J. 2001;35(4):245–56. https://doi.org/10.2343/geochemj.35.245.

Ohnuki Y, Shimizu A, Chann S, et al. Seasonal change in thick regolith hardness and water content in a dry evergreen forest in Kampong Thom Province, Cambodia. Geoderma. 2008;146(1–2):94–101. https://doi.org/10.1016/j.geoderma.2008.05.016.

Olajuyigbe S, Tobin B, Gardiner P, et al. Contribution of belowground coarse woody roots to the soil organic carbon pool. In: COST Action FP0803: Belowground carbon turnover in European forests, 2nd MC Meeting, State of the Art'. Birmensdorf, Switzerland: Swiss Federal Research Institute WSL; 2010. 26–28 Jan 2010.

Otto A, Simpson MJ. Analysis of soil organic matter biomarkers by sequential chemical degradation and gas chromatography–mass spectrometry. J Sep Sci. 2007;30(2):272–82. https://doi.org/10.1002/jssc.200600243.

Page DW, Van Leeuwen JA, Spark KM, et al. Pyrolysis characterisation of plant, humus and soil extracts from Australian catchments. J Anal Appl Pyrolysis. 2002;65(2):269–85. https://doi.org/10.1016/S0165-2370(02)00005-0.

Pankhurst CE, Pierret A, Hawke BG, et al. Microbiological and chemical properties of soil associated with macropores at different depths in a red-duplex soil in NSW Australia. Plant Soil. 2002;238(1):11–20. https://doi.org/10.1023/A:1014289632453.

Pekkala Y, Kuivasaari T, Gonçalves R, et al. Review of industrial minerals in Mozambique. Geol Soc Finland Spec Paper. 2008;48:289–306.

Prescott CE. Litter decomposition: what controls it and how can we alter it to sequester more carbon in forest soils? Biogeochem. 2010;101(1):133–49. https://doi.org/10.1007/s10533-010-9439-0.

Raeside JD. Loess deposits of the South Island, New Zealand, and soils formed on them. New Zealand J Geol Geophys. 1964;7(4):811–38. https://doi.org/10.1080/00288306.1964.10428132 .

Ralph J, Hatfield RD. Pyrolysis-GC-MS characterization of forage materials. J Agric Food Chem. 1991;39(8):1426–37.

Ramaan E. The evolution and classification of soils. London, UK: W. Heffer and Sons; 1928.

Reeves JB. Near-versus mid-infrared diffuse reflectance spectroscopy for soil analysis emphasizing carbon and laboratory versus on-site analysis: where are we and what needs to be done? Geoderma. 2010;158(1):3–14. https://doi.org/10.1016/j.geoderma.2009.04.005.

Renac C, Assassi F. Formation of non-expandable 7 Å halloysite during Eocene–Miocene continental weathering at Djebel Debbagh, Algeria: a geochemical and stable-isotope study. J Sediment Res. 2009;217(1):140–53. https://doi.org/10.1016/j.sedgeo.2009.04.001.

Richter DD, Markewitz D. How deep is soil? Bio Sci. 1995;45(9):600–9. https://doi.org/10.2307/1312764.

Rittl TF, Novotny EH, Balieiro FC, et al. Negative priming of native soil organic carbon mineralization by oilseed biochars of contrasting quality. Eur J Soil Sci. 2015;66:714–21. https://doi.org/10.1111/ejss.12257.

Rossell RA, Gasparoni JC, Galantini JA. Soil organic matter evaluation. In: Lal R, editor. Assessment methods for soil carbon. Washington D.C.: Lewis Publ; 1998. p. 311–22.

Roudnew B, Seymour JR, Jeffries TC, et al. Bacterial and virus-like particle abundances in purged and unpurged groundwater depth profiles. Ground Water Monit Remediat. 2012;32(4):72–7. https://doi.org/10.1111/j.1745-6592.2011.01393.x.

Sadleir SB, Gilkes RJ. Development of bauxite in relation to parent material near Jarrahdale, Western Australia. J Geol Soc of Aus. 1976;23(4):333–44. https://doi.org/10.1080/00167617608728948.

Sáiz-Jiménez C, De Leeuw JW. Chemical characterization of soil organic matter fractions by analytical pyrolysis-gas chromatography-mass spectrometry. J Anal Appl Pyrolysis. 1986;9(2):99–119. https://doi.org/10.1016/0165-2370(86)85002-1.

Sangmanee P. Quantification and characterisation of carbon in deep kaolinitic regolith in South-Western Australia. Australia: PhD Thesis, Murdoch University; 2016.

Sangmanee P, Dell B, Harper RJ, et al. Quantification of deep soil carbon by a wet digestion technique. Soil Res. 2017;55:78–85. https://doi.org/10.1071/SR15297.

Saviano G, Violo M, Pieruccini U, et al. Kaolin deposits from the Northern sector of the Cunene anorthosite complex (Southern Angola). Clay Clay Miner. 2005;53(6):674–85. https://doi.org/10.1346/CCMN.2005.0530613.

Schirrmeister L, Störr M. The weathering of basaltic rocks in Burundi and Vietnam. Catena. 1994;21(2):243–56. https://doi.org/10.1016/0341-8162(94)90015-9.

Schollenberger CJ. A rapid approximate method for determining soil organic matter. Soil Sci. 1927;24:65–8.

Schulten H-R, Schnitzer M. The chemistry of soil organic nitrogen: a review. Biol Fert Soils. 1997;26(1):1–15. https://doi.org/10.1007/s003740050335.

Shadkami F, Helleur R. Recent applications in analytical thermochemolysis. J Anal Appl Pyrolysis. 2010;89(1):2–16. https://doi.org/10.1016/j.jaap.2010.05.007.

Shen Y, Chapelle FH, Strom EW, et al. Origins and bioavailability of dissolved organic matter in groundwater. Biogeochem. 2015;122(1):61–78. https://doi.org/10.1007/s10533-014-0029-4.

Singh B, Gilkes RJ. Properties of soil kaolinites from South-Western Australia. J Soil Sci. 1992;43(4):645–67. https://doi.org/10.1111/j.1365-2389.1992.tb00165.x.

Soil Science Society of America. Glossary of soil science terms. Madison, Wisconsin: Soil Science Society; 1979.

Sollins P, Homann P, Caldwell BA. Stabilization and destabilization of soil organic matter: mechanisms and controls. Geoderma. 1996;74(1):65–105. https://doi.org/10.1016/S0016-7061(96)00036-5.

Stewart AD, Anand RR, Balkau J. Source of anomalous gold concentrations in termite nests, Moolart Well, Western Australia: implications for exploration. Geochem Explor Env A. 2012;12(4):327–37. https://doi.org/10.1144/geochem2012-126.

Stuut J-B, Smalley I, O'Hara-Dhand K. Aeolian dust in Europe: African sources and European deposits. Quat Int. 2009;198(1):234–45. https://doi.org/10.1016/j.quaint.2008.10.007.

Szota C, Veneklaas EJ, Koch JM, et al. Root architecture of jarrah (*Eucalyptus marginata*) trees in relation to post-mining deep ripping in Western Australia. Restor Ecol. 2007;15:S65–73. https://doi.org/10.1111/j.1526-100X.2007.00294.x.

Trumbore S. Age of soil organic matter and soil respiration: radiocarbon constraints on belowground C dynamics. Ecol Appl. 2000;10(2):399–411. https://doi.org/10.1890/1051-0761(2000)010[0399:AOSOMA]2.0.CO;2.

Tyurin IV. A new modification of the volumetric method of determining soil 344 organic matter by means of chromic acid. Pochvovedenie. 1931;26:36–47.

van Bergen PF, Nott CJ, Bull ID, et al. Organic geochemical studies of soils from the Rothamsted classical experiments-IV. Preliminary results from a study of the effect of soil pH on organic matter decay. Org Geochem. 1998;29(5):1779–95. https://doi.org/10.1016/S0146-6380(98)00188-0.

Van Praagh B, Yen A. National recovery plan for the Giant Gippsland Earthworm Megascolides Australis. East Melbourne: Department of Sustainability and Environment; 2010.

Vasiljević DA, Marković SB, Hose TA, et al. Loess–palaeosol sequences in China and Europe: common values and geoconservation issues. Catena. 2014;117:108–18. https://doi.org/10.1016/j.catena.2013.06.005.

Viscarra Rossel RA. Fine-resolution multiscale mapping of clay minerals in Australian soils measured with near infrared spectra. J Geophys Res Earth Surf. 2011;116(F04023):1–15. https://doi.org/10.1029/2011JF001977.

Viscarra Rossel RA, Walvoort DJJ, McBratney AB, et al. Visible, near infrared, mid infrared or combined diffuse reflectance spectroscopy for simultaneous assessment of various soil properties. Geoderma. 2006;131(1–2):59–75. https://doi.org/10.1016/j.geoderma.2005.03.007.

Wang Y, Ma S, Liu Z. Vertical distribution and influencing factors of soil water content within 21-m profile on the Chinese loess plateau. Geoderma. 2013;193–194:300–10. https://doi.org/10.1016/j.geoderma.2012.10.011.

Wang Y, Shao M, Zhang C, Liu Z, et al. Soil organic carbon in deep profiles under Chinese continental monsoon climate and its relations with land uses. Ecol Eng. 2015;82:361–7. https://doi.org/10.1016/j.ecoleng.2015.05.004.

Ward P, Wocheslander R, Ferris D, et al. Soil carbon storage in the root zone of a perennial grass pasture. In: Proceedings of the 17th ASA conference. Hobart, Australia; 2015.

Warnsloh JM. Geology and mineral industry of Sierra Leone. Geologic country report with emphasis on diamonds, gold and titanium. 2011:52.

Weng Z, Van Zwieten L, Singh BP, et al. Plant-biochar interactions drive the negative priming of soil organic carbon in an annual ryegrass field system. Soil Biol Biochem. 2015;90:111–21. https://doi.org/10.1016/j.soilbio.2015.08.005.

White AF, Bullen TD, Schulz MS, et al. Differential rates of feldspar weathering in granitic regoliths. Geochim Cosmochim Acta. 2001;65(6):847–69. https://doi.org/10.1016/S0016-7037(00)00577-9.

Wouatong A, Kitagawa R, Takeno S, et al. Morphological transformation of kaolin minerals from granite saprolite in the Western part of Cameroon. Clay Sci. 1996;10(1):67–81. https://doi.org/10.11362/jcssjclayscience1960.10.67.

Yu W, Wang R, Zhang Q, et al. Mineralogical and geochemical evolution of the Fusui bauxite deposit in Guangxi, South China: from the original Permian orebody to a quaternary Salento-type deposit. J Geochem Explor. 2014;146:75–88. https://doi.org/10.1016/j.gexplo.2014.07.020.

Zhang L, Luo Y, Yu G, et al. Estimated carbon residence times in three forest ecosystems of Eastern China: applications of probabilistic inversion. J Geophys Res Biogeosci. 2010;115(G01010):1–14. https://doi.org/10.1029/2009JG001004.

Zhang Y, Yao S, Mao J, et al. Chemical composition of organic matter in a deep soil changed with a positive priming effect due to glucose addition as investigated by 13C NMR spectroscopy. Soil Biol Biochem. 2015;85:137–44. https://doi.org/10.1016/j.soilbio.2015.03.013.

Chapter 15
Live Subsoils: Tropical Regolith and Biota Interactions

Carlos Ernesto Gonçalves Reynaud Schaefer and Fábio Soares de Oliveira

Abstract The effects of biota on the regolith occur at scales that range from the smallest organisms (bacteria) to the largest living ones (big trees). In this chapter, we examined how biological agents, especially plant roots, ants, and termites, play a key role in regolith and subsoil formation. We highlighted the processes of pedoturbation, microaggregate genesis, chemical reactions in the rhizosphere, and soil morphological organization. The biological turnover of soils through the action of termites and ants is discussed, emphasizing its importance in the long-term evolution of tropical soils and subsoils.

Keywords Ants · Bioturbation · Latosols · Regolith · Rhizosphere · Termites

15.1 Introduction

Terrestrial soils and associated subsoils/regoliths are living entities, representing highly complex systems. Since the emergence of plant life on land, at circa 400 MA B.P., all basic geochemical processes responsible for regolith formation have been, continuously and progressively, completely refashioned by the long-term effects of biota, not only by direct effects (roots, galleries, biochannels) but also by mediating biochemical processes at the surface, capable of exerting a deep influence into the subsoil.

The chemical reactions based on simplistic equations driving weathering (e.g. carbonic acid and H^+ from water hydrolysis) are usually presented in textbooks, with little regard to biotic weathering. The main reason for this negligence is the complexity of biochemical/organic chemistry processes leading to biological weathering, so that its understanding is beyond the scope of simplified regolith/weathering studies.

C. E. G. R. Schaefer (✉)
Soils Department, Federal University of Viçosa, Viçosa, MG, Brazil
e-mail: carlos.schaefer@ufv.br

F. S. de Oliveira
Geography Department, Federal University of Minas Gerais, Belo Horizonte, MG, Brazil

T. S. d. Oliveira, R. W. Bell (eds.), *Subsoil Constraints for Crop Production*,
https://doi.org/10.1007/978-3-031-00317-2_15

However, biological weathering was recognized as a key factor since the early days of pedology. Since the seminal works of Charles Darwin (1844, 1881) that clearly showed how the humus-rich topsoils on limestones, or more technically mollic epipedon (Johnson 1993), was created by worm activity, both biochemically and biomechanically, much has been published on the biomantle concept, across all latitudes, and under the action of various agents. The basic principle advocated by Darwin, and later broadened, was the simple movement of particles against the gravity law, by soil ingestion at lower depths, followed by deposition on the surface as faecal castings, representing a slow process of upward translocation of fine soil material. In consequence, any larger particles unworked by the biological agent, such as gravels and artefacts that cannot be ingested or moved, will slowly sink, eventually producing a biomantle as deep as the lower zone of faunal bioturbation, and usually separated by a stone-layer at its lower boundary.

As pointed out by Johnson (2002), Darwin just missed the opportunity of coining and defining specific terms, like bioturbation, biomechanical processes, textural differentiation, biomantle, mollic epipedon, stone line, and artefact line, which came much later. Different observations pointing to the same overall process came from biogeographers (von Ihering 1882), biologists-naturalists (Drummond 1888; Seton 1904), anthropologists (Holmes 1893), entomologists (Gounelle 1896), geologists (Keilhack 1899; Shaler 1891), geomorphologists (Passarge 1904), pedologists (Hilgard 1906), and marine sedimentologists (Davison 1891). Reading of these classics are, even today, illuminating for a renewed perspective on the biomantle, especially in the tropics.

The truth is that soil functioning is substantially affected by soil organisms in many effective ways, and from many perspectives, including as the basis for, and maintenance of, sustainable agricultural fertility (Kibblewhite et al. 2008). The soil biota is conceived of as the 'biological engine of the Earth' (Ritz et al. 2004), driving and modulating many of the key process that occur in soils and subsoils.

The effects of biota on the regolith occur at scales from 10^{-9} to 10^{2} m (Field and Little 2008), corresponding with the activities of the smallest organisms (bacteria) to the largest living ones (big trees). Hence, biological reorganization of the regolith operates at various scales, simultaneously.

In this chapter, we examine how biological agents play a key role in regolith/subsoil formation, emphasizing the combined influences of vegetation and soil faunal dwellers, especially ants and termites, in the long-term evolution of tropical soils and subsoils.

15.2 Importance of Soil Fauna and Vegetation in Regolith Formation

Since the early days of pedology (Dokuchaev 1883), soil fauna was considered to play a key role in soil genesis. Dokuchaev, the father of soil science, emphasized that the soil is a biogeocenosis, in which fauna occupies an integral part (Rode 1962).

This is mirrored in the subsequent works of Kubiena (1948) and Slager (1966). On the other hand, termite ecologists also recognized the importance of these organisms to soil formation (Pendleton and Sharasuvanas 1942; Nye 1955).

The role of biota in the genesis of regoliths was reported through its influence on several attributes, with emphasis on the origin and evolution of soil structures (Martin 1945; Harris et al. 1966; Tisdale and Oades 1979; Fregonezi et al. 2001), especially the homogeneous microgranular type (Lee and Wood 1971; Garnier-Sillam et al. 1985; Eschenbrenner 1986; Kooyman and Onck 1987; Lavelle et al. 1992; Black and Okwakol 1997; Dangerfield et al. 1998; Jungerius et al. 1999; Schaefer 2001; Reatto et al. 2009), nutrient availability and physicochemical properties (Lal 1988; Black and Okwakol 1997; Holt et al. 1998; Schaefer et al. 2004; Sarcinelli et al. 2009), changes in soil porosity and water circulation (Lavelle and Pashanasi 1989; Taylor and Brar 1991; Lee and Foster 1991; Miklós 1993; Mando and Stroosnijder 1999), organic matter accumulation, and formation of pedofeatures (Sleeman and Brewer 1972; Bullock et al. 1985; Simas et al. 2005). Zimmerman et al. (1982) calculated that approximately two third of the tropical land mass is inhabited by termites.

The most comprehensive study of termites in soil formation in Africa is that of Wielemaker (1984) in Kenya, who clearly demonstrated that such organisms are responsible for deep regoliths and Latosols, associated with B horizon structure formation. Similar results were presented by Schaefer (2001) for Brazil (Fig. 15.1), who identified the Plio-Pleistocene regoliths as related to long-term biological activity, despite the classical disregard of soil fauna in Brazilian and North America geomorphology (Ruhe 1959; Bigarella et al. 1965).

The rapid, direct transformation of subsoils into stable microgranular structured soils has been demonstrated by Schaefer (2021) in a termite nest study on a deeply weathered saprolite on gneiss near Viçosa City, Minas Gerais State, Brazil (Fig. 15.2). There, the growth of a *Cornitermes* colony was studied on exposed saprolite, after 6 years of evolution. Structural and chemical changes are summarized in Fig. 15.2 and show the fast conversion of an alteromorphic saprolite into a stable well-structured soil (Latosol), after 6 years of pedobioturbation.

The role of biota in the genesis and evolution of regolith requires a multiscale approach. Field and Little (2008) emphasize that the influence of fauna and vegetation occurs from the microscopic scale, with emphasis on the role of microorganisms in weathering and rhizosphere reactions, to the mesoscale, by biogeochemical transport promoted by plants, root growth, tree fall, and bioturbation, and in the macroscale, with the influence of vegetation phases and erosive processes, weathering, and water circulation (hydrobiogeochemical cycles). In summary, living organisms can act on weathering processes and regolith formation at different scales/processes (Fig. 15.3).

Emphasizing the microscale in a very comprehensive review, Banfield et al. (1999) demonstrated how microorganisms play a key role in weathering and mineral dissolution in the rhizosphere environment. For example, the cycling of N is basically controlled by microbes, since it is not available from silicate weathering, but is fixed from the atmosphere by symbiotic microbes with plants, or free-living

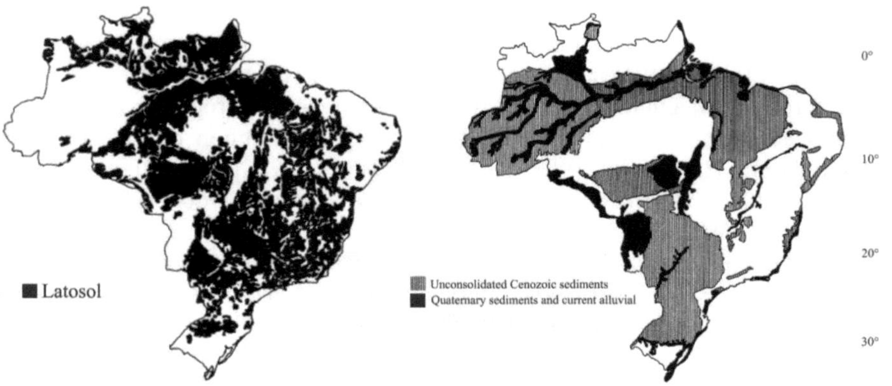

Fig. 15.1 The distribution of present-day Latosols in Brazil and the Late Cenozoic regoliths (in situ), illustrating that, except for the Upper Amazon Solimoes Formation, where Plinthosols are dominant, most areas of Latosols are underlain by this extensive Plio-Pleistocene cover related to long-term biological activity. (Schaefer 2001)

organisms. Also, mineralization and nitrification of N in the regolith are biologically mediated processes (Field and Little 2008). Soil fungi, especially mycorrhizae, are key components of the regolith and critical to plant growth. Yuan et al. (2004) showed that mycorrhizae were capable of acidifying K-interlayered vermiculite, leading to further weathering.

Hence, organic matter (live and dead) and all myriad of microorganisms are critical components of regolith weathering (Huang 2000), through the alteration of secondary and primary minerals. All scales of the three-dimensional regolith should be considered here, from molecular to microenvironmental level. However, it is in the rhizosphere microenvironment that most active biomolecules are formed, in a narrow zone of intense biological activity surrounding roots. The micro-interface surrounding roots is the zone where soil minerals, organic matter, and organisms are found, closely associated – representing the most bioactive zone in the regolith (Jones et al. 2003) (Fig. 15.4).

One key aspect of the rhizosphere that has recently received attention is mycorrhizal symbiosis, as an evolved nutrient uptake pathway for many plant species (Brundrett 2002), especially for P and K uptake (Fomina et al. 2005) and to decrease trace metal toxicity (Raapana and Field 2006). Given that mycorrhizae-infected roots have an expanded surface area, they explore a large volume of regolith, with great influence on chemical weathering.

The challenging nature of rhizosphere studies is due to the fine scale of investigation, besides the variety of biochemical interactions occurring. Novel combined techniques, such as fatty acid methyl esters (FAME) or polymerase chain reaction denaturing gradient gel electrophoresis (PRC DGGE), have been able to identify changes in microbial community structure and functioning in the rhizosphere, with promising results (Kozdrój and van Elsas 2000; Widmer et al. 2001).

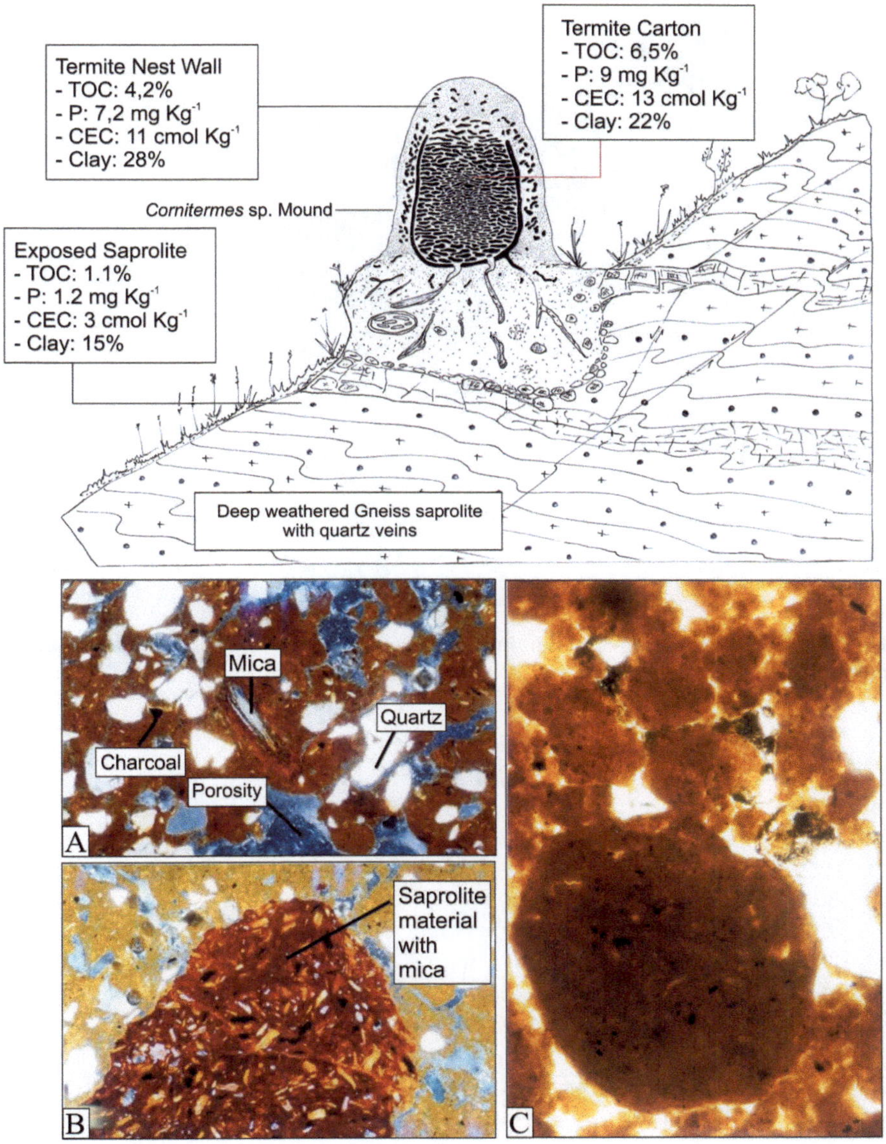

Fig. 15.2 Pedological and physical (illustration above, and optical microscope photomicrograph – OMP) transformation of an exposed saprolite into stable-structured soil promoted by termites (*Cornitermes sp.)* on a 6-years exposed road cutting near Viçosa, Minas Gerais State, Brazil. (**a**): weathered part of the external wall showing the release of micropeds following weathering; (**b**): mound-wall aggregate of saprolite nature, containing abundant mica, enclosed within a kaolinitic groundmass, and C: large termitic aggregate at the termite wall. (Adapted from Schaefer (2001))

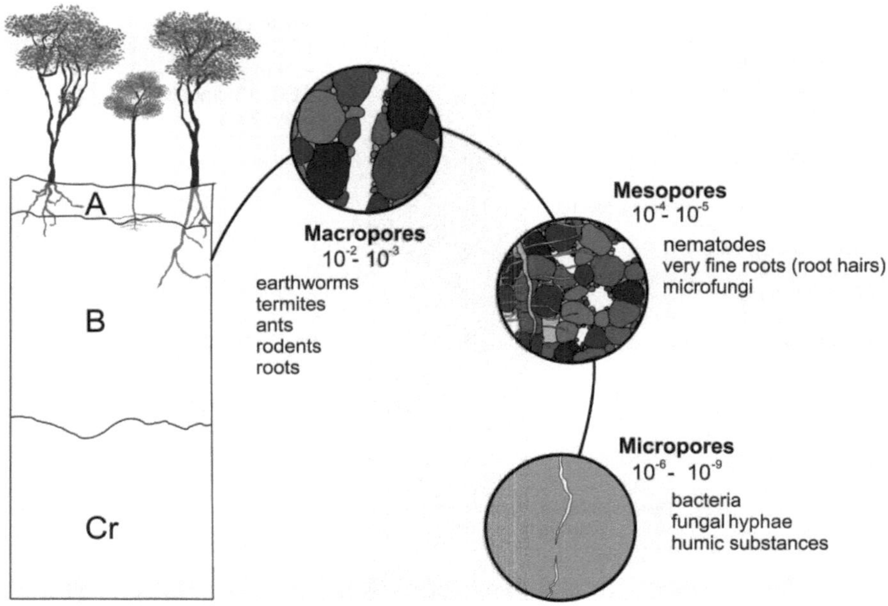

Fig. 15.3 The role for living organisms in weathering and regolith formation at different scales/processes

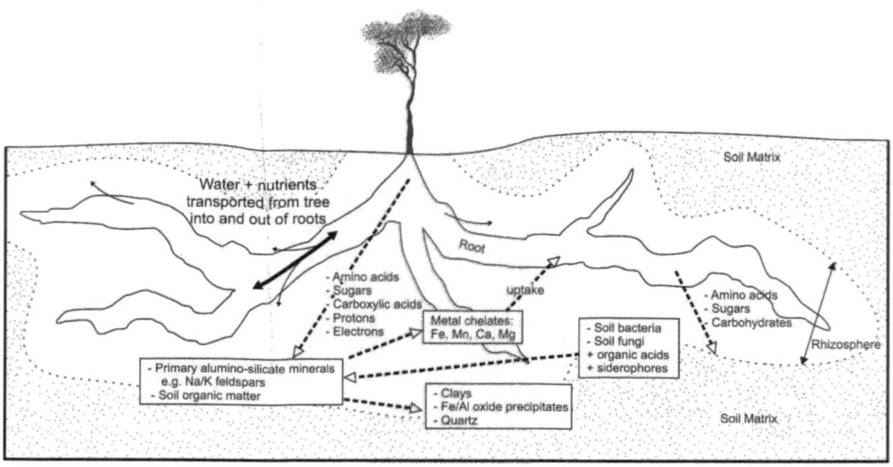

Fig. 15.4 Schematic of the rhizosphere as an important bioactive zone in the regolith. (Adapted from Field and Little 2008)

In a Latosol sequence at the Mares de Morros (Sea of Hills), Sarcinelli et al. (2009) observed significant chemical alterations on soils by termite mounds. Termite mounds showed higher pH and lower exchangeable Al than adjacent soils. A higher concentration of C, P, and N in the mounds relative to adjacent soil was associated with organic matter incorporated by termites, as faecal pellets mixed with saliva, during nest building.

In this region, termites can bring up to 7.5 m^3 ha^{-1} $year^{-1}$ or 13.0 t of soil ha^{-1} $year^{-1}$ from deeper layers to the surface during their mound-building activity, usually from the contact between B horizon and saprolites (Schaefer 2001). The mounds, once subjected to weathering processes, redistribute a relatively less weathered soil material on the surface. Sarcinelli et al. (2009) indicate that a considerable effect of termites on tropical Latosol properties exists. This is associated with their role in nutrient cycling and renewal of mineral soil brought to the surface from digging up saprolites. Dating of latosol profiles at 150–200 cm depth indicates ages between 2500 and 6000 y. BP, which are very consistent with turnover rates calculated by termite activity (Gouveia and Pessenda 2000).

Micromorphological study of the distinctive Latosol (Oxisol) microstructure, usually compared to 'coffee powder', reinforces the notion that termite activity can have an essential role on it. Thus, they should be considered as a factor in Latosol genesis. Evidence gathered from various studies corroborate termites' role as ecosystem engineers and, specifically, tropical soils engineers.

15.3 The Biological Turnover of Soils

The deeply weathered nature of tropical regoliths requires a long-term evolution under a consistent hydrological setting favourable for water infiltration to increasing depths and accounting for widespread hydrolysis of parent rocks, as well as adaptation of angiosperms to nutrient and water uptake in deep, nutrient-depleted soil.

Bioturbation is the main process leading to vertical homogenization of soils (van Bremen and Buurman 1998) and the only one that counteracts the stratification associated with soil forming factors that leads to A, E, and Bt horizons. Excavations, burrowing, and drilling, both upwards or downwards, results in a complete homogenization of Latosols, with a typical microstructure, to a given depth. The depth of Bw closely matches the depth of biological invasion by ants and termites (Schaefer 2001).

15.3.1 The Termites: Evolution, Role, Long Term Effects

Termite are very demanding with reference to climate, although they can be found in the intertropical zone from rainforest to semi-desert vegetation types, regardless of soil drainage. Some requirements are key for termite occurrence: (*i*) food availability (*ii*) high temperatures (tropical or subtropical) (*iii*) soil moisture (*iv*) soil depth for resting (*v*) clay or organic matter for building nests and (*vi*) low disturbance/low risk of inundation (Abe et al. 2000).

Soil-dwelling termites evolved since the Juro-Cretaceous age by developing unique adaptations to deal with high Al- and Fe-oxyhydroxides in soils, the insoluble products of tropical weathering, that have a high affinity for phosphorus. One such adaptation is the highly alkaline pH of the mid-gut of humus-eating termites (Bignell and Anderson 1988; Bignell and Eggleton 2000; Bignell 2006; Kappler and Brune 1999), accounting for over half of the termite species (Brune and Kuhl 1996). By digesting soil materials under high pH, they can dissolve P bound to Al and Fe oxides, which is otherwise totally unavailable under natural soil conditions. The increasing P availability had a by-product: the partial dissolution of kaolinite also ingested by termites, ending up in increased Si losses by leaching. In the long term, we suggest that soil-eating termites had a role not only on the organic matter stability of tropical soils, which is related to the recently found ability of termites to degrade lignin (Breznak and Brune 1994; Brune et al. 1995), but also on the mineral stability of tropical soils, by enhancing Si losses and decreasing kaolinite crystallinity. This is why tropical Latosols are unique.

Studies on kaolinite crystallinity in tropical soils strongly corroborate this postulate. An abrupt and enigmatic change in kaolinite crystals, which is found across the few centimetres that separate the B horizons from the underlying saprolites (Varajão et al. 2001; Schaefer 2001; Schaefer et al. 2002, 2008), suggests that some biological process in that zone affects kaolinite stability. The kaolinite of saprolite, where no soil ingestion has yet occurred, are much larger and more crystalline, compared with the overlying low-crystalline Bw kaolinites, usually separated by a stone line.

Brune and Kuhl (1996) suggest that termites developed high pH in their gut to enable the digestion of soil bacteria and make polyphenolic compounds that were soluble and unable to bind peptide nitrogen compounds, but P uptake would be even more important in nutritional terms. Sodium/potassium hydroxides are the best extracting media of phosphate from soil. Passing the alkaline mid-gut, kaolinite, and other silicates would be severely attacked, leaving a colloidal silica fraction mixed with partially dissolved kaolinite to be acidified in the following digesting tract, the acid hindgut. By doing so, soil ingested will be eventually less crystalline, poorer in P, and with more humified organic compounds, before it is excreted.

Termite can account for 2% of the world's carbon dioxide emission and 4% of the atmospheric methane (Sanderson 1996), representing an herbivory that equals that of vertebrates (Wood and Sands 1978). A single termite individual eats 2 kg of soil year^{-1}, while a colony can consume >38,000 kg (Apicotermes). Their density varies, but under the Atlantic Forest, 300–600 nests ha^{-1} are common, whereas under Cerrado (savanna), 900–1200 nests ha^{-1}, densities that illustrate how large their influence is upon the soils.

Since termites first evolved from primitive cockroach-like Triassic species (Emerson 1955), the relation between Latosols, laterites, and termites is broadly supported by palaeontology, since Latosols and Laterites became widespread by the Juro-Cretaceous, when most present-day termites were already in existence, and rapidly spreading throughout the tropics.

Tree roots remove silicon from the B horizon already transformed by termite ingestion and add it to the surface, keeping the kaolinite equilibrium, whereas gibbsite (bauxite) is formed in deeper subsoil, not reached by roots and termite galleries (Oliveira et al. 2014).

Silicon released into soil solution would have different fates under contrasting climates: in the semiarid zones, like northern Australia, free Si may have produced amorphous widespread silcretes (Si-duricrust), by cementing Si into a hard matrix. Conversely, with increasing Si losses in the humid tropics, trees would concentrate Si by cycling at the surface so as to promote kaolinite neoformation in the B horizon, leaving the Si-depleted saprolite richer in bauxite. This would explain why bauxites are overlain by kaolinitic soils, as described by Lucas et al. (1993) in Amazonia, Mateus et al. (2017) in Minas Gerais, and Oliveira et al. (2013) in Central Brazil.

The formation of bauxite-rich Latosols is consistent with Schaefer (2001), who has linked termites to the formation of Latosols in Brazil. It would also explain the large deposits of bauxite (aluminium hydroxide) formed from different parent materials. Chronologically, the long-term evolution of the weathering can be illustrated below. In the incipient phase, there are increasing rates of regolith formation, mediated by growing biomass, greater leaching, and termite activity. In the peak phase, leaching, biomass, and regolith formation reach a maximum and then begin to decrease, progressively. With time, in the last phase, when the termite role remains prominent, the regolith attains a steady state of lower evolution with stable and residual products of alteration and very low nutrient reserves, a constant rate of aerosol inputs and very low leaching of deep pre-weathered mantles (Fig. 15.5).

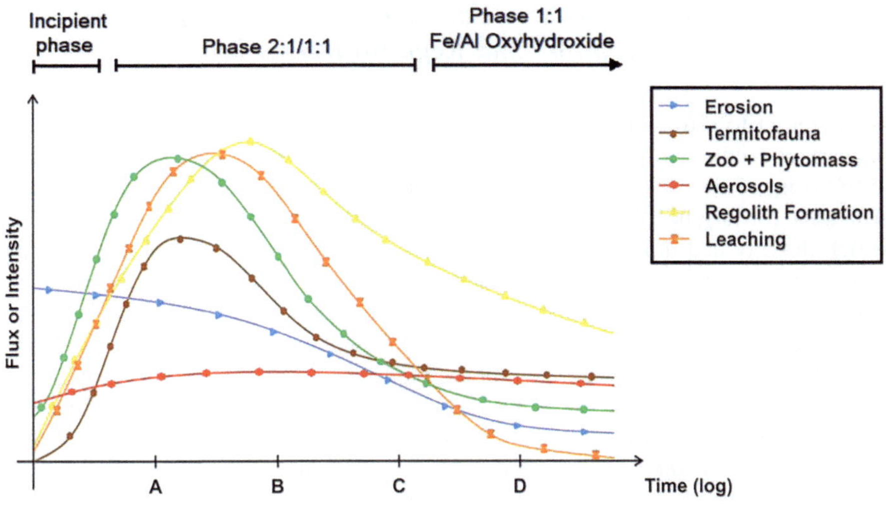

Fig. 15.5 Schematic diagram illustrating the process of weathering coupled with different ecological dynamics, partially inspired in Kronberg and Nesbitt (1981). The time periods represented by the letters A, B, C, and D on the X axis represent the timescale of, respectively, 10–10^2, 10^3–10^4, 10^5–10^6, and >10^6 years

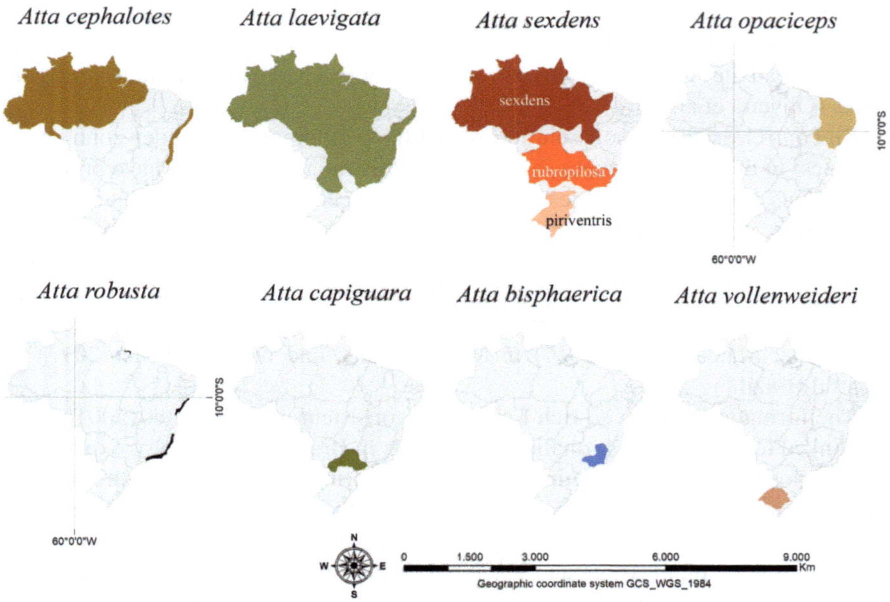

Fig. 15.6 Areas of occurrence of the main Atta ants in Brazil, showing the large spatial distribution of some species, such as *Atta cephalotes*, *Atta laevigata*, and *Atta sexdens*. Some species, like *A. cephalotes*, are exclusively found on rainforest (Amazonia and southern Bahia), whereas *Atta opaciceps* is closely associated with semi-arid areas of NE Brazil (Schaefer et al., 2021)

15.3.2 Ants and Subsoils: A Complementary Effect for Deep Latosol Formation and Regolith Deepening

Ants belong to different trophic levels, in which leafcutter and harvester ants are primary consumers of great importance in the neotropics and elsewhere (Folgarait 1998) (Fig. 15.6) and can be classified as ecosystem engineers (Jones et al. 1994). Atta ants, for instance, have a strong role in soil structure formation and processes, which directly and indirectly affect the flows of energy and material in terrestrial ecosystems, creating prominent effects on the habitats of other species (Guerra et al. 2007, Schaefer et al. 2021). Leafcutter ants can reduce 17% of the annual leaf production of a tropical forest (Cherret 1989), and one Atta nest can consume 1–2 Mg of fresh leaf material per year (Coutinho 1982).

Ant mounds have a myriad of myrmecophile inhabitants, all specialized (living on refuse piles, brood, abandoned chambers). These mounds are commonly preferred sites for plant establishment and may also allow the development of a rhizosphere-related fauna. Not only are the well-known basidiomycete fungi cultivated by leafcutter ants (Weber 1972), but other microorganisms can be also associated with ant nests, notably mycorrhizae (Friese and Allen 1993), helping P uptake in nutrient-depleted soils. Also, other non-mycorrhizal microorganisms may also favour plant productivity in richer anthill soils (McGinley et al. 1994). In general, soil and subsoil processes are strongly influenced by ant turnover, not only increasing the concentrations of nutrients in a deeply weathered and unfertile soil (Petal et al. 1977) but also collecting subsoils with finer texture, usually clay, and depositing it on the soil surface.

Soil modification by ant nest building involves physical and chemical changes in key soil properties. Ants increase soil and subsoil drainage and aeration through underground galleries. At the surface, they promote organic matter stability by organic matter-clay interactions and incorporate nutrients in subsoils by food storage (Brian 1978). Bioturbation effects occur both in topsoil and subsoil, and the amount of subsoil translocated upwards reaches gigantic values, unmatched to any social insect. A single colony of *Atta sexdens* covered an area of 100 m^2, occupied a volume of 23 m^3 in the subsoil, with a total of 40 tonnes of translocated soil (Autori 1947). Soil brought from deep subsoils are also packed into pre-existing chambers, forming smooth-walled chambers and galleries (Eschenbrenner 1994), lined with fine particles smeared with macerated plant fragments or faecal material. Also, these cavities are infilled with a porous mixture of soil aggregates (750–2000 μm) (Fig. 15.7), sometimes mixed with cocoon sacks (Humphreys 1994).

Different ant species create contrasting mounds, with pedological impacts varying from limited, small-sized to large mounds with elliptical shapes, persisting for a long time, and made of reworked soil material, with little cementation, although exceptions occur (cemented, erosion-resistant *Camponotus* nests). The subsoil movement to the surface through ant activity can be very substantial, with the highest rate recorded of 10 t ha^{-1} y^{-1} (Paton et al. 1995), and even semi-desert ants can lead to soil turnover rates as high as 420 kg ha^{-1} y^{-1} in Australia (Briese 1982) and

Fig. 15.7 Progressive infilling (a to d) of a chamber with microaggregates (100–300 micrometers) of Latosol within a gneiss saprolite from Viçosa, MG (Schaefer et al. 2021)

2100 kg of soil ha^{-1} y^{-1} in Argentinian pastures (Folgarait 1998). Comparing global rates of faunal pedoturbation, ants scored second (5 kg m^{-2} y^{-1}) to earthworms (15 kg m^{-2} y^{-1}), but ants have a wider geographical distribution than the latter (Paton et al. 1995). The marked impact of ants in the tropics are highlighted by leafcutters, where they are the most important agents of soil modification (Alvarado et al. 1981; Cherret 1989), and a single nest of *Atta capiguara* with 1.5 m width and 5 m height was found to contain 500 kg of organic matter in Brazil (Amante 1964).

Ants also have an effect on soil nutrient immobilization and humification (Anderson and Flanagan 1989; Lavelle et al. 1992): the refuse piles are habitats with increased mineralization rates, accentuated by the pre-decomposition of the refuse by fungal action (Lugo et al. 1973). According to Folgarait (1998), physical, chemical, and vegetational effects can be recognized from ant activity, as follows:

Physical changes are mainly related to soil organic matter burial, favouring the water-holding capacity, and fine particles transferred from subsoil to the surface (Petal 1978). Ant bioturbation comprises the formation of a network of channels, chambers, and new soil aggregates (Humphreys 1994), increasing soil porosity and aeration (Denning et al. 1977) and reducing bulk density (Baxter and Hole 1967; Rogers 1972). The transformation and inversion of soil layers lead to new A horizons being developed on new materials from saprolite or deep B horizons (Alvarado et al. 1981). Lobry de Bruyn and Conacher (1990) have shown that ant bioturbation activity led to homogenization of the soil texture profile. The combination of ants and termites are key to Latosols development in Brazil (Schaefer 2001).

In terms of chemical changes, depending on ant species or soil type (McGinley et al. 1994), increasing organic matter, P, N, and pH is observed in ant mounds compared with adjacent soils (Salem and Hole 1968; Czerwinski et al. 1971; Petal 1978; Mandel and Sorenson 1982; Lugo et al. 1973), with greater effects in poorer soils (Czerwinski et al. 1971; Petal 1992). Decomposition processes carried out by fungi and bacteria are very active in ant mounds in comparison to adjacent soils. The humification process is retarded due to the decreasing Actinomycetes in ant nests (Czerwinski et al. 1971; Jakubczyk et al. 1972). Nests of *Atta colombica* in Panama showed 38 times increase in the fluxes of 13 chemical elements in comparison to surrounding forest soils (Haines 1978). In infertile environments with stress-tolerant plants, low organic matter, and complex trophic webs, ants speed up the return of nutrients retained in plant and animal tissues (Petal et al. 1977).

With reference to vegetation changes, ant mounds have a different plant composition in comparison to adjacent areas (Horvitz and Schemske 1986). Ants can disperse plant propagules (Wilson 1992) and facilitate ruderal, invasive plant species (Folgarait 1996), changing plant succession (Jonkman 1978). The enriched soil of anthills is key for vegetation development in deep Latosols in south America (Coutinho 1984; Sarmiento 1984; Medina and Silva 1990). All the above-mentioned aspects of ants' role in tropical soils and subsoils are illustrated in Fig. 15.8.

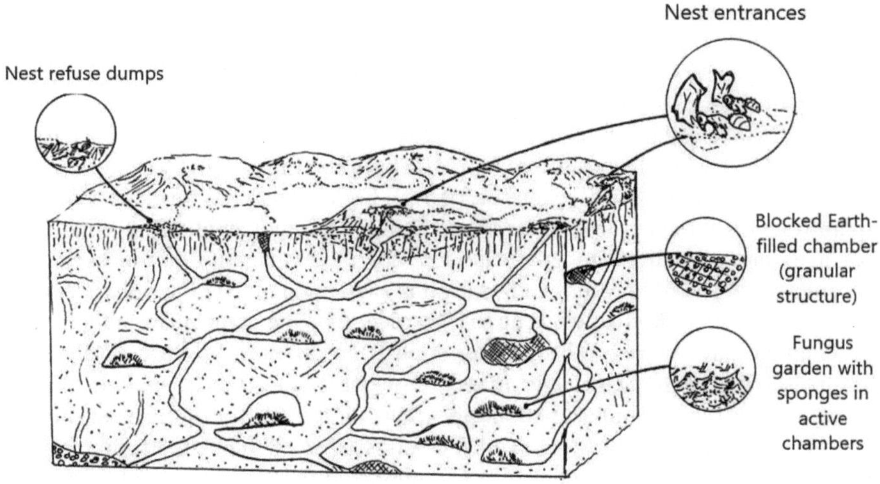

Fig. 15.8 A large nest of *Atta sexdens rubropilosa* at Coimbra (MG, Brazil), exposed on a fresh road cutting. The stone line at the bottom left marks the upper limit of pedoturbation. Colonies of *Atta sp.* have up 18 combs per hectare, resulting in 52 kg of plant harvesting daily, equivalent to three calves at the same area. Abandoned chambers are infilled with well-structure granular materials. The refuse piles at the surface represent hotspots of nutrients for plant growth. (Santos et al. 2019; Sousa-Souto et al. 2013)

15.4 Age of Soil and Biological Turnover

The age of the regolith is a subject of recent interest, since many economic ores are found on extremely old land surfaces on deep regolith. In Brazil, for example, Vasconcelos et al. (1994) dated one of the deepest Fe-rich regoliths worldwide (>400 m), from Serra dos Carajás (Amazonia), reaching ages of >120 MA BP. This is consistent with the age of Gondwana break-up that led to the Atlantic Ocean opening and resulting humid tropical climates since then. Other deep saprolites in Minas Gerais gave similar ages, dating back to the Cretaceous.

Deep regoliths, saprolites and soils together, are widespread throughout Brazil. Their ages are quite different, since most overlying Latosols are younger than 10,000 years, whereas the underlying saprolites have ages in the range of 10 MA up to 70 MA. In the Serra da Moeda (Iron Quadrangle) ironstone on itabirite, in Minas Gerais, saprolites reach more than 250 metres of depth, and more than 70 MA of age, dating back to the Late Cretaceous. The estimated erosion rate based on Be^{10} dating is one the lowest worldwide, so that differential erosion and etchplanation are the two basic processes of landform and landscape sculpturing. According to Spier et al. (2006), the ages of these weathered mantles range from early Cenozoic (Palaeocene 61 MA) to Miocene (14.2 MA). Younger saprolites can be found at the lowest surfaces, with ages up to 4.8 MA (Monteiro et al. 2014).

The radiocarbon dating of charcoal in Latosols from São Paulo state (Gouveia and Pessenda 2000) confirms the main role of soil fauna in the upward movement of saprolite materials from the deeper layers to the surface. The dating and the distribution of charcoal in the studied soils suggests the occurrence of fires between 3000 and 6000 y BP in São Paulo (Botucatu) and between 4000 and 9000 y BP in São Paulo (Jaguariúna), corresponding to dry periods (Fig. 15.9).

The carbon dating of soils can be carried out both in charcoal particles and in the humin fraction extracted from soils. According to Pessenda et al. (1998), the charcoal ages are generally similar but sometimes older than the humin ages. The Latosol ages obtained in Minas Gerais (Salitre) and São Paulo (Jaguariúna) showed that below 150 cm, the charcoal is up to 27% older than the humin. The authors state that charcoal have probably been transported from different layers by soil fauna. Results show that some organic matter movement may occur, making carbon cycling in soils more complex, with implications for best choice of the most suitable material for ^{14}C dating of soils. However, in the absence of charcoal, the humin fraction is probably the best material for ^{14}C dating, and these dates should be considered as minimum ages of the soil organic matter.

In a recent study of Latosols (red-yellow, red) and Argisols, in all parts of a strongly pedobioturbated regolith, reaching down to 4–5 metres of biological activity, Freitas et al. (2021) estimated pedobiological turnover in 100–200 cm of

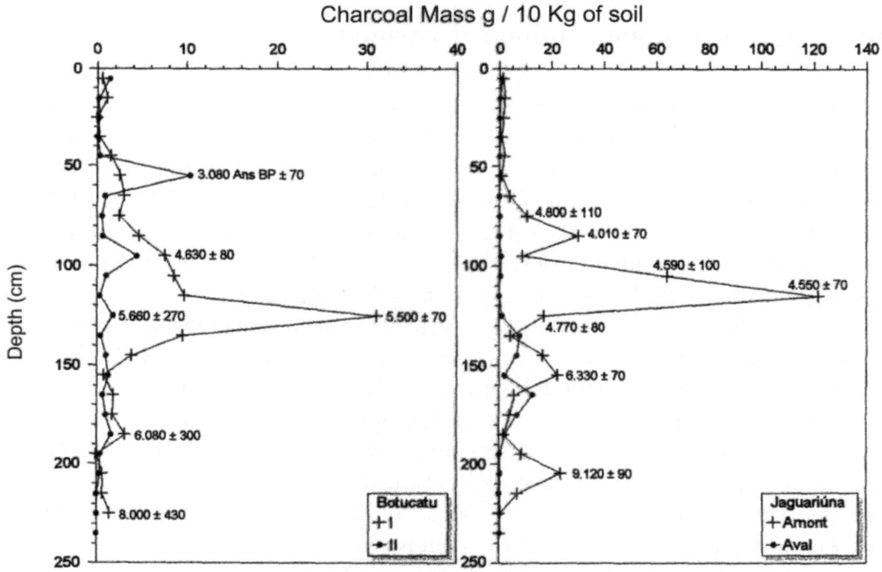

Fig. 15.9 Charcoal distribution with depth and 14C dating of Latosols from São Paulo state (Gouveia and Pessenda 2000), showing that all these Latosols were formed during the Holocene (in the last 11,000 y)

structured soils was less than 5000 years, closely matching the ^{14}C ages of charcoal or humic acids of these soils.

15.4.1 Dynamic Landscape Denudation

The biomechanical processes that Charles Darwin unveiled, demonstrating how soil biota mechanically generate, blur, or destroy regolith strata and soil layers (Darwin 1881), lost visibility in our models of landscape evolution for many reasons. For Johnson (2002), it was only a century after Darwin that a genetic language backed by a supporting theory appeared, showing the importance of such processes. It is notable that most influential earlier Earth science frameworks do not emphasize biomechanical processes (e.g. V.Dokuchaev-H. Jenny soil formation factors paradigm; W.M. Davis' geographical cycle; W. Penck-L. King-R. Ruhe backwasting-pedimentation concept), as well as other traditional approaches to geomorphology and pedology.

However, it became clear that any unified synthesis of pedogeomorphogenic processes must incorporate the concepts of biomantle, bioturbation, and pedoturbation (Johnson 2002) at different temporal and spatial scales (Schaefer 2001; Freitas et al. 2021). The dynamic denudation model provides such a synthesis, placing bioturbation on parity levels with physical, chemical, and hydrological processes. This general model has brought explanatory and predictive value in archaeology, geomorphology, and pedology (Johnson 1993, 2002).

Soon after Darwin's seminal work on bioturbation, many observations from high-ranked scientists in many different fields (e.g. von Ihering 1882; Drummond 1888; Branner 1900), including geomorphology (Passarge 1904) and pedology (Hilgard 1906), gave support for Darwin's postulates.

Despite these advances, little of this knowledge helped to change our paradigms of landscape evolution, until very recently. Hence, the main questions posed are as follows: Why until recently have biological processes been overlooked, or omitted from our introductory texts and undergraduate and graduate training, and why were biomechanical processes not a part of our conceptual traditions in geomorphology and pedology?

In our view, the answers may basically lie in an imposed bias of key authorities, negligence of relevant literature, and, not least, as Johnson (2002) tells, the lack of genetic terms and conceptual structures to showcase the processes. The idiosyncratic keepers of these disciplines have been reluctant, and even intolerant, to considering or adopting biological processes as key in landscape evolution.

In Brazil, that fact has been clearly demonstrated by the emphatic rejection by one of our eminent geomorphologists, Prof. Aziz Ab'Saber, when replying to the theory of bioturbation, ingeniously advocated by the notable French geomorphologists Andre Cailleux and Jean Tricart in the late 1950s (Cailleux and Tricart 1957). Immersed in the backwasting-pedimentation theory advocated by Penck (1953), King (1953), and Ruhe (1959), Ab'Saber could not incorporate any challenging

Fig. 15.10 Stone line formed by the redistribution of a well-structured Latosol biomantle with milky quartz pebbles and cobbles derived from the quartz veins cutting accross the green schist saprolite, at Minas Novas (MG State)

biophysicomechanical view, in spite of it being based on sound field evidence, that could undermine the conventional physicochemical views of Earth surface processes and landscape evolution. A reading of his review and criticism (Ab'Saber 1965) highlights plainly the denial of a possible interdisciplinary biomechanical approach to tropical geomorphology, strengthening the power of tradition theory of landscape evolution. This biased view has had a long tradition in Brazilian geomorphology, to this day.

The key aspect of this old controversy is the explanation for the origin of the stone line, buried beneath a friable, well-structured Latosol (Fig. 15.10). In this respect, Cailleux and Tricart (1957) postulated that termites and other soil animals have moved fine particles upwards, leaving the coarse fragments as a lag material. Due to combined bioturbation and gravity, artefacts, quartz gravels, pebbles, and large clasts – originally deposited on the surface during long phases of active erosion under dry climates – gradually sink to the top of the stone line after the onset of the wetter climate. To support the observation, they noticed that all soil particles in the overlying soil above the stone line have grain sizes that match, exactly, the maximum size of termite-transported material. But they also considered that colluvial transport could be responsible for the redistribution of the pedoturbated soil.

In rejecting the field observation, Ab'Saber (1965) pointed out that 'termites would not be capable of moving *per ascendum*, regularly, continuously, an entire layer' of well-structured soil, so creating the stone lines. Nevertheless, following his arguments disputing the biomechanical theory, we noticed his recognition of the importance of settling down this controversy in his illuminating words:

> *O estudo das linhas de cascalhos inumados das vertentes do Brasil oriental não é um mero capricho de pesquisadores para um problema marginal. Longe disso, é uma investigação*

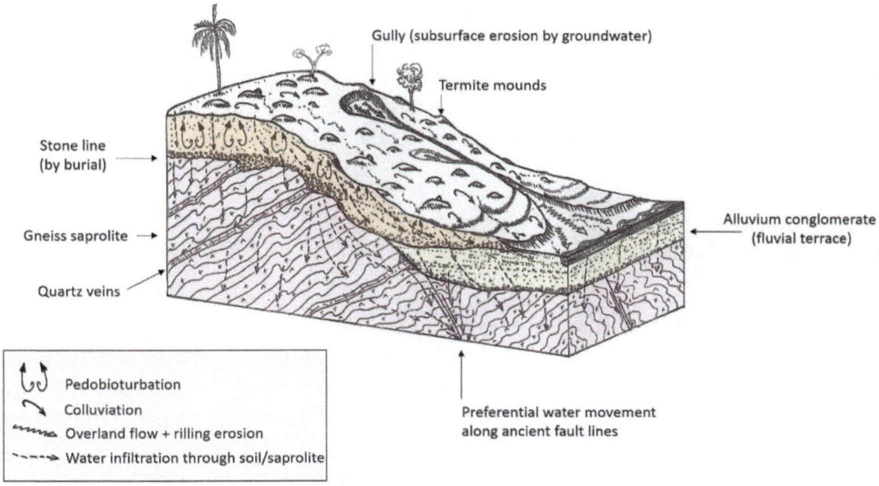

Fig. 15.11 Schematic illustration of a tropical slope in the Atlantic Forest biome on gneiss terrain, interpreted by the dynamic denudation model

> *intercientífica que pode conduzir à explicação definitiva dos principais aspectos morfogenéticos das paisagens tropicais úmidas do Brasil, e caracterizar a cronologia de eventos paleoclimáticos finais.*[1]

We agree. To address this need for integration, Johnson proposed a dynamic denudation model, here adopted. The dynamic denudation framework integrates and emphasizes theories of geomorphology and pedology and hydrology to propose a universal definition of soil. In order to offer an example of a typical tropical slope for discussion, we present a dynamic denudation interpretation of one showcase (Fig. 15.11).

Chemical denudation and deep weathering following the alteration of biotite and feldspar are enhanced by the bedding and foliation coupled with preferential capture of percolating water by fault lines, diaclases, and faunal galleries. The *per ascendum* turnover of pedobioturbated soil by termites, ants, and other soil-dwelling animals creates a thick residual mantle of well-structured soil, overlying the pedoplasmation zone and a stone line, derived from intersecting quartz veins that are the source of gravels, resistant to weathering and erosion. Both colluvial (soil creep) and sheet erosional processes occur, and colluvial foot slopes are widespread. Overland flow accounts for sheet erosion and riling, whereas gullies depend on subsurface erosion by groundwater and soil collapse. Mass wasting sediments are retained in the valleys downslope and transported by streams. The streams cut and

[1] Researchers do not study the buried stonelines of the slopes of eastern Brazil as a whimsical concession to a marginal problem. On the contrary, it is an inter-scientific investigation that can lead to the definitive explanation of the main morphogenetic aspects of the humid tropical landscapes of Brazil and to characterize the chronology of final palaeoclimatic events.

deepen their valleys in response to the depth of weathering and the stream's capacity for transporting the resulting load of eroded particles. The whole slope is strongly subjected to a biogeomorphological control that drives the increasing stability of the tropical soil mantles, triggering further water intake by a huge network of minute biopores and large biochannels (see Schaefer et al. 2021).

15.5 Final Remarks

The role of biota in Brazilian tropical regoliths is remarkable and points to a long-term, million-year-old process of coupled evolution of plants, soil fauna, microorganisms, weathering, and denudation. This combination led to the formation of a pre-weathered mantle of great depths, with ages that date back to Late Cretaceous (70 MA), or older.

The regolith may be envisaged as the balance between resistance to change due to physicochemical weathering of the substrates, the supply of nutrients to biomass, the protection against erosion and denudation, and tectonic and structural stability of the landscape.

In this sense, Brazilian subsoils are among the most deeply altered in the world and leached to the point of extremely low current solubilization rates. The upper part of the tropical regoliths is predominantly associated with a latosolization syndrome, whose age is basically Late Quaternary, that acted on pre-weathered materials that date back to Late Cretaceous. This process counterbalances the effects of extreme leaching of subsoils, with a new microaggregated order of structural organization at the surface, that allowed the adaptation of plants and animals to thrive, despite the low supply of nutrients, by exploring a large volume of porous soil.

Once formed, Latosols favour the soil fauna dwellers to move particles from within the subsoils to the surface, promoting the surface enrichment with clay and minor amounts of nutrients. Based on closed nutrient cycles and possible materials additions from aerosols, the tropical landscape remains favourable for high primary biomass productions, despite the very low status of chemical fertility.

References

Ab'Saber. Revisão dos conhecimentos sobre o horizonte subsuperficial de cascalhos inumados do Brasil Oriental. Boletim Geográfico. 1965;184:68–84.

Abe T, Bignell DE, Higashi M. Termites: evolution, sociality, symbioses. Ecology: Kluwer Academic Publishers, Dordrecht; 2000.

Alvarado L, Izquierdo JA, Enecoiz MA. Eficacia del tratamiento de semillas de maíz con carbofuran sobre larvas de *Diloboderus abderus* (Sturm). In: Actas Congresso Nacional de Maíz, 2. Pergamino; 1981. p. 168–77.

Amante E. Nota prévia sobre a estrutura do ninho de uma nova formiga saúva (*Atta* sp) (Hymenop. Formicidae). Biológico. 1964;30:96–7.

Anderson JM, Flanagan PW. Biological processes regulating organic matter dynamics in tropical soils. In: Coleman DC, Oades JM, Uehara G, editors. Dynamics of soil organic matter in tropical ecosystems. Honolulu: University of Hawaii Press; 1989. p. 97–123.
Autori M. Combate a formiga saúva. Biológico. 1947;13:196–9.
Banfield JF, Barker WW, Welch AS, Taunton A. Biological impact on mineral dissolution: application of the lichen model to understanding mineral weathering in the rhizosphere. Proc Nat Academy Sci Unit Stat Am. 1999;96:3404–11. https://doi.org/10.1073/pnas.96.7.3404.
Baxter FP, Hole FD. Ant (*Formica cinerea*) pedoturbation in a prairie soil. Soil Sci Soc Am Proc. 1967;31:425–8. https://doi.org/10.2136/sssaj1967.03615995003100030036x.
Bigarella JJ, Mousinho MR, Silva JX. Considerações a respeito da evolução das vertentes. Bol Paran de Geog. 1965;16(17):85–116.
Bignell DE. Termites as soil engineers and soil processors. In: König H, Varma A, editors. Intestinal microorganisms of termites and other invertebrates. Berlim: Springer; 2006. p. 183–220. https://doi.org/10.1007/3-540-28185-1_8.
Bignell DE, Anderson JH. Determination of pH and oxygen status in the guts of lower and higher termites. J Ins Phys. 1988;26:183–8. https://doi.org/10.1016/0022-1910(80)90079-7.
Bignell DE, Eggleton P. Termites in ecosystems. In: Abe T, Bignell DE, Higashi M, editors. Termites: evolution, sociality, symbioses, ecology. Dordrecht: Kluwer Academic Publishers; 2000. p. 363–87. https://doi.org/10.1007/978-94-017-3223-9_17.
Black HIJ, Okwakol MJN. Agricultural intensification, soil biodiversity and agroecosystem function in the tropics: the role of termites. App Soil Eco. 1997;6:37–53. https://doi.org/10.1016/S0929-1393(96)00149-7.
Branner JC. Ants as geologic agents in the tropics. J Geol. 1900;8:151–3.
Breznak JA, Brune A. Role of microorganisms in the digestion of lignocellulose by termites. Annu Rev Entomol. 1994;39:453–87. https://doi.org/10.1146/annurev.en.39.010194.002321.
Brian MV. Production ecology of ants and termites. IBP 13. Cambridge, UK: Cambridge University Press; 1978.
Briese DT. The effect of ants on the soil of a semi-arid saltbush habitat. Insects Sociaux. 1982;29:375–86. https://doi.org/10.1007/BF02228765.
Brundrett MC. Coevolution of roots and mycorrhizas of land plants. New Phytol. 2002;154:275–304. https://doi.org/10.1046/j.1469-8137.2002.00397.x.
Brune A, Kuhl M. pH profiles of the extremely alkaline hind guts of soil - feeding termites (isoptera: Termitidae) determined by micro pH electrodes. J Ins Phys. 1996;42:1121–7. https://doi.org/10.1016/S0022-1910(96)00036-4.
Brune A, Miambi E, Breznak JA. Roles of oxygen and intestinal microflora in the metabolism of lignin-derived phenylpropanoids and other monoaromatic compounds by termites. Ap Env Micro. 1995;61:2688–95. https://doi.org/10.1128/aem.61.7.2688-2695.1995.
Bullock P, Fedoroff N, Jongerius A, Stoops G, Tursina T. Handbook for soil thin section description. Wolverhampton: Waine Research Publications; 1985.
Cailleux A, Tricart J. Zones Phytogeographiques et Morphoclimatiques du Quaternaire, au Brésil. Paris: Centre Researche Societé de Biogeographie; 1957. p. 7–13.
Cherret JM. Ecological concepts. Blackwell Scientific Publications; 1989.
Coutinho LM. Aspectos ecológicos da saúva no cerrado: os murundus de terra, as características psamofiticas das espécies de sua vegetação e a sua invasão pelo Capim Gordura. Rev Bras de Biol. 1982;42(1):147–53.
Coutinho LM. Aspectos ecologicos da saúva no cerrado: a saúva, as queimadas e sua possivel relacão na ciclagem de nutrientes minerais. Bol Zool Univ S Paulo. 1984;8:1–9.
Czerwinski Z, Jakubczyk H, Petal J. Influence of ant hills on the meadow soils. Pedobiologia. 1971;1:277–85.
Dangerfield JM, McCarthy TS, Ellery WN. The mound-building termite *Macrotermes michaelseni* as an ecosystem engineer. J Trop Ecol. 1998;14:507–20. https://doi.org/10.1017/S0266467498000364.
Darwin C. On the origin of mould. Gardener's Chronicle Agricult Gazette, April 6. 1844:218.

Darwin C. The formation of vegetable mould through the action of worms. London; J Murray. Facsimilies republished in 1982 and 1985. Chicago: University of Chicago Press; 1881.
Davison C. On the amount of sand brought up by lobworms to the surface. Geological Magazine (NS). 1891;8:489–93. https://doi.org/10.1017/S0016756800187497.
Denning JL, Hole FD, Bousma J. Effects of Formica cinerea on a wetland soil on west blue mound, Wisconsin. In: Dewitt CB, Soloway E, editors. Wetlands ecology: values and impacts. Madison: University Wisconsin; 1977.
Dokuchaev VV. The Russian Chernozem. St. Petersburg: A report to the Free Economics Society; 1883.
Drummond H. On the termite as the tropical analogue of the earthworm. Proc Royal Soc Edinburgh. 1888;13:137–46. https://doi.org/10.1017/S0370164600001905.
Emerson AE. Geographical origin and dispersions of termite genera. Fieldiana: Zool. 1955;37:465–521. https://doi.org/10.5962/bhl.title.2783.
Eschenbrenner V. Contribution des termites a la microagregation des sols tropicaux. Cah Orstom, Ser Pedol. 1986;22:397–408.
Eschenbrenner V. The influence of fungus-cultivating ants (hymenoptera, Forrnicidae, Attini) on the morphology of andosols in Martinique. In: Ringrose-Voase A, Humphreys GS, editors. Soil micromorphology: studies in management and genesis. Developments in soil science. Amsterdan: Elsevier; 1994. p. 405–10.
Field J, Little D. Regolith and biota. In: Scott KM, Pain CF, editors. Regolith geoscience. Melbourne: CSIRO Publishing; 2008. p. 175–217.
Folgarait PJ. Latitudinal variation in myrmecophytic Cecropia. Association for Tropical Biology and Ecological Society of America combined meeting. 10–14 august. Rhode Island: Providence; 1996. p. 143.
Folgarait PJ. Ant biodiversity and its relationship to ecosystem functioning: a review. Biodivers Conserv. 1998;7:1221–44. https://doi.org/10.1023/A:1008891901953.
Fomina M, Hillier S, Charnock JM, Melville K, Alexander IJ, Gadd GM. Role of oxalic acid overexcretion in transformations of toxic metal minerals by Beauveria Caledonica. App Env Microb. 2005;71:371–81. https://doi.org/10.1128/AEM.71.1.371-381.2005.
Fregonezi GAF, Brossard M, Guimarães MF, Medina CC. Nodificações morfológicas e físicas de um Latossolo argiloso sob pastagens. Rev Bras de Ciên do Solo. 2001;25(4):1017–27. https://doi.org/10.1590/S0100-06832001000400024.
Freitas DF, Ker J, Silva Filho LA, Pereira TTC, Souza OGFF, Schaefer CEGR. Pedogeomorphology and paleoenvironmental implications of large termite mounds at the Brazilian semiarid landscape. Geomorphology. 2021;(34):107762. https://doi.org/10.1016/j.geomorph.2021.107762.
Friese CF, Allen MF. The interaction of harvester ants and VA mycorrhizal fungi in a patchy environment: the effects of mound structure on fungal dispersion and establishment. Funct Ecol. 1993;7:13–20. https://doi.org/10.2307/2389862.
Garnier-Sillam E, Villemin G, Toutain F, Renoux J. Formation de micro-agrégats organo-minéraux dans les féces de termites. Sc de la vie. 1985;301:213–8.
Gounelle E. Transport de terres effectue´ par des fourmis au Bre´ sil [Transport of soil by ants in Brazil]. Bulletin de la Societe´ Entomologique de France. 1896;14:332–3.
Gouveia SEM, Pessenda LCR. Datation par le C-14 de charbons inclus dans le sol pour l'étude du rôle de la remontée biologique de matière et du colluvionnement dans la formation de latosols de l'état de São Paulo, Brésil. Comptes Rendus de L'académie Des Sciences Paris. 2000;330:133–8. https://doi.org/10.1016/S1251-8050(00)00114-2.
Guerra MBB, Schaefer CEGR, Sousa-Souto L. Chemical characteristics of nest refuse of *Atta sexdens rubropilosa* (hymenoptera: Formicidae) reared with different substrates. Rev Bras Ciênc Solo. 2007;31(5):1185–9. https://doi.org/10.1590/S0100-06832007000500033.
Haines BL. Element and energy flows through colonies of the leaf-cutting ant, Atta colombica, in Panama. Biotropica. 1978;10:270–7. https://doi.org/10.2307/2387679.
Harris RF, Chesters G, Allen ON. Dynamics of soil aggregation. Adv Agron. 1966;18:107–69. https://doi.org/10.1016/S0065-2113(08)60649-5.
Hilgard EW. Soils. London: Macmillan; 1906.
Holmes WH. Vestiges of early man in Minnesota. American Geologist. 1893;1:219–40.

Holt JA, Coventry RJ, Sinclair DF. Some aspects of the biology and pedological significance of mound-building termites in a red and yellow earth landscape near charters towers, North Queensland. Aust JSoil Res. 1998;18:97–109. https://doi.org/10.1071/SR9800097.
Horvitz CC, Schemske DW. Seed dispersal of a neotropical myrmecochore: variation in removal rates and dispersal distances. Biotropica. 1986;18(4):319–23. https://doi.org/10.2307/2388575.
Huang PM. Abiotic catalysis. In: Sumner MR, editor. Handbook of soil science. Boca Raton, Florida: CRC Press; 2000.
Humphreys GS. Bioturbation, biofabrics and the biomantle: an example from the Sydney Basin. In: Ringrose-Voace AJ, Humphreys GS, editors. Soil micromorphology: studies in management and genesis, vol. 22. Amsterdam: Developments in Soil Science, Elsevier; 1994. p. 421–36. https://doi.org/10.1016/S0166-2481(08)70431-8.
Jakubczyk H, Czerwinski Z, Petal J. Ants as agents of soil habitat changes, Ekol. Polska. 1972;20:153–61.
Johnson DL. Dynamic denudation evolution of tropical, subtropical and temperate landscapes with three tiered soils: toward a general theory of landscape evolution. Quat Int. 1993;17:67–78. https://doi.org/10.1016/1040-6182(93)90082-Q.
Johnson DL. Darwin would be proud: bioturbation, dynamic denudation, and the power of theory in science. Geoarchaeology. 2002;17:7–40. https://doi.org/10.1002/gea.10001.
Jones CG, Lawton JH, Shachak M. Organisms as ecosystem engineers. Oikos. 1994;69:373–86. https://doi.org/10.2307/3545850.
Jones DT, Susilo FX, Bignell DE, Hardiwinoto S, Gillison NA, Eggleton P. Termite assemblage collapse along a land-use intensification gradient in lowland central Sumatra. Indonesia J Appl Ecol. 2003;40:380–91. https://doi.org/10.1046/j.1365-2664.2003.00794.x.
Jonkman JCM. Nest of the leaf-cutter ant Atta vollenweideri as accelerators of succession in pastures. Zeit fur Ang Ento. 1978;86:25–34. https://doi.org/10.1111/j.1439-0418.1978.tb01907.x.
Jungerius A, van den Ancker JAM, Mucher HJ. The contribution of termites to the microgranular structure of soils on the Uasin Gishu plateau, Kenya. Catena. 1999;34:349–63. https://doi.org/10.1016/S0341-8162(98)00106-4.
Kappler A, Brune A. Influence of gut alkalinity and oxygen status on mobilization and size class distribution of humic acids in the hind gut of soil feeding termites. App Soil Eco. 1999;13:219–29. https://doi.org/10.1016/S0929-1393(99)00035-9.
Keilhack K. Uber die bodenbildende Thatigkeit der insekten [On soil formation by insects]. Z Dtsch Geol Ges. 1899;51:138–41.
Kibblewhite M, Ritz K, Swift M. Soil health in agricultural systems. Phil Trans R Soc. 2008;362:685–701. https://doi.org/10.1098/rstb.2007.2178.
King LC. Canons of landscape evolution. Geol Soc Am Bull. 1953;64:721–52. https://doi.org/10.1130/0016-7606(1953)64[721:COLE]2.0.CO;2.
Kooyman C, Onck RFM. Distribution of termite (isoptera) species in Southwestern Kenya in relation to land use and the morphology of their galleries. Bio Fert of Soils. 1987;3:69–73. https://doi.org/10.1007/BF00260581.
Kozdrój J, van Elsas JD. Response of the bacterial community to root exudates in soil polluted with heavy metals assessed by molecular and cultural approaches. Soil Biol Biochem. 2000;32:1405–17. https://doi.org/10.1016/S0038-0717(00)00058-4.
Kronberg BI, Nesbitt HW. Quantification of weathering, soil geochemistry and soil fertility. J Soil Sci. 1981;32:453–9. https://doi.org/10.1111/j.1365-2389.1981.tb01721.x.
Kubiena W. Entwicklungslehre des Bodens. Wien: Springer; 1948.
Lal R. Erodibility and erosivity. In: Lal R, editor. Soil erosion research methods. Ankeny: Soil and Water Conservation Society; 1988. p. 141–60.
Lavelle P, Blanchart E, Martin A, Spain AV, Martin S. Impact of soil fauna on the properties of soils in the humid tropics. Madison: SSSA (Special Publica); 1992. p. 29. https://doi.org/10.2136/sssaspecpub29.c9.

Lavelle P, Pashanasi B. Soil macrofauna and land management in Peruvian Amazonia (Yurimaguas, Loreto). Pedobiologia. 1989;33:283–29.
Lee KE, Foster RC. Soil fauna and soil structure. Aust J Soil Res. 1991;6:745–75. https://doi.org/10.1071/SR9910745.
Lee KE, Wood TG. Termites and soils. London: Academic Press; 1971.
Lobry de Bruyn LA, Conacher AJ. The role of termites and ants in soil modification: a review. Aust J Soil Res. 1990;28:55–93. https://doi.org/10.1071/SR9900055.
Lucas Y, Luizao FJ, Chauvel A, Rouiller J, Nahon D. The relation between biological activity of the rain forest and mineral composition of soils. Science. 1993;260:521–3. https://doi.org/10.1126/science.260.5107.521.
Lugo A, Farnworth E, Pool D, Jerez P, Kaufman G. The impact of the leafcutter ant Atta colombica on the energy flow of a tropical wet forest. Ecology. 1973;54:1292–301. https://doi.org/10.2307/1934191.
Mandel RD, Sorenson CJ. The role of the western harvester ant (*Pogonomyrmex occidentalis*) in soil formation. Soil Sci Soc Amer J. 1982;46:785–8.
Mando A, Stroosnijder L. The biological and physical role of mulch in the rehabilitation of crusted soil in Sahel. Soil Use Manag. 1999;15:123–7. https://doi.org/10.1111/j.1475-2743.1999.tb00075.x.
Martin JP. Microorganisms and soil aggregation: I. origin and nature of some of the aggregating substances. Soil Sci. 1945;59:163–74.
Mateus ACC, Oliveira FS, Varajao AFDC, Soares CCV. Genesis of soils from bauxite in Southeastern Brazil: resilification as a soil-forming process. Rev Bras de Cien do Solo. 2017;41:1–18. https://doi.org/10.1590/18069657rbcs20160507.
McGinley MA, Dhillion SS, Neumann JC. Environmental heterogeneity and seedling establishment: ant-plant-microbe interactions. Funct Ecol. 1994;8:607–15.
Medina E, Silva JF. The savannas of Northern South America: a steady regulated by water interactions on a background of low nutrient availability. J Biogeogr. 1990;17:403–13. https://doi.org/10.2307/2845370.
Miklós AW. Biodinâmica da paisagem: organização e funcionamento biopedológico. Bol do Inst Biod de Desen Rural. 1993;11:9–16.
Monteiro HS, Vasconcelos PMP, Farley KA, Spier CA, Mello CL. (U-Th)/he geochronology of goethite and the origin and evolution of Cangas. Geoch et Cosmoc Acta. 2014;131:267–89. https://doi.org/10.1016/j.gca.2014.01.036.
Nye PH. Some soil-forming processes in the humid tropics -the action of the soil fauna. J Soil Sc. 1955;6(1):73–83. https://doi.org/10.1111/j.1365-2389.1954.tb02171.x.
Oliveira FS, Varajão AFDC, Varajão CAC, Boulangé B. A comparison of properties of clay minerals in isalteritic and in degraded facies. Clay Miner. 2013;48:697–711. https://doi.org/10.1180/claymin.2013.048.5.03.
Oliveira FS, Varajão AFDC, Varajão CAC, Boulangé B, Schaefer CEGR. The role of biological agents in the microstructural and mineralogical transformations in aluminium lateritic deposit in Central Brazil. Geoderma. 2014;226–227:250–9. https://doi.org/10.1016/j.geoderma.2014.02.012.
Passarge S. Die Kalahari [The Kalahari]. Berlin: Dietrich Reimer (Ernst Vohsen); 1904.
Paton TR, Humphreys GS, Mitchell PB. Soils - a new global view. London: UCL Press; 1995.
Penck W. Morphological analysis of landforms; a contribution to physical geology (translated by H. Czech & K. Cumming-Boswell). London: Macmillan; 1953.
Pendleton RL, Sharasuvanas S. Analyses and profile of some laterites and soils with iron concretions of Thailand. Soil Sci. 1942;54:1–26. https://doi.org/10.1097/00010694-194207000-00001.
Pessenda LCR, Valencia EPE, Aravena R, Telles ECC, Boulet R. Paleoclimate studies in Brazil using carbon isotopes in soils. In: Wasserman JC, Silva-Filho E, Villasboas R, editors. Environmental geochemistry in the tropics. Berlin: Springer; 1998. p. 7–16. https://doi.org/10.1007/BFb0010902.

Petal J. The role of ants in ecosystems. In: Brian MV, editor. Production and ecology of ants and termites. Cambridge University Press; 1978. p. 293–325.

Petal J. The role of ants in nutrient cycling in forest ecosystems. In: Billen J, editor. Biology and evolution of social insects. Leuven: Leuven University Press; 1992. p. 167–70.

Petal J, Nowak E, Jakubczyk H, Czerwinski Z. Effect of ants and earthworms on soil habitat modification. Ecol Bull. 1977;25:501–3.

Raapana JAW, Field JB. Macrozamia communis as a biogeochemical compounds by ericoid mycorrhizal fungi derived from heavy metal polluted sites. Soil Biol Biochem. 2006;35:133–41.

Reatto A, Bruand A, Silva EM, Guégan R, Cousin I, Brossard M, Martins ES. Shrinkage of microaggregates in Brazilian latosols during drying: significance of the clay content, mineralogy and hydric stress history. Eur J Soil Sci. 2009;60(6):1106–16. https://doi.org/10.1111/j.1365-2389.2009.01189.x.

Ritz K, McHugh M, Harris JA. Biological diversity and function in soils: contemporary perspectives and implications in relation to the formulation of effective indicators. In: Francaviglia R, editor. Agricultural soil erosion and soil biodiversity: developing indicators for policy analyses. Paris: OECD; 2004. p. 563–72.

Rode AA. Soil Science. Jerusalem: Israel program for scientific translation; 1962.

Rogers RW. Soil surface lichens in arid and sub- arid South-Eastern Australia. II. Phytosociology and geographic zonation. Aust J Bot. 1972;20:215–27. https://doi.org/10.1071/BT9720215.

Ruhe RV. Stonelines in soils. Soil Sci. 1959;87:223–31.

Salem MZ, Hole FD. Ant (formica exsectoides) pedoturbation in a forest soil. Soil Sci Soc Am Proc. 1968;32:563–7. https://doi.org/10.2136/sssaj1968.03615995003200040039x.

Sanderson MG. Biomass of termites and their emissions of methane and carbon dioxide, a global data base. Global Biog. 1996;10:543–57. https://doi.org/10.1029/96GB01893.

Santos RS, Mecenas HHB, Sousa-Souto L. Nest refuse of Atta opaciceps (hymenoptera: Formicidae) increases plant biomass and diversity during the regrowth of herbaceous. App Soil Ecol. 2019;133:160–5. https://doi.org/10.1016/j.apsoil.2018.10.002.

Sarcinelli TS, Schaefer CEGR, Lynch LS, Arato HD, Viana JHM, Albuquerque Filho MR, Gonçalves TT. Chemical, physical and micromorphological properties of termite mounds and adjacent soils along a toposequence in zona da Mata, Minas Gerais state, Brazil. Catena. 2009;76:107–13. https://doi.org/10.1016/j.catena.2008.10.001.

Sarmiento G. The ecology of neotropical savannas. Cambridge: Harvard University; 1984.

Schaefer CEGR. Brazilian latosols and their B horizon microstructure as long-term biotic constructs. Aust J Soil Res. 2001;39:909–26. https://doi.org/10.1071/SR00093.

Schaefer CEGR, Fabris JD, Ker JC. Minerals in the clay fraction of Brazilian latosols (Oxisols): a review. Clay Miner. 2008;43:1–18. https://doi.org/10.1180/claymin.2008.043.1.11.

Schaefer CEGR, Gilkes RJ, Fernandes RBA. EDS/SEM study on microaggregates of Brazilian latosols, in relation to P adsorption and clay fraction attributes. Geoderma. 2004;123:69–81. https://doi.org/10.1016/j.geoderma.2004.01.033.

Schaefer CEGR, Gomes LP, Henriques RJ, Gorsani RG, Santos MFS, Fernandes DPS. Interplays betweenAtta ants (Formicidae: Attini), soils and environmental properties in the Brazilian Neotropics: a preliminaryassessment. Revista Brasileira de Ciência do Solo. 2021;(45):e0210073. https://doi.org/10.36783/18069657rbcs20210073.

Schaefer CEGR, Ker JC, Gilkes RJ, Campos JC, Costa LM, Saadi A. Pedogenesis on the uplands of the Diamantina plateau, Minas Gerais, Brazil: a chemical and micropedological study. Geoderma. 2002;107:243–69. https://doi.org/10.1016/S0016-7061(01)00151-3.

Seton ET. Master plowman of the west. Century Magazine. 1904;68:300–7.

Shaler NS. The origin and nature of soils. US Geolog Surv 12th Ann Rep 1890–1891. 1891;1:213–345.

Simas FNB, Schaefer CEGR, Fernandes Filho EI, Chagas AC, Brandão PC. Chemistry, mineralogy and micropedology of highland soils on crystalline rocks of Serra da Mantiqueira, Southeastern Brazil. Geoderma. 2005;125:187–201. https://doi.org/10.1016/j.geoderma.2004.07.013.

Slager S. Morphological studies of some cultivated soils. Thesis Wageningen Agr Res Reports Pudoc. 1966;670:22.

Sleeman JR, Brewer R. Microstructures of some australian termite nests. Pedobiologia. 1972;12:347–73.

Sousa-Souto L, Viana-Junior AB, Nascimento ES. Spatial distribution of Acromyrmex balzani (Emery) (hymenoptera: Formicidae: Attini) nests using two sampling methods. Sociobiology. 2013;60:162–8.

Spier CA, Vasconcelos PM, Oliveira SMB. 40Ar/39Ar geochronological constraints on the evolution of lateritic iron deposits in the Quadrilátero Ferrífero, Minas Gerais. Brazil Chem Geol. 2006;234:79–104. https://doi.org/10.1016/j.chemgeo.2006.04.006.

Taylor HM, Brar GS. Effect of soil compaction on root development. Soil Till Res. 1991;19(2–3):111–9. https://doi.org/10.1016/0167-1987(91)90080-H.

Tisdale JM, Oades JM. Stabilization of soil aggregates by the root systems of ryegrass. Aust J Soil Res. 1979;17(3):429–41. https://doi.org/10.1071/SR9790429.

van Bremen N, Buurman P. Soil formation. Dordrecht, The Netherlands: Kluwer Academic Publishers; 1998.

Varajão AFDC, Gilkes RJ, Hart RD. The relationships between kaolinite crystal properties and the origin of materials for a Brazilian kaolin deposit. Clay Clay Min. 2001;49:44–59. https://doi.org/10.1346/CCMN.2001.0490104.

Vasconcelos PM, Renne PR, Brimhall GH, Becker TA. Direct dating of weathering phenomena by 40Ar/39Ar and K-Ar analysis of supergene K-Mn oxides. Geochim Cosmochim Acta. 1994;58(6):1635–65. https://doi.org/10.1016/0016-7037(94)90565-7.

von Ihering H. Uber schichtenbildung durch ameisen [on the formation of strata by ants]. Neues Jahrbuch fu ̈r Mineralogie. Geologie und Palaeontologie. 1882;1:156–7.

Weber NA. Gardening ants, the attines. Philadelphia: Memoirs of the American Philosophical Society; 1972.

Widmer F, Fliessbach A, Laczko E, Schulze-Aurich J, Zeyer J. Assessing soil biological characteristics: a comparison of bulk soil community DNA-, PLFA-, and biolog-analyses. Soil Biol Biochem. 2001;33:1029–36. https://doi.org/10.1016/S0038-0717(01)00006-2.

Wielemaker WG. Soil formation by termites, a study in the Kisii area, Kenya. PhD Thesis. Netherlands: Wageningen University; 1984.

Wilson EO. Social insects as dominant organisms. In: Billen J, editor. Biology and evolution of social insects. Belgium: Leuven University Press; 1992. p. 1–7.

Wood TG, Sands WA. The role of termites in ecosystems. In: Brian MV, editor. Production ecology of ants and termites. Cambridge University Press; 1978.

Yuan L, Huang J, Li X, Christie P. Biological mobilization of potassium from clay minerals by ectomycorrhizal fungi and eucalypt seedling roots. Plant Soil. 2004;262:351–61. https://doi.org/10.1023/B:PLSO.0000037055.67646.97.

Zimmerman PR, Greenberg JP, Wandiga SO, Crutzen P. Termites: a potentially large source of atmospheric methane, carbon dioxide, and molecular hydrogen. Sci. 1982;218:563–5. https://doi.org/10.1126/science.218.4572.563.

Chapter 16
Subsoil Constraints for Crop Production: Recent Advances, New Technologies, and Priorities for Further Research

Richard Willian Bell and Teogenes Senna de Oliveira

Abstract The existence, prevalence, and severity of subsoil constraints for crop production globally are under recognized and under-reported. Subsoil constraints (acidity, acid sulphate horizons, alkalinity, compaction, deep sand layers, gravel layers, high-density horizons, pans, pathogens, salinity, sodicity, waterlogged horizons) may be natural features of soil profiles or induced by land use and management practices. The subsoil in this chapter is considered to be the layers of the root zone below the depth of sampling for soil analysis, which typically corresponds to soil below 10–25 cm depth, depending on the soil sampling conventions of the region. Tropical regions, in particular (in Africa, Asia, Northern Australia, and Latin America), contain large areas of deeply weathered profiles that commonly have hostile subsoils that constrain root growth. The main consequence of subsoil constraints is that water and nutrients contained in subsoils are not accessed or efficiently utilized, and hence crops fail to reach their yield potential. Even when best management practices are applied to the topsoil, yield of crops is depressed by subsoil constraints. Crops may acquire up to 75% of N, 85% of P, and 70% of K uptake from the subsoil if root growth is not constrained. Technologies to sense, identify, map digitally, and ameliorate subsoil constraints represent a promising frontier for soil management, with the potential to substantially lift crop productivity in many parts of the world.

Keywords Acidity · Acidification · Alkalinity · Biopores · Compaction · Deeply weathered profiles · Digital soil mapping · Gravel-rich soils · Nutrient deficiency · Root penetration · Salinity · Sand · Sodicity · Soil-borne pathogens · Soil sensing

---

R. W. Bell (✉)
Centre for Sustainable Farming Systems, Food Futures Institute, Murdoch University, Murdoch, WA, Australia
e-mail: r.bell@murdoch.edu.au

T. S. de Oliveira
Soils Department, Federal University of Viçosa, Viçosa, MG, Brazil
e-mail: teo@ufv.br

T. S. d. Oliveira, R. W. Bell (eds.), *Subsoil Constraints for Crop Production*,
https://doi.org/10.1007/978-3-031-00317-2_16

16.1 Extent and Severity of Subsoil Constraints

The full extent of the subsoil constraints that limit crop production is not known. Indeed, a recent meta-analysis of 1146 journal papers published in prominent soil journals during the period 1989–2019 found that 73% did not even mention soil depth, and in the remainder, 27 cm was the average soil depth studied (Yost and Hartemink 2020). This approximates to the 20 cm plough layer conventionally considered to be the topsoil. In addition, there is no global mapping or quantification of subsoil constraints apart from the following: subsoil acidity (pH <4.5 in water), which it is estimated to cover 3300 Mha (Sumner and Noble 2003), and one billion hectares of subsoils that are salt-affected by the combination of irrigation and dryland salinity, sodicity, and alkalinity (Beltrán 2016). Also, globally, 29% of soils are estimated to contain >15% gravel in the subsoil (Scanlan et al. 2022), and soils with <15% clay in the subsoil are estimated to cover 5% of global arable soils or 0.9 M km^2.

In some specific areas, such as the south-west of Western Australia, subsoil constraints due to boron toxicity, acidity, alkalinity, compaction susceptibility, impenetrable layers, and water storage have been mapped (Van Gool 2016). In Brazil, the extent of subsoil constraints, such as soil acidity, nutrient deficiency, and water retention, can only be estimated by inference from previous mapping studies, but the scale of maps is a limitation to the spatial resolution with which subsoil constraints can be mapped. There is an opportunity for future soil investigations to systematically collect more data on subsoil constraints, for example, by using rapid soil analysis techniques such as NIR and MIR spectroscopy and ion-selective electrodes together with digital soil mapping protocols (Ringrose-Voase et al. 2019) to map subsoil properties. For example, digital techniques using predictive algorithms were used to predict contemporary soil organic carbon to 1 m depth in Southern Brazil (Bonfatti et al. 2016).

An example of the mapping of the extent, cause, and effect of root-restricting layers in agricultural soils is reported from Germany (Schneider and Don 2019a). The root restrictions considered were bedrock depth, rock fragments, cementation, compactness, sandy subsoil, anoxia, and acidity. Overall, 71% of soils had root-restricting subsoil layers (10–100 cm depth). Physical constraint (compaction) was the most common cause of root restriction, affecting 51% of crop land and 32% of grasslands, and most of physical restrictions (73%) was attributed to inherent geological or pedological features (soil strength, rock fragments, bedrock, acidity, anoxia). Land use practices (heavy machinery on cropped land) explained the remaining 27%. The German study is noteworthy in showing that it is possible to produce subsoil constraints maps at farm scale and to map induced subsoil constraints that arise from land management practices.

Subsoil constraints maps will be easiest to derive for naturally occurring soil chemical and physical constraints. However, much of the point source and derived spatial data on which soil mapping is based was collected decades ago and hence will not reflect current properties or induced subsoil constraints such as machinery-induced compaction or acidification that result from land management practices in more recent times. A global estimate of the various subsoil constraints will help to ensure that proper resources are allocated to developing cost-effective management tools and practices to alleviate their effects on crop production and to facilitate transfer of technologies developed in one region to another.

16.2 Deeply Weathered Soils

Subsoil constraints (physical and chemical) are commonly associated with deeply weathered soils that developed on deep regolith. Schaefer et al. (2022) highlight the regions globally with deeply weathered regolith, most of them occurring in tropical South America, South and Southeast Asia, and Africa and known as regolith-dominated terrains (González-Álvarez et al. 2016). Deep kaolinitic profiles are mostly in the range 10–60 m depth, while some have reached 200 m depth in tropical and subtropical regions of the Americas, Africa, and Oceania (Sangmanee et al. 2022). Deeply weathered subsoils occur also as paleoweathering features in temperate climates in southern Australia, central China, Russia, Europe, and Turkey (Schaefer et al. 2022).

Deeply weathered soil profiles have properties that reflect the imprint of past climates in addition to that of current climates, but both contribute to the genesis of natural subsoil constraints or to the expression of the induced constraints under crop management. Hence, the extensive research in southern Australia on hostile subsoils in the last two decades is a useful case study of subsoil constraints on agricultural land (Van Gool et al. 2007) as is the study reported for Germany (Schneider and Don 2019a, b). Deeply weathered regoliths in temperate regions such as southern Australia have been exposed to different climates to those in Brazil and a range of different subsoil constraints prevail, such as dense horizons, sodicity, salinity, alkalinity, and B toxicity (McDonald et al. 2012). The present chapter synthesizes much of the learning about the range of subsoil constraints and management practices to alleviate them based mostly on studies carried out in Brazil and Australia. Application of this learning could be particularly relevant for boosting the productivity of soils elsewhere in the world where deeply weathered subsoils also occur (Schaefer et al. 2022).

Broadly, the regoliths in the tropical zone can be divided into those involving complete desilification and those involving partial desilification (Schaefer et al. 2022). The former is dominated by Al- and or Fe-oxyhydroxides, while the latter are dominated by kaolinite mineralogy. The tropical soil profiles that have undergone complete desilification can be further divided, depending on whether the weathering environment involves continuous humid conditions or alternate wet-dry conditions. The former weathering environment leads to the deep well-structured Oxisol profiles, while the latter leads to formation of Fe segregations and/or ferricrete in the solum. Within the kaolinitic regolith types, under wet, poorly drained conditions, smectite minerals can be found, producing soils with quite distinct chemical and physical properties. The dominant soil mineralogy dictates the aggregate stability and their susceptibility to soil compaction or erosion. Deep, well-structured Oxisol profiles in Brazil with dominant oxidic composition (especially gibbsitic) are physically and mechanically more stable than kaolinitic Oxisols (Ferreira et al. 1999).

The profiles formed in the deeply weathered soils appear to be profoundly influenced by biota (Schaefer and Oliveira 2022; Schaefer et al. 2022). Biota plays a key role in subsoil formation in deep profiles of Brazil through the combined influences of vegetation and soil fauna, particularly ants and termites (Schaefer and Oliveira 2022). Biota is concurrently transforming and reorganizing the regolith across these scales.

Approximately two thirds of the tropical land mass are inhabited by termites, and deep regoliths can be strongly modified by them (Zimmerman et al. 1982). Termites were responsible for deep regoliths and the stable B horizon aggregate structure of Oxisol soils in Kenya (Wielemaker 1984). A similar conclusion was drawn by Schaefer (2001) for deep Oxisol soils with well-aggregated B horizons in Brazil. Termites and ants change soil profiles by several processes: bioturbation that homogenizes the solum, developing stable microgranular soil structure, altered nutrient availability, changes in soil porosity and soil hydrology, and organic matter accumulation (Schaefer and Oliveira 2022). Earthworms are highly mobile and also have a crucial role in soil development and functioning. Their activity creates biopores in the subsoil while adding organic matter and changing the soil environment at depth. These benefits improve soil structure, water regulation, nutrient cycling, and climate regulation (Blouin et al. 2013) and can be expressed in deeply weathered soils over the long term.

A different influence of biota on the properties of deeply weathered soils has been advanced by Verboom and Pate (2006a, b) from their studies in south-west Australia. Their phytotarium concept argues that roots of the dominant woody taxa, together with associated microorganisms, bioengineer soils through rhizosphere activity that has shaped key pedogenic processes. Proteaceae taxa produce cluster roots to enhance the acquisition of limited amounts of insoluble P in weathered soil materials. Over time, the release of P from Fe-bound forms alters the soil mineralogy to produce Fe-rich segregations and Fe-coated root channels (Verboom and Pate 2006b). The soils under the Proteaceae vegetation typically are dominated by ferruginous gravels and/or ferricrete. By contrast, adjacent soils under Myrtaceae taxa have a distinctly different profile form: ferruginous segregations and ferricrete are absent, and instead the soil profiles have a distinct texture-contrast form with a clay-rich, sodic B horizon that often contains calcrete segregations. Verboom and Pate (2006b) suggest that the vegetation type has over time changed the pedogenic processes leading to different profile types under the same climate. From these examples, we conclude that more extensive investigation is needed on the role of biota in pedogenesis of the deeply weathered soils, especially to understand their role in intensifying or alleviating subsoil constraints.

16.3 Field and Farm-Scale Variability and Diagnosis

Subsoil constraints can be highly variable in expression at farm-field scale (Wong and Asseng 2007; Wong et al. 2008). Hence, to quantify their effects on crop yield and to apply treatments that are targeted to areas with specific subsoil constraints, field-scale mapping is needed. Advanced techniques such as proximal sensing through EM mapping and gamma radiometrics are potential field-scale tools for rapid mapping of subsoil constraints. In deeply weathered landscapes of south-west Australia, field-scale variability between areas with subsoil ferruginous gravels and those with deep sands, which present quite different subsoil constraints, can be identified with these techniques (Wong et al. 2008).

At field scale, a simpler diagnosis of subsoil constraints can be achieved using an exposed soil pit (Fig. 16.1). This approach is a valuable learning context, where farmers come together with a soil adviser to collectively examine the features that represent subsoil constraints of that field, how the site relates to other soils locally, and to discuss approaches that have been found to effectively ameliorate the various constraints. There are many examples of the

Fig. 16.1 A group of farmers gathered with a soil adviser around of trench dug in south-west Australia to expose the subsoil and discuss the evidence of subsoil constraints and the effective means of amelioration. The main diagnostic tools used are the following: the presence of layers with high soil strength, evidence of restricted root growth, and unused subsoil water plus field assessment of subsoil pH using an indicator dye

methodology used in Australia including in the cotton industry: https://www.dpi.nsw.gov.au/agriculture/soils/guides/soilpak-series/soilpak. Yield mapping may also help to define the areas in a field with subsoil constraints but without necessarily identifying the underlying cause and mechanisms. Wong and Asseng (2006) illustrated how subsoil acidity and shallow depth to a cemented gravel layer could variably limit yield potential across a field and among a series of seasons.

The impacts of subsoil constraints are dependent on land use and whether the land is managed for dryland production in water-limited environments (Vance and Milroy 2022) or for irrigated crop production. In Germany, wheat yields were most constrained by subsoil constraints on sandy profiles and in low rainfall seasons, since the crop in those conditions is most constrained by the lack of root access to stored subsoil water (Schneider and Don 2019a). Water-limited yield potential has been a powerful diagnostic approach for identifying and quantifying the impact of subsoil constraints on yield in southern Australia (French and Schulz 1984). Yield potential of wheat in this environment was achieved when crops converted 1 mm of rainfall to 20 kg of grain ha^{-1} (e.g. Oliver et al. 2008).

Modelling is another option to predict the effects of subsoil constraints on crop yield over time due to climate variability (and climate change). Yield potential simulated by Yield Prophet®, an online version of the APSIM systems model (Hochman et al. 2009), is an example. Hochman et al. (2009) suggest that 80% of water-limited yield potential is comparable to the attainable farmer yield. In their study in eastern Australia on rainfed wheat cropping, elite farmers were reaching water use efficiency of ~15 kg of grain mm^{-1} ha^{-1}. By contrast, average crop yields achieved only 40–50% of water-limited yield potential. Hence, in these rainfed environments, most crops fail to use all the stored water in profile and due to a range of constraints, including those in the subsoil, fail to utilize a large portion of the water available from rainfall. Increased exploitation of subsoil stored water in the grain filling stage of wheat growth can substantially boost yield, since the conversion efficiency of this late season water is 50–60 kg of grain mm^{-1} ha^{-1} compared to the season-average conversion of 15–20 kg mm^{-1} ha^{-1} (Kirkegaard et al. 2007).

The influence of subsoil acidity on wheat yield is illustrated by Fig. 16.2. Independent of the topsoil pH (0–10 cm), the wheat yield was depressed by subsoil (10–20 cm) $pHCaCl_2$ below 4.7. At subsoil $pHCaCl_2$ of 4.1, there was a consistent 0.7 t ha^{-1} yield penalty compared to pH 4.7 or above in the subsoil, regardless of topsoil pH. It is likely that subsoil acidity is limiting yields in many parts of the world even when best management practices are used for topsoil management. Alternatively, as shown in Germany, subsoil acidity may cause farmers to choose a less profitable land use such as perennial pastures due to the low crop productivity on soils with subsoil acidity (Schneider and Don 2019a).

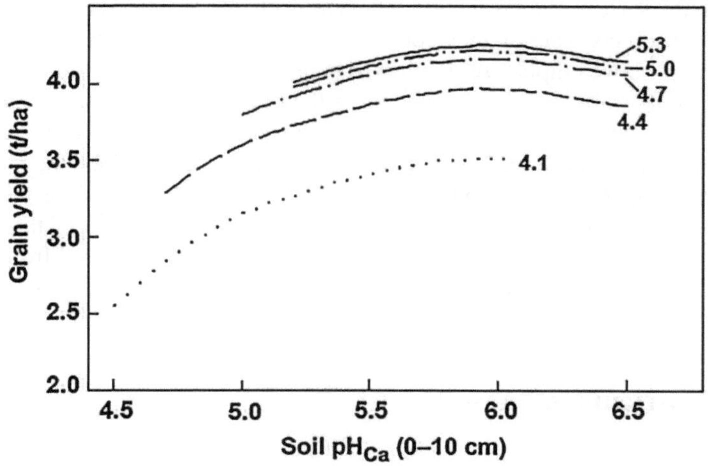

Fig. 16.2 Depression in wheat yield in southern New South Wales due to subsurface soil (10–20 cm) acidity (< $pHCaCl_2$ 4.7) independent of the topsoil pH (0–10 cm), which ranges from 6.5 to 5 or less. (From Scott et al. 1999)

16.4 Multiple Subsoil Constraints

Much of the research on subsoil constraints examined a single constraint. However, commonly multiple subsoil constraints occur on a particular soil. For example, in south-west Australia, 43% of alkaline shallow duplex soils have more than three subsoil constraints in the same profile, from combinations of subsoil compaction, high alkalinity, salinity, high sodicity, micronutrient deficiencies, and high boron (Van Gool 2016). Focussing research on a single limiting factor may fail to produce significant crop production benefits. For example, considerable research was conducted on developing improved barley and wheat cultivars for production on soils with high boron in the subsoil (Cartwright et al. 1984). However, the released wheat cultivars produced no increase in yield relative to standard cultivars (McDonald et al. 2010, 2012). This is attributed to the fact that the soils with high boron also generally have subsoil limitations due to high pH, salinity, high sodicity, micronutrient deficiencies, and perhaps high bulk density. Since the boron-tolerant cultivars had no better adaptation to the other constraints than the standard cultivars, their yield was similarly constrained.

Another example of multiple subsoil constraints is the Cerrado region of Brazil, which is dominated by deep Oxisols, with acidity and nutritional constraints in subsoils. Liming does not correct chemical limitations (low pH, high exchangeable Al^{3+}, and high potential acidity) below the depth of incorporation. Gypsum can alleviate subsoil acidity effects on roots while increasing availability of Ca, Mg, and S to crops. However, K and N leaching can increase, while plant-available P is limited to the surface soil layer. After alleviating chemical subsoil constraints in the Oxisols, subsoil compaction may continue to limit crop development.

The cases above emphasize that it will be necessary where multiple constraints co-occur on the same soil to alleviate more than one subsoil constraint rather than treating individual constraints in isolation. Hence, while chapters of the present book (Oliveira and Bell 2022) focus on specific subsoil constraints, they also refer to integrated approaches for managing subsoil constraints, including the genetic tolerance of crop cultivars to subsoil constraints. In the remainder of this chapter, we highlight key learning about managing subsoil constraints derived from preceding chapters while emphasizing approaches that address multiple constraints.

16.5 Nutrients in Subsoil

The significance of the subsoil for nutrient supply to crops is generally not well understood even though two thirds of the soil profiles' available nutrient stock is in the subsoil (Ma et al. 2022). Most soil sampling for nutrients is restricted to the upper 10–20 cm depth only so that the levels of available nutrients deeper in the profile are not recorded. However, 8–75% of the N, 3–85% of the P, and < 3–70% of K taken up by a crop is from the subsoil (Kautz et al. 2013). Indeed, several hundred kg of nitrate-N ha^{-1} can be retained in the subsoil of tropical acid soil profiles (Wong et al. 1990).

The wide global spread of Conservation Agriculture involving minimal soil disturbance and retention of crop residues accentuates nutrient concentration near the soil surface (Ma et al. 2022). In rainfed environments where topsoil drying occurs during the growing season, roots have decreased access to the nutrient-rich topsoil. The plant-available levels of subsoil nutrients and root access to the subsoil nutrients assume much greater importance in these cases.

The stock of subsoil nutrients may be relatively large, but the significance of these reserves of nutrients depends on whether enough roots have access to the subsoil. Hence, one of the research gaps is to identify the factors that determine a crop's dependence on subsoil nutrients and their likely responsiveness to deep placement of fertilizer. In general, nutrient acquisition from the subsoil varies with soil water regime, soil type, the existence of other subsoil constraints, and crop species (Ma et al. 2022).

Subsoils can play a significant role in long-term crop nutrition by storing nutrients to be recycled to the topsoil (Kautz et al. 2013). In sands and very porous Oxisols, there is potential for recovery of leached nutrients if deep-rooted species are included in the crop rotation. On deep sands in south-west Australia, deep-rooted lupin crop was able to increase plant-available K in the topsoil relative to shallow rooted sub-clover pasture (Edwards 1993).

Deep placement of fertilizer or organic materials could play a role in increasing crop production (Ma et al. 2022). Depending on the machinery used, depths of placement range from 15 to 40 cm. Crop yield increases with subsoil placement of

N, P, K, Mn, and Zn have been shown in a range of crops species. However, deep fertilizer and organic matter placement is not yet a well-proven technology for improved crop nutrition: more investigation is needed to understand the effects of crop species, soil type, and climatic constraints on crop nutrition and yield response (Ma et al. 2022). Especially for deeply weathered soil profiles, it is important to develop soil management systems that enhance nutrient retention and root growth in subsoils. Increasing organic matter at depth will increase CEC, enhance nutrient release by mineralization, decrease nutrient leaching, and enhance subsoil water retention. The increase of organic matter in subsoils of weathered deep soil occurs by root growth in subsoil or by dissolved organic matter (DOM). Although DOM comprises only a small part of soil organic matter, it is considered the major carbon source in subsoils (Leinemann et al. 2016).

Micronutrient limitations in subsoil restrict root growth, and the limited soil volume explored increases the susceptibility to drought (Nable and Webb 1993). Deep placement of fertilizer may alleviate micronutrient deficiency (Cu, Zn, Fe, and Mn) in soils under both conventional and no-tillage areas, as reported for Mn in canola fields in Australia (Brennan and Bolland 2015) and soybeans in Brazil (Tanaka et al. 1992).

In summary, subsoil nutrients are an asset that could be unlocked, but awareness of their significance as a subsoil resource is still low compared with other subsoil constraints.

16.6 Subsoil Acidity and Acidification

Identification and correction of soil acidity can convert large areas into highly productive land for crop production as in the Cerrado region of Brazil. However, under the no-tillage cropping systems, the surface application of lime is slow to ameliorate subsoil acidity, so gypsum application has been used to decrease subsoil Al^{3+} toxicity due to its greater dissolution than lime and the faster migration of soluble Ca^{2+} and SO_4^{2-} into the subsoil (Butterly et al. 2022). There is potential for this technology or some adaptation of it, to be used more widely across regions with subsoil acidity. Surface application of lime without incorporation by soil tillage has only a slow effect on subsoil pH, when the topsoil pH_{CaCl2} is 5.5 or above, and no effect, if topsoil pH_{CaCl2} is <5.5 (Whitten et al. 2000). Gypsum application plus liming can be an alternative to reduce subsoil acidity (Anderson et al. 2020). However, even with gypsum application, stratification of nutrients, lime, and organic matter in the surface soil layer will remain. This has led in recent years to on-farm renovation of fields with subsoil acidity by deep incorporation of lime to 30 cm or more (Gazey et al. 2014). By contrast, nutrient stratification does not occur in areas under conventional tillage systems.

In the surface layer of topsoil, where organic matter accumulates, increase in pH can occur due to the uptake of nitrate and due to the alkaline organic residues of crops returned to the topsoil. By contrast, uptake of an excess of cations relative to anions in the subsurface soil increases its acidity. The stratification of acidity within the profile can substantially limit root growth in cropping or pasture systems in the 10–30 cm layer which exposes crops to drought, even when the deeper parts of the profile still store substantial amounts of water. The rate at which pH stratification occurs depends not only on the rate of acidification by the land use system but also on the pH buffering capacity of soils. On sands and sandy duplex soils, the rate of subsoil pH change is quite rapid due to low pH buffering capacity (Noble et al. 2000).

Small pH differences are not easy to detect by analytical methods even when there is a well-planned strategy for monitoring crop or pasture paddock pH. To support monitoring, there is a need for qualified laboratories with standard methods and effective quality control of their results. Tools are needed to estimate how much acidification happens at farm level and the impact of crop and soil management and of amelioration methods on rates of acidification.

While most emphasis has been on lime, dolomite, gypsum, and calcium oxide for alleviation of soil acidity, there is evidence that alkaline organic amendments can play a role in neutralizing subsoil acidity (Tang and Yu 1999). Nanoparticles of alkaline products may also increase the mobility of neutralizing materials into the subsoil, but definitive research is not yet done. Butterly et al. (2022) propose that the combined incorporation of lime and organic materials could be the most effective approach for subsoil acidity amelioration, since in addition to pH increase, improved soil physical properties and nutrient supply would stimulate subsoil-root proliferation. Some industrial organic residues, such as alkaline cellulose and paper residues, can be pelletized or granulated with lime, gypsum, and nutrients and deep applied by conventional soil preparation or even deep subsoiling.

Selection and breeding for acid-tolerant species and cultivars to mitigate the effects of subsoil acidity has received much attention worldwide (Ryan 2018). However, it has been argued that acidification continues while growing acid-tolerant cultivars or species and ultimately drives subsoil pH to even lower values that will become toxic even for tolerant cultivars or more difficult to correct. This is only partly true. Butterly et al. (2022) point out that tolerant cultivars can be part of the process of ameliorating subsoil acidity. Firstly, by penetrating subsoils, the roots of tolerant cultivars can increase uptake of nitrate from the subsoil that would otherwise leach and add to the subsoil acidification. The uptake of nitrate can cause rhizosphere alkalinization if anion uptake exceeds cation uptake. Indeed, due to anion exchange capacity, the subsoils of acid soils may already contain significant stores of nitrate (Wong and Wittwer 2009). Secondly, the roots of acid-tolerant cultivars increase soil organic matter in the old root channels or excrete organic acids into the subsoil rhizosphere, which can complex Al^{3+} and alleviate toxicity. Hence, the use of Al tolerant cultivars is more than a stopgap measure to maintain crop profitability on acid soils: those cultivars can play a constructive role in subsoil acidity amelioration. However, it is not known how significant this process is in either slowing or stopping subsoil acidification. A clear demonstration of the long-term value of this approach would shift the perception of the value of the plant breeding approach for acid soil tolerance. To select for effective acid-tolerant cultivars, it is necessary to consider the tolerance not just for soil acidity but also for other subsoil constraints, such as soil compaction and nutrient deficiency in subsoil.

There are many approaches, mostly related to the nitrogen cycle in soils, that slow the rate of soil acidification such as the following: decreasing the application of ammonium-based N fertilizers, slowing the rate of nitrification, reducing the extent of NO_3^- leaching by decreasing the periods under legume pasture or fallow, applying the 4Rs principles for N fertilizer management to maximize N-use efficiency, selecting crops and their sequencing to maximize NO_3^- uptake, and minimizing the export of alkalinity as farm products. However, Butterley et al. (2022) point out that these approaches mostly tackle topsoil acidity rather than subsoil acidity. Indeed, they may have the opposite or no effect on subsoil acidity. Mechanisms for slowing the rates of subsoil acidity are limited to cropping patterns that have low cation uptake relative to anions and to biologically altering the cation/anion balance to favour subsoil anion (especially NO_3^-) uptake. Of these options, maximizing NO_3^- uptake from the subsoils has most potential, but the challenge will be to retain NO_3^- in the subsoil long enough for root absorption while avoiding the leaching of NO_3^- from the subsoil. On acid soils, there may be sufficient anion exchange capacity in the subsoil to retain significant NO_3^- (e.g. Wong et al. 1990; Wong and Wittwer 2009). If these approaches are insufficient, then either high (and often uneconomic) surface lime rates or direct incorporation of lime into the subsoil are the approaches available, but the impact of these treatments on mineralization of N and leaching of nitrate need to be checked. Given the practical and economic challenges with ameliorating acid subsoils, emphasis should be on slowing and/or avoiding subsoil acidification in farming systems.

16.7 Salinity, Alkalinity, and Sodicity

High soil salinity, alkalinity, and sodicity in the subsoil are limiting factors for plant yield that restrict the ability of the plant roots to absorb water and nutrients stored in subsoils, especially in rainfed agriculture (Rengasamy et al. 2022). Subsoil salinity and chloride constraints may be less significant in well-watered or irrigated agriculture than for rainfed or dryland crop production (Dang et al. 2008). During periods of topsoil dryness, the roots explore deeper layers, where they may encounter saline constraints or salinity combined with chloride, sodicity, alkalinity, and excess B (Sadras et al. 2003; Dang et al. 2008). An important symptom of subsoil constraints due to salinity and sodicity is that at harvest time, significant volumes of stored water remain in the subsoil due to the inability of roots to extract that water (Rengasamy et al. 2003; Dang et al. 2008). In the case of rainfed agriculture, negative impacts may be more intense in drought years when the soil water storage is lower and the evaporative demand is higher.

Historically, research on salinity has focussed on irrigation and on topsoil properties and management. The issue of irrigation salinity in topsoil and subsoil will increase in importance over time as high-quality water sources are increasingly diverted into higher value uses or uses in cities (UNESCO 2017). By contrast, compromised water and water from waste streams may be more available to agriculture for irrigation. The value of this water for irrigation will depend on the development of strategies to avoid accumulation of salt or other contaminants and the avoidance of higher sodicity or alkalinity in the soil profile, including in the subsoil.

In dryland agriculture, salinity also arises from groundwater sources of salt, particularly where a change in land use or vegetation type alters water balance and increases the rate of groundwater recharge within a water catchment (Clarke et al. 2002). This form of salinity is first expressed as rising saline groundwater into the subsoil and later by accumulation of salt at the soil surface. The impact of waterlogging in the root zone of saline soils is underestimated. The combination of waterlogging and salinity has more profound damaging effects on the growth of most agricultural plants than either stress individually (Barrett-Lennard and Shabala 2013).

Transient salinity is another distinctive form of dryland salinity that has only recently been recognized (Rengasamy et al. 2022). There are 5.8 M km^2 globally of sodic soils with dispersive soil layers, which have the potential for transient salinity (Bui et al. 1998). Transient salinity lowers water use efficiency in dryland agriculture due to temporal decreases in the osmotic potential of the soil solution. Transient salinity can occur when there is a combination of low rainfall and dispersive clay layers below the roots zone (Rengasamy 2002). Together, these properties restrict salt leaching below the root zones in semi-arid regions. Due to degraded soil structure, dispersive subsoils restrict water flow and salt leaching, leading to salinization of root zone layers including the subsoil. However, the consequences for plant growth vary with seasonal rainfall, leaching, water use by plants, and water evaporation from the soil surface. In addition, the decrease of the osmotic potential of soil water, which decreases the non-limiting plant-available water range, may also be associated with high concentrations of soluble boron, carbonates, and microelements, such as Al, Mn, and Fe, which can also be toxic to crops. Hence, transient salinity often involves multiple constraints so that a focus on alleviating any single constraint is unlikely to be effective.

Confusion has existed in the literature about the definition of sodic soils and in the recommended management practices for such soils, because dispersiveness on sodic soils varies with many soil factors such as mineralogy, organic matter, exchangeable cations, electrolyte concentration of soil solution, and pH (Rengasamy et al. 2022). Sodic soils have been defined by exchangeable Na percentage (ESP), or the Na absorption ratio (SAR). Threshold values vary from 6% to 15% ESP. However, the critical property is usually not sodicity per se but rather the dispersive tendency of the soils. A new concept of "net dispersive charge" has been proposed to account for the amount of clay dispersed from a dispersive soil based on the difference between dispersive charge and flocculating charge (Rengasamy et al. 2016).

When the flocculating charge in a saline soil is equal to or greater than the dispersive charge, the soil structural effects of sodicity will be minimal, and the effects on crop growth are largely explained by osmotic effects and ion toxicity. When the flocculating charge is lower than the dispersive charge, soil physical properties will be degraded with damaging effects on crop growth. The concept of net dispersive charge of dispersive soils considers the concentrations of all cations unlike SAR, which ignores the dispersive effects of K, and the flocculating effect of both Mg and Ca. Based on the concept of net dispersive charge, there is a need to examine a broad range of dispersive soils to derive appropriate threshold values for dispersion and appropriate soil management practices for different classes of sodic soil. These investigations should include assessments of subsoils and their relationship to transient salinity.

While gypsum is commonly recommended as an ameliorant for sodic soils, there are still inadequate guidelines on when to expect responses to its application especially for subsoil sodicity (Zoca and Penn 2017). Amelioration of sodic, dense subsoils by the deep incorporation of organic materials (30–40 cm) has shown promising results in southeast Australia (Gill et al. 2008; Sale et al. 2019) as a strategy to restore degraded soils (Celestina et al. 2019). The organic materials, applied as pellets of lucerne shoots or poultry litter at 20 t ha^{-1}, were in fact more effective than incorporation of gypsum or deep ripping (Gill et al. 2008). The 60% increase in wheat yield was attributed to 50 mm extra water extraction from below 40 cm depth due to increased root growth and to increased N uptake, which enabled leaves of wheat to remain green for longer. The incorporation of lucerne or poultry litter pellets at a depth of 30–40 cm almost doubled the macroporosity from <10% to >18%, decreased the volumetric water content retained at −1500 kPa, and caused a 50-fold increase on saturated hydraulic conductivity in the subsoil layer (Gill et al. 2009). The yield responses persisted for at least 4 years (Sale et al. 2019).

In alkaline soils, much of the research has been on bicarbonate toxicity (pH 8–9), carbonate toxicity (> pH 9), Fe deficiency (pH 8–9), and boron deficiency and toxicity (Rengasamy et al. 2022). Another constraint when soil pH increases above 9 is attributed to toxicity due to the formation of soluble anionic aluminate species (Brautigan et al. 2012); however, the veracity of this conclusion is contested by Kopittke et al. (2016). In addition, with increasing pH, monovalent ions, particularly Na^+, increase in soil solution while concentrations of divalent Ca^{2+} and Mg^2 decrease, leading to greater clay dispersion and soil structural degradation. Soil structural degradation exacerbates transient waterlogging, and the anoxic conditions in the subsoil can trigger microelement toxicities (such as Mn, Fe, Al, and B) due to the combination of increasing concentrations in soil solutions and/or reduced exclusion by roots (Setter et al. 2009). When alkaline soils are salinized by chloride and sulphate salts of sodium, soil pH is considerably reduced and toxicity due to carbonates and elements (Mn, Al, and Fe) is diminished, and osmotic effects become dominant.

16.8 Soil Physical Constraints

Physical subsoil constraints are linked with chemical properties and biological processes (Oliveira and Fernandes 2022). Physical restrictions in the subsoil can be natural or induced. The most well-studied physical constraint is compaction caused by pressure from the weight of machines on the soil surface and the shear force of implements on soils. Mechanization of all phases of crop production is a reality in many countries and landscapes. To achieve greater economic efficiency, there is a trend towards heavier machines with higher soil surface pressure.

Controlled traffic farming is a rapidly evolving technology for minimizing subsoil compaction in mechanized crop production. There is a need to extend this technology to perennial crops where multiple wheel passes (for fertilization, chemical application, harvesting, etc.) between the tree lines (Fig. 16.3) compact the surface and subsurface soil and decrease rates of water infiltration and lateral root expansion. This can happen not only on sandy soils but also on fine-textured soils. It may be exacerbated under irrigated agricultural systems (localized, sprinkler, or surface inundation), since the soil water content will often be higher in the tramlines, making the soil more prone to compaction. In vegetables crops, the timing of operations is critical so as to achieve a high-value product, which often results in machinery operating in wet soils. Avoiding soil compaction and the degradation of soil structure in these systems remains a challenge.

The most common correction for subsoil compaction is deep subsoiling (~ 60 cm) or deep ploughing at ~40 cm, but first, it is necessary to identify the depth of the compacted layer. However, periodic elimination of a compacted layer by any implement runs the risk of creating a new compacted layer at a greater depth, especially as machinery weight tends to increase over time. With the trend towards heavier machinery and more mechanized operations in each crop, it is essential to better predict subsoil compaction susceptibility and to incorporate this kind of information in soil use planning as demonstrated in the study of Schneider and Don (2019a, b).

Fig. 16.3 Compacted surface tramlines in perennial crop (*Malpighia punicifolia* L.) production under drip irrigation

The risk of compacted subsoil layers occurs even in oxidic Ferralsols, which are characterized by very well-developed, stable microgranular structure. However, under intensive vegetable systems (at least two crops per year) and central pivot irrigation, compaction is occurring. Within the aggregates, there is increased porosity but fewer large, elongated pores, and less connected pores (Pereira 2020).

16.9 Sand and Gravel-Rich Subsoils

A major limitation of soils with either gravel or sand subsoils is their limited capacity to store and supply water and nutrients for root uptake. Even when roots can grow into these gravelly and sandy subsoils, the amount of stored water and nutrients that can be accessed is limited. Root growth in these subsoils can be hampered further by subsoil acidity and compaction.

At <15% content, gravels are not considered to significantly influence edaphic soil properties: indeed, if the <2 mm fraction has sand or loamy sand texture, increases in gravel up to 20–30% can slightly increase root depth (Bablola and Lal 1977). However, with progressive increases in gravel content, water storage declines. The implications for crops and pastures depend on whether the gravel layer is penetrated by roots (Scanlan et al. 2022). However, even with penetrable gravel layers, root depth declines with increasing gravel content. Due to the low plant-available water content (PAWC) of soils with gravel subsoil, decreases in root depth can profoundly hamper crop growth. For example, in sandy profiles with gravel subsoil in south-west Australia, the cumulative PAWC to 1 m depth was 55–105 mm, but if root depth was restricted to 30 cm depth, the PAWC was only 30–35 mm (Scanlan et al. 2022).

Like profiles with gravel subsoil, sand-textured subsoils severely limit PAWC (Scanlan et al. 2022). In the World Reference Base (FAO Soils Portal 1998), the deep sands or Arenosols can have up to 16% clay. However, across the range in sand content from 0 to 16%, there are major changes in water and nutrient availability and implications for soil management. Profiles with <2% clay in the subsoil will behave very differently to those with 10–16% clay in terms of water storage, nutrient leaching, susceptibility to compaction, acidification, and crop yields.

Sands are susceptible to subsoil compaction particularly when repeatedly trafficked by heavy machinery. Deep sands in south-west Australia can develop up to 3 MPa penetration resistance within 20 cm depth that stops root elongation (Hall et al. 2020). Subsoil compaction also develops in deep sands in Northeast Thailand due the repeated wet puddling of soils for rice cultivation (Bruand et al. 2004). Deep ripping to 50 cm or more removes the compaction but is only effective for 2–3 years before a similar level of compaction redevelops, due to the continuation of trafficking with heavy machinery or re-settling of sand particles. Controlled traffic farming, whereby the wheel tracks of all vehicles follow the same tramlines for all operations, can preserve the benefits of deep ripping for deep root penetration on deep sands. This confines the compaction of soil to the tramlines, which cover <15% of the land area: the remaining land generally maintains a level of penetration resistance that is not limiting to root penetration.

A possible solution to the water and nutrient limitations of subsoils in deep sands is to increase their content of high surface-reactivity materials such as clay. Clay addition and incorporation to 30–40 cm depth is a semi-permanent treatment to alleviate the low soil water and nutrient retention of the sandy subsoil (Hall et al. 2010; Betti et al. 2015). In addition, biochar incorporation into sandy subsoils to 75 cm depth at 100 and 200 t ha^{-1} increased water retention from 8.0 to 11.0 and 13.9% (vol/vol), respectively (Bruun et al. 2014).

16.10 Biological Subsoil Constraints

Biological constraints in subsoil may arise from infections of roots by soil pathogens, especially by those that prefer anoxic or hypoxic environments. However, Moura et al. (2022) uncovered relatively few studies, which examined the activity and significance of soil-borne pathogens in subsoils. This remains a research gap. Indeed, in a recent meta-analysis of published soils papers, the depth in soil biology research was on average only 18 cm (Yost and Hartemink 2020).

In the absence of specific research, the impact of soil-borne diseases in the subsoil can be predicted from general principles. Firstly, soil-borne diseases infect roots and are propagated on remnant plant roots that are most abundant in the upper 10–20 cm of soils (Costa and Coutinho 2022). Secondly, soil organic matter levels, which have a strong influence on soil microbial activity, are greatest in the topsoil and decline with depth. Hence, the capacity of non-disease organisms in the soil to compete with soil-borne diseases and suppress their activity may be lower in the subsoil than topsoil. In addition, the value of organic matter added to topsoil and incorporated in topsoil may have limited effect on activity of soil-borne diseases in subsoils. The shift towards no-tillage crop establishment increases the stratification of organic matter close to the soil surface. However, there is evidence that where no-tillage increases soil organic matter in the 0–30 cm layer, it also increases soil organic matter, albeit by modest amounts, in the 30–60 cm layer (Sun et al. 2020). The converse was also true: where soil organic matter in the 0–30 cm layer decreased, it also decreased in the 30–60 cm layer. Earlier concerns that increase in topsoil organic matter levels after adoption of no-tillage were negated by decline in the subsurface layers (Powlson et al. 2014) are not universally the case (Sun et al. 2020). Finally, those soil-borne diseases which are suppressed by tillage, such as rhizoctonia or nematodes, would remain unaffected in the subsoils below the direct effect of soil tillage disturbance on crop roots.

There has also been limited study on the subsoil activity of beneficial soil organisms (Moreira et al. 2022). However, the subsoil and even deeper layers of regolith are not devoid of soil biological activity. De Araujo Pereira et al. (2018) reported arbuscular mycorrhizal fungal (AMF) spores and root colonization in pure and mixed stands of *Eucalyptus grandis* and *Acacia mangium* to 8.0 m depth in south-eastern Brazil. In this study, the diversity of AMF species declined with depth. Similarly, the diversity of ectomycorrhizal fungi (EMF) and the abundance of external hyphae were reported to decrease with soil depth (Shahin et al. 2013). In Norway spruce, the number of EM root tips was higher in the mineral soil horizons to a depth of 0.75 m than in the surface organic-rich layer (Scattolin et al. 2008). Wang et al. (2014) found that the proportion of high efficiency mineral-dissolving bacteria increased with soil depth in the upper 1 m of a soil profile in a grassland. Brewer et al. (2019) also found that some taxa of bacteria and archaea increased with depth in soils even through overall diversity declined. Yost and Hartemink (2020) cite several studies reported over 100 years ago in which the authors reported measurable and significant bacterial counts at depths up to 6 m.

The extent to which alleviation of subsoil constraints increases root activity and density in subsoils and the repercussions for beneficial microorganisms needs to be better understood, especially for rainfed agriculture. It would also be prudent to increase the focus on pathogenic root infections in the subsoil to ensure that the gains from alleviating chemical and physical constraints are not negated by increased subsoil-root disease.

Ants, termites, earthworms, and other soil fauna have the capacity for subsoil mixing, creation of biopores, adding organic matter, and changing the soil environment at depth (Schaefer and Oliveira 2022). However, their role in crop production soils is not well-understood, and technologies to harness their activity have not yet been developed.

16.11 Root Penetration Rates and Biopores

Subsoil constraints may not be significant for species that produce shallow roots. Conversely, investing in alleviation of subsoil constraints is only justifiable if root systems have the capacity to grow deeply into the subsoil. Many crop species have 50% or more of the roots within 8–20 cm depth, but 95% of all roots occur within 64–138 cm depth (Fan et al. 2016). Globally, tree roots on average are 7 ± 1.2 m deep. By contrast, for herbaceous species, the average is 2.1 ± 0.2 m (Jackson et al. 1996). Hence, Thorup-Kristiansen et al. (2020) argue that selecting deep-rooted species represents the greatest potential for exploiting subsoil resources. While the maximum depth of roots is of interest, there is a need for more information demonstrating the extent of root function with depth together with convenient methods to select genetically superior materials from crop cultivar/varieties/hybrids and advanced lines. The importance of subsoil constraints depends on the proportion of crop water uptake and nutrient uptake from subsoil and the relative efficiency of different crop cultivar/varieties/hybrids (Vance and Milroy 2022; Ma et al. 2022).

The duration of root occupancy at depth may be quite limited for annual crops (Thorup-Kristiansen et al. 2020). The rate of transfer of water across the soil-root interface is particularly important in the case of many annual crops (Vance and Milroy 2022). For annual crops, root occupancy of deep subsoil layers may occur late in the growing season, which limits the length of time when roots can absorb subsoil water (Thorup-Kristansen et al. 2020). This trend is exacerbated by breeding shorter duration crops. Roots of early maturing crops have less time to grow deeper. For example, winter wheat, which grew for 6 months longer than spring wheat, had roots twice as deep (Thorup-Kristensen et al. 2009). By contrast, perennial crops continuously access the subsoil for water and nutrients, especially under rainfed systems. Under irrigation, the roots concentrate in the surface layer (Guohua et al. 2009).

The optimum depth for roots depends on where water and nutrients are stored in the profile and whether the availability of those resources changes with time. Practices to enhance higher water storage in subsoil could be beneficial. Increased topsoil macroporosity, for example, reduces water evaporation by interrupting capillary flow of water and slowing heat transfer by air present in the superficial layer, reducing soil water evaporation. When these properties develop at the end of the wet season, this practice can favour higher water levels in subsoil. By contrast in a no-tillage context, the stratification of nutrients and also organic matter facilitate root abundance near the soil surface. While plant-available P is immobile in most soils and stratified with highest concentrations close to the surface, N and water are more mobile. For P foraging, shallow roots concentrated in the topsoil are advantageous (Lynch and Brown 2001). By contrast for N and water, Lynch (2013) proposed a root ideotype for maize based on roots traits that fit the criteria of being steep, cheap, and deep. While plants need to invest some resources in acquiring N and water early in their growth from shallow soil layers, the mobility of those resources means that over time it is advantageous for the roots to grow deeper. One of the root traits that was advantageous for the uptake of N, P, and K when levels were suboptimal in soils was the development of root cortical aerenchyma that decreases the respiratory cost of growing deep roots (Potsma and Lynch 2011; Schneider et al. 2017).

An obvious limitation to efforts to encourage deeper root growth and activity is the ability to accurately measure accurately root properties, especially those related to water uptake, nutrient acquisition, soil biology, and soil carbon (Thorup-Kristensen et al. 2020). Traditional methods for assessing roots by digging, recovering, washing, and measuring are tedious and expensive, which limits the number of samples collected, the frequency of sampling, and the depth of sampling. Improved methods for assessing subsoil roots, which overcome the above limitations, are needed. Electrical resistivity tomography (ERT), which can provide non-invasive estimates of root depth of woody species (Mary et al. 2018), may provide such capability for annual crops in future. Changes in electrical resistivity were related to uptake of subsoil water by apple trees (Cassiani et al. 2016). Three-dimensional ERT was also able to quantify root biomass in herbaceous plant species (Amato et al. 2008).

Biopores are a critical feature of subsoils for root penetration, especially when the subsoil has high bulk density. However, more study is needed to understand the role of biopores in subsoils, and effective methods are needed for quantifying their presence, abundance, and function. Even though their density may be low, a significant proportion of roots occupy biopores in subsoils (Vance and Milroy 2022). According to Stirzaker et al. (1996), in soil with high strength, the frequency of roots in pores was three to four times higher than might have been expected based on random root distribution. Biopores in subsoils are created by successive generations of sinker/tap roots or the activity of soil-dwelling invertebrates.

While there is good evidence for the importance of soil biopores for root extension in soils with high strength, the impact on access to soil resources is likely to differ among species, depending on the mode of contact between a root and the pore wall (Vance and Milroy 2022). The seminal roots of barley make contact with the pore wall mainly through long root hairs. By contrast, the taproot of oilseed rape grows vertically down the pores and makes contact with the pore wall via the lateral roots. More research is needed to establish how a wider diversity of species and genotypes within a species utilize pores in high-strength soils. The key questions relate to determining what traits determine the ability of roots to grow into available pores and to access resources particularly in zones of high soil strength, or in otherwise hostile subsoils.

Larger macropores, formed by root channels, are prevalent in forests, where roots must access water at depth during periods of drought (Dell et al. 1983). Not all soil pores contain roots, with only 5% of subsoil pores in a wheat field containing roots compared to 20% in surface soil (White and Kirkegaard 2010). Crop rotation can be used to enhance the formation of macropores. For example, when chicory was used in rotation with barley and winter oilseed rape, the number of biopores was increased (Perkons et al. 2014). More systematic study is needed on the influence of cultivation methods, traffic management, crop rotation, cover management, and soil chemical amendment on the rate of development and persistence of biopores of different sizes.

Biopores are commonly nutrient enriched around their walls through the accumulation of organic matter (Ma et al. 2022). Also, the organic matter that earthworms incorporate into biopores creates habitats for microorganisms, which can facilitate SOM turnover in the subsoil (Hoang et al. 2017). Subsoil biopores may have more labile P than in the bulk soil (Bauke et al. 2017). The extent to which microorganisms cohabit with roots in biopores at depths below 2 m has not been explored.

Establishing cover crops on amended soils is advocated in Brazil to improve the physical condition of the soil profile. The continuous biopores formed by roots or macrofauna increase macroporosity and maximize water infiltration, which is critical to achieve amelioration of subsoil layers. Gramineous crops (bamboo, sugar cane, forages, etc.) with very dense fibrous roots and also trees (e.g. eucalyptus roots) in rotation with annuals crops can promote subsoil biopores, root growth, organic matter accumulation, and access to subsoil water and transfer at the soil-root interface.

16.12 Conclusions

Large areas of land globally suffer from subsoil constraints, but these are not well defined in terms of location, severity, and area affected. The area and location of subsoil constraints, both from natural causes and those induced by crop and farm management practices, can be predicted by digital soil mapping. Subsoil constraints are common on deeply weathered soils, including the deep regoliths that are common in tropical regions. Many individual subsoil constraints depress crop production. However, multiple constraints are also common. In such cases, alleviation of any one of those constraints may fail to generate significant benefit because other constraints remain.

The main consequence of subsoil constraints is that water and nutrients stored in subsoils are not efficiently utilized, and hence, crops fail reach their yield potential. Even when best management practices are applied to the topsoil, yield of crops is depressed by subsoil constraints. Crops may acquire up to 75% of N, 85% of P uptake, and up to 70% of K uptake from the subsoil, if root growth is not constrained. Technologies to identify and ameliorate subsoil constraints represent a promising frontier for crop and soil management with the potential to achieve a substantial lift in crop productivity in many parts of the world. The present chapter and companion chapters in this book have identified knowledge gaps and opportunities for further research and technology development to alleviate the effects of subsoil constraints on crop production.

The significance of subsoil constraints and their alleviation need to be highlighted in training for university graduates, field technicians, and farmers. Moreover, methodologies, such as SubVESS (Ball et al., 2015), need to be taught so that soil managers have the skills to assess susceptibility to induced subsoil constraints and to identify natural constraints as well as being able to match appropriate mitigation and amelioration techniques to particular subsoil constraints. Capacity development should emphasize avoidance of the causes of subsoil degradation rather than relying only on interventions, such as subsoiling to alleviate it.

References

Amato M, Basso B, Celano G, et al. In situ detection of tree root distribution and biomass by multi-electrode resistivity imaging. Tree Physiol. 2008;28:1441–8. https://doi.org/10.1093/treephys/28.8.1441.

Anderson GC, Pathan S, Easton J, Hall DJM, Sharma R. Short and long term effects of lime and gypsum application on acid soils in a water-limited environment: 1. Grain yield response and nutrient concentration. Agronomy. 2020;10:1213. https://doi.org/10.3390/agronomy10081213.

Babalola O, Lal R. Subsoil gravel horizon and maize root growth. I. Gravel concentration and bulk density effects. Plant Soil. 1977;46:337–46. https://doi.org/10.1007/BF00010090.

Ball BC, Batey T, Guimaraes RML, et al. The numeric visual evaluation of subsoil structure (SubVESS) under agricultural production. Soil Tillage Res. 2015;148:85–96. https://doi.org/10.1016/j.still.2014.12.005.

Barrett-Lennard EG, Shabala SN. The waterlogging/salinity interaction in higher plants revisited – focusing on the hypoxia-induced disturbance to K^+ homeostasis. Funct Plant Biol. 2013;40:872–82. https://doi.org/10.1071/FP12235.
Bauke SL, von Sperber C, Siebers N, et al. Biopore effects on phosphorus biochemistry in subsoils. Soil Biol Biochem. 2017;111:157–65. https://doi.org/10.1016/j.soilbio.2017.04.012.
Beltrán JM. Integrated approach to address salinity problems in irrigated agriculture. In: Gheyi HR, Dias NS, Lacerda CF, et al., editors. Manejo da salinidade na agricultura: Estudos básicos e aplicados. 2nd ed. Fortaleza: INCT Sal; 2016. p. 3–6.
Betti G, Grant C, Churchman G, Murray R. Increased profile wettability in texture-contrast soils from clay delving: case studies in South Australia. Soil Res. 2015;53:125–36. https://doi.org/10.1071/SR14133.
Blouin M, Hodson ME, Delgado EA, et al. A review of earthworm impact on soil function and ecosystem services. Eur J Soil Sci. 2013;64:161–82. https://doi.org/10.1111/ejss.12025.
Bonfatti BR, Hartemink AE, Giasson E, et al. Digital mapping of soil carbon in a viticultural region of Southern Brazil. Geoderma. 2016;261:204–21. https://doi.org/10.1016/J.GEODERMA.2015.07.016.
Brautigan DJ, Rengasamy P, Chittleborough DJ. Aluminium speciation and phytotoxicity in alkaline soils. Plant Soil. 2012;360:187–96. https://doi.org/10.1007/s11104-012-1232-5.
Brennan RF, Bolland MDA. Manganese deficiency in canola induced by liming to ameliorate soil acidity. J Plant Nutr. 2015;38:886–903. https://doi.org/10.1080/01904167.2014.964362.
Brewer TE, Aronson EL, Arogyaswamy K, et al. Ecological and genomic attributes of novel bacterial taxa that thrive in subsurface soil horizons. mBio. 2019;10:e01318-19. https://doi.org/10.1128/mBio.01318-19.
Bruand A, Hartmann C, Ratana-Anupap S, et al. Composition, fabric, and porosity of an Arenic Haplustalf of Northeast Thailand. Soil Sci Soc Am J. 2004;68:185–93. https://doi.org/10.2136/sssaj2004.0185.
Bruun EW, Petersen CT, Hansen E, et al. Biochar amendment to coarse sandy subsoil improves root growth and increases water retention. Soil Use Manag. 2014;30:09–118. https://doi.org/10.1111/sum.12102.
Bui EN, Krogh L, Lavado RS, et al. Distribution of sodic soils: the world scene. In: Sumner ME, Naidu R, editors. Sodic soils: Distribution, properties, management and environmental consequences. New York: Oxford University Press; 1998. p. 19–33.
Butterly CR, Amado TJC, Tang C. Soil acidity and acidification. In: Oliveira TS, Bell RW, editors. Subsoil constraints for crop production. 1st ed. Springer; 2022. p. 53–91.
Cartwright B, Zarcinas BA, Mayfield AH. Toxic concentrations of boron in a red-brown earth at Gladstone, South Australia. Aust J Soil Res. 1984;22:261–72. https://doi.org/10.1071/SR9840261.
Cassiani G, Boaga J, Rossi M, et al. Soil-plant interaction monitoring: small scale example of an apple orchard in Trentino, north-eastern Italy. Sci Total Environ. 2016;543:851–61. https://doi.org/10.1016/j.scitotenv.2015.03.113.
Celestina C, Hunt J, Sale PWG, et al. Attribution of crop yield responses to application of organic amendments: a critical review. Soil Till Res. 2019;186:135–45. https://doi.org/10.1016/j.still.2018.10.002.
Clarke CJ, George RJ, Bell RW, et al. Dryland salinity in southwestern Australia: its origins, remedies, and future research directions. Aust J Soil Res. 2002;40:93–113. https://doi.org/10.1071/SR01028.
Costa MCG, Coutinho IAC. Root systems of agricultural crops and their response to physical and chemical soil constraints. In: Oliveira TS, Bell RW, editors. Subsoil constraints for crop production. 1st ed. Springer; 2022. p. 225–61.
Dang YP, Dalal RC, Mayer DG, et al. High subsoil chloride concentrations reduce soil water extraction and crop yield on Vertosols in North-Eastern Australia. Aust J Soil Res. 2008;59:321–30. https://doi.org/10.1071/AR07192.

de Araujo Pereira AP, Santana MC, Bonfim JA, et al. Digging deeper to study the distribution of mycorrhizal arbuscular fungi along the soil profile in pure and mixed *Eucalyptus grandis* and *Acacia mangium* plantations. Appl Soil Ecol. 2018;128:1–11. https://doi.org/10.1016/j.apsoil.2018.03.015.
Dell B, Bartle J, Tacey W. Root occupation and root channels of jarrah forest subsoils. Aust J Bot. 1983;31(6):615–27. https://doi.org/10.1071/BT9830615.
Edwards NK. Distribution of potassium in the soil profile of a sandplain soil under pasture species. Plant Soil. 1993;155–156:407–10. https://doi.org/10.1007/BF00025069.
Fan J, McConkey B, Wang H, et al. Root distribution by depth for temperate agricultural crops. Field Crops Res. 2016;189:68–74. https://doi.org/10.1016/j.fcr.2016.02.013.
FAO Soils Portal: World reference base. http://www.fao.org/soils-portal/data-hub/soil-classification/world-reference-base/en/ (1998). Accessed 29 Nov 2020
Ferreira MM, Fernandes B, Curi N. Mineralogia da fração argila e estrutura de latossolos da região sudeste do Brasil. Rev Bras Ci Solo. 1999;23:507–14. https://doi.org/10.1590/S0100-06831999000300003.
French RJ, Schultz JE. Water use efficiency of wheat in a Mediterranean-type environment. I. The relation between yield, water use and climate. Aust J Agric Res. 1984;35:743–64. https://doi.org/10.1071/AR9840743.
Gazey C, Davies S, Master R. Soil acidity: a guide for WA farmers and consultants, Bulletin 4858. Perth: Department of Agriculture and Food, Western Australia; 2014.
Gill JS, Sale PWG, Tang C. Amelioration of dense sodic subsoil using organic amendments increases wheat yield more than using gypsum in a high rainfall zone of southern Australia. Field Crops Res. 2008;107:265–75. https://doi.org/10.1016/j.fcr.2008.02.014.
Gill JS, Sale PWG, Peries RR, et al. Changes in soil physical properties and crop root growth in dense sodic subsoil following incorporation of organic amendments. Field Crops Res. 2009;114:137–46. https://doi.org/10.1016/j.fcr.2009.07.018.
González-Álvarez I, Boni M, Anand RR. Mineral exploration in regolith-dominated terrains: global considerations and challenges. Ore Geol Rev. 2016;73:375–9. https://doi.org/10.1016/j.oregeorev.2015.11.017.
Guohua L, Kang Y, Wan S. Effect of irrigation methods on root development and profile soil water uptake in winter wheat. Irrig Sci. 2009;28:387–98. https://doi.org/10.1007/s00271-009-0200-1.
Hall DJM, Jones HR, Crabtree WL, et al. Claying and deep ripping can increase crop yields and profits on water repellent sands with marginal fertility in southern Western Australia. Aust J Soil Res. 2010;48:178–87. https://doi.org/10.1071/SR09078.
Hall DJM, Davies SL, Bell RW et al. Soil management systems to overcome multiple constraints for dryland crops on deep sands in the Mediterranean climate of the south coast of Western Australia. Agronomy-988806; 2020.
Hoang DTT, Bauke SL, Kuzyakov Y, et al. Rolling in the deep: priming effects in earthworm biopores in topsoil and subsoil. Soil Biol Biochem. 2017;114:59–71. https://doi.org/10.1016/j.soilbio.2017.06.021.
Hochman Z, Holzworth D, Hunt JR. Potential to improve on-farm wheat yield and WUE in Australia. Crop Past Sci. 2009;60:708–16. https://doi.org/10.1071/CP09064.
Jackson RB, Canadell J, Ehleringer JR, Mooney HA, Sala OE, Schulze ED. A global analysis of root distributions for terrestrial biomes. Oecologia. 1996;108:389–411.
Kautz T, Amelung W, Ewert F, et al. Nutrient acquisition from arable subsoils in temperate climates: a review. Soil Biol Biochem. 2013;57:1003–22. https://doi.org/10.1016/j.soilbio.2012.09.014.
Kirkegaard JA, Lilley JM, Howe GN, et al. Impact of subsoil water use on wheat yield. Aust J Agric Res. 2007;58:303–15. https://doi.org/10.1071/AR06285.
Kopittke PM, Menzies NW, Wang P, et al. Kinetics and nature of aluminium rhizotoxic effects: a review. J Exp Bot. 2016;67:4451–67. https://doi.org/10.1093/jxb/erw233.
Leinemann T, Mikutta R, Kalbitz K, et al. Small scale variability of vertical water and dissolved organic matter fluxes in sandy Cambisol subsoils as revealed by segmented suction plates. Biogeochemistry. 2016;131:1–15. https://doi.org/10.1007/s10533-016-0259-8.

Lynch JP. Steep, cheap and deep: an ideotype to optimize water and N acquisition by maize root systems. Ann Bot. 2013;112:347–57. https://doi.org/10.1093/aob/mcs293.

Lynch JP, Brown KM. Topsoil foraging – an architectural adaptation of plants to low phosphorus availability. Plant Soil. 2001;237:225–37. https://doi.org/10.1023/A:1013324727040.

Ma Q, Bell RW, Mattiello E. Nutrient acquisition with particular reference to subsoil constraints. In: Oliveira TS, Bell RW, editors. Subsoil constraints for crop production. 1st ed. Springer; 2022. p. 289–321.

Mary B, Peruzzo L, Boaga J, et al. Small-scale characterization of vine plant root water uptake via 3-D electrical resistivity tomography and mise-à-la-masse method. Hydrol Earth Syst Sci. 2018;22:5427–44. https://doi.org/10.5194/hess-22-5427-2018.

McDonald GK, Eglinton JK, Barr AR. Assessment of the agronomic value of QTL on chromosomes 2H and 4H linked to tolerance to boron toxicity in barley (Hordeum vulgare L.). Plant Soil. 2010;326:275–90. https://doi.org/10.1007/s11104-009-0006-1.

McDonald GK, Taylor JD, Verbyla A, et al. Assessing the importance of subsoil constraints to yield of wheat and its implications for yield improvement. Crop Pasture Sci. 2012;263:1043–65. https://doi.org/10.1071/CP12244.

Moreira BC, Prates P Jr, Dell B, et al. Roots and beneficial interactions with soil microbes. In: Oliveira TS, Bell RW, editors. Subsoil constraints for crop production. 1st ed. Springer; 2022. p. 263–87.

Moura AB, Backhouse D, Souza IT Jr, et al. Soilborne pathogens. In: Oliveira TS, Bell RW, editors. Subsoil constraints for crop production. 1st ed. Springer; 2022. p. 199–224.

Nable RO, Webb MJ. Further evidence that zinc is required throughout the root zone for optimal plant growth and development. Plant Soil. 1993;150:247–53.

Noble AD, Gillman GP, Ruaysoongnern S. A cation exchange index for assessing degradation of acid soil by further acidification under permanent agriculture in the tropics. Eur J Soil Sci. 2000;51:233–43. https://doi.org/10.1046/j.1365-2389.2000.00313.x.

Oliveira TS, Bell RW, editors. Subsoil constraints for crop production. 1st ed. Springer; 2022.

Oliveira TS, Fernandes RBA. Physical subsoil constraints of agricultural and forestry land. In: Oliveira TS, Bell RW, editors. Subsoil constraints for crop production. 1st ed. Springer; 2022. p. 125–60.

Oliver Y, Robertson MJ, Stone P. Modification to the French and. Schultz formula to account for soil type and within-season rainfall. In: Unkovich EM, editor. Proceedings of the 14th Australian agronomy conference. September 2008, Adelaide, South Australia. Gosford: The Regional Institute Ltd; 2008. http://www.agronomyaustraliaproceedings.org/images/sampledata/2008/concurrent/assessing-yield-potential/5827_oliverym.pdf.

Pereira MTJJR. Impactos do cultivo intensivo de olerícolas em longo prazo na mineralogia, química e estrutura de Latossolo oxídico. PhD Thesis, Universidade Federal de Viçosa; 2020.

Perkons U, Kautz T, Uteau D, et al. Root-length densities of various annual crops following crops with contrasting root systems. Soil Tillage Res. 2014;137:50–7. https://doi.org/10.1016/j.still.2013.11.005.

Postma JA, Lynch JP. Root cortical aerenchyma enhances the growth of maize on soils with suboptimal availability of nitrogen, phosphorus, and potassium. Plant Physiol. 2011;156:1190–201. https://doi.org/10.1104/pp.111.175489.

Powlson DS, Stirling CM, Jat ML, et al. Limited potential of no-tillage agriculture for climate change mitigation. Nat Clim Chang. 2014;4:678–83. https://doi.org/10.1038/nclimate2292.

Rengasamy P. Transient salinity and subsoil constraints to dryland farming in Australian sodic soils: an overview. Aust J Exp Agric. 2002;42:351–61. https://doi.org/10.1071/EA01111.

Rengasamy P, Chittleborough D, Helyar K. Root-zone constraints and plant-based solutions for dryland salinity. Plant Soil. 2003;257:249–60. https://doi.org/10.1023/A:1027326424022.

Rengasamy P, Tavakkoli E, McDonald GK. Exchangeable cations and clay dispersion: net dispersive charge, a new concept for dispersive soil. Eur J Soil Sci. 2016;67:659–65. https://doi.org/10.1111/ejss.12369.

Rengasamy P, Lacerda CF, Gheyi HR. Salinity, sodicity and alkalinity. In: Oliveira TS, Bell RW, editors. Subsoil constraints for crop production. 1st ed. Springer; 2022. p. 83–107.

Ringrose-Voase AJ, Grealish GJ, Thomas M, et al. Four pillars of digital land resource mapping to address information and capacity shortages in developing countries. Geoderma. 2019;352:299–313. https://doi.org/10.1016/j.geoderma.2017.10.014.

Ryan PR. Assessing the role of genetics for improving the yield of Australia's major grain crops on acid soils. Crop Pasture Sci. 2018;69:242–64. https://doi.org/10.1071/CP17310.

Sadras V, Baldock J, Roget D, et al. Measuring and modelling yield and water budget components of wheat crops in coarse-textured soils with chemical constraints. Field Crops Res. 2003;84:241–60. https://doi.org/10.1016/S0378-4290(03)00093-5.

Sale PW, Gill JS, Peries RR, et al. Crop responses to subsoil manuring. I. Results in South-Western Victoria from 2009 to 2012. Crop Pasture Sci. 2019;70:44–54. https://doi.org/10.1071/CP18115.

Sangmanee P, Dell B, Harper RJ, et al. Deep soil carbon – characteristics and measurement with particular bearing on kaolinitic profiles. In: Oliveira TS, Bell RW, editors. Subsoil constraints for crop production. 1st ed. Springer; 2022. p. 347–72.

Scanlan CA, Holmes KW, Bell RW. Sand and gravel subsoils. In: Oliveira TS, Bell RW, editors. Subsoil constraints for crop production. 1st ed. Springer; 2022. p. 179–198.

Scattolin L, Montecchio L, Mosca E, et al. Vertical distribution of the ectomycorrhizal community in the top soil of Norway spruce stands. Eur J For Res. 2008;127:347–57. https://doi.org/10.1007/s10342-008-0209-7.

Schaefer CEGR. Brazilian latosols and their B horizon microstructure as long-term biotic constructs. Aust J Soil Res. 2001;39:909–26. https://doi.org/10.1071/SR00093.

Schaefer C, Oliveira FS. Chapter 15: Live subsoils: tropical regolith and biota interactions. In: Oliveira TS, Bell RW, editors. Subsoil constraints for crop production. 1st ed. Springer; 2022. p. 373–96.

Schaefer CEGR, Souza JJLL, Oliveira FS, et al. The geological, geomorphological, climatic and hydrological background of tropical regoliths and hostile subsoils: The Brazilian landmass. In: Oliveira TS, Bell RW, editors. Subsoil constraints for crop production. 1st ed. Springer; 2022. p. 11–52.

Schneider F, Don A. Root-restricting layers in German agricultural soils. Part I: extent and cause. Plant Soil. 2019a;442:433–51. https://doi.org/10.1007/s11104-019-04185-9.

Schneider F, Don A. Root-restricting layers in German agricultural soils. II: adaptation and melioration strategies. Plant Soil. 2019b;442:419–32. https://doi.org/10.1007/s11104-019-04186-8.

Schneider HM, Postma JA, Wojciechowski T, et al. Root cortical senescence improves growth under suboptimal availability of N, P, and K. Plant Phys. 2017;174:2333–23247. https://doi.org/10.1104/pp.17.00648.

Scott BJ, Conyers MK, Poile GJ, et al. Reacidification and reliming effects on soil properties and wheat yield. Aust J Exp Agric. 1999;39:849–56. https://doi.org/10.1071/EA99040.

Setter TL, Waters I, Sharma SK, et al. Review of wheat improvement for waterlogging tolerance in Australia and India: the importance of anaerobiosis and element toxicities associated with different soils. Ann Bot. 2009;103:221–35. https://doi.org/10.1093/aob/mcn137.

Shahin O, Martin-St Paul N, Rambal S, et al. Ectomycorrhizal fungal diversity in Quercus ilex Mediterranean woodlands: variation among sites and over soil depth profiles in hyphal exploration types, species richness and community composition. Symbiosis. 2013;61:1–12. https://doi.org/10.1007/s13199-013-0252-0.

Stirzaker RJ, Passioura JB, Wilms Y. Soil structure and plant growth: impact of bulk density and biopores. Plant Soil. 1996;185:151–62. https://doi.org/10.1007/BF02257571.

Sumner ME, Noble AD. Soil acidification: the world story. In: Rengel Z, editor. Handbook of soil acidity. New York: Marcel Dekker; 2003. p. 1–28.

Sun W, Canadell JG, Yu L, et al. Climate drives global soil carbon sequestration and crop yield changes under conservation agriculture. Glob Chang Biol. 2020;26:3325–35. https://doi.org/10.1111/gcb.15001.

Tanaka RT, Mascarenhas HA, Bulisani EA. Deficiência de manganês em soja induzida por excesso de calcário. Pesq Agropec Bras. 1992;27:247–50.

Tang C, Yu Q. Impact of chemical composition of legume residues and initial soil pH on pH change of a soil after residue incorporation. Plant Soil. 1999;215(1):29–38. https://doi.org/10.1023/A:1004704018912.

Thorup-Kristensen K, Salmerón M, Loges R. Winter wheat roots grow twice as deep as spring wheat roots, is this important for N uptake and N leaching losses? Plant Soil. 2009;322:101–14. https://doi.org/10.1007/s11104-009-9898-z.

Thorup-Kristensen K, Halberg N, Nicolaisen M, et al. Review. Digging deeper for agricultural resources, the value of deep rooting. Trends Plant Sci. 2020;25:406–17. https://doi.org/10.1016/j.tplants.2019.12.007.

UNESCO. The United Nations world water development report 2017: wastewater the untapped resource. Paris: UNESCO; 2017.

van Gool D. Identifying soil constraints that limit wheat yield in south-West Western Australia, Resource management technical report 399. Perth: Department of Agriculture and Food, Western Australia; 2016.

van Gool D, Evans D, Vernon L, et al. Managing hostile subsoils in the high rainfall zone of South-Western Australia, Resource management technical report 321. Department of Agriculture and Food, Western Australia, Perth; 2007.

Vance WH, Milroy SP. Water acquisition by roots from the subsoil: impact of physical constraints on the dynamics of water capture. In: Oliveira TS, Bell RW, editors. Subsoil constraints for crop production. 1st ed. Springer; 2022. p. 323–45.

Verboom WH, Pate JS. Bioengineering of soil profiles in semiarid ecosystems: the 'phytotarium' concept. A review. Plant Soil. 2006a;289:71–102. https://doi.org/10.1007/s11104-006-9073-8.

Verboom WH, Pate JS. Evidence of active biotic influences in pedogenetic processes. Case studies from semiarid ecosystems of South-West Western Australia. Plant Soil. 2006b;289:103–21. https://doi.org/10.1007/s11104-006-9075-6.

Wang Q, Cheng C, He L, et al. Characterization of depth-related changes in bacterial communities involved in mineral weathering along a mineral-rich soil profile. Geomicrobiol J. 2014;31:43–444. https://doi.org/10.1080/01490451.2013.848248.

White RG, Kirkegaard JA. The distribution and abundance of wheat roots in a dense, structured subsoil – implications for water uptake. Plant Cell Environ. 2010;33:133–48. https://doi.org/10.1111/j.1365-3040.2009.02059.x.

Whitten MG, Wong MTF, Rate AW. Amelioration of subsurface acidity in the south-west of Western Australia: downward movement and mass balance of surface-incorporated lime after 2-15 years. Aust J Soil Res. 2000;38(3):711–28. https://doi.org/10.1071/sr99054.

Wielemaker WG. Soil formation by termites, a study in the Kisii area, Kenya. PhD Thesis, Wageningen University, Netherlands; 1984.

Wong MTF, Asseng S. Determining the causes of spatial and temporal variability of wheat yields at sub-field scale using a new method of upscaling a crop model. Plant Soil. 2006;283:203–15. https://doi.org/10.1007/s11104-006-0012-5.

Wong MTF, Asseng S. Yield and environmental benefits of ameliorating subsoil constraints under variable rainfall in a Mediterranean environment. Plant Soil. 2007;297:29–42. https://doi.org/10.1007/s11104-007-9316-3.

Wong MTF, Wittwer K. Positive charge discovered across Western Australian wheatbelt soils challenges key soil and nitrogen management assumptions. Aust J Soil Res. 2009;47:127–35. https://doi.org/10.1071/SR08098.

Wong MTF, Hughes R, Rowell DL. The retention of nitrate in acid soils from the tropics. Soil Use Manag. 1990;6:72–4. https://doi.org/10.1111/j.1475-2743.1990.tb00805.x.

Wong MTF, Asseng S, Robertson MJ, et al. Mapping subsoil acidity and shallow soil across a field with information from yield maps, geophysical sensing, and the grower. Prec Agric. 2008;9:3–15. https://doi.org/10.1007/s11119-008-9052-6.

Yost JL, Hartemink AE. How deep is the soil studied – an analysis of four soil science journals. Plant Soil. 2020;452:5–18. https://doi.org/10.1007/s11104-020-04550-z.

Zimmerman PR, Greenberg JP, Wandiga SO, et al. Termites: a potentially large source of atmospheric methane, carbon dioxide, and molecular hydrogen. Science. 1982;218:563–5. https://doi.org/10.1126/science.218.4572.563.

Zoca SM, Penn C. An important tool with no instruction manual: a review of gypsum use in agriculture. Adv Agron. 2017;144:1–44. https://doi.org/10.1016/bs.agron.2017.03.001.

Index

T. S. d. Oliveira, R. W. Bell (eds.), *Subsoil Constraints for Crop Production*,
https://doi.org/10.1007/978-3-031-00317-2

E

F

G

N

O

P

Q

R

S

MIX
Papier aus verantwortungsvollen Quellen
Paper from responsible sources
FSC® C105338

If you have any concerns about our products,
you can contact us on
ProductSafety@springernature.com

In case Publisher is established outside the EU,
the EU authorized representative is:
Springer Nature Customer Service Center GmbH
Europaplatz 3, 69115 Heidelberg, Germany

Printed by Libri Plureos GmbH
in Hamburg, Germany